# IMAGE and VIDEO COMPRESSION for MULTIMEDIA ENGINEERING

## Fundamentals, Algorithms, and Standards

# IMAGE PROCESSING SERIES

Series Editor: Phillip A. Laplante

## *Forthcoming Titles*

**Adaptive Image Processing: A Computational Intelligence Perspective**
Ling Guan, Hau-San Wong, and Stuart William Perry

**Shape Analysis and Classification: Theory and Practice**
Luciano da Fontoura Costa and Roberto Marcondes Cesar, Jr.

# IMAGE and VIDEO COMPRESSION for MULTIMEDIA ENGINEERING

## Fundamentals, Algorithms, and Standards

### Yun Q. Shi

New Jersey Institute of Technology
Newark, NJ

### Huifang Sun

Mitsubishi Electric Information Technology Center
America Advanced Television Laboratory
New Providence, NJ

**CRC Press**

**Boca Raton London New York Washington, D.C.**

## Library of Congress Cataloging-in-Publication Data

Shi, Yun Q.
    Image and video compression for multimedia engineering : fundamentals, algorithms, and standards / Yun Q. Shi, Huifang Sun.
        p.  cm.
    Includes bibliographical references and index.
    ISBN 0-8493-3491-8 (alk. paper)
    1.Multimedia systems. 2. Image compressions. 3. Video compression.
    I. Sun, Huifnag.  II. Title.
    QA76.575.S555 1999
    006.7—dc21
                                                            99-047137
                                                                 CIP

This book contains information obtained from authentic and highly regarded sources. Reprinted material is quoted with permission, and sources are indicated. A wide variety of references are listed. Reasonable efforts have been made to publish reliable data and information, but the author and the publisher cannot assume responsibility for the validity of all materials or for the consequences of their use.

Neither this book nor any part may be reproduced or transmitted in any form or by any means, electronic or mechanical, including photocopying, microfilming, and recording, or by any information storage or retrieval system, without prior permission in writing from the publisher.

The consent of CRC Press LLC does not extend to copying for general distribution, for promotion, for creating new works, or for resale. Specific permission must be obtained in writing from CRC Press LLC for such copying.

Direct all inquiries to CRC Press LLC, 2000 N.W. Corporate Blvd., Boca Raton, Florida 33431.

**Trademark Notice:** Product or corporate names may be trademarks or registered trademarks, and are used only for identification and explanation, without intent to infringe.

*Figures and tables from ISO reproduced with the permission of the International Organization for Standardization. These standards can be obtained from any member body or directly from the Central Secretariat, ISO, Case postale 56, 1211 Geneva 20, Switzerland.*

# Preface

It is well known that in the 1960s the advent of the semiconductor computer and the space program swiftly brought the field of digital image processing into public focus. Since then the field has experienced rapid growth and has entered into every aspect of modern technology. Since the early 1980s, digital image sequence processing has been an attractive research area because an image sequence, as a collection of images, may provide more information than a single image frame. The increased computational complexity and memory space required for image sequence processing are becoming more attainable. This is due to more advanced, achievable computational capability resulting from the continuing progress made in technologies, especially those associated with the VLSI industry and information processing.

In addition to image and image sequence processing in the digitized domain, facsimile transmission has switched from analog to digital since the 1970s. However, the concept of high definition television (HDTV) when proposed in the late 1970s and early 1980s continued to be analog. This has since changed. In the U.S., the first digital system proposal for HDTV appeared in 1990. The Advanced Television Standards Committee (ATSC), formed by the television industry, recommended the digital HDTV system developed jointly by the seven Grand Alliance members as the standard, which was approved by the Federal Communication Commission (FCC) in 1997. Today's worldwide prevailing concept of HDTV is digital. Digital television (DTV) provides the signal that can be used in computers. Consequently, the marriage of TV and computers has begun. Direct broadcasting by satellite (DBS), digital video disks (DVD), video-on-demand (VOD), video games, and other digital video related media and services are available now, or soon will be.

As in the case of image and video transmission and storage, audio transmission and storage through some media have changed from analog to digital. Examples include entertainment audio on compact disks (CD) and telephone transmission over long and medium distances. Digital TV signals, mentioned above, provide another example since they include audio signals. Transmission and storage of audio signals through some other media are about to change to digital. Examples of this include telephone transmission through local area and cable TV.

Although most signals generated from various sensors are analog in nature, the switching from analog to digital is motivated by the superiority of digital signal processing and transmission over their analog counterparts. The principal advantage of the digital signal is its robustness against various noises. Clearly, this results from the fact that only binary digits exist in digital format and it is much easier to distinguish one state from the other than to handle analog signals.

Another advantage of being digital is ease of signal manipulation. In addition to the development of a variety of digital signal processing techniques (including image, video, and audio) and specially designed software and hardware that may be well known, the following development is an example of this advantage. The digitized information format, i.e., the bitstream, often in a compressed version, is a revolutionary change in the video industry that enables many manipulations which are either impossible or very complicated to execute in analog format. For instance, video, audio, and other data can be first compressed to separate bitstreams and then combined to form a signal bitstream, thus providing a multimedia solution for many practical applications. Information from different sources and to different devices can be multiplexed and demultiplexed in terms of the bitstream. Bitstream conversion in terms of bit rate conversion, resolution conversion, and syntax conversion becomes feasible. In digital video, content-based coding, retrieval, and manipulation and the ability to edit video in the compressed domain become feasible. All system-timing signals in the digital systems can be included in the bitstream instead of being transmitted separately as in traditional analog systems.

The digital format is well suited to the recent development of modern telecommunication structures as exemplified by the Internet and World Wide Web (WWW). Therefore, we can see that digital computers, consumer electronics (including television and video games), and telecommunications networks are combined to produce an information revolution. By combining audio, video, and other data, multimedia becomes an indispensable element of modern life. While the pace and the future of this revolution cannot be predicted, one thing is certain: this process is going to drastically change many aspects of our world in the next several decades.

One of the enabling technologies in the information revolution is digital data compression, since the digitization of analog signals causes data expansion. In other words, storage and/or transmission of digitized signals require more storage space and/or bandwidth than the original analog signals.

The focus of this book is on image and video compression encountered in multimedia engineering. Fundamentals, algorithms, and standards are the three emphases of the book. It is intended to serve as a senior/graduate-level text. Its material is sufficient for a one-semester or one-quarter graduate course on digital image and video coding. For this purpose, at the end of each chapter there is a section of exercises containing problems and projects for practice, and a section of references for further reading.

Based on this book, a short course entitled "Image and Video Compression for Multimedia," was conducted at Nanyang Technological University, Singapore in March and April, 1999. The response to the short course was overwhelmingly positive.

# Authors

**Dr. Yun Q. Shi** has been a professor with the Department of Electrical and Computer Engineering at the New Jersey Institute of Technology, Newark, NJ since 1987. Before that he obtained his B.S. degree in Electronic Engineering and M.S. degree in Precision Instrumentation from the Shanghai Jiao Tong University, Shanghai, China and his Ph.D. in Electrical Engineering from the University of Pittsburgh. His research interests include motion analysis from image sequences, video coding and transmission, digital image watermarking, computer vision, applications of digital image processing and pattern recognition to industrial automation and biomedical engineering, robust stability, spectral factorization, multidimensional systems and signal processing. Prior to entering graduate school, he worked in a radio factory as a design and test engineer in digital control manufacturing and in electronics.

He is the author or coauthor of about 90 journal and conference proceedings papers in his research areas and has been a formal reviewer of the *Mathematical Reviews* since 1987, an IEEE senior member since 1993, and the chairman of Signal Processing Chapter of IEEE North Jersey Section since 1996. He was an associate editor for *IEEE Transactions on Signal Processing* responsible for Multidimensional Signal Processing from 1994 to 1999, the guest editor of the special issue on Image Sequence Processing for the *International Journal of Imaging Systems and Technology*, published as Volumes 9.4 and 9.5 in 1998, one of the contributing authors in the area of Signal and Image Processing to the *Comprehensive Dictionary of Electrical Engineering*, published by the CRC Press LLC in 1998. His biography has been selected by Marquis Who's Who for inclusion in the 2000 edition of *Who's Who in Science and Engineering*.

**Dr. Huifang Sun** received the B.S. degree in Electrical Engineering from Harbin Engineering Institute, Harbin, China, and the Ph.D. in Electrical Engineering from University of Ottawa, Ottawa, Canada. In 1986 he jointed Fairleigh Dickinson University, Teaneck, NJ as an assistant professor and was promoted to an associate professor in electrical engineering. From 1990 to 1995, he was with the David Sarnoff Research Center (Sarnoff Corp.) in Princeton as a member of technical staff and later promoted to technology leader of Digital Video Technology where his activities included MPEG video coding, AD-HDTV, and Grand Alliance HDTV development. He joined the Advanced Television Laboratory, Mitsubishi Electric Information Technology Center America (ITA), New Providence, NJ in 1995 as a senior principal technical staff and was promoted to deputy director in 1997 working in advanced television development and digital video processing. He has been active in MPEG video standards for many years and holds 10 U.S. patents with several pending. He has authored or coauthored more than 80 journal and conference papers and obtained the 1993 best paper award of IEEE Transactions on Consumer Electronics, and 1997 best paper award of International Conference on Consumer Electronics. For his contributions to HDTV development, he obtained the 1994 Sarnoff technical achievement award. He is currently the associate editor of *IEEE Transactions on Circuits and Systems for Video Technology*.

# Acknowledgments

We are pleased to express our gratitude here for the support and help we received in the course of writing this book.

The first author thanks his friend and former colleague, Dr. C. Q. Shu, for fruitful technical discussions related to some contents of the book. Sincere thanks also are directed to several of his friends and former students, Drs. J. N. Pan, X. Xia, S. Lin, and Y. Shi, for their technical contributions and computer simulations related to some subjects of the book. He is grateful to Ms. L. Fitton for her English editing of 11 chapters, and to Dr. Z. F. Chen for her help in preparing many graphics.

The second author expresses his appreciation to his colleagues, Anthony Vetro and Ajay Divakaran, for fruitful technical discussion related to some contents of the book and for proofreading nine chapters. He also extends his appreciation to Dr. Xiaobing Lee for his help in providing some useful references, and to many friends and colleagues of the MPEGers who provided wonderful MPEG documents and tutorial materials that are cited in some chapters of this book. He also would like to thank Drs. Tommy Poor, Jim Foley, and Toshiaki Sakaguchi for their continuing support and encouragement.

Both authors would like to express their deep appreciation to Dr. Z. F. Chen for her great help in formatting all the chapters of the book. They also thank Dr. F. Chichester for his help in preparing the book.

Special thanks go to the editor-in-chief of the Image Processing book series of CRC Press, Dr. P. Laplante, for his constant encouragement and guidance. Help from the editors at CRC Press, N. Konopka, M. Mogck, and other staff, is appreciated.

The first author acknowledges the support he received associated with writing this book from the Electrical and Computer Engineering Department at the New Jersey Institute of Technology. In particular, thanks are directed to the department chairman, Professor R. Haddad, and the associate chairman, Professor K. Sohn. He is also grateful to the Division of Information Engineering and the Electrical and Electronic Engineering School at Nanyang Technological University (NTU), Singapore for the support he received during his sabbatical leave. It was in Singapore that he finished writing the manuscript. In particular, thanks go to the dean of the school, Professor Er Meng Hwa, and the division head, Professor A. C. Kot. With pleasure, he expresses his appreciation to many of his colleagues at the NTU for their encouragement and help. In particular, his thanks go to Drs. G. Li and J. S. Li, and Dr. G. A. Bi. Thanks are also directed to many colleagues, graduate students, and some technical staff from industrial companies in Singapore who attended the short course which was based on this book in March/April 1999 and contributed their enthusiastic support and some fruitful discussion.

Last but not least, both authors thank their families for their patient support during the course of the writing. Without their understanding and support we would not have been able to complete this book.

Yun Q. Shi
Huifang Sun

# Content and Organization of the Book

The entire book consists of 20 chapters which can be grouped into four sections:

I. Fundamentals,
II. Still Image Compression,
III. Motion Estimation and Compensation, and
IV. Video Compression.

In the following, we summarize the aim and content of each chapter and each part, and the relationships between some chapters and between the four parts.

Section I includes the first six chapters. It provides readers with a solid basis for understanding the remaining three parts of the book. In Chapter 1, the practical needs for image and video compression is demonstrated. The feasibility of image and video compression is analyzed. Specifically, both statistical and psychovisual redundancies are analyzed and the removal of these redundancies leads to image and video compression. In the course of the analysis, some fundamental characteristics of the human visual system are discussed. Visual quality measurement as another important concept in the compression is addressed in both subjective and objective quality measures. The new trend in combining the virtues of the two measures also is presented. Some information theory results are presented as the final subject of the chapter.

Quantization, as a crucial step in lossy compression, is discussed in Chapter 2. It is known that quantization has a direct impact on both the coding bit rate and quality of reconstructed frames. Both uniform and nonuniform quantization are covered. The issues of quantization distortion, optimum quantization, and adaptive quantization are addressed. The final subject discussed in the chapter is pulse code modulation (PCM) which, as the earliest, best-established, and most frequently applied coding system normally serves as a standard against which other coding techniques are compared.

Two efficient coding schemes, differential coding and transform coding (TC), are discussed in Chapters 3 and 4, respectively. Both techniques utilize the redundancies discussed in Chapter 1, thus achieving data compression. In Chapter 3, the formulation of general differential pulse code modulation (DPCM) systems is described first, followed by discussions of optimum linear prediction and several implementation issues. Then, delta modulation (DM), an important, simple, special case of DPCM, is presented. Finally, application of the differential coding technique to interframe coding and information-preserving differential coding are covered.

Chapter 4 begins with the introduction of the Hotelling transform, the discrete version of the optimum Karhunen and Loeve transform. Through statistical, geometrical, and basis vector (image) interpretations, this introduction provides a solid understanding of the transform coding technique. Several linear unitary transforms are then presented, followed by performance comparisons between these transforms in terms of energy compactness, mean square reconstruction error, and computational complexity. It is demonstrated that the discrete cosine transform (DCT) performs better than others, in general. In the discussion of bit allocation, an efficient adaptive scheme is presented using thresholding coding devised by Chen and Pratt in 1984, which established a basis for the international still image coding standard, Joint Photographic (image) Experts Group (JPEG). The comparison between DPCM and TC is given. The combination of these two techniques (hybrid transform/waveform coding), and its application in image and video coding also are described.

The last two chapters in the first part cover some coding (codeword assignment) techniques. In Chapter 5, two types of variable-length coding techniques, Huffman coding and arithmetic coding, are discussed. First, an introduction to some basic coding theory is presented, which can be viewed as a continuation of the information theory results presented in Chapter 1. Then the Huffman code, as an optimum and instantaneous code, and a modified version are covered. Huffman coding is a systematic procedure for encoding a source alphabet with each source symbol having an occurrence probability. As a block code (a fixed codeword having an integer number of bits is assigned to a source symbol), it is optimum in the sense that it produces minimum coding redundancy. Some limitations of Huffman coding are analyzed. As a stream-based coding technique, arithmetic coding is distinct from and is gaining more popularity than Huffman coding. It maps a string of source symbols into a string of code symbols. Free of the integer-bits-per-source-symbol restriction, arithmetic coding is more efficient. The principle of arithmetic coding and some of its implementation issues are addressed.

While the two types of variable-length coding techniques introduced in Chapter 5 can be classified as fixed-length to variable-length coding techniques, both run-length coding (RLC) and dictionary coding, discussed in Chapter 6, can be classified as variable-length to fixed-length coding techniques. The discrete Markov source model (another portion of the information theory results) that can be used to characterize 1-D RLC, is introduced at the beginning of Chapter 6. Both 1-D RLC and 2-D RLC are then introduced. The comparison between 1-D and 2-D RLC is made in terms of coding efficiency and transmission error effect. The digital facsimile coding standards based on 1-D and 2-D RLC are introduced. Another focus of Chapter 6 is on dictionary coding. Two groups of adaptive dictionary coding techniques, the LZ77 and LZ78 algorithms, are presented and their applications are discussed. At the end of the chapter, a discussion of international standards for lossless still image compression is given. For both lossless bilevel and multilevel still image compression, the respective standard algorithms and their performance comparisons are provided.

Section II of the book (Chapters 7, 8, and 9) is devoted to still image compression. In Chapter 7, the international still image coding standard, JPEG, is introduced. Two classes of encoding: lossy and lossless; and four modes of operation: sequential DCT-based mode, progressive DCT-based mode, lossless mode, and hierarchical mode are covered. The discussion in the first part of the book is very useful in understanding what is introduced here for JPEG.

Due to its higher coding efficiency and superior spatial and quality scalability features over the DCT coding technique, the discrete wavelet transform (DWT) coding has been adopted by JPEG-2000 still image coding standards as the core technology. Chapter 8 begins with an introduction to wavelet transform (WT), which includes a comparison between WT and the short-time Fourier transform (STFT), and presents WT as a unification of several existing techniques known as filter bank analysis, pyramid coding, and subband coding. Then the DWT for still image coding is discussed. In particular, the embedded zerotree wavelet (EZW) technique and set partitioning in hierarchical trees (SPIHT) are discussed. The updated JPEG-2000 standard activity is presented.

Chapter 9 presents three nonstandard still image coding techniques: vector quantization (VQ), fractal, and model-based image coding. All three techniques have several important features such as very high compression ratios for certain kinds of images, and very simple decoding procedures. Due to some limitations, however, they have not been adopted by the still image coding standards. On the other hand, the facial model and face animation technique have been adopted by the MPEG-4 video standard.

Section III, consisting of Chapters 10 through 14, addresses the motion estimation and motion compensation — key issues in modern video compression. In this sense, Section III is a prerequisite to Section IV, which discusses various video coding standards. The first chapter in Section III, Chapter 10, introduces motion analysis and compensation in general. The chapter begins with the concept of imaging space, which characterizes all images and all image sequences in temporal and spatial domains. Both temporal and spatial image sequences are special proper subsets of the imaging space. A single image becomes merely a specific cross section of the imaging space. Two techniques in video compression utilizing interframe correlation, both developed in the late 1960s and early 1970s, are presented. Frame replenishment is relatively simpler in modeling and implementation. However, motion compensated coding achieves higher coding efficiency and better quality in reconstructed frames with a 2-D displacement model. Motion analysis is then viewed from the signal processing perspective. Three techniques in motion analysis are briefly discussed. They are block matching, pel recursion, and optical flow, which are presented in detail in Chapters 11, 12, and 13, respectively. Finally, other applications of motion compensation to image sequence processing are discussed.

Chapter 11 addresses the block matching technique, which presently is the most frequently used motion estimation technique. The chapter first presents the original block matching technique proposed by Jain and Jain. Several different matching criteria and search strategies are then discussed. A thresholding multiresolution block matching algorithm is described in some detail so as to provide an insight into the technique. Then, the limitations of block matching techniques are analyzed, from which several new improvements are presented. They include hierarchical block matching, multigrid block matching, predictive motion field segmentation, and overlapped block matching. All of these techniques modify the nonoverlapped, equally spaced, fix-sized, small rectangular block model proposed by Jain and Jain in some way so that the motion estimation is more accurate and has fewer block artifacts and less overhead side information.

The pel recursive technique is discussed in Chapter 12. First, determination of 2-D displacement vectors is converted via the use of the displaced frame difference (DFD) concept to a minimization problem. Second, descent methods in optimization theory are discussed. In particular, the steepest descent method and Newton-Raphson method are addressed in terms of algorithm, convergence, and implementation issues such as selection of step-size and initial value. Third, the first pel recursive techniques proposed by Netravali and Robbins are presented. Finally, several improvement algorithms are described.

Optical flow, the third technique in motion estimation for video coding, is covered in Chapter 13. First, some fundamental issues in motion estimation are addressed. They include the difference and relationships between 2-D motion and optical flow, the aperture problem, and the ill-posed nature of motion estimation. The gradient-based and correlation-based approaches to optical flow determination are then discussed in detail. For the former, the Horn and Schunck algorithm is illustrated as a representative technique and some other algorithms are briefly introduced. For the latter, the Singh method is introduced as a representative technique. In particular, the concepts of conservation information and neighborhood information are emphasized. A correlation-feedback algorithm is presented in detail to provide an insight into the correlation technique. Finally, multiple attributes for conservation information are discussed.

Chapter 14, the last chapter in Section III, provides a further discussion and summary of 2-D motion estimation. First, a few features common to all three major techniques discussed in Chapters 11, 12, and 13 are addressed. They are the aperture and ill-posed inverse problems, conservation and neighborhood information, occlusion and disocclusion, rigid and nonrigid motion. Second, a variety of different classifications of motion estimation techniques is presented. Frequency domain methods are discussed as well. Third, a performance comparison between the three major techniques in motion estimation is made. Finally, the new trends in motion estimation are presented.

Section IV, discussing various video coding standards, is covered in Chapters 15 through 20. Chapter 15 presents fundamentals of video coding. First, digital video representation is discussed. Second, the rate distortion function of the video signal is covered — the fourth portion of the information theory results presented in this book. Third, various digital video formats are discussed. Finally, the current digital image/video coding standards are summarized. The full names and abbreviations of some organizations, the completion time, and the major features of various image/video coding standards are listed in two tables.

Chapter 16 is devoted to video coding standards MPEG-1/2, which are the most widely used video coding standards at the present. The basic technique of MPEG-1/2 is a full-motion-compensated DCT and DPCM hybrid coding algorithm. The features of MPEG-1 (including layered data structure) and the MPEG-2 enhancements (including field/frame modes for supporting the interlaced video input and scalability extension) are described. Issues of rate control, optimum mode decision, and multiplexing are discussed.

Chapter 17 presents several application examples of MPEG-1/2 video standards. They are the ATSC DTV standard approved by the FCC in the U.S., transcoding, the down-conversion decoder, and error concealment. Discussion of these applications can enhance the understanding and mastering of MPEG-1/2 standards. Some research work is reported that may be helpful for graduate students to broaden their knowledge of digital video processing — an active research field.

Chapter 18 presents the MPEG-4 video standard. The predominant feature of MPEG-4, content-based manipulation, is emphasized. The underlying concept of audio/visual objects (AVOs) is introduced. The important functionalities of MPEG-4: content-based interactivity (including bitstream editing, synthetic and natural hybrid coding [SNHC]), content-based coding efficiency, and universal access (including content-based scalability), are discussed. Since neither MPEG-1 nor MPEG-2 includes synthetic video and content-based coding, the most important application of MPEG-4 is in a multimedia environment.

Chapter 19 introduces ITU-T video coding standards H.261 and H.263, which are utilized mainly for videophony and videoconferencing. The basic technical details of H.261, the earliest video coding standard, are presented. The technical improvements by which H.263 achieves high coding efficiency are discussed. Features of H.263+, H.263++, and H.26L are presented.

Chapter 20 covers the systems part of MPEG — multiplexing/demultiplexing and synchronizing the coded audio and video as well as other data. Specifically, MPEG-2 systems and MPEG-4 systems are introduced. In MPEG-2 systems, two forms: Program Stream and Transport Stream, are described. In MPEG-4 systems, some multimedia application related issues are discussed.

# Contents

## Section I Fundamentals

**Chapter 1    Introduction**
1.1    Practical Needs for Image and Video Compression..................................................4
1.2    Feasibility of Image and Video Compression ........................................................4
    1.2.1    Statistical Redundancy ..............................................................................4
    1.2.2    Psychovisual Redundancy.........................................................................9
1.3    Visual Quality Measurement ................................................................................18
    1.3.1    Subjective Quality Measurement...........................................................19
    1.3.2    Objective Quality Measurement ............................................................20
1.4    Information Theory Results...................................................................................24
    1.4.1    Entropy......................................................................................................24
    1.4.2    Shannon's Noiseless Source Coding Theorem.....................................25
    1.4.3    Shannon's Noisy Channel Coding Theorem .......................................26
    1.4.4    Shannon's Source Coding Theorem ......................................................27
    1.4.5    Information Transmission Theorem........................................................27
1.5    Summary .................................................................................................................27
1.6    Exercises..................................................................................................................28
References ..........................................................................................................................28

**Chapter 2    Quantization**
2.1    Quantization and the Source Encoder .................................................................31
2.2    Uniform Quantization ...........................................................................................33
    2.2.1    Basics .........................................................................................................33
    2.2.2    Optimum Uniform Quantizer.................................................................37
2.3    Nonuniform Quantization .....................................................................................40
    2.3.1    Optimum (Nonuniform) Quantization ..................................................42
    2.3.2    Companding Quantization ......................................................................43
2.4    Adaptive Quantization ...........................................................................................45
    2.4.1    Forward Adaptive Quantization .............................................................47
    2.4.2    Backward Adaptive Quantization ..........................................................48
    2.4.3    Adaptive Quantization with a One-Word Memory .............................48
    2.4.4    Switched Quantization ............................................................................48
2.5    PCM .........................................................................................................................49
2.6    Summary ..................................................................................................................50
2.7    Exercises...................................................................................................................52
References ..........................................................................................................................52

**Chapter 3    Differential Coding**
3.1    Introduction to DPCM ...........................................................................................55
    3.1.1    Simple Pixel-to-Pixel DPCM..................................................................55
    3.1.2    General DPCM Systems ..........................................................................58
3.2    Optimum Linear Prediction ..................................................................................60

    3.2.1    Formulation .................................................................................................60
    3.2.2    Orthogonality Condition and Minimum Mean Square Error.......................61
    3.2.3    Solution to Yule-Walker Equations.............................................................62
3.3    Some Issues in the Implementation of DPCM.......................................................62
    3.3.1    Optimum DPCM System...............................................................................62
    3.3.2    1-D, 2-D, and 3-D DPCM ...........................................................................63
    3.3.3    Order of Predictor........................................................................................64
    3.3.4    Adaptive Prediction......................................................................................64
    3.3.5    Effect of Transmission Errors......................................................................65
3.4    Delta Modulation ...................................................................................................65
3.5    Interframe Differential Coding ..............................................................................68
    3.5.1    Conditional Replenishment..........................................................................68
    3.5.2    3-D DPCM ...................................................................................................69
    3.5.3    Motion-Compensated Predictive Coding.....................................................71
3.6    Information-Preserving Differential Coding...........................................................71
3.7    Summary ................................................................................................................72
3.8    Exercises.................................................................................................................73
References .......................................................................................................................73

**Chapter 4        Transform Coding**

4.1    Introduction ...........................................................................................................75
    4.1.1    Hotelling Transform.....................................................................................75
    4.1.2    Statistical Interpretation ...............................................................................77
    4.1.3    Geometrical Interpretation ...........................................................................78
    4.1.4    Basis Vector Interpretation...........................................................................79
    4.1.5    Procedures of Transform Coding.................................................................80
4.2    Linear Transforms..................................................................................................80
    4.2.1    2-D Image Transformation Kernel ..............................................................80
    4.2.2    Basis Image Interpretation...........................................................................83
    4.2.3    Subimage Size Selection..............................................................................84
4.3    Transforms of Particular Interest ..........................................................................84
    4.3.1    Discrete Fourier Transform (DFT) ..............................................................85
    4.3.2    Discrete Walsh Transform (DWT)...............................................................86
    4.3.3    Discrete Hadamard Transform (DHT).........................................................87
    4.3.4    Discrete Cosine Transform (DCT) ..............................................................88
    4.3.5    Performance Comparison.............................................................................92
4.4    Bit Allocation ........................................................................................................95
    4.4.1    Zonal Coding................................................................................................95
    4.4.2    Threshold Coding.........................................................................................96
4.5    Some Issues..........................................................................................................102
    4.5.1    Effect of Transmission Errors....................................................................102
    4.5.2    Reconstruction Error Sources ....................................................................102
    4.5.3    Comparison Between DPCM and TC .........................................................103
    4.5.4    Hybrid Coding............................................................................................103
4.6    Summary ..............................................................................................................103
4.7    Exercises..............................................................................................................105
References .....................................................................................................................106

**Chapter 5        Variable-Length Coding: Information Theory Results (II)**

5.1    Some Fundamental Results....................................................................................107

5.1.1 Coding an Information Source ...........................................................................107
5.1.2 Some Desired Characteristics .........................................................................108
5.1.3 Discrete Memoryless Sources .........................................................................111
5.1.4 Extensions of a Discrete Memoryless Source .................................................112
5.2 Huffman Codes ...........................................................................................................114
5.2.1 Required Rules for Optimum Instantaneous Codes .......................................114
5.2.2 Huffman Coding Algorithm ............................................................................115
5.3 Modified Huffman Codes ............................................................................................117
5.3.1 Motivation .......................................................................................................117
5.3.2 Algorithm ........................................................................................................118
5.3.3 Codebook Memory Requirement ....................................................................118
5.3.4 Bounds on Average Codeword Length ............................................................119
5.4 Arithmetic Codes ........................................................................................................119
5.4.1 Limitations of Huffman Coding ......................................................................120
5.4.2 Principle of Arithmetic Coding .......................................................................120
5.4.3 Implementation Issues .....................................................................................125
5.4.4 History .............................................................................................................126
5.4.5 Applications .....................................................................................................127
5.5 Summary ......................................................................................................................127
5.6 Exercises......................................................................................................................128
References .............................................................................................................................129

**Chapter 6    Run-Length and Dictionary Coding: Information Theory Results (III)**
6.1 Markov Source Model ..................................................................................................131
6.1.1 Discrete Markov Source ..................................................................................131
6.1.2 Extensions of a Discrete Markov Source ........................................................133
6.1.3 Autoregressive (AR) Model ............................................................................133
6.2 Run-Length Coding (RLC) ..........................................................................................134
6.2.1 1-D Run-Length Coding ..................................................................................134
6.2.2 2-D Run-Length Coding ..................................................................................135
6.2.3 Effect of Transmission Error and Uncompressed Mode..................................138
6.3 Digital Facsimile Coding Standards ............................................................................139
6.4 Dictionary Coding........................................................................................................140
6.4.1 Formulation of Dictionary Coding .................................................................140
6.4.2 Categorization of Dictionary-Based Coding Techniques ...............................140
6.4.3 Parsing Strategy ..............................................................................................141
6.4.4 Sliding Window (LZ77) Algorithms................................................................142
6.4.5 LZ78 Algorithms..............................................................................................145
6.5 International Standards for Lossless Still Image Compression ....................................149
6.5.1 Lossless Bilevel Still Image Compression ......................................................150
6.5.2 Lossless Multilevel Still Image Compression .................................................150
6.6 Summary ......................................................................................................................151
6.7 Exercises......................................................................................................................152
References .............................................................................................................................153

# Section II   Still Image Compression

**Chapter 7    Still Image Coding Standard: JPEG**
7.1 Introduction .................................................................................................................157
7.2 Sequential DCT-Based Encoding Algorithm................................................................159

7.3    Progressive DCT-Based Encoding Algorithm ...............................................163
7.4    Lossless Coding Mode ...............................................................................164
7.5    Hierarchical Coding Mode .........................................................................166
7.6    Summary ...................................................................................................167
7.7    Exercises ...................................................................................................167
References ........................................................................................................167

**Chapter 8       Wavelet Transform for Image Coding**
8.1    Review of the Wavelet Transform .............................................................169
       8.1.1    Definition and Comparison with Short-Time Fourier Transform .......169
       8.1.2    Discrete Wavelet Transform ..........................................................172
8.2    Digital Wavelet Transform for Image Compression ...................................174
       8.2.1    Basic Concept of Image Wavelet Transform Coding .......................174
       8.2.2    Embedded Image Wavelet Transform Coding Algorithms ................176
8.3    Wavelet Transform for JPEG-2000 ...........................................................179
       8.3.1    Introduction of JPEG-2000 ...........................................................179
       8.3.2    Verification Model of JPEG-2000 ..................................................180
8.4    Summary ...................................................................................................182
8.5    Exercises ...................................................................................................182
References ........................................................................................................183

**Chapter 9       Nonstandard Image Coding**
9.1    Introduction ..............................................................................................185
9.2    Vector Quantization ...................................................................................186
       9.2.1    Basic Principle of Vector Quantization ..........................................186
       9.2.2    Several Image Coding Schemes with Vector Quantization ...............189
       9.2.3    Lattice VQ for Image Coding ........................................................191
9.3    Fractal Image Coding .................................................................................193
       9.3.1    Mathematical Foundation ..............................................................193
       9.3.2    *IFS*-Based Fractal Image Coding ....................................................195
       9.3.3    Other Fractal Image Coding Methods .............................................197
9.4    Model-Based Coding ..................................................................................197
       9.4.1    Basic Concept ...............................................................................197
       9.4.2    Image Modeling ............................................................................198
9.5    Summary ...................................................................................................198
9.6    Exercises ...................................................................................................198
References ........................................................................................................199

# Section III   Motion Estimation and Compression

**Chapter 10     Motion Analysis and Motion Compensation**
10.1    Image Sequences .......................................................................................203
10.2    Interframe Correlation ...............................................................................205
10.3    Frame Replenishment ................................................................................208
10.4    Motion-Compensated Coding .....................................................................209
10.5    Motion Analysis ........................................................................................211
       10.5.1   Biological Vision Perspective .........................................................212
       10.5.2   Computer Vision Perspective .........................................................212
       10.5.3   Signal Processing Perspective ........................................................213

10.6    Motion Compensation for Image Sequence Processing ....................................214
        10.6.1  Motion-Compensated Interpolation ...........................................214
        10.6.2  Motion-Compensated Enhancement ............................................215
        10.6.3  Motion-Compensated Restoration .............................................217
        10.6.4  Motion-Compensated Down-Conversion ....................................217
10.7    Summary ...........................................................................................217
10.8    Exercises ............................................................................................218
References ....................................................................................................219

**Chapter 11    Block Matching**

11.1    Nonoverlapped, Equally Spaced, Fixed Size, Small Rectangular Block Matching ..........221
11.2    Matching Criteria ...............................................................................222
11.3    Searching Procedures ..........................................................................224
        11.3.1  Full Search ...............................................................................224
        11.3.2  2-D Logarithm Search ...............................................................224
        11.3.3  Coarse-Fine Three-Step Search ..................................................226
        11.3.4  Conjugate Direction Search .......................................................226
        11.3.5  Subsampling in the Correlation Window ....................................227
        11.3.6  Multiresolution Block Matching ................................................227
        11.3.7  Thresholding Multiresolution Block Matching ...........................229
11.4    Matching Accuracy .............................................................................234
11.5    Limitations with Block Matching Techniques .......................................235
11.6    New Improvements .............................................................................236
        11.6.1  Hierarchical Block Matching ....................................................236
        11.6.2  Multigrid Block Matching .........................................................238
        11.6.3  Predictive Motion Field Segmentation ......................................242
        11.6.4  Overlapped Block Matching ......................................................244
11.7    Summary ...........................................................................................245
11.8    Exercises ............................................................................................247
References ....................................................................................................248

**Chapter 12    PEL Recursive Technique**

12.1    Problem Formulation ..........................................................................251
12.2    Descent Methods ................................................................................252
        12.2.1  First-Order Necessary Conditions..............................................252
        12.2.2  Second-Order Sufficient Conditions ..........................................253
        12.2.3  Underlying Strategy ..................................................................253
        12.2.4  Convergence Speed ...................................................................255
        12.2.5  Steepest Descent Method ..........................................................256
        12.2.6  Newton-Raphson's Method .......................................................257
        12.2.7  Other Methods ..........................................................................258
12.3    Netravali-Robbins Pel Recursive Algorithm .........................................258
        12.3.1  Inclusion of a Neighborhood Area .............................................259
        12.3.2  Interpolation.............................................................................259
        12.3.3  Simplification............................................................................259
        12.3.4  Performance ..............................................................................260
12.4    Other Pel Recursive Algorithms ..........................................................260
        12.4.1  The Bergmann Algorithm (1982)................................................260
        12.4.2  The Bergmann Algorithm (1984)................................................260
        12.4.3  The Cafforio and Rocca Algorithm ............................................261
        12.4.4  The Walker and Rao Algorithm .................................................261

12.5 Performance Comparison...........................................................................................261
12.6 Summary ..............................................................................................................262
12.7 Exercises...............................................................................................................262
References ........................................................................................................................263

**Chapter 13     Optical Flow**

13.1 Fundamentals .......................................................................................................265
    13.1.1  2-D Motion and Optical Flow..................................................................265
    13.1.2  Aperture Problem ....................................................................................266
    13.1.3  Ill-Posed Inverse Problem .......................................................................267
    13.1.4  Classification of Optical Flow Techniques ..............................................269
13.2 Gradient-Based Approach....................................................................................269
    13.2.1  The Horn and Schunck Method................................................................269
    13.2.2  Modified Horn and Schunck Method .......................................................273
    13.2.3  The Lucas and Kanade Method ................................................................275
    13.2.4  The Nagel Method....................................................................................276
    13.2.5  The Uras, Girosi, Verri, and Torre Method ..............................................276
13.3 Correlation-Based Approach................................................................................276
    13.3.1  The Anandan Method................................................................................277
    13.3.2  The Singh Method.....................................................................................278
    13.3.3  The Pan, Shi, and Shu Method.................................................................281
13.4 Multiple Attributes for Conservation Information ..............................................293
    13.4.1  The Weng, Ahuja, and Huang Method .....................................................294
    13.4.2  The Xia and Shi Method...........................................................................296
13.5 Summary ..............................................................................................................300
13.6 Exercises...............................................................................................................301
References ........................................................................................................................302

**Chapter 14     Further Discussion and Summary on 2-D Motion Estimation**

14.1 General Characterization......................................................................................305
    14.1.1  Aperture Problem .....................................................................................305
    14.1.2  Ill-Posed Inverse Problem .......................................................................305
    14.1.3  Conservation Information and Neighborhood Information......................306
    14.1.4  Occlusion and Disocclusion.....................................................................306
    14.1.5  Rigid and Nonrigid Motion......................................................................307
14.2 Different Classifications.......................................................................................308
    14.2.1  Deterministic Methods vs. Stochastic Methods ......................................308
    14.2.2  Spatial Domain Methods vs. Frequency Domain Methods ......................308
    14.2.3  Region-Based Approaches vs. Gradient-Based Approaches ....................311
    14.2.4  Forward vs. Backward Motion Estimation ..............................................312
14.3 Performance Comparison Among Three Major Approaches...............................313
    14.3.1  Three Representatives ..............................................................................313
    14.3.2  Algorithm Parameters...............................................................................314
    14.3.3  Experimental Results and Observations ..................................................314
14.4 New Trends ..........................................................................................................315
    14.4.1  DCT-Based Motion Estimation................................................................315
14.5 Summary ..............................................................................................................318
14.6 Exercises...............................................................................................................319
References ........................................................................................................................319

# Section IV  Video Compression

**Chapter 15      Fundamentals of Digital Video Coding**

15.1   Digital Video Representation ...................................................................................323
15.2   Information Theory Results (IV): Rate Distortion Function of Video Signal ...................324
15.3   Digital Video Formats ...........................................................................................327
15.4   Current Status of Digital Video/Image Coding Standards .........................................328
15.5   Summary ..............................................................................................................331
15.6   Exercises ...............................................................................................................331
References .......................................................................................................................332

**Chapter 16      Digital Video Coding Standards — MPEG-1/2 Video**

16.1   Introduction ..........................................................................................................333
16.2   Features of MPEG-1/2 Video Coding .....................................................................333
        16.2.1   MPEG-1 Features ...................................................................................334
        16.2.2   MPEG-2 Enhancements .........................................................................340
16.3   MPEG-2 Video Encoding ......................................................................................346
        16.3.1   Introduction ...........................................................................................346
        16.3.2   Preprocessing .........................................................................................346
        16.3.3   Motion Estimation and Motion Compensation .......................................347
16.4   Rate Control .........................................................................................................350
        16.4.1   Introduction of Rate Control ...................................................................350
        16.4.2   Rate Control of Test Model 5 (TM5) for MPEG-2 ...................................350
16.5   Optimum Mode Decision ......................................................................................354
        16.5.1   Problem Formation ................................................................................354
        16.5.2   Procedure for Obtaining the Optimal Mode ...........................................357
        16.5.3   Practical Solution with New Criteria for the Selection of Coding Mode ..........359
16.6   Statistical Multiplexing Operations on Multiple Program Encoding ........................360
        16.6.1   Background of Statistical Multiplexing Operation ....................................360
        16.6.2   VBR Encoders in StatMux ......................................................................362
        16.6.3   Research Topics of StatMux ....................................................................363
16.7   Summary ..............................................................................................................365
16.8   Exercises ...............................................................................................................365
References .......................................................................................................................366

**Chapter 17      Application Issues of MPEG-1/2 Video Coding**

17.1   Introduction ..........................................................................................................367
17.2   ATSC DTV Standards .............................................................................................367
        17.2.1   A Brief History .......................................................................................367
        17.2.2   Technical Overview of ATSC Systems .....................................................368
17.3   Transcoding with Bitstream Scaling ........................................................................371
        17.3.1   Background .............................................................................................371
        17.3.2   Basic Principles of Bitstream Scaling ......................................................373
        17.3.3   Architectures of Bitstream Scaling ..........................................................374
        17.3.4   Analysis .................................................................................................378
17.4   Down-Conversion Decoder ....................................................................................379
        17.4.1   Background .............................................................................................379
        17.4.2   Frequency Synthesis Down-Conversion ...................................................381

17.4.3 Low-Resolution Motion Compensation ...................................................383
17.4.4 Three-Layer Scalable Decoder.............................................................385
17.4.5 Summary of Down-Conversion Decoder...........................................388
17.4.6 DCT-to-Spatial Transformation............................................................388
17.4.7 Full-Resolution Motion Compensation in Matrix Form ...............389
17.5 Error Concealment ...........................................................................................391
17.5.1 Background...................................................................................................391
17.5.2 Error Concealment Algorithms .............................................................392
17.5.3 Algorithm Enhancements ........................................................................397
17.5.4 Summary of Error Concealment ...........................................................400
17.6 Summary .............................................................................................................400
17.7 Exercises..............................................................................................................401
References .....................................................................................................................401

**Chapter 18     MPEG-4 Video Standard: Content-Based Video Coding**

18.1 Introduction .......................................................................................................403
18.2 MPEG-4 Requirements and Functionalities.................................................404
18.2.1 Content-Based Interactivity.....................................................................404
18.2.2 Content-Based Efficient Compression ...............................................404
18.2.3 Universal Access .......................................................................................405
18.2.4 Summary of MPEG-4 Features..............................................................405
18.3 Technical Description of MPEG-4 Video......................................................406
18.3.1 Overview of MPEG-4 Video...................................................................406
18.3.2 Motion Estimation and Compensation ...............................................407
18.3.3 Texture Coding ...........................................................................................409
18.3.4 Shape Coding ..............................................................................................413
18.3.5 Sprite Coding...............................................................................................416
18.3.6 Interlaced Video Coding..........................................................................417
18.3.7 Wavelet-Based Texture Coding..............................................................417
18.3.8 Generalized Spatial and Temporal Scalability.................................418
18.3.9 Error Resilience .........................................................................................419
18.4 MPEG-4 Visual Bitstream Syntax and Semantics .....................................420
18.5 MPEG-4 Video Verification Model ...............................................................421
18.5.1 VOP-Based Encoding and Decoding Process .................................422
18.5.2 Video Encoder ...........................................................................................422
18.5.3 Video Decoder ...........................................................................................426
18.6 Summary .............................................................................................................427
18.7 Exercises..............................................................................................................427
Reference........................................................................................................................427

**Chapter 19     ITU-T Video Coding Standards H.261 and H.263**

19.1 Introduction .......................................................................................................429
19.2 H.261 Video-Coding Standard........................................................................429
19.2.1 Overview of H.261 Video-Coding Standard......................................429
19.2.2 Technical Detail of H.261 .......................................................................430
19.2.3 Syntax Description .....................................................................................432
19.3 H.263 Video-Coding Standard........................................................................433
19.3.1 Overview of H.263 Video Coding ........................................................433
19.3.2 Technical Features of H.263 ...................................................................434
19.4 H.263 Video-Coding Standard Version 2 .....................................................439

      19.4.1  Overview of H.263 Version 2 ................................................................439
      19.4.2  New Features of H.263 Version 2.........................................................439
19.5  H.263++ Video Coding and H.26L ...............................................................446
19.6  Summary ........................................................................................................447
19.7  Exercises........................................................................................................447
References .................................................................................................................447

**Chapter 20     MPEG System — Video, Audio, and Data Multiplexing**

20.1  Introduction ...................................................................................................449
20.2  MPEG-2 System ............................................................................................450
      20.2.1  Major Technical Definitions in MPEG-2 System Document..................450
      20.2.2  Transport Streams...............................................................................451
      20.2.3  Transport Stream Splicing...................................................................456
      20.2.4  Program Streams .................................................................................458
      20.2.5  Timing Model and Synchronization ....................................................459
20.3  MPEG-4 System ............................................................................................462
      20.3.1  Overview and Architecture..................................................................462
      20.3.2  Systems Decoder Model ......................................................................464
      20.3.3  Scene Description................................................................................465
      20.3.4  Object Description Framework ............................................................466
20.4  Summary ........................................................................................................466
20.5  Exercises........................................................................................................466
References .................................................................................................................467

**Index** ......................................................................................................................469

# Dedication

*To beloved Kong Wai Shih and Wen Su,*
*Yi Xi Li and Shu Jun Zheng,*
*Xian Hong Li,*
*and*
*To beloved Xuedong, Min, Yin, Andrew, and Haixin*

# Section I

## Fundamentals

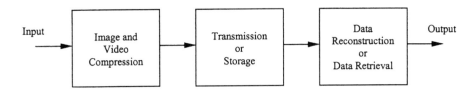

**FIGURE 1.1**    Image and video compression for visual transmission and storage.

## 1.1   PRACTICAL NEEDS FOR IMAGE AND VIDEO COMPRESSION

Needless to say, visual information is of vital importance if human beings are to perceive, recognize, and understand the surrounding world. With the tremendous progress that has been made in advanced technologies, particularly in very large scale integrated (VLSI) circuits, and increasingly powerful computers and computations, it is becoming more than ever possible for video to be widely utilized in our daily lives. Examples include videophony, videoconferencing, high definition TV (HDTV), and the digital video disk (DVD), to name a few.

Video as a sequence of video frames, however, involves a huge amount of data. Let us take a look at an illustrative example. Assume the present switch telephone network (PSTN) modem can operate at a maximum bit rate of 56,600 bits per second. Assume each video frame has a resolution of 288 by 352 (288 lines and 352 pixels per line), which is comparable with that of a normal TV picture and is referred to as common intermediate format (CIF). Each of the three primary colors RGB (red, green, blue) is represented for 1 pixel with 8 bits, as usual, and the frame rate in transmission is 30 frames per second to provide a continuous motion video. The required bit rate, then, is $288 \times 352 \times 8 \times 3 \times 30 = 72,990,720$ bps. Therefore, the ratio between the required bit rate and the largest possible bit rate is about 1289. This implies that we have to compress the video data by at least 1289 times in order to accomplish the transmission described in this example. Note that an audio signal has not yet been accounted for yet in this illustration.

With increasingly complex video services such as 3-D movies and 3-D games, and high video quality such as HDTV, advanced image and video data compression is necessary. It becomes an enabling technology to bridge the gap between the required huge amount of video data and the limited hardware capability.

## 1.2   FEASIBILITY OF IMAGE AND VIDEO COMPRESSION

In this section we shall see that image and video compression is not only a necessity for the rapid growth of digital visual communications, but it is also feasible. Its feasibility rests with two types of redundancies, i.e., statistical redundancy and psychovisual redundancy. By eliminating these redundancies, we can achieve image and video compression.

### 1.2.1   STATISTICAL REDUNDANCY

Statistical redundancy can be classified into two types: interpixel redundancy and coding redundancy. By interpixel redundancy we mean that pixels of an image frame and pixels of a group of successive image or video frames are not statistically independent. On the contrary, they are correlated to various degrees. (Note that the differences and relationships between image and video sequences are discussed in Chapter 10, when we begin to discuss video compression.) This type of interpixel correlation is referred to as interpixel redundancy. Interpixel redundancy can be divided into two categories, spatial redundancy and temporal redundancy. By coding redundancy we mean the statistical redundancy associated with coding techniques.

### 1.2.1.1  Spatial Redundancy

Spatial redundancy represents the statistical correlation between pixels within an image frame. Hence it is also called intraframe redundancy.

It is well known that for most properly sampled TV signals the normalized autocorrelation coefficients along a row (or a column) with a one-pixel shift is very close to the maximum value of 1. That is, the intensity values of pixels along a row (or a column) have a very high autocorrelation (close to the maximum autocorrelation) with those of pixels along the same row (or the same column), but shifted by a pixel. This does not come as a surprise because most of the intensity values change continuously from pixel to pixel within an image frame except for the edge regions. This is demonstrated in Figure 1.2. Figure 1.2(a) is a normal picture — a boy and a girl in a park, and is of a resolution of 883 by 710. The intensity profiles along the 318th row and the 262nd column are depicted in Figures 1.2(b) and (c), respectively. For easy reference, the positions of the 318th row and 262nd column in the picture are shown in Figure 1.2(d). That is, the vertical axis represents intensity values, while the horizontal axis indicates the pixel position within the row or the column. These two plots (shown in Figures 1.2(b) and 1.2(c)) indicate that intensity values often change gradually from one pixel to the other along a row and along a column.

The study of the statistical properties of video signals can be traced back to the 1950s. Knowing that we must study and understand redundancy in order to remove redundancy, Kretzmer designed some experimental devices such as a picture autocorrelator and a probabiloscope to measure several statistical quantities of TV signals and published his outstanding work in (Kretzmer, 1952). He found that the autocorrelation in both horizontal and vertical directions exhibits similar behaviors, as shown in Figure 1.3. Autocorrelation functions of several pictures with different complexities were measured. It was found that from picture to picture, the shape of the autocorrelation curves ranges from remarkably linear to somewhat exponential. The central symmetry with respect to the vertical axis and the bell-shaped distribution, however, remains generally the same. When the pixel shifting becomes small, it was found that the autocorrelation is high. This "local" autocorrelation can be as high as 0.97 to 0.99 for one- or two-pixel shifting. For very detailed pictures, it can be from 0.43 to 0.75. It was also found that autocorrelation generally has no preferred direction.

The Fourier transform of autocorrelation, the power spectrum, is known as another important function in studying statistical behavior. Figure 1.4 shows a typical power spectrum of a television signal (Fink, 1957; Connor et al., 1972). It is reported that the spectrum is quite flat until 30 kHz for a broadcast TV signal. Beyond this line frequency the spectrum starts to drop at a rate of around 6 dB per octave. This reveals the heavy concentration of video signals in low frequencies, considering a nominal bandwidth of 5 MHz.

Spatial redundancy implies that the intensity value of a pixel can be *guessed* from that of its neighboring pixels. In other words, it is not necessary to represent each pixel in an image frame independently. Instead, one can predict a pixel from its neighbors. Predictive coding, also known as differential coding, is based on this observation and is discussed in Chapter 3. The direct consequence of recognition of spatial redundancy is that by removing a large amount of the redundancy (or utilizing the high correlation) within an image frame, we may save a lot of data in representing the frame, thus achieving data compression.

### 1.2.1.2  Temporal Redundancy

Temporal redundancy is concerned with the statistical correlation between pixels from successive frames in a temporal image or video sequence. Therefore, it is also called interframe redundancy.

Consider a temporal image sequence. That is, a camera is fixed in the 3-D world and it takes pictures of the scene one by one as time goes by. As long as the time interval between two consecutive pictures is short enough, i.e., the pictures are taken densely enough, we can imagine that the similarity between two neighboring frames is strong. Figures 1.5(a) and (b) show, respectively,

## Row Profile

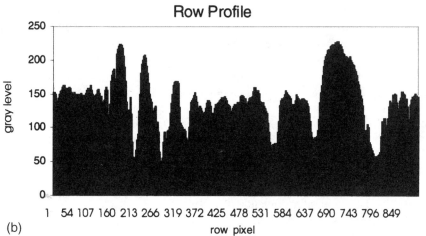

(b)

## Column Profile

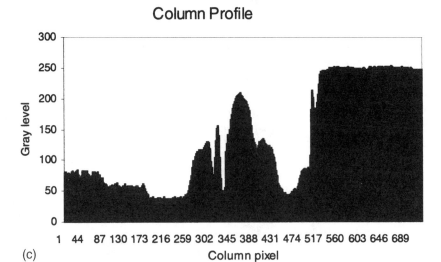

(c)

**FIGURE 1.2**  (a) A picture of "Boy and Girl," (b) Intensity profile along 318th row, (c) Intensity profile along 262nd column, (d) Positions of 318th row and 262nd column.

**FIGURE 1.2** (continued)

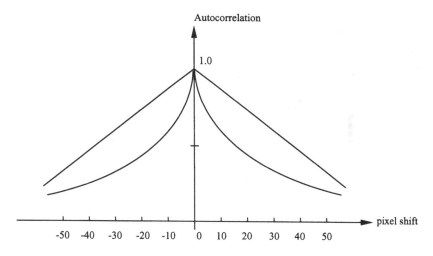

**FIGURE 1.3** Autocorrelation in the horizontal direction for some pictures. (After Kretzmer, 1952.)

the 21st and 22nd frames of the "Miss America" sequence. The frames have a resolution of 176 by 144. Among the total of 25,344 pixels, only 3.4% change their gray value by more than 1% of the maximum gray value (255 in this case) from the 21st frame to the 22nd frame. This confirms an observation made in (Mounts, 1969): for a videophone-like signal with moderate motion in the scene, on average, less than 10% of pixels change their gray values between two consecutive frames by an amount of 1% of the peak signal. The high interframe correlation was reported in (Kretzmer, 1952). There, the autocorrelation between two adjacent frames was measured for two typical motion-picture films. The measured autocorrelations are 0.80 and 0.86. In summary, pixels within successive frames usually bear a strong similarity or correlation.

As a result, we may predict a frame from its neighboring frames along the temporal dimension. This is referred to as interframe predictive coding and is discussed in Chapter 3. A more precise hence, more efficient interframe predictive coding scheme, which has been in development sin

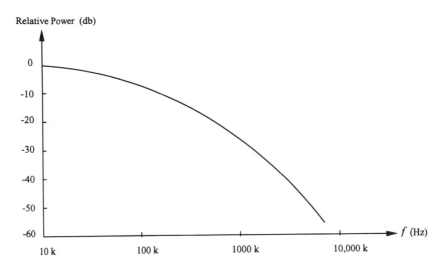

**FIGURE 1.4** Typical power spectrum of a TV broadcast signal. (Adapted from Fink, D.G., *Television Engineering Handbook,* McGraw-Hill, New York, 1957.)

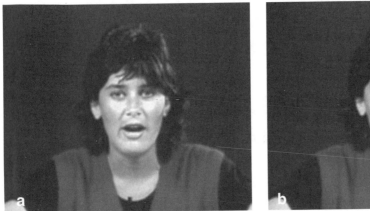

**FIGURE 1.5**   (a) The 21st frame, and (b) 22nd frame of the "Miss America" sequence.

the 1980s, uses motion analysis. That is, it considers that the changes from one frame to the next are mainly due to the motion of some objects in the frame. Taking this motion information into consideration, we refer to the method as motion compensated predictive coding. Both interframe correlation and motion compensated predictive coding are covered in detail in Chapter 10.

Removing a large amount of temporal redundancy leads to a great deal of data compression. At present, all the international video coding standards have adopted motion compensated predictive coding, which has been a vital factor to the increased use of digital video in digital media.

### 1.2.1.3  Coding Redundancy

As we discussed, interpixel redundancy is concerned with the correlation between pixels. That is, some information associated with pixels is redundant. The psychovisual redundancy, which is discussed in the next subsection, is related to the information that is psychovisually redundant, i.e., to which the HVS is not sensitive. Hence, it is clear that both the interpixel and psychovisual redundancies are somehow associated with some information contained in the image and video. Eliminating these redundancies, or utilizing these correlations, by using fewer bits to represent the information results in image and video data compression. In this sense, the coding redundancy is different. It has nothing to do with information redundancy but with the representation of information, i.e., coding itself. To see this, let us take a look at the following example.

**TABLE 1.1**
**An Illustrative Example**

| Symbol | Occurrence Probability | Code 1 | Code 2 |
|--------|------------------------|--------|--------|
| $a_1$ | 0.1 | 000 | 0000 |
| $a_2$ | 0.2 | 001 | 01 |
| $a_3$ | 0.5 | 010 | 1 |
| $a_4$ | 0.05 | 011 | 0001 |
| $a_5$ | 0.15 | 100 | 001 |

One illustrative example is provided in Table 1.1. The first column lists five distinct symbols that need to be encoded. The second column contains occurrence probabilities of these five symbols. The third column lists code 1, a set of codewords obtained by using uniform-length codeword assignment. (This code is known as the natural binary code.) The fourth column shows code 2, in which each codeword has a variable length. Therefore, code 2 is called the variable-length code. It is noted that the symbol with a higher occurrence probability is encoded with a shorter length. Let us examine the efficiency of the two different codes. That is, we will examine which one provides a shorter average length of codewords. It is obvious that the average length of codewords in code 1, $L_{avg,1}$, is three bits. The average length of codewords in code 2, $L_{avg,2}$, can be calculated as follows.

$$L_{avg,2} = 4 \times 0.1 + 2 \times 0.2 + 1 \times 0.5 + 4 \times 0.05 + 3 \times 0.15 = 1.95 \; bits \; per \; symbol \qquad (1.1)$$

Therefore, it is concluded that code 2 with variable-length coding is more efficient than code 1 with natural binary coding.

From this example, we can see that for the same set of symbols different codes may perform differently. Some may be more efficient than others. For the same amount of information, code 1 contains some redundancy. That is, some data in code 1 are not necessary and can be removed without any effect. Huffman coding and arithmetic coding, two variable-length coding techniques, will be discussed in Chapter 5.

From the study of coding redundancy, it is clear that we should search for more efficient coding techniques in order to compress image and video data.

## 1.2.2 PSYCHOVISUAL REDUNDANCY

While interpixel redundancy inherently rests in image and video data, psychovisual redundancy originates from the characteristics of the human visual system (HVS).

It is known that the HVS perceives the outside world in a rather complicated way. Its response to visual stimuli is not a linear function of the strength of some physical attributes of the stimuli, such as intensity and color. HVS perception is different from camera sensing. In the HVS, visual information is not perceived equally; some information may be more important than other information. This implies that if we apply fewer data to represent less important visual information, perception will not be affected. In this sense, we see that some visual information is psychovisually redundant. Eliminating this type of psychovisual redundancy leads to data compression.

In order to understand this type of redundancy, let us study some properties of the HVS. We may model the human vision system as a cascade of two units (Lim, 1990), as depicted in Figure 1.6.

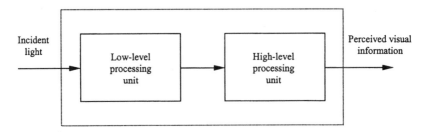

**FIGURE 1.6**   A two-unit cascade model of the human visual system (HVS).

The first one is a low-level processing unit which converts incident light into a neural signal. The second one is a high-level processing unit, which extracts information from the neural signal. While much research has been carried out to investigate low-level processing, high-level processing remains wide open. The low-level processing unit is known as a nonlinear system (approximately logarithmic, as shown below). While a great body of literature exists, we will limit our discussion only to video compression-related results. That is, several aspects of the HVS which are closely related to image and video compression are discussed in this subsection. They are luminance masking, texture masking, frequency masking, temporal masking, and color masking. Their relevance in image and video compression is addressed. Finally, a summary is provided in which it is pointed out that all of these features can be unified as one: differential sensitivity. This seems to be the most important feature of human visual perception.

### 1.2.2.1   Luminance Masking

Luminance masking concerns the brightness perception of the HVS, which is the most fundamental aspect among the five to be discussed here. Luminance masking is also referred to as *luminance dependence* (Connor et al., 1972), and *contrast masking* (Legge and Foley, 1980, Watson, 1987). As pointed in (Legge and Foley, 1980), the term *masking* usually refers to a destructive interaction or interference among stimuli that are closely coupled in time or space. This may result in a failure in detection, or errors in recognition. Here, we are mainly concerned with the detectability of one stimulus when another stimulus is present simultaneously. The effect of one stimulus on the detectability of another, however, does not have to decrease detectability. Indeed, there are some cases in which a low-contrast masker increases the detectability of a signal. This is sometimes referred to as *facilitation*, but in this discussion we only use the term masking.

Consider the monochrome image shown in Figure 1.7. There, a uniform disk-shaped object with a gray level (intensity value) $I_1$ is imposed on a uniform background with a gray level $I_2$. Now the question is under what circumstances can the disk-shaped object be discriminated from the background by the HVS? That is, we want to study the effect of one stimulus (the background in this example, the masker) on the detectability of another stimulus (in this example, the disk). Two extreme cases are obvious. That is, if the difference between the two gray levels is quite large, the HVS has no problem with discrimination, or in other words the HVS notices the object from the background. If, on the other hand, the two gray levels are the same, the HVS cannot identify the existence of the object. What we are concerned with here is the critical threshold in the gray level difference for discrimination to take place.

If we define the threshold $\Delta I$ as such a gray level difference $\Delta I = I_1 - I_2$ that the object can be noticed by the HVS with a 50% chance, then we have the following relation, known as *contrast sensitivity function*, according to Weber's law:

$$\frac{\Delta I}{I} \approx constant \qquad (1.2)$$

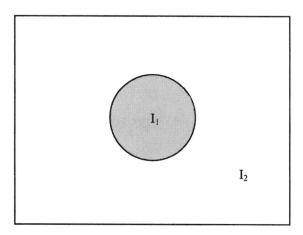

**FIGURE 1.7**    A uniform object with gray level $I_1$ imposed on a uniform background with gray level $I_2$.

where the constant is about 0.02. Weber's law states that for a relatively very wide range of I, the threshold for discrimination, $\Delta I$, is directly proportional to the intensity $I$. The implication of this result is that when the background is bright, a larger difference in gray levels is needed for the HVS to discriminate the object from the background. On the other hand, the intensity difference required could be smaller if the background is relatively dark. It is noted that Equation 1.2 implies a logarithmic response of the HVS, and Weber's law holds for all other human senses as well.

Further research has indicated that the luminance threshold $\Delta I$ increases more slowly than is predicted by Weber's law. Some more accurate contrast sensitivity functions have been presented in the literature. In (Legge and Foley, 1980), it was reported that an exponential function replaces the linear relation in Weber's law. The following exponential expression is reported in (Watson, 1987).

$$\Delta I = I_0 \cdot \max\left\{1, \left(\frac{I}{I_0}\right)^\alpha\right\}, \qquad (1.3)$$

where $I_0$ is the luminance detection threshold when the gray level of the background is equal to zero, i.e., $I = 0$, and $\alpha$ is a constant, approximately equal to 0.7.

Figure 1.8 shows a picture uniformly corrupted by additive white Gaussian noise (AWGN). It can be observed that the noise is more visible in the dark areas than in the bright areas if comparing, for instance, the dark portion and the bright portion of the cloud above the bridge. This indicates that noise filtering is more necessary in the dark areas than in the bright areas. The lighter areas can accommodate more additive noise before the noise becomes visible. This property has found application in embedding digital watermarks (Huang and Shi, 1998).

The direct impact that luminance masking has on image and video compression is related to quantization, which is covered in detail in the next chapter. Roughly speaking, quantization is a process that converts a continuously distributed quantity into a set of many finitely distinct quantities. The number of these distinct quantities (known as quantization levels) is one of the keys in quantizer design. It significantly influences the resulting bit rate and the quality of the reconstructed image and video. An effective quantizer should be able to minimize the visibility of quantization error. The contrast sensitivity function provides a guideline in analysis of the visibility of quantization error. Therefore, it can be applied to quantizer design. Luminance masking suggests a nonuniform quantization scheme that takes the contrast sensitivity function into consideration. One such example was presented in (Watson, 1987).

**FIGURE 1.8** The Burrard bridge in Vancouver. (a) Original picture (courtesy of Minhuai Shi). (b) Picture uniformly corrupted by additive white Gaussian noise.

### 1.2.2.2  Texture Masking

Texture masking is sometimes also called *detail dependence* (Connor et al., 1972), *spatial masking* (Netravali and Presada, 1977; Lim, 1990), or *activity masking* (Mitchell et al., 1997). It states that the discrimination threshold increases with increasing picture detail. That is, the stronger the texture, the larger the discrimination threshold. In Figure 1.8, it can be observed that the additive random noise is less pronounced in the strongly textured area than in the smooth area if comparing, for instance, the dark portion of the cloud (the upper-right corner of the picture) with the water area (the lower right corner of the picture). This is a confirmation of texture masking.

In Figure 1.9(b), the number of quantization levels decreases from 256, as in Figure 1.9(a), to 16. That is, we use only four bits instead of eight bits to represent the intensity value for each pixel.

**FIGURE 1.9** Christmas in Winorlia. (a) Original. (b) Four-bit quantized. (c) Improved IGS quantized with four bits.

The unnatural contours caused by coarse quantization can be noticed in the relatively uniform regions, compared with Figure 1.9(a). This phenomenon was first noted in (Goodall, 1951) and is called *false contouring* (Gonzalez and Woods, 1992). Now we see that the false contouring can be explained by using texture masking, since texture masking indicates that the human eye is more sensitive to the smooth region than to the textured region, where intensity exhibits a high variation. A direct impact on image and video compression is that the number of quantization levels, which affects the bit rate significantly, should be adapted according to the intensity variation of image regions.

### 1.2.2.3 Frequency Masking

While the above two characteristics are picture dependent in nature, frequency masking is picture independent. It states that the discrimination threshold increases with frequency increase. It is also referred to as *frequency dependence*.

**FIGURE 1.9** (continued)

Frequency masking can be well illustrated in using Figure 1.9 above. In Figure 1.9(c), high-frequency random noise has been added to the original image before quantization. This method is referred to as the improved gray-scale (IGS) quantization (Gonzalez and Woods, 1992). With the same number of quantization levels, 16, as in Figure 1.9(b), the picture quality of Figure 1.9(c) is improved dramatically compared with that of Figure 1.9(b): the annoying false contours have disappeared despite an increase in the root mean square value of the total noise in Figure 1.9(c). This is due to the fact that the low-frequency quantization error is converted to the high-frequency noise, and that the HVS is less sensitive to the high-frequency content. We thus see, as pointed out in (Connor, 1972), that our human eyes function like a low-pass filter.

Owing to frequency masking in the transform domain, say, the discrete cosine transform (DCT) domain, we can drop some high-frequency coefficients with small magnitudes to achieve data compression without noticeably affecting the perception of the HVS. This leads to what is called transform coding, which is discussed in Chapter 4.

### 1.2.2.4  Temporal Masking

Temporal masking is another picture-independent feature of the HVS. It states that it takes a while for the HVS to adapt itself to the scene when the scene changes abruptly. During this transition the HVS is not sensitive to details. The masking takes place both before and after the abrupt change. It is called forward temporal masking if it happens after the scene change. Otherwise, it is referred to backward temporal masking (Mitchell et al., 1997).

This implies that one should take temporal masking into consideration when allocating data in image and video coding.

### 1.2.2.5  Color Masking

Digital color image processing is gaining increasing popularity due to the wide application of color images in modern life. As mentioned at the beginning of the discussion about psychovisual redundancy, we are not going to cover all aspects of the perception of the HVS. Instead, we cover only those aspects related to psychovisual redundancy, thus to image and video compression. Therefore, our discussion here on color perception is by no means exhaustive.

In physics, it is known that any visible light corresponds to an electromagnetic spectral distribution. Therefore, a color, as a sensation of visible light, is an energy with an intensity as well as a set of wavelengths associated with the electromagnetic spectrum. Obviously, intensity is an attribute of visible light. The composition of wavelengths is another attribute: chrominance. There are two elements in the chrominance attribute: *hue* and *saturation*. The hue of a color is characterized by the dominant wavelength in the composition. Saturation is a measure of the purity of a color. A pure color has a saturation of 100%, whereas white light has a saturation of 0.

**RGB model** — The red-green-blue (RGB) primary color system is the best known of several color systems. This is due to the following feature of the human perception of color. The color-sensitive area in the HVS consists of three different sets of cones and each set is sensitive to the light of one of the three primary colors: red, green, and blue. Consequently, any color sensed by the HVS can be considered as a particular linear combination of the three primary colors. Many research studies are available, the CIE (Commission Internationale de l'Eclairage) chromaticity diagram being a well-known example. These results can be easily found in many classic optics and digital image processing texts.

The RGB model is used mainly in color image acquisition and display. In color signal processing including image and video compression, however, the luminance-chrominance color system is more efficient and, hence, widely used. This has something to do with the color perception of the HVS. It is known that the HVS is more sensitive to green than to red, and is least sensitive to blue. An equal representation of red, green, and blue leads to inefficient data representation when the HVS is the ultimate viewer. Allocating data only to the information that the HVS can perceive, on the other hand, can make video coding more efficient.

Luminance is concerned with the perceived brightness, while chrominance is related to the perception of hue and saturation of color. Roughly speaking, the luminance-chrominance representation agrees more with the color perception of the HVS. This feature makes the luminance-chrominance color models more suitable for color image processing. A good example was presented in (Gonzalez and Woods, 1992), about histogram equalization. It is well known that applying histogram equalization can bring out some details originally hidden in dark regions. Applying histogram equalization to the RGB components separately can certainly achieve the goal. In doing so, however, the chrominance elements, hue and saturation, have been changed, thus leading to color distortion. With a luminance-chrominance model, histogram equalization can be applied to the luminance component only. Hence, the details in the dark regions are brought out, whereas the chrominance elements remain unchanged, producing no color distortion. With the luminance component Y serving as a black-white signal, a luminance-chrominance color model offers compatibility with black and white TV systems. This is another virtue of luminance-chrominance color models.

It is known that a nonlinear relationship (basically a power function) exists between electrical signal magnitude and light intensity for both cameras and CRT-based display monitors (Haskell et al., 1997). That is, the light intensity is a linear function of the signal voltage raised to the power of $\gamma$. It is a common practice to correct this nonlinearity before transmission. This is referred to as gamma correction. The gamma-corrected RGB components are denoted by R', G', and B', respectively. They are used in the discussion on various color models. For the sake of notational brevity, we simply use R, G, and B instead of R', G' and B' in the following discussion, while keeping the gamma-correction in mind.

Discussed next are several different luminance-chrominance color models: HSI, YUV, YCbCr, and YDbDr.

**HSI model** — In this model, I stands for the intensity component, H for the hue component, and S for saturation. One merit of this color system is that the intensity component is decoupled from the chromatic components. As analyzed above, this decoupling usually facilitates color image processing tasks. Another merit is that this model is closely related to the way the HVS perceives color pictures. Its main drawback is the complicated conversion between RGB and HSI models. A detailed derivation of the conversion may be found in (Gonzalez and Woods, 1992). Because of this complexity, the HSI model is not used in any TV systems.

**YUV model** — In this model, Y denotes the luminance component, and U and V are the two chrominance components. The luminance Y can be determined from the RGB model via the following relation:

$$Y = 0.299\,R + 0.587\,G + 0.114\,B \tag{1.4}$$

It is noted that the three weights associated with the three primary colors, R, G, and B, are not the same. Their different magnitudes reflect different responses of the HVS to different primary colors.

Instead of being directly related to hue and saturation, the other two chrominance components, U and V, are defined as color differences as follows.

$$U = 0.492(B - Y) \tag{1.5}$$

$$V = 0.877(R - Y) \tag{1.6}$$

In this way, the YUV model lowers computational complexity. It has been used in PAL (Phase Alternating Line) TV systems. Note that PAL is an analog composite color TV standard and is used in most European countries, some Asian countries, and Australia. By composite systems, we mean both the luminance and chrominance components of the TV signals are multiplexed within the same channel. For completeness, an expression of YUV in terms of RGB is listed below.

$$\begin{pmatrix} Y \\ U \\ V \end{pmatrix} = \begin{pmatrix} 0.299 & 0.587 & 0.114 \\ -0.147 & -0.289 & 0.436 \\ 0.615 & -0.515 & -0.100 \end{pmatrix} \begin{pmatrix} R \\ G \\ B \end{pmatrix} \tag{1.7}$$

**YIQ model** — This color space has been utilized in NTSC (National Television Systems Committee) TV systems for years. Note that NTSC is an analog composite color TV standard and is used in North America and Japan. The Y component is still the luminance. The two chrominance components are the linear transformation of the U and V components defined in the YUV model. Specifically,

$$I = -0.545U + 0.839V \tag{1.8}$$

$$Q = 0.839U + 0.545V \tag{1.9}$$

Substituting the U and V expressed in Equations 1.4 and 1.5 into the above two equations, we can express YIQ directly in terms of RGB. That is,

$$\begin{pmatrix} Y \\ I \\ Q \end{pmatrix} = \begin{pmatrix} 0.299 & 0.587 & 0.114 \\ 0.596 & -0.275 & -0.321 \\ 0.212 & -0.523 & 0.311 \end{pmatrix} \begin{pmatrix} R \\ G \\ B \end{pmatrix} \tag{1.10}$$

**YDbDr model** — The YDbDr model is used in the SECAM (Sequential Couleur a Memoire) TV system. Note that SECAM is used in France, Russia, and some eastern European countries. The relationship between YDbDr and RGB appears below.

$$\begin{pmatrix} Y \\ Db \\ Dr \end{pmatrix} = \begin{pmatrix} 0.299 & 0.587 & 0.114 \\ -0.450 & -0.883 & 1.333 \\ -1.333 & 1.116 & -0.217 \end{pmatrix} \begin{pmatrix} R \\ G \\ B \end{pmatrix} \qquad (1.11)$$

That is,

$$Db = 3.059U \qquad (1.12)$$

$$Dr = -2.169V \qquad (1.13)$$

**YCbCr model** — From the above, we can see that the U and V chrominance components are differences between the gamma-corrected color B and the luminance Y, and the gamma-corrected R and the luminance Y, respectively. The chrominance component pairs I and Q, and Db and Dr are both linear transforms of U and V. Hence they are very closely related to each other. It is noted that U and V may be negative as well. In order to make chrominance components nonnegative, the Y, U, and V are scaled and shifted to produce the YCbCr model, which is used in the international coding standards JPEG and MPEG. (These two standards are covered in Chapters 7 and 16, respectively.)

$$\begin{pmatrix} Y \\ Cb \\ Cr \end{pmatrix} = \begin{pmatrix} 0.257 & 0.504 & 0.098 \\ -0.148 & -0.291 & 0.439 \\ 0.439 & -0.368 & -0.071 \end{pmatrix} \begin{pmatrix} R \\ G \\ B \end{pmatrix} + \begin{pmatrix} 16 \\ 128 \\ 128 \end{pmatrix} \qquad (1.14)$$

### 1.2.2.6  Color Masking and Its Application in Video Compression

It is well-known that the HVS is much more sensitive to the luminance component than to the chrominance components. Following Van Ness and Bouman (1967) and Mullen (1985), there is a figure in Mitchell et al. (1997) to quantitatively illustrate the above statement. A modified version is shown in Figure 1.10. There, the abscissa represents spatial frequency in the unit of cycles per degree (cpd), while the ordinate is the contrast threshold for a just detectable change in the sinusoidal testing signal. Two observations are in order. First, for each of the three curves, i.e., curves for the luminance component Y and the opposed-color chrominance, the contrast sensitivity increases when spatial frequency increases, in general. This agrees with frequency masking discussed above. Second, for the same contrast threshold, we see that the luminance component corresponds to a much higher spatial frequency. This is an indication that the HVS is much more sensitive to luminance than to chrominance. This statement can also be confirmed, perhaps more easily, by examining those spatial frequencies at which all three curves have data available. Then we can see that the contrast threshold of luminance is much lower than that of the chrominance components.

The direct impact of color masking on image and video coding is that by utilizing this psychovisual feature we can allocate more bits to the luminance component than to the chrominance components. This leads to a common practice in color image and video coding: using full resolution for the intensity component, while using a 2 by 1 subsampling both horizontally and vertically for the two chrominance components. This has been adopted in related international coding standards, which will be discussed in Chapter 16.

### 1.2.2.7  Summary: Differential Sensitivity

In this subsection we discussed luminance masking, texture masking, frequency masking, temporal masking, and color masking. Before we enter the next subsection, let us summarize what we have discussed so far.

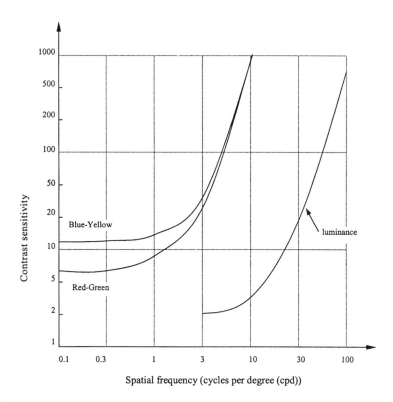

**FIGURE 1.10**  Contrast sensitivity vs. spatial frequency. (Revised from Van Ness and Bouman [1967] and Mullen [1985].)

We see that luminance masking, also known as contrast masking, is of fundamental importance among several types of masking. It states that the sensitivity of the eyes to a stimulus depends on the intensity of another stimulus. Thus it is a differential sensitivity. Both the texture (detail or activity) and frequency of another stimulus significantly influence this differential sensitivity. The same mechanism exists in color perception, where the HVS is much more sensitive to luminance than to chrominance. Therefore, we conclude that differential sensitivity is the key in studying human visual perception. These features can be utilized to eliminate psychovisual redundancy, and thus compress image and video data.

It is also noted that variable quantization, which depends on activity and luminance in different regions, seems to be reasonable from a data compression point of view. Its practical applicability, however, is somehow questionable. That is, some experimental work does not support this expectation (Mitchell et al., 1997).

It is noted that this differential sensitivity feature of the HVS is common to human perception. For instance, there is also forward and backward temporal masking in human audio perception.

## 1.3  VISUAL QUALITY MEASUREMENT

As the definition of image and video compression indicates, image and video quality is an important factor in dealing with image and video compression. For instance, in evaluating two different compression methods we have to base the evaluation on some definite image and video quality. When both methods achieve the same quality of reconstructed image and video, the one that requires less data is considered to be superior to the other. Alternatively, with the same amount of data the method providing a higher-quality reconstructed image or video is considered the better method. Note that here we have not considered other performance criteria, such as computational complexity.

Surprisingly, however, it turns out that the measurement of image and video quality is not straightforward. There are two types of visual quality assessments. One is objective assessment (using electrical measurements), and the other is subjective assessment (using human observers). Each has its merits and drawbacks. A combination of these two methods is now widely utilized in practice. In this section we first discuss subjective visual quality measurement, followed by objective quality measurement.

## 1.3.1  SUBJECTIVE QUALITY MEASUREMENT

It is natural that the visual quality of reconstructed video frames should be judged by human viewers if they are to be the ultimate receivers of the data (see Figure 1.1). Therefore, the subjective visual quality measure plays an important role in visual communications.

In subjective visual quality measurement, a set of video frames is generated with varying coding parameters. Observers are invited to subjectively evaluate the visual quality of these frames. Specifically, observers are asked to rate the pictures by giving some measure of picture quality. Alternatively, observers are requested to provide some measure of impairment to the pictures. A five-scale rating system of the degree of impairment, used by Bell Laboratories, is listed below (Sakrison, 1979). It has been adopted as one of the standard scales in CCIR Recommendation 500-3 (CCIR, 1986). Note that CCIR is now ITU-R (International Telecommunications Union — Recommendations).

1. Impairment is not noticeable
2. Impairment is just noticeable
3. Impairment is definitely noticeable, but not objectionable
4. Impairment is objectionable
5. Impairment is extremely objectionable

In regard to the subjective evaluation, there are a few things worth mentioning. In most applications there is a whole array of pictures simultaneously available for evaluation. These pictures are generated with different encoding parameters. By keeping some parameters fixed while making one parameter (or a subset of parameters) free to change, the resulting quality rating can be used to study the effect of the one parameter (or the subset of parameters) on encoding. An example using this method for studying the effect of varying numbers of quantization levels on image quality can be found in (Gonzalez and Woods, 1992).

Another possible way to study the effect is to identify pictures with the same subjective quality measure from the whole array of pictures. From this subset of test pictures we can produce, in the encoding parameter space, isopreference curves that can be used to study the effect of the parameter(s) under investigation. An example using this method to study the effect of varying both image resolution and numbers of quantization levels on image quality can be found in (Huang, 1965).

In this rating, a whole array of pictures is usually divided into columns, with each column sharing some common conditions. The evaluation starts within each column with a pairwise comparison. This is because a pairwise comparison is relatively easy for the eyes. As a result, pictures in one column are arranged in an order according to visual quality, and quality or impairment measures are then assigned to the pictures in that one column. After each column has been rated, a unification between columns is necessary. That is, different columns need to have a unified quality measurement. As pointed out in (Sakrison, 1979), this task is not easy since it means we may need to equate impairment that results from different types of errors.

One thing can be understood from the above discussion: subjective evaluation of visual quality is costly. It needs a large number of pictures and observers. The evaluation takes a long time because human eyes are easily fatigued and bored. Some special measures have to be taken in order to arrive at an accurate subjective quality measurement. Examples in this regard include averaging

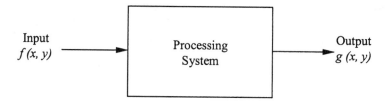

**FIGURE 1.11**    An image processing system.

subjective ratings and taking their deviation into consideration. For further details on subjective visual quality measurement, readers may refer to Sakrison (1979), Hidaka and Ozawa (1990), and Webster et al. (1993).

### 1.3.2    OBJECTIVE QUALITY MEASUREMENT

In this subsection, we first introduce the concept of signal-to-noise ratio (SNR), which is a popularly utilized objective quality assessment. Then we present a promising new objective visual quality assessment technique based on human visual perception.

#### 1.3.2.1    Signal-to-Noise Ratio

Consider Figure 1.11, where $f(x, y)$ is the input image to a processing system. The system can be a low-pass filter, a subsampling system, or a compression system. It can even represent a process in which additive white Gaussian noise corrupts the input image. The $g(x, y)$ is the output of the system. In evaluating the quality of $g(x, y)$, we define an error function $e(x, y)$ as the difference between the input and the output. That is,

$$e(x, y) = f(x, y) - g(x, y) \tag{1.15}$$

The mean square error is defined as $E_{ms}$:

$$E_{ms} = \frac{1}{MN} \sum_{x=0}^{M-1} \sum_{y=0}^{N-1} e(x, y)^2 \tag{1.16}$$

where $M$ and $N$ are the dimensions of the image in the horizontal and vertical directions. Note that it is sometimes denoted by $MSE$. The root mean square error is defined as $E_{rms}$:

$$E_{rms} = \sqrt{E_{ms}} \tag{1.17}$$

It is sometimes denoted by $RMSE$.

As noted earlier, $SNR$ is widely used in objective quality measurement. Depending whether mean square error or root mean square error is used, the $SNR$ may be called the mean square signal-to-noise ratio, $SNR_{ms}$, or the root mean square signal-to-noise ratio, $SNR_{rms}$. We have

$$SNR_{ms} = 10 \log_{10} \left( \frac{\sum_{x=0}^{M-1} \sum_{y=0}^{N-1} g(x, y)^2}{MN \cdot E_{ms}} \right), \tag{1.18}$$

and

$$SNR_{rms} = \sqrt{SNR_{ms}} \qquad (1.19)$$

In image and video data compression, another closely related term, *PSNR* (peak signal-to-noise ratio), which is essentially a modified version of $SNR_{ms}$, is widely used. It is defined as follows.

$$PSNR = 10\log_{10}\left(\frac{255^2}{E_{ms}}\right) \qquad (1.20)$$

The interpretation of the *SNRs* is that the larger the *SNR* ($SNR_{ms}$, $SNR_{rms}$, or *PSNR*) the better the quality of the processed image, $g(x, y)$; that is, the closer the processed image $g(x,y)$ is to the original image $f(x, y)$. This seems to be correct. However, from our above discussion about the features of the HVS, we know that the HVS does not respond to visual stimuli in a straightforward way. Its low-level processing unit is known to be nonlinear. Several masking phenomena exist. Each confirms that the visual perception of the HVS is not simple. It is worth noting that our understanding of the high-level processing unit of the HVS is far from complete. Therefore, we may understand that the *SNR* does not always provide us with reliable assessments of image quality. One good example is presented in Section 1.2.2.3, which uses the IGS quantization technique to achieve high compression (using only four bits for quantization instead of the usual eight bits) without introducing noticeable false contouring. In this case, the subjective quality is improved, and the *SNR* decreases due to the additive high-frequency random noise. Another example, drawn from our discussion about the masking phenomena, is that some additive noise in bright areas or in highly textured regions may be masked, while some minor artifacts in dark and uniform regions may turn out to be quite annoying. In this case, the *SNR* cannot truthfully reflect visual quality, as well.

On the one hand, we see that objective quality measurement does not always provide reliable picture quality assessment. On the other hand, however, its implementation is much faster and easier than that of the subjective quality measurement. Furthermore, objective assessment is repeatable. Owing to these merits, objective quality assessment is still widely used despite this drawback.

It is noted that combining subjective and objective assessments has been a common practice in international coding-standard activity.

### 1.3.2.2  Objective Quality Measurement Based on Human Visual Perception

Introduced here is a new development in visual quality assessment, which is an objective quality measurement based on human visual perception (Webster et al., 1993). Since it belongs to the category of objective assessment, it possesses virtues such as repeatability and fast and easy implementation. Because it is based on human visual perception, on the other hand, its assessment of visual quality agrees closely to that of subjective assessment. In this sense the new method attempts to combine the merits of the two different types of assessment.

**Motivation** — Visual quality assessment is best conducted via the subjective approach since in this case the HVS is the ultimate viewer. The implementation of subjective assessment, however, is time-consuming, costly, and lacks repeatability. On the other hand, although not always accurate, objective assessment is fast, easy, and repeatable. The motivation here is to develop an objective quality measurement system such that its quality assessment is very close to that obtained by using subjective assessment. In order to achieve this goal, this objective system is based on subjective assessment. That is, it uses the rating achieved via subjective assessment as a criterion to search for new objective measurements so as to have the objective rating as close to the subjective one as possible.

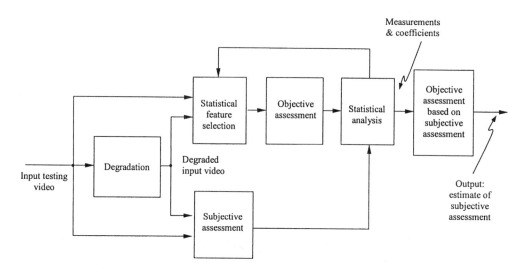

**FIGURE 1.12** Block diagram of objective assessment based on subjective assessment.

**Methodology** — The derivation of the objective quality assessment system is shown in Figure 1.12. The input testing video goes through a degradation block, resulting in degraded input video. The degradation block, or impairment generator, includes various video compression codecs (coder-decoder pairs) with bit rates ranging from 56 kb/sec to 45 Mb/sec and other video operations. The input video and degraded input video form a pair of testing videos, which is sent to a subjective assessment block as well as a statistical feature selection block.

A normal subjective visual quality assessment as introduced in the previous subsection is performed in the subjective assessment block, which involves a large panel of observers, e.g., 48 observers in Webster et al. (1993). In the statistical feature selection block, a variety of statistical operations are conducted and various statistical features are selected. Examples cover Sobel filtering, Laplacian operator, first-order differencing, moment calculation, fast Fourier transform, etc. Statistical measurements are then selected based on these statistical operations and features. An objective assessment is formed as follows:

$$\hat{s} = a_0 + \sum_{i=1}^{l} a_i n_i,$$ (1.21)

where $\hat{s}$ denotes the output rating of the object assessment, or simply the objective measure, which is supposed to be a good estimate of the corresponding subjective score. The $n_i$, $i = 1, \cdots, l$ are selected objective measurements. The $a_0$, $a_i$, $i = 1, \cdots, l$ are coefficients in the linear model of the objective assessment.

The results of the objective assessment and subjective assessment are applied to a statistical analysis block. In the statistical analysis block, the objective assessment rating is compared with that of the subjective assessment. The result of the comparison is fed back to the statistical feature selection block. The statistical measurements obtained in the statistical feature selection block are examined according to their performance in the assessment. A statistical measurement is regarded to be good if it can reduce by a significant amount of the difference between the objective assessment and the subjective assessment. The best measurement is determined via an exhaustive search among the various measurements. Note that the coefficients in Equation 1.21 are also examined in the statistical analysis block in a similar manner to that used for the measurements.

The measurements and coefficients determined after iterations result in an optimal objective assessment via Equation 1.21, which is finally passed to the last block as the output of the system. The whole process will become much clearer below.

*Results*

The results reported by Webster (1993) are introduced here.

**Information features** — As mentioned in Section 1.2.2, differential sensitivity is a key in human visual perception. Two selected features: perceived spatial information (the amount of spatial detail) and perceived temporal information (the amount of temporal luminance variation), involve pixel differencing. Spatial information (SI) is defined as shown below.

$$SI(f_n) = STD_s \{Sobel(f_n)\} \tag{1.22}$$

where $STD_s$ stands for the standard deviation operator in the spatial domain, $Sobel$ denotes the Sobel operation, and $f_n$ represents the $n$th video frame. Temporal information (TI) is defined similarly:

$$TI(f_n) = STD_s \{\Delta f_n\} \tag{1.23}$$

where $\Delta f_n = f_n - f_{n-1}$, i.e., the successive frame difference.

*Determined Measurements*

The parameter $l$ in Equation 1.21 is chosen as three. That is

$$\hat{s} = a_0 + a_1 n_1 + a_2 n_2 + a_3 n_3 \tag{1.24}$$

The measurements $n_1$, $n_2$, and $n_3$ are formulated based on the above-defined information features, SI and TI, as follows:

1. Measurement $n_1$:

$$n_1 = RMS_t \left( 5.81 \left| \frac{SI(of_n) - SI(df_n)}{SI(of_n)} \right| \right) \tag{1.25}$$

where $RMS_t$ represents the root mean square value taken over the time dimension, and $of_n$ and $df_n$ denote the original $n$th frame and the degraded $n$th frame, respectively. It is observed that $n_1$ is a measure of the relative change in the spatial information between the original frame and the degraded frame.

2. Measurement $n_2$:

$$n_2 = \Im_t \{0.108 \cdot MAX \{[TI(of_n) - TI(df_n)], 0\}\} \tag{1.26}$$

where

$$\Im_t \{y_t\} = STD_t \{CONV(y_t, [-1, 2, -1])\} \tag{1.27}$$

where $STD_t$ denotes the standard deviation operator with respect to time, and $CONV$ indicates the convolution operation between its two arguments. It is understood that temporal information, TI, measures temporal luminance variation (temporal motion) and the convolution kernel, [-1, 2, -1], enhances the variation due to its high-pass filter nature. Therefore, $n_2$ measures the difference of TI between the original and graded frames.

3. Measurement $n_3$:

$$n_3 = MAX_t \left\{ 4.23 \cdot \log_{10} \left( \frac{TI(df_n)}{TI(of_n)} \right) \right\} \tag{1.28}$$

where $MAX_t$ indicates the taking of the maximum value over time. Therefore, measurement $n_3$ responds to the ratio between the temporal information of the degraded video and that of the original video. Distortions such as block artifacts (discussed in Chapter 11) and motion jerkiness (discussed in Chapter 10), which occur in video coding, will cause $n_3$ to be large.

**Objective estimator** — The least square error procedure is applied to testing video sequences with measurements $n_i$, $i = 1,2,3$, determined above, to minimize the difference between the rating scores obtained from the subjective assessment and the objective assessment, resulting in the estimated coefficients $a_0$ and $a_i$, $i = 1,2,3$. Consequently, the objective assessment of visual quality $\hat{s}$ becomes

$$\hat{s} = 4.77 - 0.992 n_1 - 0.272 n_2 - 0.356 n_3 \tag{1.29}$$

**Reported experimental results** — It was reported that the correlation coefficient between the subjective assessment score and the objective assessment score (an estimate of the subjective score) is in the range of 0.92 to 0.94. It is noted that a set of 36 testing scenes containing various amounts of spatial and temporal information was used in the experiment. Hence, it is apparent that quite good performance was achieved. Though there is surely room for further improvement, this work does open a new and promising way to assess visual quality by combining subjective and objective approaches. Since it is objective it is fast and easy; and because it is based on the subjective measurement, it is more accurate in terms of the high correlation to human perception. Theoretically, the spatial information measure and temporal information measure defined on differencing are very important. They reflect the most important aspect of human visual perception.

## 1.4   INFORMATION THEORY RESULTS

In the beginning of this chapter it was noted that the term information is considered one of the fundamental concepts in image and video compression. We will now address some information theory results. In this section, the measure of information and the entropy of an information source are covered first. We then introduce some coding theorems, which play a fundamental role in studying image and video compression.

### 1.4.1   ENTROPY

Entropy is a very important concept in information theory and communications. So is it in image and video compression. We first define the information content of a source symbol. Then we define entropy as average information content per symbol for a discrete memoryless source.

#### 1.4.1.1   Information Measure

As mentioned at the beginning of this chapter, information is defined as knowledge, fact, and news. It can be measured quantitatively. The carriers of information are symbols. Consider a symbol with an occurrence probability p. Its information content (i.e., the amount of information contained in the symbol), I, is defined as follows.

$$I = \log_2 \frac{1}{p} \quad bits \quad or \quad I = -\log_2 p \quad bits \tag{1.30}$$

where the *bit* is a contraction of *binary unit*. In the above equations we set the base of the logarithmic function to equal 2. It is noted that these results can be easily converted as follows for the case where the r-ary digits are used for encoding. Hence, from now on, we restrict our discussion to binary encoding.

$$I = -\log_r 2 \cdot \log_2 p \quad bits \tag{1.31}$$

According to Equation 1.30, the information contained within a symbol is a logarithmic function of its occurrence probability. The smaller the probability, the more information the symbol contains. This agrees with common sense. The occurrence probability is somewhat related to the uncertainty of the symbol. A small occurrence probability means large uncertainty. In this way, we see that the information content of a symbol is about the uncertainty of the symbol. It is noted that the information measure defined here is valid for both equally probable symbols and nonequally probable symbols (Lathi, 1998).

### 1.4.1.2   Average Information per Symbol

Now consider a discrete memoriless information source. By discreteness, we mean the source is a countable set of symbols. By memoriless, we mean the occurrence of a symbol in the set is independent of that of its preceding symbol. Take a look at a source of this type that contains $m$ possible symbols: $\{s_i, i = 1,2\cdots,m\}$. The corresponding occurrence probabilities are denoted by $\{p_i, i = 1,2,\cdots,m\}$. According to the discussion above, the information content of a symbol $s_i$, $I_i$, is equal to $I_i = -\log_2 p_i$ bits. Entropy is defined as the average information content per symbol of the source. Obviously, the entropy, $H$, can be expressed as follows.

$$H = -\sum_{i=1}^{m} p_i \log_2 p_i \quad bits \tag{1.32}$$

From this definition, we see that the entropy of an information source is a function of occurrence probabilities. It is straightforward to show that the entropy reaches the maximum when all symbols in the set are equally probable.

### 1.4.2   SHANNON'S NOISELESS SOURCE CODING THEOREM

Consider a discrete, memoriless, stationary information source. In what is called source encoding, a *codeword* is assigned to each symbol in the source. The number of bits in the codeword is referred to as the length of the codeword. The average length of codewords is referred to as the bit rate, expressed in the unit of bits per symbol.

Shannon's noiseless source coding theorem states that for a discrete, memoriless, stationary information source, the minimum bit rate required to encode a symbol, on average, is equal to the entropy of the source. This theorem provides us with a lower bound in source coding. Shannon showed that the lower bound can be achieved when the *encoding delay* extends to infinity. By encoding delay, we mean the encoder waits and then encodes a certain number of symbols at once. Fortunately, with finite encoding delay, we can already achieve an average codeword length fairly close to the entropy. That is, we do not have to actually sacrifice bit rate much to avoid long encoding delay, which involves high computational complexity and a large amount of memory space.

Note that the discreteness assumption is not necessary. We assume a discrete source simply because digital image and video are the focus in this book. Stationarity assumption is necessary in deriving the noiseless source coding theorem. This assumption may not be satisfied in practice. Hence, Shannon's theorem is a theoretical guideline only. There is no doubt, however, that it is a fundamental theoretical result in information theory.

In summary, the noiseless source coding theorem, Shannon's first theorem, which was published in his celebrated paper (Shannon, 1948), is concerned with the case where both the channel and the coding system are noise free. The aim under these circumstances is coding compactness. The more compact it is, the better the coding. This theorem specifies the lower bound, which is the source entropy, and how to reach this lower bound.

One way to evaluate the efficiency of a coding scheme is to determine its *efficiency* with respect to the lower bound, i.e., entropy. The efficiency $\eta$ is defined as follows.

$$\eta = \frac{H}{L_{avg}}$$ (1.33)

where $H$ is entropy, and $L_{avg}$ denotes the average length of the codewords in the code. Since the entropy is the lower bound, the efficiency never exceeds the unity, i.e., $\eta \leq 1$. The same definition can be generalized to calculate the relative efficiency between two codes. That is

$$\eta = \frac{L_{avg,1}}{L_{avg,2}}$$ (1.34)

where $L_{avg,1}$ and $L_{avg,2}$ represent the average codeword length for code 1 and code 2, respectively. We usually put the larger of the two in the denominator, and $\eta$ is called the efficiency of code 2 with respect to code 1. A complementary parameter of coding efficiency is coding *redundancy*, $\zeta$, which is defined as

$$\zeta = 1 - \eta$$ (1.35)

### 1.4.3 SHANNON'S NOISY CHANNEL CODING THEOREM

If a code has an efficiency of $\eta = 1$, i.e., it reaches the lower bound of source encoding, then coding redundancy is $\zeta = 0$. Now consider a noisy transmission channel. In transmitting the coded symbol through the noisy channel, the received symbols may be erroneous due to the lack of redundancy. On the other hand, it is well known that by adding redundancy (e.g., parity check bits) some errors occurring during the transmission over the noisy channel may be corrected or identified. In the latter, the coded symbols are then resent. In this way, we see that adding redundancy may combat noise.

Shannon's noisy channel coding theorem states that it is possible to transmit symbols over a noisy channel without error if the bit rate is below a *channel capacity,* C. That is

$$R < C$$ (1.36)

where $R$ denotes the bit rate. The channel capacity is determined by the noise and signal power.

In conclusion, the noisy channel coding theorem, Shannon's second theorem (Shannon, 1948), is concerned with a noisy, memoriless channel. By memoriless, we mean the channel output corresponding to the current input is independent of the output corresponding to previous input symbols. Under these circumstances, the aim is reliable communication. To be error free, the bit rate cannot exceed channel capacity. That is, channel capacity sets an upper bound on the bit rate.

### 1.4.4   SHANNON'S SOURCE CODING THEOREM

As seen in the previous two subsections, the noiseless source coding theorem defines the lowest possible bit rate for noiseless source coding and noiseless channel transmission; whereas the noisy channel coding theorem defines the highest possible coding bit rate for error-free transmission. Therefore, both theorems work for reliable (no error) transmission. In this subsection, we continue to deal with discrete memoriless information sources, but we discuss the situation in which lossy coding is encountered. As a result, distortion of the information source takes place. For instance, quantization, which is covered in the next chapter, causes information loss. Therefore, it is concluded that if an encoding procedure involves quantization, then it is lossy coding. That is, errors occur during the coding process, even though the channel is error free. We want to find the lower bound of the bit rate for this case.

The source coding theorem (Shannon, 1948) states that for a given distortion $D$, there exists a rate distortion function $R(D)$ (Berger, 1971), which is the minimum bit rate required to transmit the source with distortion less than or equal to $D$. That is, in order to have distortion not larger than $D$, the bit rate $R$ must satisfy the following condition:

$$R \geq R(D) \tag{1.37}$$

A more detailed discussion about this theorem and the rate distortion function is given in Chapter 15, when we introduce video coding.

### 1.4.5   INFORMATION TRANSMISSION THEOREM

It is clear that by combining the noisy channel coding theorem and the source coding theorem we can derive the following relationship:

$$C \geq R(D) \tag{1.38}$$

This is called the information transmission theorem (Slepian, 1973). It states that if the channel capacity of a noisy channel, $C$, is larger than the rate distortion function $R(D)$, then it is possible to transmit an information source with distortion $D$ over a noisy channel.

## 1.5   SUMMARY

In this chapter, we first discussed the necessity for image and video compression. It is shown that image and video compression becomes an enabling technique in today's exploding number of digital multimedia applications. Then, we show that the feasibility of image and video compression rests in redundancy removal. Two types of redundancies: statistical redundancy and psychovisual redundancy are studied. Statistical redundancy comes from interpixel correlation and coding redundancy. By interpixel correlation, we mean correlation between pixels either located in one frame (spatial or intraframe redundancy) or pixels located in successive frames (temporal or interframe redundancy). Coding redundancy is related to coding technique. Psychovisual redundancy is based on the features (several types of masking phenomena) of human visual perception. That is, visual information is not perceived equally from the human visual point of view. In this sense, some information is psychovisually redundant.

The visual quality of the reconstructed image and video is a crucial criterion in the evaluation of the performance of visual transmission or storage systems. Both subjective and objective assessments are discussed. A new and promising objective technique based on subjective assessment is introduced. Since it combines the merits of both types of visual quality assessment, it achieves a quite satisfactory performance. The selected statistical features reveal some possible mechanism of the human visual perception. Further study in this regard would be fruitful.

In the last section, we introduced some fundamental information theory results, relevant to image and video compression. The results introduced include information measurement, entropy, and several theorems. All the theorems assume discrete, memoriless, and stationary information sources. The noiseless source coding theorem points out that the entropy of an information source is the lower bound of the coding bit rate that a source encoder can achieve. The source coding theorem deals with lossy coding applied in a noise-free channel. It states that for a given distortion, $D$, there is a rate distortion function, $R(D)$. When the bit rate in the source coding is greater than $R(D)$, the reconstructed source at the receiving end may satisfy the fidelity requirement defined by $D$. The noisy channel coding theorem states that, in order to achieve error-free performance, the source coding bit rate must be smaller than the channel capacity. Channel capacity is a function of noise and signal power. The information transmission theorem combines the noisy channel coding theorem and the source coding theorem. It states that it is possible to have a reconstructed waveform at the receiving end, satisfying the fidelity requirement corresponding to distortion $D$ if the channel capacity, $C$, is larger than the rate distortion function $R(D)$. Though some of the assumptions on which these theorems were developed may not be valid in complicated practical situations, these theorems provide important theoretical limits for image and video coding. They can also be used for evaluation of the performance of different coding techniques.

## 1.6   EXERCISES

**1-1.** Using your own words, define spatial and temporal redundancy, and psychovisual redundancy, and state the impact they have on image and video compression.

**1-2.** Why is differential sensitivity considered the most important feature in human visual perception?

**1-3.** From the description of the newly developed objective assessment technique based on subjective assessment, discussed in Section 1.3, what points do you think are related to and support the statement made in Exercise 1-2?

**1-4.** Interpret Weber's law using your own words.

**1-5.** What is the advantage possessed by color models that decouple the luminance component from chrominance components.

**1-6.** Why has the HIS model not been adopted by any TV systems?

**1-7.** What is the problem with the objective visual quality measure of PSNR?

## REFERENCES

Berger, T. *Rate Distortion Theory*, Englewood Cliffs, NJ, Prentice-Hall, 1971.

CCIR Recommendation 500-3, Method for the subjective assessment of the quality of television pictures, Recommendations and Reports of the CCIR, 1986, XVIth Plenary Assembly, Volume XI, Part 1.

Connor, D. J., R. C. Brainard, and J. O. Limb, Interframe coding for picture transmission, *Proc. IEEE,* 60(7), 779-790, 1972.

Fink, D. G. *Television Engineering Handbook*, New York, McGraw-Hill, 1957, Sect. 10.7.

Goodall, W. M. Television by pulse code modulation, *Bell Syst. Tech. J.,* 33-49, 1951.

Gonzalez, R. C. and R. E. Woods, *Digital Image Processing*, Reading, MA, Addison-Wesley, 1992.

Haskell, B. G., A. Puri, and A. N. Netravali, *Digital Video: An Introduction to MPEG-2,* Chapman and Hall, New York, 1997.

Hidaka, T. and K. Ozawa, Subjective assessment of redundancy-reduced moving images for interactive application: test methodology and report, *Signal Process. Image Commun.,* 2, 201-219, 1990.

Huang, T. S. PCM picture transmission, *IEEE Spectrum,* 2(12), 57-63, 1965.

Huang, J. and Y. Q. Shi, Adaptive image watermarking scheme based on visual masking, *IEE Electron. Lett.,* 34(8), 748-750, 1998.

Kretzmer, E. R. Statistics of television signal, *Bell Syst. Tech. J.,* 31(4), 751-763, 1952.

Lathi, B. P. *Modern Digital and Analog Communication Systems,* 3rd ed., Oxford University Press, New York, 1998.

Legge, G. E. and J. M. Foley, Contrast masking in human vision, *J. Opt. Soc. Am.,* 70(12), 1458-1471, 1980.

Lim, J. S. *Two-Dimensional Signal and Image Processing,* Englewood Cliffs, NJ, Prentice Hall, 1990.

Mitchell, J. L., W. B. Pennebaker, C. E. Fogg, and D. J. LeGall, *MPEG Video Compression Standard,* Chapman and Hall, New York, 1997.

Mounts, F. W. A video encoding system with conditional picture-element replenishment, *Bell Syst. Tech. J.,* 48(7), 2545-2554, 1969.

Mullen, K.T. The contrast sensitivity of human color vision to red-green and blue-yellow chromatic gratings, *J. Physiol.,* 359, 381-400, 1985.

Netravali, A. N. and B. Prasada, Adaptive quantization of picture signals using spatial masking, *Proc. IEEE,* 65, 536-548, 1977.

Sakrison, D. J. Image coding applications of vision model, in *Image Transmission Techniques,* W. K. Pratt (Ed.), 21-71, New York, Academic Press, 1979.

Seyler, A. J. The coding of visual signals to reduce channel-capacity requirements, *IEE Monogr.,* 533E, July 1962.

Seyler, A. J. Probability distributions of television frame difference, *Proc. IREE (Australia),* 26, 335, 1965.

Shannon, C. E. A mathematical theory of communication, *Bell Syst. Tech. J.,* 27, 379-423 (Part I), 1948; 623-656 (Part II), 1948.

Slepian, D. *Key Papers in the Development of Information Theory,* Slepian, D. (Ed.), IEEE Press, New York, 1973.

Van Ness, F. I. and M. A. Bouman, Spatial modulation transfer in the human eye, *J. Opt. Soc. Am.,* 57(3), 401-406, 1967.

Watson, A. B. Efficiency of a model human image code, *J. Opt. Soc. Am., A,* 4(12), 2401-2417, 1987.

Webster, A. A., C. T. Jones, and M. H. Pinson, An objective video quality assessment system based on human perception, *Proc. Human Vision, Visual Processing and Digital Display IV,* J. P. Allebach and B. E. Rogowitz (Eds.), SPIE, 1913, 15-26, 1993.

# 2 Quantization

After the introduction to image and video compression presented in Chapter 1, we now address several fundamental aspects of image and video compression in the remaining chapters of Section I. Chapter 2, the first chapter in the series, concerns quantization. Quantization is a necessary component in lossy coding and has a direct impact on the bit rate and the distortion of reconstructed images or videos. We discuss concepts, principles and various quantization techniques which include uniform and nonuniform quantization, optimum quantization, and adaptive quantization.

## 2.1 QUANTIZATION AND THE SOURCE ENCODER

Recall Figure 1.1, in which the functionality of image and video compression in the applications of visual communications and storage is depicted. In the context of visual communications, the whole system may be illustrated as shown in Figure 2.1. In the transmitter, the input analog information source is converted to a digital format in the A/D converter block. The digital format is compressed through the image and video source encoder. In the channel encoder, some redundancy is added to help combat noise and, hence, transmission error. Modulation makes digital data suitable for transmission through the analog channel, such as air space in the application of a TV broadcast. At the receiver, the counterpart blocks reconstruct the input visual information. As far as storage of visual information is concerned, the blocks of channel, channel encoder, channel decoder, modulation, and demodulation may be omitted, as shown in Figure 2.2. If input and output are required to be in the digital format in some applications, then the A/D and D/A converters are omitted from the system. If they are required, however, other blocks such as encryption and decryption can be added to the system (Sklar, 1988). Hence, what is conceptualized in Figure 2.1 is a fundamental block diagram of a visual communication system.

In this book, we are mainly concerned with source encoding and source decoding. To this end, we take it a step further. That is, we show block diagrams of a source encoder and decoder in Figure 2.3. As shown in Figure 2.3(a), there are three components in source encoding: transformation, quantization, and codeword assignment. After the transformation, some form of an input information source is presented to a quantizer. In other words, the transformation block decides which types of quantities from the input image and video are to be encoded. It is not necessary that the original image and video waveform be quantized and coded: we will show that some formats obtained from the input image and video are more suitable for encoding. An example is the difference signal. From the discussion of interpixel correlation in Chapter 1, it is known that a pixel is normally highly correlated with its immediate horizontal or vertical neighboring pixel. Therefore, a better strategy is to encode the difference of gray level values between a pixel and its neighbor. Since these data are highly correlated, the difference usually has a smaller dynamic range. Consequently, the encoding is more efficient. This idea is discussed in Chapter 3 in detail.

Another example is what is called transform coding, which is addressed in Chapter 4. There, instead of encoding the original input image and video, we encode a transform of the input image and video. Since the redundancy in the transform domain is greatly reduced, the coding efficiency is much higher compared with directly encoding the original image and video.

Note that the term transformation in Figure 2.3(a) is sometimes referred to as *mapper* and *signal processing* in the literature (Gonzalez and Woods, 1992; Li and Zhang, 1995). Quantization refers to a process that converts input data into a set of finitely different values. Often, the input data to a quantizer are continuous in magnitude.

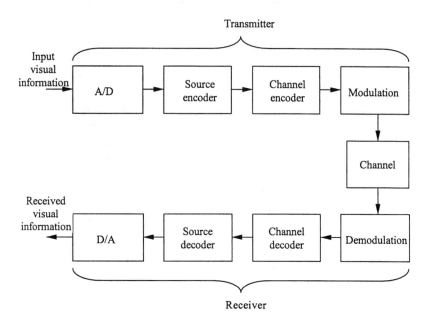

**FIGURE 2.1**    Block diagram of a visual communication system.

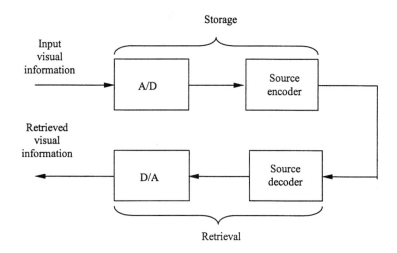

**FIGURE 2.2**    Block diagram of a visual storage system.

Hence, quantization is essentially discretization in magnitude, which is an important step in the lossy compression of digital image and video. (The reason that the term lossy compression is used here will be shown shortly.) The input and output of quantization can be either scalars or vectors. The quantization with scalar input and output is called *scalar quantization*, whereas that with vector input and output is referred to as *vector quantization*. In this chapter we discuss scalar quantization. Vector quantization will be addressed in Chapter 9.

After quantization, codewords are assigned to the many finitely different values from the output of the quantizer. Natural binary code (NBC) and variable-length code (VLC), introduced in Chapter 1, are two examples of this. Other examples are the widely utilized entropy code (including Huffman code and arithmetic code), dictionary code, and run-length code (RLC) (frequently used in facsimile transmission), which are covered in Chapters 5 and 6.

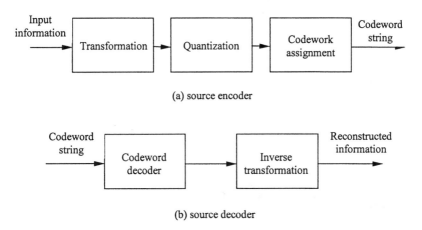

(a) source encoder

(b) source decoder

**FIGURE 2.3**   Block diagram of a source encoder and a source decoder.

The source decoder, as shown in Figure 2.3(b), consists of two blocks: codeword decoder and inverse transformation. They are counterparts of the codeword assignment and transformation in the source encoder. Note that there is no block that corresponds to quantization in the source decoder. The implication of this observation is the following. First, quantization is an irreversible process. That is, in general there is no way to find the original value from the quantized value. Second, quantization, therefore, is a source of information loss. In fact, quantization is a critical stage in image and video compression. It has significant impact on the distortion of reconstructed image and video as well as the bit rate of the encoder. Obviously, coarse quantization results in more distortion and a lower bit rate than fine quantization.

In this chapter, uniform quantization, which is the simplest yet the most important case, is discussed first. Nonuniform quantization is covered after that, followed by optimum quantization for both uniform and nonuniform cases. Then a discussion of adaptive quantization is provided. Finally, pulse code modulation (PCM), the best established and most frequently implemented digital coding method involving quantization, is described.

## 2.2   UNIFORM QUANTIZATION

Uniform quantization is the simplest and most popular quantization technique. Conceptually, it is of great importance. Hence, we start our discussion on quantization with uniform quantization. Several fundamental concepts of quantization are introduced in this section.

### 2.2.1   Basics

This subsection concerns several basic aspects of uniform quantization. These are some fundamental terms, quantization distortion, and quantizer design.

#### 2.2.1.1   Definitions

Take a look at Figure 2.4. The horizontal axis denotes the input to a quantizer, while the vertical axis represents the output of the quantizer. The relationship between the input and the output best characterizes this quantizer; this type of configuration is referred to as the input-output characteristic of the quantizer. It can be seen that there are nine intervals along the $x$-axis. Whenever the input falls in one of the intervals, the output assumes a corresponding value. The input-output characteristic of the quantizer is staircase-like and, hence, clearly nonlinear.

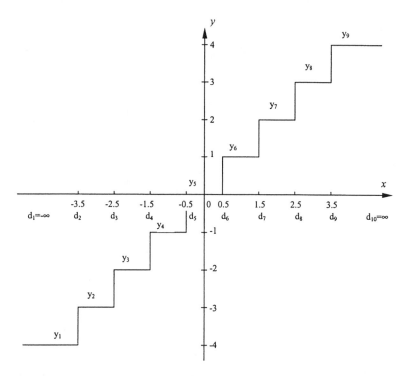

**FIGURE 2.4**   Input-output characteristic of a uniform midtread quantizer.

The end points of the intervals are called *decision levels*, denoted by $d_i$ with $i$ being the index of intervals. The output of the quantization is referred to as the *reconstruction level* (also known as *quantizing level* [Musmann, 1979]), denoted by $y_i$ with $i$ being its index. The length of the interval is called the *step size* of the quantizer, denoted by $\Delta$. With the above terms defined, we can now mathematically define the function of the quantizer in Figure 2.4 as follows.

$$y_i = Q(x) \quad if \quad x \in \left(d_i, d_{i+1}\right) \tag{2.1}$$

where $i = 1, 2, \cdots, 9$ and $Q(x)$ is the output of the quantizer with respect to the input $x$.

It is noted that in Figure 2.4, $\Delta = 1$. The decision levels and reconstruction levels are evenly spaced. It is a uniform quantizer because it possesses the following two features.

1. Except for possibly the right-most and left-most intervals, all intervals (hence, decision levels) along the $x$-axis are uniformly spaced. That is, each inner interval has the same length.
2. Except for possibly the outer intervals, the reconstruction levels of the quantizer are also uniformly spaced. Furthermore, each inner reconstruction level is the arithmetic average of the two decision levels of the corresponding interval along the $x$-axis.

The uniform quantizer depicted in Figure 2.4 is called *midtread* quantizer. Its counterpart is called a *midrise* quantizer, in which the reconstructed levels do not include the value of zero. A midrise quantizer having step size $\Delta = 1$ is shown in Figure 2.5. Midtread quantizers are usually utilized for an odd number of reconstruction levels and midrise quantizers are used for an even number of reconstruction levels.

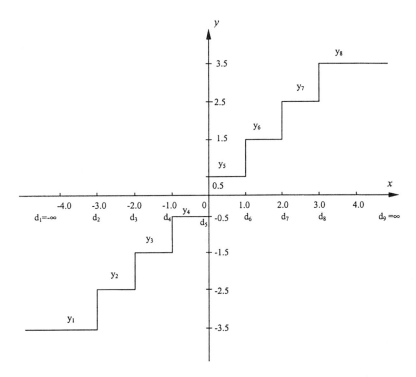

**FIGURE 2.5**  Input-output characteristic of a uniform midrise quantizer.

Note that the input-output characteristic of both the midtread and midrise uniform quantizers as depicted in Figures 2.4 and 2.5, respectively, is odd symmetric with respect to the vertical axis $x = 0$. In the rest of this chapter, our discussion develops under this symmetry assumption. The results thus derived will not lose generality since we can always subtract the statistical mean of input x from the input data and thus achieve this symmetry. After quantization, we can add the mean value back.

Denote by $N$ the total number of reconstruction levels of a quantizer. A close look at Figure 2.4 and 2.5 reveals that if $N$ is even, then the decision level $d_{(N/2)+1}$ is located in the middle of the input x-axis. If $N$ is odd, on the other hand, then the reconstruction level $y_{(N+1)/2} = 0$. This convention is important in understanding the design tables of quantizers in the literature.

### 2.2.1.2  Quantization Distortion

The source coding theorem presented in Chapter 1 states that for a certain distortion $D$, there exists a rate distortion function $R(D)$, such that as long as the bit rate used is larger than $R(D)$ then it is possible to transmit the source with a distortion smaller than $D$. Since we cannot afford an infinite bit rate to represent an original source, some distortion in quantization is inevitable. In other words, we can say that since quantization causes information loss irreversibly, we encounter *quantization error* and, consequently, an issue: how do we evaluate the quality or, equivalently, the distortion of quantization. According to our discussion on visual quality assessment in Chapter 1, we know that there are two ways to do so: subjective evaluation and objective evaluation.

In terms of subjective evaluation, in Section 1.3.1 we introduced a five-scale rating adopted in CCIR Recommendation 500-3. We also described the false contouring phenomenon, which is caused by coarse quantization. That is, our human eyes are more sensitive to the relatively uniform regions in an image plane. Therefore an insufficient number of reconstruction levels results in annoying false contours. In other words, more reconstruction levels are required in relatively uniform regions than in relatively nonuniform regions.

In terms of objective evaluation, in Section 1.3.2 we defined mean square error (*MSE*) and root mean square error (*RMSE*), signal-to-noise ratio (*SNR*), and peak signal-to-noise ratio (*PSNR*). In dealing with quantization, we define quantization error, $e_q$, as the difference between the input signal and the quantized output:

$$e_q = x - Q(x),\qquad(2.2)$$

where $x$ and $Q(x)$ are input and quantized output, respectively. Quantization error is often referred to as *quantization noise*. It is a common practice to treat input $x$ as a random variable with a probability density function (*pdf*) $f_x(x)$. Mean square quantization error, $MSE_q$, can thus be expressed as

$$MSE_q = \sum_{i=1}^{N} \int_{d_i}^{d_{i+1}} \left(x - Q(x)\right)^2 f_x(x)dx\qquad(2.3)$$

where $N$ is the total number of reconstruction levels. Note that the outer decision levels may be $-\infty$ or $\infty$, as shown in Figures 2.4 and 2.5. It is clear that when the *pdf*, $f_x(x)$, remains unchanged, fewer reconstruction levels (smaller $N$) result in more distortion. That is, coarse quantization leads to large quantization noise. This confirms the statement that quantization is a critical component in a source encoder and significantly influences both bit rate and distortion of the encoder. As mentioned, the assumption we made above that the input-output characteristic is odd symmetric with respect to the $x = 0$ axis implies that the mean of the random variable, $x$, is equal to zero, i.e., $E(x) = 0$. Therefore the mean square quantization error $MSE_q$ is the variance of the quantization noise equation, i.e., $MSE_q = \sigma_q^2$.

The quantization noise associated with the midtread quantizer depicted in Figure 2.4 is shown in Figure 2.6. It is clear that the quantization noise is signal dependent. It is observed that, associated with the inner intervals, the quantization noise is bounded by $\pm 0.5\Delta$. This type of quantization noise is referred to as *granular noise*. The noise associated with the right-most and the left-most

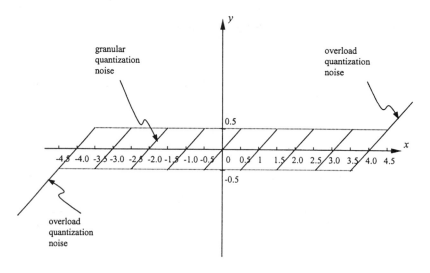

**FIGURE 2.6**   Quantization noise of the uniform midtread quantizer shown in Figure 2.4.

intervals are unbounded as the input $x$ approaches either $-\infty$ or $\infty$. This type of quantization noise is called *overload noise*. Denoting the mean square granular noise and overload noise by $MSE_{q,g}$ and $MSE_{q,o}$, respectively, we then have the following relations:

$$MSE_q = MSE_{q,g} + MSE_{q,o} \qquad (2.4)$$

and

$$MSE_{q,g} = \sum_{i=2}^{N-1} \int_{d_i}^{d_{i+1}} (x - Q(x))^2 \, f_X(x)dx \qquad (2.5)$$

$$MSE_{q,o} = 2\int_{d_1}^{d_2} (x - Q(x))^2 \, f_X(x)dx \qquad (2.6)$$

### 2.2.1.3  Quantizer Design

The design of a quantizer (either uniform or nonuniform) involves choosing the number of reconstruction levels, $N$ (hence, the number of decision levels, $N+1$), and selecting the values of decision levels and reconstruction levels (deciding where to locate them). In other words, the design of a quantizer is equivalent to specifying its input-output characteristic.

The *optimum* quantizer design can be stated as follows. For a given probability density function of the input random variable, $f_X(x)$, determine the number of reconstruction levels, $N$, choose a set of decision levels $\{d_i, i = 1, \cdots, N+1\}$ and a set of reconstruction levels $\{y_i, i = 1, \cdots, N\}$ such that the mean square quantization error, $MSE_q$, defined in Equation 2.3, is minimized.

In the uniform quantizer design, the total number of reconstruction levels, $N$, is usually given. According to the two features of uniform quanitzers described in Section 2.2.1.1, we know that the reconstruction levels of a uniform quantizer can be derived from the decision levels. Hence, only one of these two sets is independent. Furthermore, both decision levels and reconstruction levels are uniformly spaced except possibly the outer intervals. These constraints together with the symmetry assumption lead to the following observation: There is in fact only one parameter that needs to be decided in uniform quantizer design, which is the step size $\Delta$. As to the optimum uniform quantizer design, a different *pdf* leads to a different step size.

### 2.2.2  OPTIMUM UNIFORM QUANTIZER

In this subsection, we first discuss optimum uniform quantizer design when the input $x$ obeys uniform distribution. Then, we cover optimum uniform quantizer design when the input $x$ has other types of probabilistic distributions.

### 2.2.2.1  Uniform Quantizer with Uniformly Distributed Input

Let us return to Figure 2.4, where the input-output characteristic of a nine reconstruction-level midtread quantizer is shown. Now, consider that the input $x$ is a uniformly distributed random variable. Its input-output characteristic is shown in Figure 2.7. We notice that the new characteristic is restricted within a finite range of $x$, i.e., $-4.5 \leq x \leq 4.5$. This is due to the definition of uniform distribution. Consequently, the overload quantization noise does not exist in this case, which is shown in Figure 2.8.

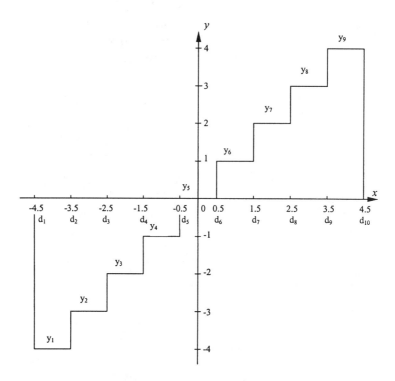

**FIGURE 2.7** Input-output characteristic of a uniform midtread quantizer with input $x$ having uniform distribution in [-4.5, 4.5].

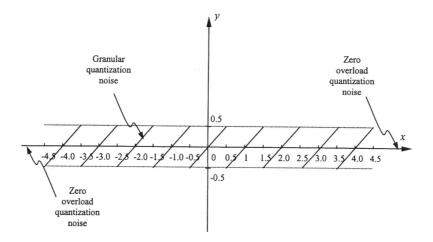

**FIGURE 2.8** Quantization noise of the quantizer shown in Figure 2.7.

The mean square quantization error is found to be

$$MSE_q = N \int_{d_1}^{d_2} \left( x - Q(x) \right)^2 \frac{1}{N\Delta} dx$$

(2.7)

$$MSE_q = \frac{\Delta^2}{12}$$

This result indicates that if the input to a uniform quantizer has a uniform distribution and the number of reconstruction levels is fixed, then the mean square quantization error is directly proportional to the square of the quantization step size. Or, in other words, the root mean square quantization error (the standard deviation of the quantization noise) is directly proportional to the quantization step. The larger the step size, the larger (according to square law) the mean square quantization error. This agrees with our previous observation: coarse quantization leads to large quantization error.

As mentioned above, the mean square quantization error is equal to the variance of the quantization noise, i.e., $MSE_q = \sigma_q^2$. In order to find the signal-to-noise ratio of the uniform quantization in this case, we need to determine the variance of the input $x$. Note that we assume the input $x$ to be a zero mean uniform random variable. So, according to probability theory, we have

$$\sigma_x^2 = \frac{(N\Delta)^2}{12} \tag{2.8}$$

Therefore, the mean square signal-to-noise ratio, $SNR_{ms}$, defined in Chapter 1, is equal to

$$SNR_{ms} = 10\log_{10}\frac{\sigma_x^2}{\sigma_q^2} = 10\log_{10} N^2. \tag{2.9}$$

Note that here we use the subscript $ms$ to indicate the signal-to-noise ratio in the mean square sense, as defined in the previous chapter. If we assume $N = 2^n$, we then have

$$SNR_{ms} = 20\log_{10} 2^n = 6.02n \quad dB. \tag{2.10}$$

The interpretation of the above result is as follows. If we use the natural binary code to code the reconstruction levels of a uniform quantizer with a uniformly distributed input source, then every increased bit in the coding brings out a 6.02-dB increase in the $SNR_{ms}$. An equivalent statement can be derived from Equation 2.7. That is, whenever the step size of the uniform quantizer decreases by a half, the mean square quantization error decreases four times.

### 2.2.2.2 Conditions of Optimum Quantization

The conditions under which the mean square quantization error $MSE_q$ is minimized were derived (Lloyd, 1982; Max, 1960) for a given probability density function of the quantizer input, $f_X(x)$.

The mean square quantization error $MSE_q$ was given in Equation 2.3. The necessary conditions for optimum (minimum mean square error) quantization are as follows. That is, the derivatives of $MSE_q$ with respect to the $d_i$ and $y_i$ have to be zero.

$$(d_i - y_{i-1})^2 f_x(d_i) - (d_i - y_i)^2 f_x(d_i) = 0 \qquad i = 2,\cdots,N \tag{2.11}$$

$$-\int_{d_i}^{d_i+1} (x - y_i) f_x(x)dx = 0 \qquad i = 1,\cdots,N \tag{2.12}$$

The sufficient conditions can be derived accordingly by involving the second-order derivatives (Max, 1960; Fleischer, 1964). The symmetry assumption of the input-output characteristic made earlier holds here as well. These sufficient conditions are listed below.

1.  $x_1 = -\infty$ and $x_{N+1} = +\infty$                       (2.13)

2.  $\displaystyle\int_{d_i}^{d_{i+1}} (x - y_i) f_x(x) dx = 0 \qquad i = 1, 2, \cdots, N$    (2.14)

3.  $d_i = \dfrac{1}{2}(y_{i-1} + y_i) \qquad i = 2, \cdots, N$           (2.15)

Note that the first condition is for an input $x$ whose range is $-\infty < x < \infty$. The interpretation of the above conditions is that each decision level (except for the outer intervals) is the arithmetic average of the two neighboring reconstruction levels, and each reconstruction level is the centroid of the area under the probability density function $f_x(x)$ and between the two adjacent decision levels.

Note that the above conditions are general in the sense that there is no restriction imposed on the *pdf*. In the next subsubsection, we discuss the optimum uniform quantization when the input of quantizer assumes different distributions.

### 2.2.2.3  Optimum Uniform Quantizer with Different Input Distributions

Let's return to our discussion on the optimum quantizer design whose input has uniform distribution. Since the input has uniform distribution, the outer intervals are also finite. For uniform distribution, Equation 2.14 implies that each reconstruction level is the arithmetic average of the two corresponding decision levels. Considering the two features of a uniform quantizer, presented in Section 2.2.1.1, we see that a uniform quantizer is optimum (minimizing the mean square quantization error) when the input has uniform distribution.

When the input $x$ is uniformly distributed in [-1,1], the step size $\Delta$ of the optimum uniform quantizer is listed in Table 2.1 for the number of reconstruction levels, $N$, equal to 2, 4, 8, 16, and 32. From the table, we notice that the $MSE_q$ of the uniform quantization with a uniformly distributed input decreases four times as $N$ doubles. As mentioned in Section 2.2.2.1, this is equivalent to an increase of $SNR_{ms}$ by 6.02 dB as $N$ doubles.

The derivation above is a special case, i.e., the uniform quantizer is optimum for a uniformly distributed input. Normally, if the probability density function is not uniform, the optimum quantizer is not a uniform quantizer. Due to the simplicity of uniform quantization, however, it may sometimes be desirable to design an optimum uniform quantizer for an input with an other-than-uniform distribution.

Under these circumstances, however, Equations 2.13, 2.14, and 2.15 are not a set of simultaneous equations one can hope to solve with any ease. Numerical procedures were suggested to solve for design of optimum uniform quantizers. Max derived uniform quantization step size $\Delta$ for an input with a Gaussian distribution (Max, 1960). Paez and Glisson (1972) found step size $\Delta$ for Laplacian- and Gamma-distributed input signals. These results are listed in Table 2.1. Note that all three distributions have a zero mean and unit standard deviation. If the mean is not zero, only a shift in input is needed when applying these results. If the standard deviation is not unity, the tabulated step size needs to be multiplied by the standard deviation. The theoretical $MSE$ is also listed in Table 2.1. Note that the subscript $q$ associated with $MSE$ has been dropped from now on in the chapter for the sake of notational brevity as long as it does not cause confusion.

## 2.3  NONUNIFORM QUANTIZATION

It is not difficult to see that, except for the special case of the uniformly distributed input variable $x$, the optimum (minimum $MSE$, also denoted sometimes by $MMSE$) quantizers should be nonuniform.

## TABLE 2.1
## Optimal Symmetric Uniform Quantizer for Uniform Gaussian, Laplacian, and Gamma Distributions [a]

| N | Uniform | | | Gaussian | | | Laplacian | | | Gamma | | |
|---|---|---|---|---|---|---|---|---|---|---|---|---|
| | $d_i$ | $y_i$ | MSE | $d_i$ | $y_i$ | MSE | $d_i$ | $y_i$ | MSE | $d_i$ | $y_i$ | MSE |
| 2 | −1.000 | −0.500 | 8.33 | −1.596 | −0.798 | 0.363 | −1.414 | −0.707 | 0.500 | −1.154 | −0.577 | 0.668 |
| | 0.000 | 0.500 | ×10⁻² | 0.000 | 0.798 | | 0.000 | 0.707 | | 0.000 | 0.577 | |
| | **1.000** | | | **1.596** | | | **1.414** | | | **1.154** | | |
| 4 | −1.000 | −0.750 | 2.08 | −1.991 | −1.494 | 0.119 | −2.174 | −1.631 | 1.963 | −2.120 | −1.590 | 0.320 |
| | −0.500 | −0.250 | ×10⁻² | −0.996 | −0.498 | | −1.087 | −0.544 | ×10⁻¹ | −1.060 | −0.530 | |
| | 0.000 | 0.250 | | 0.000 | 0.498 | | 0.000 | 0.544 | | 0.000 | 0.530 | |
| | **0.500** | 0.750 | | **0.996** | 1.494 | | **1.087** | 1.631 | | **1.060** | 1.590 | |
| | 1.000 | | | 1.991 | | | 2.174 | | | 2.120 | | |
| 8 | −1.000 | −0.875 | 5.21 | −2.344 | −2.051 | 3.74 | −2.924 | −2.559 | 7.17 | −3.184 | −2.786 | 0.132 |
| | −0.750 | −0.625 | ×10⁻³ | −1.758 | −1.465 | ×10⁻² | −2.193 | −1.828 | ×10⁻² | −2.388 | −1.990 | |
| | −0.500 | −0.375 | | −1.172 | −0.879 | | −1.462 | −1.097 | | −1.592 | −1.194 | |
| | −0.250 | −0.125 | | −0.586 | −0.293 | | −0.731 | −0.366 | | −0.796 | −0.398 | |
| | 0.000 | 0.125 | | 0.000 | 0.293 | | 0.000 | 0.366 | | 0.000 | 0.398 | |
| | **0.250** | 0.375 | | **0.586** | 0.879 | | **0.731** | 1.097 | | **0.796** | 1.194 | |
| | 0.500 | 0.625 | | 1.172 | 1.465 | | 1.462 | 1.828 | | 1.592 | 1.990 | |
| | 0.750 | 0.875 | | 1.758 | 2.051 | | 2.193 | 2.559 | | 2.388 | 2.786 | |
| | 1.000 | | | 2.344 | | | 2.924 | | | 3.184 | | |
| 16 | −1.000 | −0.938 | 1.30 | −2.680 | −2.513 | 1.15 | −3.648 | −3.420 | 2.54 | −4.320 | −4.050 | 5.01 |
| | −0.875 | −0.813 | ×10⁻³ | −2.345 | −2.178 | ×10⁻² | −3.192 | −2.964 | ×10⁻² | −3.780 | −3.510 | ×10⁻² |
| | −0.750 | −0.688 | | −2.010 | −1.843 | | −2.736 | −2.508 | | −3.240 | −2.970 | |
| | −0.625 | −0.563 | | −1.675 | −1.508 | | −2.280 | −2.052 | | −2.700 | −2.430 | |
| | −0.500 | −0.438 | | −1.340 | −1.173 | | −1.824 | −1.596 | | −2.160 | −1.890 | |
| | −0.375 | −0.313 | | −1.005 | −0.838 | | −1.368 | −1.140 | | −1.620 | −1.350 | |
| | −0.250 | −0.188 | | −0.670 | −0.503 | | −0.912 | −0.684 | | −1.080 | −0.810 | |
| | −0.125 | −0.063 | | −0.335 | −0.168 | | −0.456 | −0.228 | | −0.540 | −0.270 | |
| | 0.000 | 0.063 | | 0.000 | 0.168 | | 0.000 | 0.228 | | 0.000 | 0.270 | |
| | **0.125** | 0.188 | | **0.335** | 0.503 | | **0.456** | 0.684 | | **0.540** | 0.810 | |
| | 0.250 | 0.313 | | 0.670 | 0.838 | | 0.912 | 1.140 | | 1.080 | 1.350 | |
| | 0.375 | 0.438 | | 1.005 | 1.173 | | 1.368 | 1.596 | | 1.620 | 1.890 | |
| | 0.500 | 0.563 | | 1.340 | 1.508 | | 1.824 | 2.052 | | 2.160 | 2.430 | |
| | 0.625 | 0.688 | | 1.675 | 1.843 | | 2.280 | 2.508 | | 2.700 | 2.970 | |
| | 0.750 | 0.813 | | 2.010 | 2.178 | | 2.736 | 2.964 | | 3.240 | 3.510 | |
| | 0.875 | 0.938 | | 2.345 | 2.513 | | 3.192 | 3.420 | | 3.780 | 4.050 | |
| | 1.000 | | | 2.680 | | | 3.648 | | | 4.320 | | |

*Note:* The uniform distribution is between [−1,1], the other three distributions have zero mean and unit variance. The numbers in bold type are the step sizes.

[a] Data from (Max, 1960; Paez and Glisson, 1972).

Consider a case in which the input random variable obeys the Gaussian distribution with a zero mean and unit variance, and the number of reconstruction levels is finite. We naturally consider that having decision levels more densely located around the middle of the $x$-axis, $x = 0$ (high-probability density region), and choosing decision levels more coarsely distributed in the range far away from the center of the $x$-axis (low-probability density region) will lead to less *MSE*. The strategy adopted here is analogous to the superiority of variable-length code over fixed-length code discussed in the previous chapter.

### 2.3.1  OPTIMUM (NONUNIFORM) QUANTIZATION

Conditions for optimum quantization were discussed in Section 2.2.2.2. With some constraints, these conditions were solved in a closed form (Panter and Dite, 1951). The equations characterizing these conditions, however, cannot be solved in a closed form in general. Lloyd and Max proposed an iterative procedure to numerically solve the equations. The optimum quantizers thus designed are called Lloyd-Max quantizers.

**TABLE 2.2**
**Optimal Symmetric Quantizer for Uniform, Gaussian, Laplacian, and Gamma Distributions**[a]

| N | Uniform $d_i$ | Uniform $y_i$ | Uniform MSE | Gaussian $d_i$ | Gaussian $y_i$ | Gaussian MSE | Laplacian $d_i$ | Laplacian $y_i$ | Laplacian MSE | Gamma $d_i$ | Gamma $y_i$ | Gamma MSE |
|---|---|---|---|---|---|---|---|---|---|---|---|---|
| 2 | −1.000 | −0.500 | 8.33 | −∞ | −0.799 | 0.363 | −∞ | −0.707 | 0.500 | −∞ | −0.577 | 0.668 |
|   | 0.000 | 0.500 | ×10⁻² | 0.000 | 0.799 | | 0.000 | 0.707 | | 0.000 | 0.577 | |
|   | 1.000 | | | ∞ | | | ∞ | | | ∞ | | |
| 4 | −1.000 | −0.750 | 2.08 | −∞ | −1.510 | 0.118 | −∞ | −1.834 | 1.765 | −∞ | −2.108 | 0.233 |
|   | −0.500 | −0.250 | ×10⁻² | −0.982 | −0.453 | | −1.127 | −0.420 | ×10⁻¹ | −1.205 | −0.302 | |
|   | 0.000 | 0.250 | | 0.000 | 0.453 | | 0.000 | 0.420 | | 0.000 | 0.302 | |
|   | 0.500 | 0.750 | | −0.982 | 1.510 | | 1.127 | 1.834 | | 1.205 | 2.108 | |
|   | 1.000 | | | ∞ | | | ∞ | | | ∞ | | |
| 8 | −1.000 | −0.875 | 5.21 | −∞ | −2.152 | 3.45 | −∞ | −3.087 | 5.48 | −∞ | −3.799 | 7.12 |
|   | −0.750 | −0.625 | ×10⁻³ | −1.748 | −1.344 | ×10⁻² | −2.377 | −1.673 | ×10⁻² | −2.872 | −1.944 | ×10⁻² |
|   | −0.500 | −0.375 | | −1.050 | −0.756 | | −1.253 | −0.833 | | −1.401 | −0.859 | |
|   | −0.250 | −0.125 | | −0.501 | −0.245 | | −0.533 | −0.233 | | −0.504 | −0.149 | |
|   | 0.000 | 0.125 | | 0.000 | 0.245 | | 0.000 | 0.233 | | 0.000 | 0.149 | |
|   | 0.250 | 0.375 | | 0.501 | 0.756 | | 0.533 | 0.833 | | 0.504 | 0.859 | |
|   | 0.500 | 0.625 | | 1.050 | 1.344 | | 1.253 | 1.673 | | 1.401 | 1.944 | |
|   | 0.750 | 0.875 | | 1.748 | 2.152 | | 2.377 | 3.087 | | 2.872 | 3.799 | |
|   | 1.000 | | | ∞ | | | ∞ | | | ∞ | | |
| 16 | −1.000 | −0.938 | 1.30 | −∞ | −2.733 | 9.50 | −∞ | −4.316 | 1.54 | −∞ | −6.085 | 1.96 |
|   | −0.875 | −0.813 | ×10⁻³ | −2.401 | −2.069 | ×10⁻³ | −3.605 | −2.895 | ×10⁻² | −5.050 | −4.015 | ×10⁻² |
|   | −0.750 | −0.688 | | −1.844 | −1.618 | | −2.499 | −2.103 | | −3.407 | −2.798 | |
|   | −0.625 | −0.563 | | −1.437 | −1.256 | | −1.821 | −1.540 | | −2.372 | −1.945 | |
|   | −0.500 | −0.438 | | −1.099 | −0.942 | | −1.317 | −1.095 | | −1.623 | −1.300 | |
|   | −0.375 | −0.313 | | −0.800 | −0.657 | | −0.910 | −0.726 | | −1.045 | −0.791 | |
|   | −0.250 | −0.188 | | −0.522 | −0.388 | | −0.566 | −0.407 | | −0.588 | −0.386 | |
|   | −0.125 | −0.063 | | −0.258 | −0.128 | | −0.266 | −0.126 | | −0.229 | −0.072 | |
|   | 0.000 | 0.063 | | 0.000 | 0.128 | | 0.000 | 0.126 | | 0.000 | 0.072 | |
|   | 0.125 | 0.188 | | 0.258 | 0.388 | | 0.266 | 0.407 | | 0.229 | 0.386 | |
|   | 0.250 | 0.313 | | 0.522 | 0.657 | | 0.566 | 0.726 | | 0.588 | 0.791 | |
|   | 0.375 | 0.438 | | 0.800 | 0.942 | | 0.910 | 1.095 | | 1.045 | 1.300 | |
|   | 0.500 | 0.563 | | 1.099 | 1.256 | | 1.317 | 1.540 | | 1.623 | 1.945 | |
|   | 0.625 | 0.688 | | 1.437 | 1.618 | | 1.821 | 2.103 | | 2.372 | 2.798 | |
|   | 0.750 | 0.813 | | 1.844 | 2.069 | | 2.499 | 2.895 | | 3.407 | 4.015 | |
|   | 0.875 | 0.938 | | 2.401 | 2.733 | | 3.605 | 4.316 | | 5.050 | 6.085 | |
|   | 1.000 | | | ∞ | | | ∞ | | | ∞ | | |

*Note:* The uniform distribution is between [-1, 1], the other three distributions have zero mean and unit variance.

[a] Data from Lloyd, 1957, 1982; Max, 1990; and Paez, 1972.

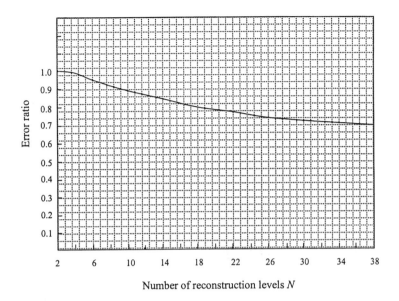

Number of reconstruction levels $N$

**FIGURE 2.9**   Ratio of the error for the optimal quantizer to the error for the optimum uniform quantizer vs. the number of reconstruction levels $N$. (Minimum mean square error for Gaussian-distributed input with a zero mean and unit variance). (Data from Max, 1960.)

The solution to optimum quantizer design for many finite reconstruction levels $N$ when input $x$ obeys Gaussian distribution was obtained (Lloyd, 1982; Max, 1960). That is, the decision levels and reconstruction levels together with theoretical minimum *MSE* and optimum *SNR* have been determined. Following this procedure, the design for Laplacian and Gamma distribution were tabulated in (Paez and Glisson, 1972). These results are contained in Table 2.2. As stated before, we see once again that uniform quantization is optimal if the input $x$ is a uniform random variable.

Figure 2.9 (Max, 1960) gives a performance comparison between optimum uniform quantization and optimum quantization for the case of a Gaussian-distributed input with a zero mean and unit variance. The abscissa represents the number of reconstruction levels, $N$, and the ordinate the ratio between the error of the optimum quantizer and the error of the optimum uniform quantizer. It can be seen that when $N$ is small, the ratio is close to one. That is, the performances are close. When $N$ increases, the ratio decreases. Specifically, when $N$ is large the nonuniform quantizer is about 20 to 30% more efficient than the uniform optimum quantizer for the Gaussian distribution with a zero mean and unit variance.

### 2.3.2   COMPANDING QUANTIZATION

It is known that a speech signal usually has a large dynamic range. Moreover, its statistical distribution reveals that very low speech volumes predominate most voice communications. Specifically, by a 50% chance, the voltage characterizing detected speech energy is less than 25% of the root mean square (rms) value of the signal. Large amplitude values are rare: only by a 15% chance does the voltage exceed the rms value (Sklar, 1988). These statistics naturally lead to the need for nonuniform quantization with relatively dense decision levels in the small-magnitude range and relatively coarse decision levels in the large-magnitude range.

When the bit rate is eight bits per sample, the following companding technique (Smith, 1957), which realizes nonuniform quantization, is found to be extremely useful. Though speech coding is not the main focus of this book, we briefly discuss the companding technique here as an alternative way to achieve nonuniform quantization.

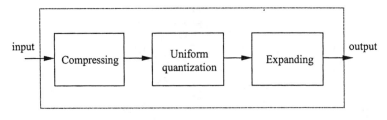

Nonuniform quantization

**FIGURE 2.10**  Companding technique in achieving quantization.

The companding technique, also known as logarithmic quantization, consists of the following three stages: compressing, uniform quantization, and expanding (Gersho, 1977), as shown in Figure 2.10. It first compresses the input signal with a logarithmic characteristic, and then it quantizes the compressed input using a uniform quantizer. Finally, the uniformly quantized results are expanded inversely. An illustration of the characteristics of these three stages and the resultant nonuniform quantization are shown in Figure 2.11.

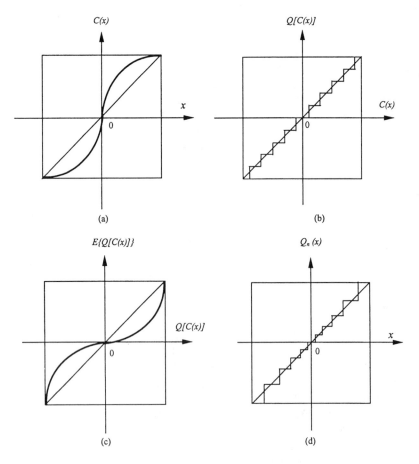

**FIGURE 2.11**  Characteristics of companding techniques. (a) Compressing characteristic. (b) Uniform quantizer characteristic. (c) Expanding characteristic. (d) Nonuniform quantizer characteristic.

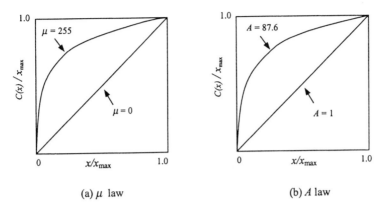

(a) $\mu$ law          (b) $A$ law

**FIGURE 2.12**  Compression characteristics.

In practice, a piecewise linear approximation of the logarithmic compression characteristic is used. There are two different ways. In North America, a $\mu$-law compression characteristic is used, which is defined as follows:

$$c(x) = x_{max} \frac{\ln\left[1 + \mu\left(|x|/x_{max}\right)\right]}{\ln(1 + \mu)} \, \text{sgn} \, x, \tag{2.16}$$

where sgn is a sign function defined as

$$\text{sgn} \, x = \begin{cases} +1 & if \quad x \geq 0 \\ -1 & if \quad x < 0 \end{cases} \tag{2.17}$$

The $\mu$-law compression characteristic is shown in Figure 2.12(a). The standard value of $\mu$ is 255. Note from the figure that the case of $\mu = 0$ corresponds to uniform quantization.

In Europe, the A-law characteristic is used. The A-law characteristic is depicted in Figure 2.12(b), and is defined as follows:

$$c(x) = \begin{cases} x_{max} \dfrac{A\left(|x|/x_{max}\right)}{1 + \ln A} \, \text{sgn} \, x & 0 < \dfrac{|x|}{x_{max}} \leq \dfrac{1}{A} \\[3mm] x_{max} \dfrac{1 + \ln\left[A\left(|x|/x_{max}\right)\right]}{1 + \ln A} \, \text{sgn} \, x & \dfrac{1}{A} < \dfrac{|x|}{x_{max}} < 1 \end{cases} \tag{2.18}$$

It is noted that the standard value of A is 87.6. The case of A = 1 corresponds to uniform quantization.

## 2.4  ADAPTIVE QUANTIZATION

In the previous section, we studied nonuniform quantization, whose motivation is to minimize the mean square quantization error $MSE_q$. We found that nonuniform quantization is necessary if the *pdf* of the input random variable $x$ is not uniform. Consider an optimum quantizer for a Gaussian-distributed input when the number of reconstruction levels $N$ is eight. Its input-output characteristic can be derived from Table 2.2 and is shown in Figure 2.13. This plot reveals that the decision levels are densely located in the central region of the $x$-axis and coarsely elsewhere. In other words, the

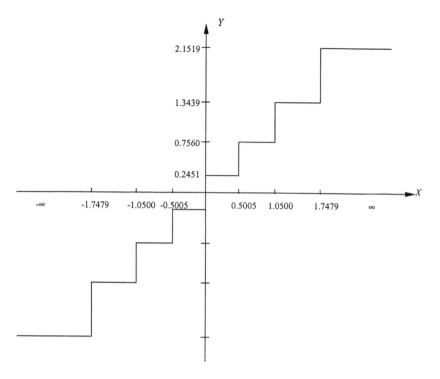

**FIGURE 2.13**  Input-output characteristic of the optimal quantizer for Gaussian distribution with zero mean, unit variance, and N = 8.

decision levels are densely distributed in the region having a higher probability of occurrence and coarsely distributed in other regions. A logarithmic companding technique also allocates decision levels densely in the small-magnitude region, which corresponds to a high occurrence probability, but in a different way. We conclude that nonuniform quantization achieves minimum mean square quantization error by distributing decision levels according to the statistics of the input random variable.

These two types of nonuniform quantizers are both time-invariant. That is, they are not designed for nonstationary input signals. Moreover, even for a stationary input signal, if its *pdf* deviates from that with which the optimum quantizer is designed, then what is called *mismatch* will take place and the performance of the quantizer will deteriorate. There are two main types of mismatch. One is called variance mismatch. That is, the *pdf* of input signal is matched, while the variance is mismatched. Another type is *pdf* mismatch. Noted that these two kinds of mismatch also occur in optimum uniform quantization, since there the optimization is also achieved based on the input statistics assumption. For a detailed analysis of the effects of the two types of mismatch on quantization, readers are referred to (Jayant and Noll, 1984).

Adaptive quantization attempts to make the quantizer design adapt to the varying input statistics in order to achieve better performance. It is a means to combat the mismatch problem discussed above. By statistics, we mean the statistical mean, variance (or the dynamic range), and type of input *pdf*. When the mean of the input changes, differential coding (discussed in the next chapter) is a suitable method to handle the variation. For other types of cases, adaptive quantization is found to be effective. The price paid for adaptive quantization is processing delays and an extra storage requirement as seen below.

There are two different types of adaptive quantization: forward adaptation and backward adaptation. Before we discuss these, however, let us describe an alternative way to define quantization (Jayant and Noll, 1984). Look at Figure 2.14. Quantization can be viewed as a two-stage

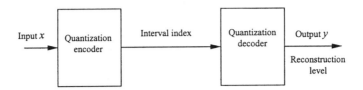

**FIGURE 2.14**   A two-stage model of quantization.

process. The first stage is the quantization encoder and the second stage is the quantization decoder. In the encoder, the input to quantization is converted to the index of an interval into which the input $x$ falls. This index is mapped to (the codeword that represents) the reconstruction level corresponding to the interval in the decoder. Roughly speaking, this definition considers a quantizer as a communication system in which the quantization encoder is on the transmitter side while the quantization decoder is on the receiver side. In this sense, this definition is broader than that for quantization defined in Figure 2.3(a).

### 2.4.1   FORWARD ADAPTIVE QUANTIZATION

A block diagram of forward adaptive quantization is shown in Figure 2.15. There, the input to the quantizer, $x$, is first split into blocks, each with a certain length. Blocks are stored in a buffer one at a time. A statistical analysis is then carried out with respect to the block in the buffer. Based on the analysis, the quantization encoder is set up, and the input data within the block are assigned indexes of respective intervals. In addition to these indexes, the encoder setting parameters, derived from the statistical analysis, are sent to the quantization decoder as *side* information. The term *side* comes from the fact that the amount of bits used for coding the setting parameter is usually a small fraction of the total amount of bits used.

Selection of the block size is a critical issue. If the size is small, the adaptation to the local statistics will be effective, but the side information needs to be sent frequently. That is, more bits are used for sending the side information. If the size is large, the bits used for side information decrease. On the other hand, the adaptation becomes less sensitive to changing statistics, and both processing delays and storage required increase. In practice, a proper compromise between the quantity of side information and the effectiveness of adaptation produces a good selection of the block size.

Examples of using the forward approach to adapt quantization to a changing input variance (to combat variance mismatch) can be found in (Jayant and Noll, 1984; Sayood, 1996).

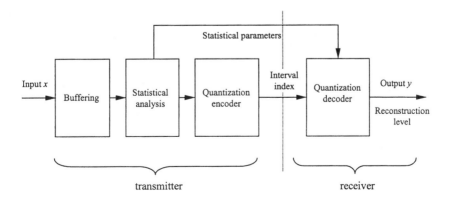

**FIGURE 2.15**   Forward adaptive quantization.

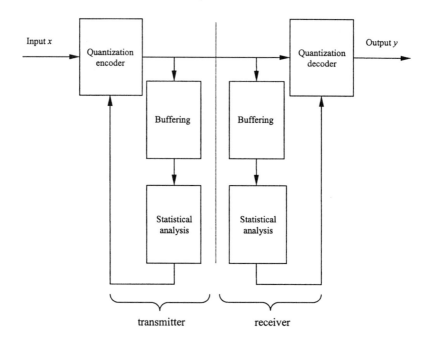

**FIGURE 2.16**   Backward adaptive quantization.

## 2.4.2   BACKWARD ADAPTIVE QUANTIZATION

Figure 2.16 shows a block diagram of backward adaptive quantization. A close look at the block diagram reveals that in both the quantization encoder and decoder the buffering and the statistical analysis are carried out with respect to the output of the quantization encoder. In this way, there is no need to send side information. The sensitivity of adaptation to the changing statistics will be degraded, however, since instead of the original input, only the output of the quantization encoder is used in the statistical analysis. That is, the quantization noise is involved in the statistical analysis.

## 2.4.3   ADAPTIVE QUANTIZATION WITH A ONE-WORD MEMORY

Intuitively, it is expected that observance of a sufficiently large number of input or output (quantized) data is necessary in order to track the changing statistics and then adapt the quantizer setting in adaptive quantization. Through an analysis, Jayant showed that effective adaptations can be realized with an explicit memory of only one word. That is, either one input sample, $x$, in forward adaptive quantization or a quantized output, $y$, in backward adaptive quantization is sufficient (Jayant, 1973).

In (Jayant, 1984), examples on step-size adaptation (with the number of total reconstruction levels larger than four) were given. The idea is as follows. If at moment $t_i$ the input sample $x_i$ falls into the outer interval, then the step size at the next moment $t_{i+1}$ will be enlarged by a factor of $m_i$ (multiplying the current step size by $m_i$, $m_i > 1$). On the other hand, if the input $x_i$ falls into an inner interval close to $x = 0$ then, the multiplier is less than 1, i.e., $m_i < 1$. That is, the multiplier $m_i$ is small in the interval near $x = 0$ and monotonically increases for an increased $x$. Its range varies from a small positive number less than 1 to a number larger than 1. In this way, the quantizer adapts itself to the input to avoid *overload* as well as *underload* to achieve better performance.

## 2.4.4   SWITCHED QUANTIZATION

This is another adaptive quantization scheme. A block diagram is shown in Figure 2.17. It consists of a bank of $L$ quantizers. Each quantizer in the bank is fixed, but collectively they form a bank

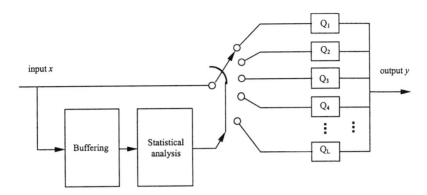

**FIGURE 2.17**   Switched quantization.

of quantizers with a variety of input-output characteristics. Based on a statistical analysis of recent input or output samples, a switch connects the current input to one of the quantizers in the bank such that the best possible performance may be achieved. It is reported that in both video and speech applications, this scheme has shown improved performance even when the number of quantizers in the bank, $L$, is two (Jayant and Noll, 1984). Interestingly, it is noted that as $L \to \infty$, the switched quantization converges to the adaptive quantizer discussed above.

## 2.5   PCM

Pulse code modulation (PCM) is closely related to quantization, the focus of this chapter. Further-more, as pointed out in (Jayant, 1984), PCM is the earliest, best established, and most frequently applied coding system despite the fact that it is the most bit-consuming digitizing system (since it encodes each pixel independently) as well as being a very demanding system in terms of the bit error rate on the digital channel. Therefore, we discuss the PCM technique in this section.

PCM is now the most important form of pulse modulation. The other forms of pulse modulation are pulse amplitude modulation (PAM), pulse width modulation (PWM), and pulse position mod-ulation (PPM), which are covered in most communications texts. Briefly speaking, pulse modulation links an analog signal to a pulse train in the following way. The analog signal is first sampled (a discretization in the time domain). The sampled values are used to modulate a pulse train. If the modulation is carried out through the amplitude of the pulse train, it is called PAM. If the modified parameter of the pulse train is the pulse width, we then have PWM. If the pulse width and magnitude are constant — only the position of pulses is modulated by the sample values — we then encounter PPM. An illustration of these pulse modulations is shown in Figure 2.18.

In PCM, an analog signal is first sampled. The sampled value is then quantized. Finally the quantized value is encoded, resulting in a bit steam. Figure 2.19 provides an example of PCM. We see that through a sampling and a uniform quantization the PCM system converts the input analog signal, which is continuous in both time and magnitude, into a digital signal (discretized in both time and magnitude) in the form of a natural binary code sequence. In this way, an analog signal modulates a pulse train with a natural binary code.

By far, PCM is more popular than other types of pulse modulation since the *code* modulation is much more robust against various noises than amplitude modulation, width modulation, and position modulation. In fact, almost all coding techniques include a PCM component. In digital image processing, given digital images usually appear in PCM format. It is known that an acceptable PCM representation of a monochrome picture requires six to eight bits per pixel (Huang, 1975). It is used so commonly in practice that its performance normally serves as a standard against which other coding techniques are compared.

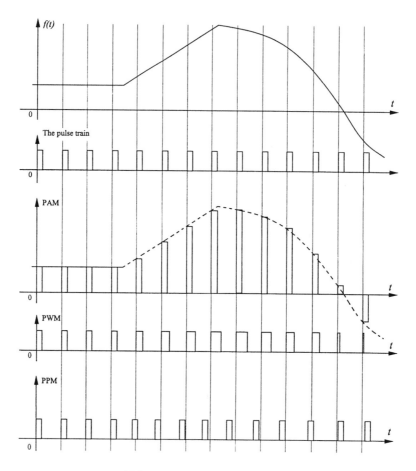

**FIGURE 2.18**   Pulse modulation.

Recall the false contouring phenomenon discussed in Chapter 1, when we discussed texture masking. It states that our eyes are more sensitive to relatively uniform regions in an image plane. If the number of reconstruction levels is not large enough (coarse quantization), then some unnatural contours will appear. When frequency masking was discussed, it was noted that by adding some high-frequency signal before quantization, the false contouring can be eliminated to a great extent. This technique is called dithering. The high-frequency signal used is referred to as a dither signal. Both false contouring and dithering were first reported in (Goodall, 1951).

## 2.6   SUMMARY

Quantization is a process in which a quantity having possibly an infinite number of different values is converted to another quantity having only finite many values. It is an important element in source encoding that has significant impact on both bit rate and distortion of reconstructed images and video in visual communication systems. Depending on whether the quantity is a scalar or a vector, quantization is called either scalar quantization or vector quantization. In this chapter we considered only scalar quantization.

Uniform quantization is the simplest and yet the most important case. In uniform quantization, except for outer intervals, both decision levels and reconstruction levels are uniformly spaced. Moreover, a reconstruction level is the arithmetic average of the two corresponding decision levels. In uniform quantization design, the step size is usually the only parameter that needs to be specified.

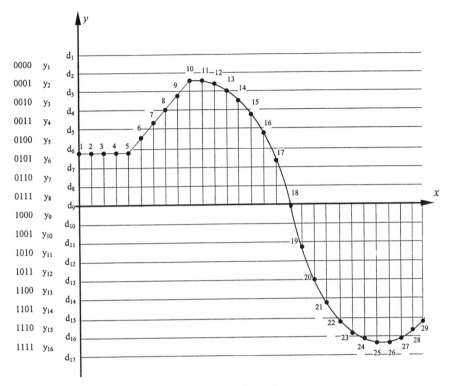

Output code (from left to right, from top to bottom):

| 0101 | 0101 | 0101 | 0101 | 0101 | 0100 | 0011 | 0011 | 0010 | 0001 | 0001 | 0001 | 0010 | 0010 | 0011 |
| 0100 | 0101 | 1000 | 1010 | 1100 | 1101 | 1110 | 1110 | 1111 | 1111 | 1111 | 1111 | 1110 | 1110 | |

**FIGURE 2.19**   Pulse code modulation (PCM).

Optimum quantization implies minimization of the mean square quantization error. When the input has a uniform distribution, uniform quantization is optimum. For the sake of simplicity, a uniform optimum quantizer is sometimes desired even when the input does not obey uniform distribution. The design under these circumstances involves an iterative procedure. The design problem in cases where the input has Gaussian, Lapacian, or Gamma distribution was solved and the parameters are available.

When the constraint of uniform quantization is removed, the conditions for optimum quantization are derived. The resultant optimum quantizer is normally nonuniform. An iterative procedure to solve the design is established and the optimum design parameters for Gaussian, Laplacian, and Gamma distribution are tabulated.

The companding technique is an alternative way to implement nonuniform quantization. Both nonuniform quantization and companding are time-invariant and hence not suitable for nonstationary input. Adaptive quantization deals with nonstationary input and combats the mismatch that occurs in optimum quantization design.

In adaptive quantization, buffering is necessary to store some recent input or sampled output data. A statistical analysis is carried out with respect to the stored recent data. Based on the analysis, the quantizer's parameters are adapted to changing input statistics to achieve better quantization performance. There are two types of adaptive quantization: forward and backward adaptive quantization. With the forward type, the statistical analysis is derived from the original input data, while with the backward type, quantization noise is involved in the analysis. Therefore, the forward technique usually achieves more effective adaptation than the backward manner. The latter, however, does not need to send quantizer setting parameters as side information to the receiver side, since the output values of the quantization encoder (based on which the statistics are analyzed and the quantizer's parameters are adapted) are available in both the transmitter and receiver sides.

Switched quantization is another type of adaptive quantization. In this scheme, a bank of fixed quantizers is utilized, each quantizer having different input-output characteristics. A statistical analysis based on recent input decides which quantizer in the bank is suitable for the present input. The system then connects the input to this particular quantizer.

Nowadays, pulse code modulation is the most frequently used form of pulse modulation due to its robustness against noise. PCM consists of three stages: sampling, quantization, and encoding. Analog signals are first sampled with a proper sampling frequency. The sampled data are then quantized using a uniform quantizer. Finally, the quantized values are encoded with natural binary code. It is the best established and most applied coding system. Despite its bit-consuming feature, it is utilized in almost all coding systems.

## 2.7   EXERCISES

**2-1.** Using your own words, define quantization and uniform quantization. What are the two features of uniform quantization?

**2-2.** What is optimum quantization? Why is uniform quantization sometimes desired, even when the input has a *pdf* different from uniform? How was this problem solved? Draw an input-output characteristic of an optimum uniform quantizer with an input obeying Gaussian *pdf* having zero mean, unit variance, and the number of reconstruction levels, $N$, equal to 8.

**2-3.** What are the conditions of optimum nonuniform quantization? From Table 2.2, what observations can you make?

**2-4.** Define variance mismatch and *pdf* mismatch. Discuss how you can resolve the mismatch problem.

**2-5.** What is the difference between forward and backward adaptive quantization? Comment on the merits and drawbacks for each.

**2-6.** What are PAM, PWM, PPM, and PCM? Why is PCM the most popular type of pulse modulation?

## REFERENCES

Fleischer, P. E. Sufficient conditions for achieving minimum distortion in quantizer, *IEEE Int. Convention Rec.,* 12, 104-111, 1964.

Gersho, A, Quantization, *IEEE Commun. Mag.,* pp. 6-29, September, 1977.

Gonzalez, R. C. and R. E. Woods, *Digital Image Processing*, Addison-Wesley, Reading, MA, 1992.

Goodall, W. M. Television by pulse code modulation, *Bell Syst. Tech. J.,* 33-49, January 1951.

Huang, T. S. PCM picture transmission, *IEEE Spectrum*, 2, 57-63, 1965.

Jayant, N. S. Adaptive delta modulation with one-bit memory, *Bell Syst. Tech. J.,* 49, 321-342, 1970.

Jayant, N. S. Adaptive quantization with one word memory, *Bell Syst. Tech. J.,* 52, 1119-1144, 1973.

Jayant, N. S. and P. Noll, *Digital Coding of Waveforms*, Prentice-Hall, Englewood Cliffs, NJ, 1984.

Li, W. and Y.-Q. Zhang, Vector-based signal processing and qunatization for image and video compression, *Proc. IEEE,* 83(2), 317-335, 1995.

Lloyd, S. P. Least squares quantization in PCM, Inst. Mathematical Statistics Meet., Atlantic City, NJ, September 1957; *IEEE Trans. Inf. Theory,* pp. 129-136, March 1982.

Max, J. Quantizing for minimum distortion, *IRE Trans. Inf. Theory,* it-6, 7-12, 1960.

Musmann, H. G. Predictive Image Coding, in *Image Transmission Techniques*, W. K. Pratt (Ed.), Academic Press, New York, 1979.

Paez, M. D. and T. H. Glisson, Minimum mean squared error qunatization in speech PCM and DPCM Systems, *IEEE Trans. Commun.,* 225-230, April 1972.

Panter, P. F. and W. Dite, Quantization distortion in pulse count modulation with nonuniform spacing of levels, *Proc. IRE*, 39, 44-48, 1951.

Sayood, K. *Introduction to Data Compression*, Morgan Kaufmann Publishers, San Francisco, CA, 1996.

Sklar, B. *Digital Communications: Fundamentals and Applications*, Prentice-Hall, Englewood Cliffs, NJ, 1988.

Smith, B. Instantaneous companding of quantized signals, *Bell Syst. Tech. J.,* 36, 653-709, 1957.

# 3 Differential Coding

Instead of encoding a signal directly, the *differential coding* technique codes the difference between the signal itself and its prediction. Therefore it is also known as *predictive coding*. By utilizing spatial and/or temporal interpixel correlation, differential coding is an efficient and yet computationally simple coding technique. In this chapter, we first describe the differential technique in general. Two components of differential coding, prediction and quantization, are discussed. There is an emphasis on (optimum) prediction, since quantization was discussed in Chapter 2. When the difference signal (also known as prediction error) is quantized, the differential coding is called differential pulse code modulation (DPCM). Some issues in DPCM are discussed, after which delta modulation (DM) as a special case of DPCM is covered. The idea of differential coding involving image sequences is briefly discussed in this chapter. More detailed coverage is presented in Sections III and IV, starting from Chapter 10. If quantization is not included, the differential coding is referred to as information-preserving differential coding. This is discussed at the end of the chapter.

## 3.1 INTRODUCTION TO DPCM

As depicted in Figure 2.3, a source encoder consists of the following three components: transformation, quantization, and codeword assignment. The transformation converts input into a format for quantization followed by codeword assignment. In other words, the component of transformation decides which format of input is to be encoded. As mentioned in the previous chapter, input itself is not necessarily the most suitable format for encoding.

Consider the case of monochrome image encoding. The input is usually a 2-D array of gray level values of an image obtained via PCM coding. The concept of spatial redundancy, discussed in Section 1.2.1.1, tells us that neighboring pixels of an image are usually highly correlated. Therefore, it is more efficient to encode the gray difference between two neighboring pixels instead of encoding the gray level values of each pixel. At the receiver, the decoded difference is added back to reconstruct the gray level value of the pixel. Since neighboring pixels are highly correlated, their gray level values bear a great similarity. Hence, we expect that the variance of the difference signal will be smaller than that of the original signal. Assume uniform quantization and natural binary coding for the sake of simplicity. Then we see that for the same bit rate (bits per sample) the quantization error will be smaller, i.e., a higher quality of reconstructed signal can be achieved. Or, for the same quality of reconstructed signal, we need a lower bit rate.

### 3.1.1 SIMPLE PIXEL-TO-PIXEL DPCM

Denote the gray level values of pixels along a row of an image as $z_i$, $i = 1, \cdots, M$, where $M$ is the total number of pixels within the row. Using the immediately preceding pixel's gray level value, $z_{i-1}$, as a prediction of that of the present pixel, $\hat{z}_i$, i.e.,

$$\hat{z}_i = z_{i-1} \qquad (3.1)$$

we then have the difference signal

$$d_i = z_i - \hat{z}_i = z_i - z_{i-1} \qquad (3.2)$$

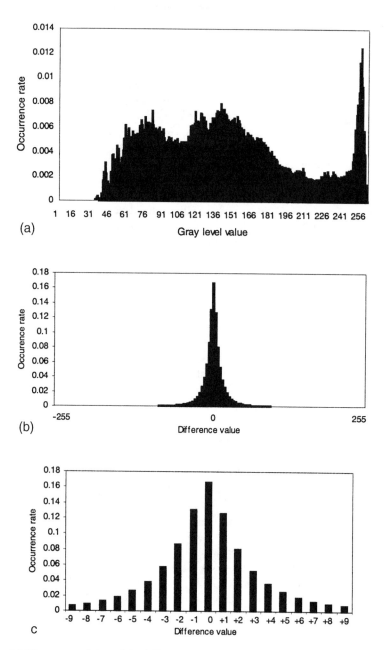

**FIGURE 3.1**  (a) Histogram of the original "boy and girl" image. (b) Histogram of the difference image obtained by using horizontal pixel-to-pixel differencing. (c) A close-up of the central portion of the histogram of the difference image.

Assume a bit rate of eight bits per sample in the quantization. We can see that although the dynamic range of the difference signal is theoretically doubled, from 256 to 512, the variance of the difference signal is actually much smaller. This can be confirmed from the histograms of the "boy and girl" image (refer to Figure 1.1) and its difference image obtained by horizontal pixel-to-pixel differencing, shown in Figure 3.1(a) and (b), respectively. Figure 3.1(b) and its close-up (c) indicate that by a rate of 42.44% the difference values fall into the range of –1, 0, and +1. In other words, the histogram of the difference signal is much more narrowly concentrated than that of the original signal.

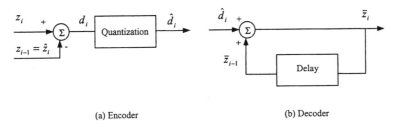

(a) Encoder                              (b) Decoder

**FIGURE 3.2**   Block diagram of a pixel-to-pixel differential coding system.

A block diagram of the scheme described above is shown in Figure 3.2. There $z_i$ denotes the sequence of pixels along a row, $d_i$ is the corresponding difference signal, and $\hat{d}_i$ is the quantized version of the difference, i.e.,

$$\hat{d}_i = Q(d_i) = d_i + e_q \tag{3.3}$$

where $e_q$ represents the quantization error. In the decoder, $\bar{z}_i$ represents the reconstructed pixel gray value, and we have

$$\bar{z}_i = \bar{z}_{i-1} + \hat{d}_i \tag{3.4}$$

This simple scheme, however, suffers from an accumulated quantization error. We can see this clearly from the following derivation (Sayood, 1996), where we assume the initial value $z_0$ is available for both the encoder and the decoder.

$$\text{as} \quad i = 1, \quad d_1 = z_1 - z_0$$

$$\hat{d}_1 = d_1 + e_{q,1} \tag{3.5}$$

$$\bar{z}_1 = z_0 + \hat{d}_1 = z_0 + d_1 + e_{q,1} = z_1 + e_{q,1}$$

Similarly, we can have

$$\text{as} \quad i = 2, \quad \bar{z}_2 = z_2 + e_{q,1} + e_{q,2} \tag{3.6}$$

and, in general,

$$\bar{z}_i = z_i + \sum_{j=1}^{i} e_{q,j} \tag{3.7}$$

This problem can be remedied by the following scheme, shown in Figure 3.3. Now we see that in both the encoder and the decoder, the reconstructed signal is generated in the same way, i.e.,

$$\bar{z}_i = \bar{z}_{i-1} + \hat{d}_i \tag{3.8}$$

and in the encoder the difference signal changes to

$$d_i = z_i - \bar{z}_{i-1} \tag{3.9}$$

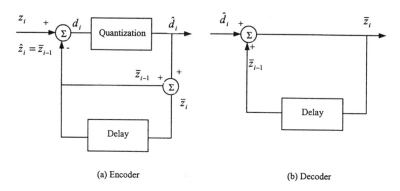

(a) Encoder                                                    (b) Decoder

**FIGURE 3.3**   Block diagram of a practical pixel-to-pixel differential coding system.

Thus, the previously reconstructed $\bar{z}_{i-1}$ is used as the predictor, $\hat{z}_i$, i.e.,

$$\hat{z}_i = \bar{z}_{i-1}.$$                                                              (3.10)

In this way, we have

$$as \quad i = 1, \quad d_1 = z_1 - z_0$$

$$\hat{d}_1 = d_1 + e_{q,1}$$                                                          (3.11)

$$\bar{z}_1 = z_0 + \hat{d}_1 = z_0 + d_1 + e_{q,1} = z_1 + e_{q,1}$$

Similarly, we have

$$as \quad i = 2, \quad d_2 = z_2 - \bar{z}_1$$

$$\hat{d}_2 = d_2 + e_{q,2}$$                                                          (3.12)

$$\bar{z}_2 = \bar{z}_1 + \hat{d}_2 = z_2 + e_{q,2}$$

In general,

$$\bar{z}_i = z_i + e_{q,i}$$                                                              (3.13)

Thus, we see that the problem of the quantization error accumulation has been resolved by having both the encoder and the decoder work in the same fashion, as indicated in Figure 3.3, or in Equations 3.3, 3.9, and 3.10.

### 3.1.2   GENERAL DPCM SYSTEMS

In the above discussion, we can view the reconstructed neighboring pixel's gray value as a prediction of that of the pixel being coded. Now, we generalize this simple pixel-to-pixel DPCM. In a general DPCM system, a pixel's gray level value is first predicted from the preceding reconstructed pixels' gray level values. The difference between the pixel's gray level value and the predicted value is then quantized. Finally, the quantized difference is encoded and transmitted to the receiver. A block

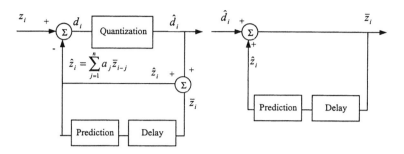

**FIGURE 3.4**  Block diagram of a general DPCM system.

diagram of this general differential coding scheme is shown in Figure 3.4, where the codeword assignment in the encoder and its counterpart in decoder are not included.

It is noted that, instead of using the previously reconstructed sample, $\bar{z}_{i-1}$, as a predictor, we now have the predicted version of $z_i$, $\hat{z}_i$, as a function of the $n$ previously reconstructed samples, $\bar{z}_{i-1}, \bar{z}_{i-2}, \cdots, \bar{z}_{i-n}$. That is,

$$\hat{z}_i = f\left(\bar{z}_{i-1}, \bar{z}_{i-2}, \cdots, \bar{z}_{i-n}\right) \tag{3.14}$$

Linear prediction, i.e., that the function $f$ in Equation 3.14 is linear, is of particular interest and is widely used in differential coding. In linear prediction, we have

$$\hat{z}_i = \sum_{j=1}^{n} a_j \bar{z}_{i-j} \tag{3.15}$$

where $a_j$ are real parameters. Hence, we see that the simple pixel-to-pixel differential coding is a special case of general differential coding with linear prediction, i.e., $n = 1$ and $a_1 = 1$.

In Figure 3.4, $d_i$ is the difference signal and is equal to the difference between the original signal, $z_i$, and the prediction $\hat{z}_i$. That is,

$$d_i = z_i - \hat{z}_i \tag{3.16}$$

The quantized version of $d_i$ is denoted by $\hat{d}_i$. The reconstructed version of $z_i$ is represented by $\bar{z}_i$, and

$$\bar{z}_i = \hat{z}_i + \hat{d}_i \tag{3.17}$$

Note that this is true for both the encoder and the decoder. Recall that the accumulation of the quantization error can be remedied by using this method.

The difference between the original input and the predicted input is called prediction error, which is denoted by $e_p$. That is,

$$e_p = z_i - \hat{z}_i \tag{3.18}$$

where the $e_p$ is understood as the prediction error associated with the index $i$. Quantization error, $e_q$, is equal to the reconstruction error or coding error, $e_r$, defined as the difference between the original signal, $z_i$, and the reconstructed signal, $\bar{z}_i$, when the transmission is error free:

$$e_q = d_i - \hat{d}_i$$

$$= \left(z_i - \hat{z}_i\right) - \left(\bar{z}_i - \hat{z}_i\right) \qquad (3.19)$$

$$= z_i - \bar{z}_i = e_r$$

This indicates that quantization error is the only source of information loss with an error-free transmission channel.

The DPCM system depicted in Figure 3.4 is also called closed-loop DPCM with feedback around the quantizer (Jayant, 1984). This term reflects the feature in DPCM structure.

Before we leave this section, let us take a look at the history of the development of differential image coding. According to an excellent early article on differential image coding (Musmann, 1979), the first theoretical and experimental approaches to image coding involving linear prediction began in 1952 at the Bell Telephone Laboratories (Oliver, 1952; Kretzmer, 1952; Harrison, 1952). The concepts of DPCM and DM were also developed in 1952 (Cutler, 1952; Dejager, 1952). Predictive coding capable of preserving information for a PCM signal was established at the Massachusetts Institute of Technology (Elias, 1955).

The differential coding technique has played an important role in image and video coding. In the international coding standard for still images, JPEG (covered in Chapter 7), we can see that differential coding is used in the lossless mode and in the DCT-based mode for coding DC coefficients. Motion-compensated (MC) coding has been a major development in video coding since the 1980s and has been adopted by all the international video coding standards such as H.261 and H.263 (covered in Chapter 19), MPEG 1 and MPEG 2 (covered in Chapter 16). MC coding is essentially a predictive coding technique applied to video sequences involving displacement motion vectors.

## 3.2  OPTIMUM LINEAR PREDICTION

Figure 3.4 demonstrates that a differential coding system consists of two major components: prediction and quantization. Quantization was discussed in the previous chapter. Hence, in this chapter we emphasize prediction. Below, we formulate an optimum linear prediction problem and then present a theoretical solution to the problem.

### 3.2.1  FORMULATION

Optimum linear prediction can be formulated as follows. Consider a discrete-time random process $z$. At a typical moment $i$, it is a random variable $z_i$. We have $n$ previous observations $\bar{z}_{i-1}, \bar{z}_{i-2}, \cdots, \bar{z}_{i-n}$ available and would like to form a prediction of $z_i$, denoted by $\hat{z}_i$. The output of the predictor, $\hat{z}_i$, is a linear function of the $n$ previous observations. That is,

$$\hat{z}_i = \sum_{j=1}^{n} a_j \bar{z}_{i-j} \qquad (3.20)$$

with $a_j, j = 1,2,\cdots,n$ being a set of real coefficients. An illustration of a linear predictor is shown in Figure 3.5. As defined above, the prediction error, $e_p$, is

$$e_p = z_i - \hat{z}_i \qquad (3.21)$$

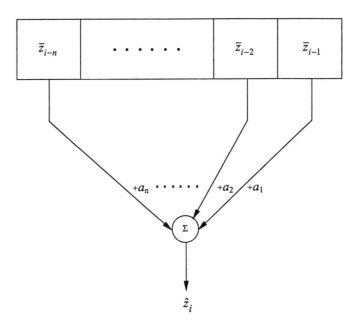

**FIGURE 3.5**  An illustration of a linear predictor.

The mean square prediction error, $MSE_p$, is

$$MSE_p = E\left[\left(e_p\right)^2\right] = E\left[\left(z_i - \hat{z}_i\right)^2\right]$$

(3.22)

The optimum prediction, then, refers to the determination of a set of coefficients $a_j$, $j = 1,2,\cdots,n$ such that the mean square prediction error, $MSE_p$, is minimized.

This optimization problem turns out to be computationally intractable for most practical cases due to the feedback around the quantizer shown in Figure 3.4, and the nonlinear nature of the quantizer. Therefore, the optimization problem is solved in two separate stages. That is, the best linear predictor is first designed ignoring the quantizer. Then, the quantizer is optimized for the distribution of the difference signal (Habibi, 1971). Although the predictor thus designed is sub-optimal, ignoring the quantizer in the optimum predictor design allows us to substitute the reconstructed $\bar{z}_{i-j}$ by $z_{i-j}$ for $j = 1,2,\cdots,n$, according to Equation 3.20. Consequently, we can apply the theory of optimum linear prediction to handle the design of the optimum predictor as shown below.

### 3.2.2  ORTHOGONALITY CONDITION AND MINIMUM MEAN SQUARE ERROR

By taking the differentiation of $MSE_p$ with respect to coefficient $a_j$s, one can derive the following necessary conditions, which are usually referred to as the *orthogonality condition*:

$$E\left[e_p \cdot z_{i-j}\right] = 0 \quad for \quad j = 1,2,\cdots,n$$

(3.23)

The interpretation of Equation 3.23 is that the prediction error, $e_p$, must be orthogonal to all the observations, which are now the preceding samples: $z_{i-j}, j = 1,2,\cdots,n$ according to our discussion in Section 3.2.1. These are equivalent to

$$R_z(m) = \sum_{j=1}^{n} a_j R_z(m-j) \quad for \quad m = 1, 2, \cdots, n \tag{3.24}$$

where $R_z$ represents the autocorrelation function of $z$. In a vector-matrix format, the above orthogonal conditions can be written as

$$
\begin{bmatrix}
R_z(1) \\
R_z(2) \\
\vdots \\
\vdots \\
R_z(n)
\end{bmatrix}
=
\begin{bmatrix}
R_z(0) & R_z(1) & \cdots & \cdots & R_z(n-1) \\
R_z(1) & R_z(0) & \cdots & \cdots & R_z(n-2) \\
\vdots & \vdots & \cdots & \cdots & \vdots \\
\vdots & \vdots & \cdots & \cdots & \vdots \\
R_z(n-1) & R_z(n) & \cdots & \cdots & R_z(0)
\end{bmatrix}
\cdot
\begin{bmatrix}
a_1 \\
a_2 \\
\vdots \\
\vdots \\
a_n
\end{bmatrix}
\tag{3.25}
$$

Equations 3.24 and 3.25 are called Yule-Walker equations.

The minimum mean square prediction error is then found to be

$$MSE_p = R_z(0) - \sum_{j=1}^{n} a_j R_z(j) \tag{3.26}$$

These results can be found in texts dealing with random processes, e.g., in (Leon-Garcia, 1994).

### 3.2.3 SOLUTION TO YULE-WALKER EQUATIONS

Once autocorrelation data are available, the Yule-Walker equation can be solved by matrix inversion. A recursive procedure was developed by Levinson to solve the Yule-Walker equations (Leon-Garcia, 1993). When the number of previous samples used in the linear predictor is large, i.e., the dimension of the matrix is high, the Levinson recursive algorithm becomes more attractive. Note that in the field of image coding the autocorrelation function of various types of video frames is derived from measurements (O'Neal, 1966; Habibi, 1971).

## 3.3 SOME ISSUES IN THE IMPLEMENTATION OF DPCM

Several related issues in the implementation of DPCM are discussed in this section.

### 3.3.1 OPTIMUM DPCM SYSTEM

Since DPCM consists mainly of two parts, prediction and quantization, its optimization should not be carried out separately. The interaction between the two parts is quite complicated, however, and thus combined optimization of the whole DPCM system is difficult. Fortunately, with the mean square error criterion, the relation between quantization error and prediction error has been found:

$$MSE_q \approx \frac{9}{2N^2} MSE_p \tag{3.27}$$

where $N$ is the total number of reconstruction levels in the quantizer (O'Neal, 1966; Musmann, 1979). That is, the mean square error of quantization is approximately proportional to the mean square error of prediction. With this approximation, we can optimize the two parts separately, as mentioned in Section 3.2.1. While the optimization of quantization was addressed in Chapter 2, the optimum predictor was discussed in Section 3.2. A large amount of work has been done on this subject. For instance, the optimum predictor for color image coding was designed and tested in (Pirsch and Stenger, 1977).

### 3.3.2   1-D, 2-D, AND 3-D DPCM

In Section 3.1.2, we expressed linear prediction in Equation 3.15. However, so far we have not really discussed how to predict a pixel's gray level value by using its neighboring pixels' coded gray level values.

Recall that a practical pixel-to-pixel differential coding system was discussed in Section 3.1.1. There, the reconstructed intensity of the immediately preceding pixel along the same scan line is used as a prediction of the pixel intensity being coded. This type of differential coding is referred to as 1-D DPCM. In general, 1-D DPCM may use the reconstructed gray level values of more than one of the preceding pixels within the same scan line to predict that of a pixel being coded. By far, however, the immediately preceding pixel in the same scan line is most frequently used in 1-D DPCM. That is, pixel A in Figure 3.6 is often used as a prediction of pixel Z, which is being DPCM coded.

Sometimes in DPCM image coding, both the decoded intensity values of adjacent pixels within the same scan line and the decoded intensity values of neighboring pixels in different scan lines are involved in the prediction. This is called 2-D DPCM. A typical pixel arrangement in 2-D predictive coding is shown in Figure 3.6. Note that the pixels involved in the prediction are restricted to be either in the lines above the line where the pixel being coded, Z, is located or on the left-hand side of pixel Z if they are in the same line. Traditionally, a TV frame is scanned from top to bottom and from left to right. Hence, the above restriction indicates that only those pixels, which have been coded and available in both the transmitter and the receiver, are used in the prediction. In 2-D system theory, this support is referred to as recursively computable (Bose, 1982). An often-used 2-D prediction involves pixels A, D, and E.

Obviously, 2-D predictive coding utilizes not only the spatial correlation existing within a scan line but also that existing in neighboring scan lines. In other words, the spatial correlation is utilized both horizontally and vertically. It was reported that 2-D predictive coding outperforms 1-D predictive coding by decreasing the prediction error by a factor of two, or equivalently, 3dB in *SNR*. The improvement in subjective assessment is even larger (Musmann, 1979). Furthermore, the transmission error in 2-D predictive image coding is much less severe than in 1-D predictive image coding. This is discussed in Section 3.6.

In the context of image sequences, neighboring pixels may be located not only in the same image frame but also in successive frames. That is, neighboring pixels along the time dimension are also involved. If the prediction of a DPCM system involves three types of neighboring pixels: those along the same scan line, those in the different scan lines of the same image frame, and those

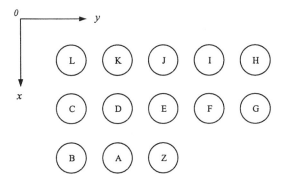

**FIGURE 3.6**   Pixel arrangement in 1-D and 2-D prediction.

in the different frames, the DPCM is then called 3-D differential coding. It will be discussed in Section 3.5.

### 3.3.3 ORDER OF PREDICTOR

The number of coefficients in the linear prediction, $n$, is referred to as the order of the predictor. The relation between the mean square prediction error, $MSE_p$, and the order of the predictor, $n$, has been studied. As shown in Figure 3.7, the $MSE_p$ decreases quite effectively as $n$ increases, but the performance improvement becomes negligible as $n > 3$ (Habibi, 1971).

### 3.3.4 ADAPTIVE PREDICTION

Adaptive DPCM means adaptive prediction and adaptive quantization. As adaptive quantization was discussed in Chapter 2, here we discuss adaptive prediction only.

Similar to the discussion on adaptive quantization, adaptive prediction can be done in two different ways: forward adaptive and backward adaptive prediction. In the former, adaptation is based on the input of a DPCM system, while in the latter, adaptation is based on the output of the DPCM. Therefore, forward adaptive prediction is more sensitive to changes in local statistics. Prediction parameters (the coefficients of the predictor), however, need to be transmitted as side information to the decoder. On the other hand, quantization error is involved in backward adaptive prediction. Hence, the adaptation is less sensitive to local changing statistics. But, it does not need to transmit side information.

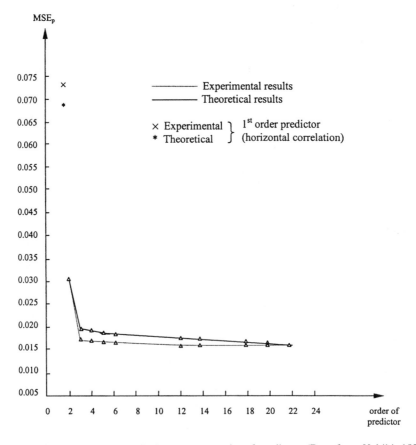

**FIGURE 3.7**   Mean square prediction error vs. order of predictor. (Data from Habibi, 1971.)

In either case, the data (either input or output) have to be buffered. Autocorrelation coefficients are analyzed, based on which prediction parameters are determined.

### 3.3.5 EFFECT OF TRANSMISSION ERRORS

Transmission errors caused by channel noise may reverse the binary bit information from 0 to 1 or 1 to 0 with what is known as *bit error probability*, or *bit error rate*. The effect of transmission errors on reconstructed images varies depending on different coding techniques.

In the case of the PCM-coding technique, each pixel is coded independently of the others. Therefore bit reversal in the transmission only affects the gray level value of the corresponding pixel in the reconstructed image. It does not affect other pixels in the reconstructed image.

In DPCM, however, the effect caused by transmission errors becomes more severe. Consider a bit reversal occurring in transmission. It causes error in the corresponding pixel. But, this is not the end of the effect. The affected pixel causes errors in reconstructing those pixels towards which the erroneous gray level value was used in the prediction. In this way, the transmission error propagates.

Interestingly, it is reported that the error propagation is more severe in 1-D differential image coding than in 2-D differential coding. This may be explained as follows. In 1-D differential coding, usually only the immediate preceding pixel in the same scan line is involved in prediction. Therefore, an error will be propagated along the scan line until the beginning of the next line, where the pixel gray level value is reinitialized. In 2-D differential coding, the prediction of a pixel gray level value depends not only on the reconstructed gray level values of pixels along the same scan line but also on the reconstructed gray level values of the vertical neighbors. Hence, the effect caused by a bit reversal transmission error is less severe than in the 1-D differential coding.

For this reason, the bit error rate required by DPCM coding is lower than that required by PCM coding. For instance, while a bit error rate less than $5 \cdot 10^{-6}$ is normally required for PCM to provide broadcast TV quality, for the same application a bit error rate less than $10^{-7}$ and $10^{-9}$ are required for DPCM coding with 2-D prediction and 1-D prediction, respectively (Musmann, 1979).

Channel encoding with an error correction capability was applied to lower the bit error rate. For instance, to lower the bit error rate from the order of $10^{-6}$ to the order of $10^{-9}$ for DPCM coding with 1-D prediction, an error correction by adding 3% redundancy in channel coding has been used (Bruders, 1978).

## 3.4  DELTA MODULATION

Delta modulation (DM) is an important, simple, special case of DPCM, as discussed above. It has been widely applied and is thus an important coding technique in and of itself.

The above discussion and characterization of DPCM systems are applicable to DM systems. This is because DM is essentially a special type of DPCM, with the following two features.

1. The linear predictor is of the first order, with the coefficient $a_1$ equal to 1.
2. The quantizer is a one-bit quantizer. That is, depending on whether the difference signal is positive or negative, the output is either $+\Delta/2$ or $-\Delta/2$.

To perceive these two features, let us take a look at the block diagram of a DM system and the input-output characteristic of its one-bit quantizer, shown in Figures 3.8 and 3.9, respectively. Due to the first feature listed above, we have:

$$\hat{z}_i = \bar{z}_{i-1} \tag{3.28}$$

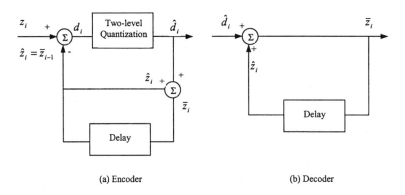

(a) Encoder                                              (b) Decoder

**FIGURE 3.8**  Block diagram of DM systems.

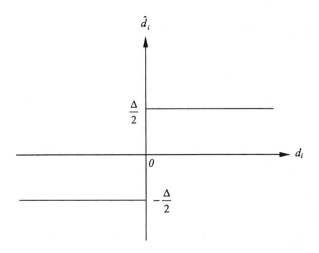

**FIGURE 3.9**  Input-output characteristic of two-level quantization in DM.

Next, we see that there are only two reconstruction levels in quantization because of the second feature. That is,

$$\hat{d}_i = \begin{cases} +\Delta/2 & if \quad z_i > \bar{z}_{i-1} \\ -\Delta/2 & if \quad z_i < \bar{z}_{i-1} \end{cases} \tag{3.29}$$

From the relation between the reconstructed value and the predicted value of DPCM, discussed above, and the fact that DM is a special case of DPCM, we have

$$\bar{z}_i = \hat{z}_i + \hat{d}_i \tag{3.30}$$

Combining Equations 3.28, 3.29, and 3.30, we have

$$\bar{z}_i = \begin{cases} \bar{z}_{i-1} + \Delta/2 & if \quad z_i > \bar{z}_{i-1} \\ \bar{z}_{i-1} - \Delta/2 & if \quad z_i < \bar{z}_{i-1} \end{cases} \tag{3.31}$$

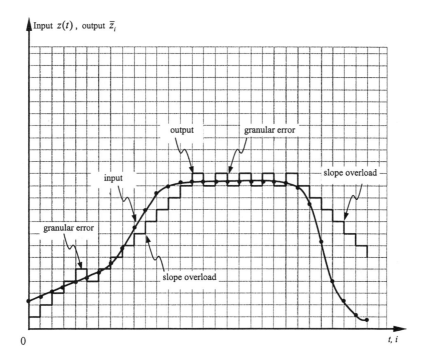

**FIGURE 3.10**   DM with fixed step size.

The above mathematical relationships are of importance in understanding DM systems. For instance, Equation 3.31 indicates that the step size $\Delta$ of DM is a crucial parameter. We notice that a large step size compared with the magnitude of the difference signal causes granular error, as shown in Figure 3.10. Therefore, in order to reduce the granular error we should choose a small step size. On the other hand, a small step size compared with the magnitude of the difference signal will lead to the overload error discussed in Chapter 2 for quantization. Since in DM systems it is the difference signal that is quantized, the overload error in DM becomes *slope overload* error, as shown in Figure 3.10. That is, it takes a while (multiple steps) for the reconstructed samples to catch up with the sudden change in input. Therefore, the step size should be large in order to avoid the slope overload. Considering these two conflicting factors, a proper compromise in choosing the step size is common practice in DM.

To improve the performance of DM, an oversampling technique is often applied. That is, the input is oversampled prior to the application of DM. By oversampling, we mean that the sampling frequency is higher than the sampling frequency used in obtaining the original input signal. The increased sample density caused by oversampling decreases the magnitude of the difference signal. Consequently, a relatively small step size can be used so as to decrease the granular noise without increasing the slope overload error. Thus, the resolution of the DM-coded image is kept the same as that of the original input (Jayant, 1984; Lim, 1990).

To achieve better performance for changing inputs, an adaptive technique can be applied in DM. That is, either input (forward adaptation) or output (backward adaptation) data are buffered and the data variation is analyzed. The step size is then chosen accordingly. If it is forward adaptation, side information is required for transmission to the decoder. Figure 3.11 demonstrates step size adaptation. We see the same input as that shown in Figure 3.10. But, the step size is now not fixed. Instead, the step size is adapted according to the varying input. When the input changes with a large slope, the step size increases to avoid the slope overload error. On the other hand, when the input changes slowly, the step size decreases to reduce the granular error.

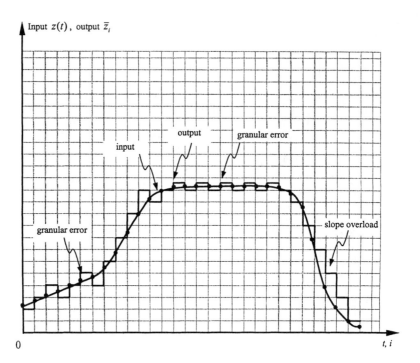

**FIGURE 3.11**   Adaptive DM.

## 3.5   INTERFRAME DIFFERENTIAL CODING

As was mentioned in Section 3.3.2, 3-D differential coding involves an image sequence. Consider a sensor located in 3-D world space. For instance, in applications such as videophony and video-conferencing, the sensor is fixed in position for a while and it takes pictures. As time goes by, the images form a temporal image sequence. The coding of such an image sequence is referred to as interframe coding. The subject of image sequence and video coding is addressed in Sections III and IV. In this section, we briefly discuss how differential coding is applied to interframe coding.

### 3.5.1   CONDITIONAL REPLENISHMENT

Recognizing the great similarity between consecutive TV frames, a conditional replenishment coding technique was proposed and developed (Mounts, 1969). It was regarded as one of the first real demonstrations of interframe coding exploiting interframe redundancy (Netravali and Robbins, 1979).

In this scheme, the previous frame is used as a reference for the present frame. Consider a pair of pixels: one in the previous frame, the other in the present frame — both occupying the same spatial position in the frames. If the gray level difference between the pair of pixels exceeds a certain criterion, then the pixel is considered a *changing* pixel. The present pixel gray level value and its position information are transmitted to the receiving side, where the pixel is replenished. Otherwise, the pixel is considered *unchanged*. At the receiver its previous gray level is repeated. A block diagram of conditional replenishment is shown in Figure 3.12. There, a frame memory unit in the transmitter is used to store frames. The differencing and thresholding of corresponding pixels in two consecutive frames can then be conducted there. A buffer in the transmitter is used to smooth the transmission data rate. This is necessary because the data rate varies from region to region within an image frame and from frame to frame within an image sequence. A buffer in the receiver is needed for a similar consideration. In the frame memory unit, the replenishment is

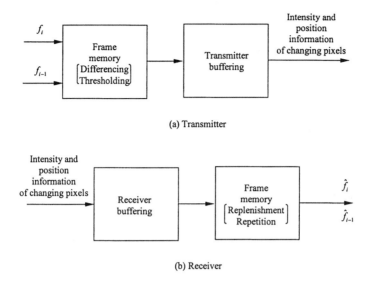

(a) Transmitter

(b) Receiver

**FIGURE 3.12**    Block diagram of conditional replenishment.

carried out for the changing pixels and the gray level values in the receiver are repeated for the unchanged pixels.

With conditional replenishment, a considerable savings in bit rate was achieved in applications such as videophony, videoconferencing, and TV broadcasting. Experiments in real time, using the head-and-shoulder view of a person in animated conversation as the video source, demonstrated an average bit rate of 1 bit/pixel with a quality of reconstructed video comparable with standard 8 bit/pixel PCM transmission (Mounts, 1969). Compared with pixel-to-pixel 1-D DPCM, the most popularly used coding technique at the time, the conditional replenishment technique is more efficient due to the exploitation of high interframe redundancy. As pointed in (Mounts, 1969), there is more correlation between television pixels along the frame-to-frame temporal dimension than there is between adjacent pixels within a signal frame. That is, the temporal redundancy is normally higher than spatial redundancy for TV signals.

Tremendous efforts have been made to improve the efficiency of this rudimentary technique. For an excellent review, readers are referred to (Haskell et al., 1972, 1979). 3-D DPCM coding is among the improvements and is discussed next.

### 3.5.2    3-D DPCM

It was soon realized that it is more efficient to transmit the gray level difference than to transmit the gray level itself, resulting in interframe differential coding. Furthermore, instead of treating each pixel independently of its neighboring pixels, it is more efficient to utilize spatial redundancy as well as temporal redundancy, resulting in 3-D DPCM.

Consider two consecutive TV frames, each consisting of an odd and an even field. Figure 3.13 demonstrates the small neighborhood of a pixel, Z, in the context. As with the 1-D and 2-D DPCM discussed before, the prediction can only be based on the previously encoded pixels. If the pixel under consideration, Z, is located in the even field of the present frame, then the odd field of the present frame and both odd and even fields of the previous frame are available. As mentioned in Section 3.3.2, it is assumed that in the even field of the present frame, only those pixels in the lines above the line where pixel Z lies and those pixels left of the Z in the line where Z lies are used for prediction.

Table 3.1 lists several utilized linear prediction schemes. It is recognized that the case of *element difference* is a 1-D predictor since the immediately preceding pixel is used as the predictor. The

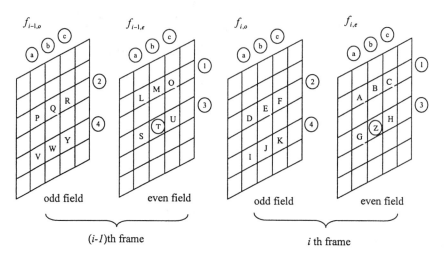

**FIGURE 3.13** Pixel arrangement in two TV frames. (After Haskell, 1979.)

field difference is defined as the arithmetic average of two immediately vertical neighboring pixels in the previous odd field. Since the odd field is generated first, followed by the even field, this predictor cannot be regarded as a pure 2-D predictor. Instead, it should be considered a 3-D predictor. The remaining cases are all 3-D predictors. One thing is common in all the cases: the gray levels of pixels used in the prediction have already been coded and thus are available in both the transmitter and the receiver. The prediction error of each changing pixel Z identified in thresholding process is then quantized and coded.

An analysis of the relationship between the entropy of moving areas (bits per changing pixel) and the speed of the motion (pixels per frame interval) in an image containing a moving mannequin's head was studied with different linear predictions, as listed in Table 3.1 in Haskell (1979). It was found that the element difference of field difference generally corresponds to the lowest entropy, meaning that this prediction is the most efficient. The frame difference and element difference correspond to higher entropy. It is recognized that, in the circumstances, transmission error will be propagated if the pixels in the previous line are used in prediction (Connor, 1973). Hence, the linear predictor should use only pixels from the same line or the same line in the previous frame when bit reversal error in transmission needs to be considered. Combining these two factors, the element difference of frame difference prediction is preferred.

**TABLE 3.1**
**Some Linear Prediction Schemes. (After Haskell, 1979).**

| | Original signal (Z) | Prediction signal ($\hat{Z}$) | Differential signal ($d_z$) |
|---|---|---|---|
| Element difference | Z | $G$ | $Z-G$ |
| Field difference | Z | $\dfrac{E+J}{2}$ | $Z - \dfrac{E+J}{2}$ |
| Frame difference | Z | $T$ | $Z-T$ |
| Element difference of frame difference | Z | $T + G - S$ | $(Z-G)-(T-S)$ |
| Line difference of frame difference | Z | $T + B - M$ | $(Z-B)-(T-M)$ |
| Element difference of field difference | Z | $T + \dfrac{E+J}{2} - \dfrac{Q+W}{2}$ | $\left(Z - \dfrac{E+J}{2}\right) - \left(T - \dfrac{Q+W}{2}\right)$ |

### 3.5.3  MOTION-COMPENSATED PREDICTIVE CODING

When frames are taken densely enough, changes in successive frames can be attributed to the motion of objects during the interval between frames. Under this assumption, if we can analyze object motion from successive frames, then we should be able to predict objects in the next frame based on their positions in the previous frame and the estimated motion. The difference between the original frame and the predicted frame thus generated and the motion vectors are then quantized and coded. If the motion estimation is accurate enough, the motion-compensated prediction error can be smaller than 3-D DPCM. In other words, the variance of the prediction error will be smaller, resulting in more efficient coding. Take motion into consideration — this differential technique is called motion compensated predictive coding. This has been a major development in image sequence coding since the 1980s. It has been adopted by all international video coding standards. A more detailed discussion is provided in Chapter 10.

## 3.6  INFORMATION-PRESERVING DIFFERENTIAL CODING

As emphasized in Chapter 2, quantization is not reversible in the sense that it causes permanent information loss. The DPCM technique, discussed above, includes quantization, and hence is lossy coding. In applications such as those involving scientific measurements, information preservation is required. In this section, the following question is addressed: under these circumstances, how should we apply differential coding in order to reduce the bit rate while preserving information?

Figure 3.14 shows a block diagram of information-preserving differential coding. First, we see that there is no quantizer. Therefore, the irreversible information loss associated with quantization does not exist in this technique. Second, we observe that prediction and differencing are still used. That is, the differential (predictive) technique still applies. Hence it is expected that the variance of the difference signal is smaller than that of the original signal, as explained in Section 3.1. Consequently, the higher-peaked histograms make coding more efficient. Third, an efficient lossless coder is utilized. Since quantizers cannot be used here, PCM with natural binary coding is not used here. Since the histogram of the difference signal is narrowly concentrated about its mean, lossless coding techniques such as an efficient Huffman coder (discussed in Chapter 5) is naturally a suitable choice here.

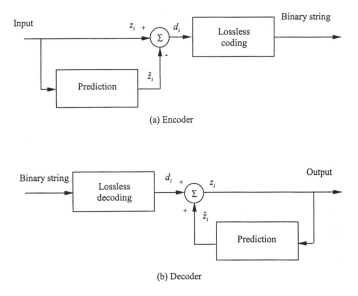

(a) Encoder

(b) Decoder

**FIGURE 3.14**  Block diagram of information-preserving differential coding.

As mentioned before, input images are normally in a PCM coded format with a bit rate of eight bits per pixel for monochrome pictures. The difference signal is therefore integer-valued. Having no quantization and using an efficient lossless coder, the coding system depicted in Figure 3.14, therefore, is an information-preserving differential coding technique.

## 3.7  SUMMARY

Rather than coding the signal itself, differential coding, also known as predictive coding, encodes the difference between the signal and its prediction. Utilizing spatial and/or temporal correlation between pixels in the prediction, the variance of the difference signal can be much smaller than that of the original signal, thus making differential coding quite efficient.

Among differential coding methods, differential pulse code modulation (DPCM) is used most widely. In DPCM coding, the difference signal is quantized and codewords are assigned to the quantized difference. Prediction and quantization are therefore two major components in the DPCM systems. Since quantization was addressed in Chapter 2, this chapter emphasizes prediction. The theory of optimum linear prediction is introduced. Here, optimum means minimization of the mean square prediction error. The formulation of optimum linear prediction, the orthogonality condition, and the minimum mean square prediction error are presented. The orthogonality condition states that the prediction error must be orthogonal to each observation, i.e., to the reconstructed sample intensity values used in the linear prediction. By solving the Yule-Walker equation, the optimum prediction coefficients may be determined.

In addition, some fundamental issues in implementing the DPCM technique are discussed. One issue is the dimensionality of the predictor in DPCM. We discussed 1-D, 2-D, and 3-D predictors. DPCM with a 2-D predictor demonstrates better performance than a 1-D predictor since 2-D DPCM utilizes more spatial correlation, i.e., not only horizontally but also vertically. As a result, a 3-dB improvement in *SNR* was reported. 3-D prediction is encountered in what is known as interframe coding. There, temporal correlation exists. 3-D DPCM utilizes both spatial and temporal correlation between neighboring pixels in successive frames. Consequently, more redundancy can be removed. Motion-compensated predictive coding is a very powerful technique in video coding related to differential coding. It uses a more advanced translational motion model in the prediction, however, and it is covered in Sections III and IV.

Another issue is the order of predictors and its effect on the performance of prediction in terms of mean square prediction error. Increasing the prediction order can lower the mean square prediction error effectively, but the performance improvement becomes insignificant after the third order.

Adaptive prediction is another issue. Similar to adaptive quantization, discussed in Chapter 2, we can adapt the prediction coefficients in the linear predictor to varying local statistics.

The last issue is concerned with the effect of transmission error. Bit reversal in transmission causes a different effect on reconstructed images depending on the type of coding technique used. PCM is known to be bit-consuming. (An acceptable PCM representation of monochrome images requires six to eight bits per pixel.) But a one-bit reversal only affects an individual pixel. For the DPCM coding technique, however, a transmission error may propagate from one pixel to the other. In particular, DPCM with a 1-D predictor suffers from error propagation more severely than DPCM with a 2-D predictor.

Delta modulation is an important special case of DPCM in which the predictor is of the first order. Specifically, the immediately preceding coded sample is used as a prediction of the present input sample. Furthermore, the quantizer has only two reconstruction levels.

Finally, an information-preserving differential coding technique is discussed. As mentioned in Chapter 2, quantization is an irreversible process: it causes information loss. In order to preserve information, there is no quantizer in this type of system. To be efficient, lossless codes such as Huffman code or arithmetic code should be used for difference signal encoding.

## 3.8   EXERCISES

**3-1.** Justify the necessity of the closed-loop DPCM with feedback around quantizers. That is, convince yourself why the quantization error will be accumulated if, instead of using the reconstructed preceding samples, we use the immediately preceding sample as the prediction of the sample being coded in DPCM.

**3-2.** Why does the overload error encountered in quantization appear to be the slope overload in DM?

**3-3.** What advantage does oversampling bring up in the DM technique?

**3-4.** What are the two features of DM that make it a subclass of DPCM?

**3-5.** Explain why DPCM with a 1-D predictor suffers from bit reversal transmission error more severely than DPCM with a 2-D predictor.

**3-6.** Explain why no quantizer can be used in information-preserving differential coding, and why the differential system can work without a quantizer.

**3-7.** Why do all the pixels involved in prediction of differential coding have to be in a recursively computable order from the point of view of the pixel being coded?

**3-8.** Discuss the similarity and dissimilarity between DPCM and motion compensated predictive coding.

## REFERENCES

Bose, N. K. *Applied Multidimensional System Theory*, Van Nostrand Reinhold, New York, 1982.

Bruders, R., T. Kummerow, P. Neuhold, and P. Stamnitz, Ein versuchssystem zur digitalen ubertragung von fernsehsignalen unter besonderer berucksichtigung von ubertragungsfehlern, Festschrift 50 Jahre Heinrich-Hertz-Institut, Berlin, 1978.

Connor, D. J. *IEEE Trans. Commun.,* com-21, 695-706, 1973.

Cutler, C. C. U. S. Patent 2,605,361, 1952.

DeJager, F. *Philips Res. Rep.*, 7, 442-466, 1952.

Elias, P. *IRE Trans. Inf. Theory*, it-1, 16-32, 1955.

Habibi, A. Comparison of nth-order DPCM encoder with linear transformations and block quantization techniques, *IEEE Trans. Commun. Technol.*, COM-19(6), 948-956, 1971.

Harrison, C. W. *Bell Syst. Tech. J.*, 31, 764-783, 1952.

Haskell, B. G., F. W. Mounts, and J. C. Candy, Interframe coding of videotelephone pictures, *Proc. IEEE*, 60, 7, 792-800, 1972.

Haskell, B. G. Frame replenishment coding of television, in *Image Transmission Techniques*, W. K. Pratt (Ed.), Academic Press, New York, 1979.

Jayant, N. S. and P. Noll, *Digital Coding of Waveforms,* Prentice-Hall, Upper Saddle River, NJ, 1984.

Kretzmer, E. R. Statistics of television signals, *Bell Syst. Tech. J.*, 31, 751-763, 1952.

Leon-Garcia, A. *Probability and Random Processes for Electrical Engineering*, 2nd ed., Addison-Wesley, Reading, MA, 1994.

Lim, J. S. *Two-Dimensional Signal and Image Processing*, Prentice-Hall, Englewood Cliffs, NJ, 1990.

Mounts, F. W. A video encoding system with conditional picture-element replenishment, *Bell Syst. Tech. J.*, 48, 7, 1969.

Musmann, H. G. Predictive Image Coding, in *Image Transmission Techniques*, W. K. Pratt (Ed.), Academic Press, New York, 1979.

Netravali, A. N. and J. D. Robbins, Motion-compensated television coding. Part I, *Bell Syst. Tech. J.*, 58, 3, 631-670, 1979.

Oliver, B. M. *Bell Syst. Tech. J.*, 31, 724-750, 1952.

O'Neal, J. B. *Bell Syst. Tech. J.*, 45, 689-721, 1966.

Pirsch, P. and L. Stenger, *Acta Electron.*, 19, 277-287, 1977.

Sayood, K. *Introduction to Data Compression*, Morgan Kaufmann, San Francisco, CA, 1996.

# 4 Transform Coding

As introduced in the previous chapter, differential coding achieves high coding efficiency by utilizing the correlation between pixels existing in image frames. Transform coding (TC), the focus of this chapter, is another efficient coding scheme based on utilization of interpixel correlation. As we will see in Chapter 7, TC has become a fundamental technique recommended by the international still image coding standard, JPEG. Moreover, TC has been found to be efficient in coding prediction error in motion-compensated predictive coding. As a result, TC was also adopted by the international video coding standards such as H.261, H.263, and MPEG 1, 2, and 4. This will be discussed in Section IV.

## 4.1 INTRODUCTION

Recall the block diagram of source encoders shown in Figure 2.3. There are three components in a source encoder: transformation, quantization, and codeword assignment. It is the transformation component that decides which format of input source is quantized and encoded. In DPCM, for instance, the difference between an original signal and a predicted version of the original signal is quantized and encoded. As long as the prediction error is small enough, i.e., the prediction resembles the original signal well (by using correlation between pixels), differential coding is efficient.

In transform coding, the main idea is that if the transformed version of a signal is less correlated compared with the original signal, then quantizing and encoding the transformed signal may lead to data compression. At the receiver, the encoded data are decoded and transformed back to reconstruct the signal. Therefore, in transform coding, the transformation component illustrated in Figure 2.3 is a transform. Quantization and codeword assignment are carried out with respect to the transformed signal, or, in other words, carried out in the transform domain.

We begin with the Hotelling transform, using it as an example of how a transform may decorrelate a signal in the transform domain.

### 4.1.1 HOTELLING TRANSFORM

Consider an $N$-dimensional vector $\vec{z}_s$. The ensemble of such vectors, $\{\vec{z}_s\}$ $s \in I$, where $I$ represents the set of all vector indexes, can be modeled by a random vector $\vec{z}$ with each of its component $z_i$ $i = 1, 2, \cdots, N$ as a random variable. That is,

$$\vec{z} = \left(z_1, z_2, \cdots, z_N\right)^T \tag{4.1}$$

where $T$ stands for the operator of matrix transposition. The mean vector of the population, $m_{\vec{z}}$, is defined as

$$m_{\vec{z}} = E[\vec{z}] = \left(m_1, m_2, \cdots, m_N\right)^T \tag{4.2}$$

where $E$ stands for the expectation operator. Note that $m_{\vec{z}}$ is an $N$-dimensional vector with the $i$th component, $m_i$, being the expectation value of the $i$th random variable component in $\vec{z}$.

$$m_i = E[z_i] \quad i = 1, 2, \cdots, N \tag{4.3}$$

The covariance matrix of the population, denoted by $C_{\bar{z}}$, is equal to

$$C_{\bar{z}} = E\left[\left(\bar{z} - m_{\bar{z}}\right)\left(\bar{z} - m_{\bar{z}}\right)^T\right].$$  (4.4)

Note that the product inside the $E$ operator is referred to as the *outer product* of the vector $(\bar{z} - m_{\bar{z}})$. Denote an entry at the $i$th row and $j$th column in the covariance matrix by $c_{i,j}$. From Equation 4.4, it can be seen that $c_{i,j}$ is the covariance between the $i$th and $j$th components of the random vector $\bar{z}$. That is,

$$c_{i,j} = E\left[\left(z_i - m_i\right)\left(z_j - m_j\right)\right] = Cov\left(z_i, z_j\right).$$  (4.5)

On the main diagonal of the covariance matrix $C_{\bar{z}}$, the element $c_{i,i}$ is the variance of the $i$th component of $\vec{z}$, $z_i$. Obviously, the covariance matrix $C_{\bar{z}}$ is a real and symmetric matrix. It is real because of the definition of random variables. It is symmetric because $Cov(z_i, z_j) = Cov(z_j, z_i)$. According to the theory of linear algebra, it is always possible to find a set of $N$ orthonormal eigenvectors of the matrix $C_{\bar{z}}$, with which we can convert the real symmetric matrix $C_{\bar{z}}$ into a fully ranked diagonal matrix. This statement can be found in texts of linear algebra, e.g., in (Strang, 1998).

Denote the set of $N$ orthonormal eigenvectors and their corresponding eigenvalues of the covariance matrix $C_{\bar{z}}$ by $\vec{e}_i$ and $\lambda_i$, $i = 1,2,\cdots,N$, respectively. Note that eigenvectors are column vectors. Form a matrix $\Phi$ such that its rows comprise the $N$ transposed eigenvectors. That is,

$$\Phi = \left(\vec{e}_1, \vec{e}_2, \cdots, \vec{e}_N\right)^T.$$  (4.6)

Now, consider the following transformation:

$$\vec{y} = \Phi\left(\vec{z} - m_{\bar{z}}\right)$$  (4.7)

It is easy to verify that the transformed random vector $\vec{y}$ has the following two characteristics:

1. The mean vector, $m_{\vec{y}}$, is a zero vector. That is,

$$m_{\bar{y}} = 0.$$  (4.8)

2. The covariance matrix of the transformed random vector $C_{\vec{y}}$ is

$$C_{\bar{y}} = \Phi C_{\bar{z}} \Phi^T = \begin{bmatrix} \lambda_1 & & & 0 \\ & \lambda_2 & & \\ & & \ddots & \\ & & & \ddots \\ 0 & & & \lambda_n \end{bmatrix}.$$  (4.9)

This transform is called the Hotelling transform (Hotelling, 1933), or eigenvector transform (Tasto, 1971; Wintz, 1972).

The inverse Hotelling transform is defined as

$$\bar{z} = \Phi^{-1}\vec{y} + m_{\bar{z}},$$  (4.10)

where $\Phi^{-1}$ is the inverse matrix of $\Phi$. It is easy to see from its formation discussed above that the matrix $\Phi$ is orthogonal. Therefore, we have $\Phi^T = \Phi^{-1}$. Hence, the inverse Hotelling transform can be expressed as

$$\vec{z} = \Phi^T \vec{y} + m_{\vec{z}}.$$ 

(4.11)

Note that in implementing the Hotelling transform, the mean vector $m_{\vec{z}}$ and the covariance matrix $C_{\vec{z}}$ can be calculated approximately by using a given set of $K$ sample vectors (Gonzalez and Woods, 1992).

$$m_{\vec{z}} = \frac{1}{K} \sum_{s=1}^{K} \vec{z}_s$$ 

(4.12)

$$C_{\vec{z}} = \frac{1}{K} \sum_{s=1}^{K} \vec{z}_s \vec{z}_s^T - m_{\vec{z}} m_{\vec{z}}^T$$ 

(4.13)

The analogous transform for continuous data was devised by Karhunen and Loeve (Karhunen, 1947; Loeve, 1948). Alternatively, the Hotelling transform can be viewed as the discrete version of the Karhunen-Loeve transform (KLT). We observe that the covariance matrix $C_{\vec{y}}$ is a diagonal matrix. The elements in the diagonal are the eigenvalues of the covariance matrix $C_{\vec{z}}$. That is, the two covariance matrices have the same eigenvalues and eigenvectors because the two matrices are similar. The fact that zero values are everywhere except along the main diagonal in $C_{\vec{y}}$ indicates that the components of the transformed vector $\vec{y}$ are uncorrelated. That is, the correlation previously existing between the different components of the random vector $\vec{z}$ has been removed in the transformed domain. Therefore, if the input is split into *blocks* and the Hotelling transform is applied blockwise, the coding may be more efficient since the data in the transformed block are uncorrelated. At the receiver, we may produce a replica of the input with an inverse transform. This basic idea behind transform coding will be further illustrated next. Note that transform coding is also referred to as block quantization (Huang, 1963).

## 4.1.2   STATISTICAL INTERPRETATION

Let's continue our discussion of the 1-D Hotelling transform. Recall that the covariance matrix of the transformed vector $\vec{y}$, $C_{\vec{y}}$, is a diagonal matrix. The elements in the main diagonal are eigenvalues of the covariance matrix $C_{\vec{y}}$. According to the definition of a covariance matrix, these elements are the variances of the components of vector $\vec{y}$, denoted by $\sigma_{y,1}^2, \sigma_{y,2}^2, \cdots, \sigma_{y,N}^2$. Let us arrange the eigenvalues (variances) in a nonincreasing order. That is, $\lambda_1 \geq \lambda_2 \geq \cdots \geq \lambda_N$. Choose an integer $L$, and $L < N$. Using the corresponding $L$ eigenvectors, $\vec{e}_1, \vec{e}_2, \cdots, \vec{e}_L$, we form a matrix $\overline{\Phi}$ with these $L$ eigenvectors (transposed) as its $L$ rows. Obviously, the matrix $\overline{\Phi}$ is of $L \times N$. Hence, using the matrix $\overline{\Phi}$ in Equation 4.7 we will have the transformed vector $\vec{y}$ of $L \times 1$. That is,

$$\vec{y} = \overline{\Phi}(\vec{z} - m_{\vec{z}}).$$ 

(4.14)

The inverse transform changes accordingly:

$$\vec{z}' = \overline{\Phi}^T \vec{y} + m_{\vec{z}}.$$ 

(4.15)

Note that the reconstructed vector $\vec{z}$, denoted by $\vec{z}'$, is still an $N \times 1$ column vector. It can be shown (Wintz, 1972) that the mean square reconstruction error between the original vector $\vec{z}$ and the reconstructed vector $\vec{z}$ is given by

$$MSE_r = \sum_{i=L+1}^{N} \sigma_{y,i}^2 . \qquad (4.16)$$

This equation indicates that the mean square reconstruction error equals the sum of variances of the discarded components. Note that although we discuss the reconstruction error here, we have not considered the quantization error and transmission error involved. Equation 4.15 implies that if, in the transformed vector $\vec{y}$, the first $L$ components have their variances occupy a large percentage of the total variances, the mean square reconstruction error will not be large even though only the first $L$ components are kept, i.e., the $(N - L)$ remaining components in the $\vec{y}$ are discarded. Quantizing and encoding only $L$ components of vector $\vec{y}$ in the transform domain lead to higher coding efficiency. This is the basic idea behind transform coding.

### 4.1.3  GEOMETRICAL INTERPRETATION

Transforming a set of statistically dependent data into another set of uncorrelated data, then discarding the insignificant transform coefficients (having small variances) illustrated above using the Hotelling transform, can be viewed as a statistical interpretation of transform coding. Here, we give a geometrical interpretation of transform coding. For this purpose, we use 2-D vectors instead of $N$-D vectors.

Consider a binary image of a car in Figure 4.1(a). Each pixel in the shaded object region corresponds to a 2-D vector with its two components being coordinates $z_1$ and $z_2$, respectively. Hence, the set of all pixels associated with the object forms a population of vectors. We can determine its mean vector and covariance matrix using Equations 4.12 and 4.13, respectively. We can then apply the Hotelling transform by using Equation 4.7. Figure 4.1(b) depicts the same object after the application of the Hotelling transform in the $y_1$-$y_2$ coordinate system. We notice that the origin of the new coordinate system is now located at the centroid of the binary object. Furthermore, the new coordinate system is aligned with the two eigenvectors of the covariance matrix $C_{\vec{z}}$.

As mentioned, the elements along the main diagonal of $C_{\vec{y}}$ (two eigenvalues of the $C_{\vec{y}}$ and $C_{\vec{z}}$) are the two variances of the two components of the $\vec{y}$ population. Since the covariance matrix

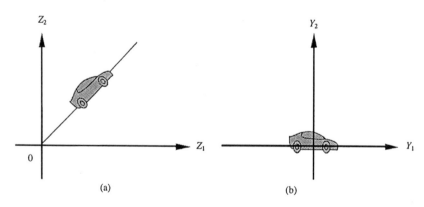

FIGURE 4.1   (a) A binary object in the $z_1$-$z_2$ coordinate system. (b) After the Hotelling transform, the object is aligned with its principal axes.

$C_{\vec{y}}$ is a diagonal matrix, the two components are uncorrelated after the transform. Since one variance (along the $y_1$ direction) is larger than the other (along the $y_2$ direction), it is possible for us to achieve higher coding efficiency by ignoring the component associated with the smaller variance without too much sacrifice of the reconstructed image quality.

It is noted that the alignment of the object with the eigenvectors of the covariance matrix is of importance in pattern recognition (Gonzalez and Woods, 1992).

### 4.1.4  BASIS VECTOR INTERPRETATION

Basis vector expansion is another interpretation of transform coding. For simplicity, in this sub-section we assume a zero mean vector. Under this assumption, the Hotelling transform and its inverse transform become

$$\vec{y} = \Phi\vec{z} \tag{4.17}$$

$$\vec{z} = \Phi^T\vec{y} \tag{4.18}$$

Recall that the row vectors in the matrix $\Phi$ are the transposed eigenvectors of the covariance matrix $C_{\vec{z}}$. Therefore, Equation 4.18 can be written as

$$\vec{z} = \sum_{i=1}^{N} y_i\vec{e}_i. \tag{4.19}$$

In the above equation, we can view vector $\vec{z}$ as a linear combination of *basis vectors* $\vec{e}_i$, $i = 1,2,\cdots,N$. The components of the transformed vector $\vec{y}$, $y_i$, $i = 1,2,\cdots,N$ serve as coefficients in the linear combination, or weights in the weighted sum of basis vectors. The coefficient $y_i$, $i = 1,2,\cdots,N$ can be produced according to Equation 4.17:

$$y_i = \vec{e}_i^T\vec{z}. \tag{4.20}$$

That is, $y_i$ is the *inner product* between vectors $\vec{e}_i$ and $\vec{z}$. Therefore, the coefficient $y_i$ can be interpreted as the amount of correlation between the basis vector $\vec{e}_i$ and the original signal $\vec{z}$.

In the Hotelling transform the coefficients $y_i$, $i = 1,2,\cdots,N$ are uncorrelated. The variance of $y_i$ can be arranged in a nonincreasing order. For $i > L$, the variance of the coefficient becomes insignificant. We can then discard these coefficients without introducing significant error in the linear combination of basis vectors and achieve higher coding efficiency.

In the above three interpretations of transform coding, we see that the linear unitary transform can provide the following two functions:

1. Decorrelate input data; i.e., transform coefficients are less correlated than the original data, and
2. Have some transform coefficients more significant than others (with large variance, eigenvalue, or weight in basis vector expansion) such that transform coefficients can be treated differently: some can be discarded, some can be coarsely quantized, and some can be finely quantized.

**Note** that the definition of *unitary* transform is given shortly in Section 4.2.1.3.

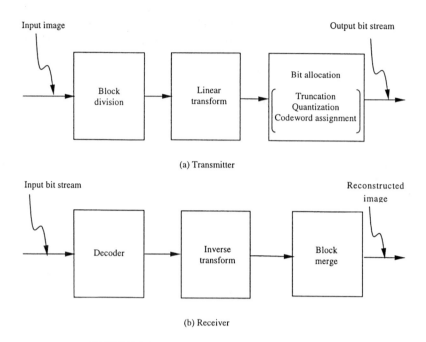

**FIGURE 4.2**   Block diagram of transform coding.

### 4.1.5   Procedures of Transform Coding

Prior to leaving this section, we summarize the procedures of transform coding. There are three steps in transform coding as shown in Figure 4.2. First, the input data (frame) are divided into blocks (subimages). Each block is then linearly transformed. The transformed version is then truncated, quantized, and encoded. These last three functions, which are discussed in Section 4.4, can be grouped and termed as bit allocation. The output of the encoder is a bitstream.

In the receiver, the bitstream is decoded and then inversely transformed to form reconstructed blocks. All the reconstructed blocks collectively produce a replica of the input image.

## 4.2   LINEAR TRANSFORMS

In this section, we first discuss a general formulation of a linear unitary 2-D image transform. Then, a basis image interpretation of TC is given.

### 4.2.1   2-D Image Transformation Kernel

There are two different ways to handle image transformation. In the first way, we convert a 2-D array representing a digital image into a 1-D array via row-by-row stacking, for example. That is, from the second row on, the beginning of each row in the 2-D array is cascaded to the end of its previous row. Then we transform this 1-D array using a 1-D transform. After the transformation, we can convert the 1-D array back to a 2-D array. In the second way, a 2-D transform is directly applied to the 2-D array corresponding to an input image, resulting in a transformed 2-D array. These two ways are essentially the same. It can be straightforwardly shown that the difference between the two is simply a matter of notation (Wintz, 1972). In this section, we use the second way to handle image transformation. That is, we work on 2-D image transformation.

Assume a digital image is represented by a 2-D array $g(x, y)$, where $(x, y)$ is the coordinates of a pixel in the 2-D array, while $g$ is the gray level value (also often called intensity or brightness) of the pixel. Denote the 2-D transform of $g(x, y)$ by $T(u, v)$, where $(u, v)$ is the coordinates in the transformed domain. Assume that both $g(x, y)$ and $T(u, v)$ are a square 2-D array of $N \times N$; i.e., $0 \le x, y, u, v \le N - 1$.

The 2-D forward and inverse transforms are defined as

$$T(u,v) = \sum_{x=0}^{N-1} \sum_{y=0}^{N-1} g(x,y) f(x,y,u,v) \tag{4.21}$$

and

$$g(x,y) = \sum_{u=0}^{N-1} \sum_{v=0}^{N-1} T(u,v) i(x,y,u,v) \tag{4.22}$$

where $f(x, y, u, v)$ and $i(x, y, u, v)$ are referred to as the forward and inverse *transformation kernels*, respectively.

A few characteristics of transforms are discussed below.

### 4.2.1.1   Separability

A transformation kernel is called separable (hence, the transform is said to be separable) if the following conditions are satisfied.

$$f(x,y,u,v) = f_1(x,u) f_2(y,v), \tag{4.23}$$

and

$$i(x,y,u,v) = i_1(x,u) i_2(y,v). \tag{4.24}$$

Note that a 2-D separable transform can be decomposed into two 1-D transforms. That is, a 2-D transform can be implemented by a 1-D transform rowwise followed by another 1-D transform columnwise. That is,

$$T_1(x,v) = \sum_{y=0}^{N-1} g(x,y) f_2(y,v), \tag{4.25}$$

where $0 \le x, v \le N - 1$, and

$$T(u,v) = \sum_{x=0}^{N-1} T_1(x,v) f_1(x,u), \tag{4.26}$$

where $0 \le u, v \le N - 1$. Of course, the 2-D transform can also be implemented in a reverse order with two 1-D transforms, i.e., columnwise first, followed by rowwise. The counterparts of Equations 4.25 and 4.26 for the inverse transform can be derived similarly.

#### 4.2.1.2  Symmetry

The transformation kernel is symmetric (hence, the transform is symmetric) if the kernel is separable and the following condition is satisfied:

$$f_1(y,v) = f_2(y,v). \tag{4.27}$$

That is, $f_1$ is functionally equivalent to $f_2$.

#### 4.2.1.3  Matrix Form

If a transformation kernel is symmetric (hence, separable) then the 2-D image transform discussed above can be expressed compactly in the following matrix form. Denote an *image matrix* by $G$ and $G = \{g_{i,j}\} = \{g(i-1, j-1)\}$. That is, a typical element (at the $i$th row and $j$th column) in the matrix $G$ is the pixel gray level value in the 2-D array $g(x, y)$ at the same geometrical position. Note that the subtraction of one in the notation $g(i-1, j-1)$ comes from Equations 4.21 and 4.22. Namely, the indexes of a square 2-D image array are conventionally defined from 0 to $N$-1, while the indexes of a square matrix are from 1 to $N$. Denote the *forward transform matrix* by $F$ and $F = \{f_{i,j}\} = \{f_1(i-1, j-1)\}$. We then have the following matrix form of a 2-D transform:

$$T = F^T GF \tag{4.28}$$

where $T$ on the left-hand side of the equation denotes the matrix corresponding to the transformed 2-D array in the same fashion as that used in defining the $G$ matrix. The inverse transform can be expressed as

$$G = I^T TI \tag{4.29}$$

where the matrix $I$ is the *inverse transform matrix* and $I = \{i_{j,k}\} = \{i_1 (j-1, k-1)\}$. The forward and inverse transform matrices have the following relation:

$$I = F^{-1} \tag{4.30}$$

Note that all of the matrices defined above, $G$, $T$, $F$, and $I$ are of $N \times N$.

It is known that the discrete Fourier transform involves complex quantities. In this case, the counterparts of Equations 4.28, 4.29, and 4.30 become Equations 4.31, 4.32, and 4.33, respectively:

$$T = F^{*T} GF \tag{4.31}$$

$$G = I^{*T} TI \tag{4.32}$$

$$I = F^{-1} = F^{*T} \tag{4.33}$$

where $*$ indicates complex conjugation. Note that the transform matrices $F$ and $I$ contain complex quantities and satisfy Equation 4.33. They are called unitary matrices and the transform is referred to as a unitary transform.

### 4.2.1.4  Orthogonality

A transform is said to be orthogonal if the transform matrix is orthogonal. That is,

$$F^T = F^{-1} \qquad (4.34)$$

Note that an orthogonal matrix (orthogonal transform) is a special case of a unitary matrix (unitary transform), where only real quantities are involved. We will see that all the 2-D image transforms, presented in Section 4.3, are separable, symmetric, and unitary.

### 4.2.2  BASIS IMAGE INTERPRETATION

Here we study the concept of *basis images* or *basis matrices*. Recall that we discussed basis vectors when we considered the 1-D transform. That is, the components of the transformed vector (also referred to as the transform coefficients) can be interpreted as the coefficients in the basis vector expansion of the input vector. Each coefficient is essentially the amount of correlation between the input vector and the corresponding basis vector. The concept of basis vectors can be extended to basis images in the context of 2-D image transforms.

Recall that the 2-D inverse transform introduced at the beginning of this section is defined as

$$g(x,y) = \sum_{u=0}^{N-1} \sum_{v=0}^{N-1} T(u,v) i(x,y,u,v) \qquad (4.35)$$

where $0 \leq x, y \leq N - 1$. This equation can be viewed as a *component* form of the inverse transform. As defined above in Section 4.2.1.3, the whole image $\{g(x, y)\}$ is denoted by the image matrix $G$ of $N \times N$. We now denote the "image" formed by the inverse transformation kernel $\{i(x, y, u, v), 0 \leq x, y \leq N - 1\}$ as a 2-D array $I_{u,v}$ of $N \times N$ for a specific pair of $(u, v)$ with $0 \leq u, v \leq N - 1$. Recall that a digital image can be represented by a 2-D array of gray level values. In turn the 2-D array can be arranged into a matrix. Namely, we treat the following three: a digital image, a 2-D array (with proper resolution), and a matrix (with proper indexing), interchangeably. We then have

$$I_{u,v} = \begin{bmatrix} i(0,0,u,v) & i(0,1,u,v) & \cdots & \cdots & i(0,N-1,u,v) \\ i(1,0,u,v) & i(1,1,u,v) & \cdots & \cdots & i(1,N-1,u,v) \\ \vdots & \vdots & \cdots & \cdots & \vdots \\ \vdots & \vdots & \cdots & \cdots & \vdots \\ i(N-1,0,u,v) & i(N-1,1,u,v) & \cdots & \cdots & i(N-1,N-1,u,v) \end{bmatrix} \qquad (4.36)$$

The 2-D array $I_{u,v}$ is referred to as a basis image. There are $N^2$ basis images in total since $0 \leq u,v \leq N - 1$. The inverse transform expressed in Equation 4.35 can then be written in a *collective* form as

$$G = \sum_{u=0}^{N-1} \sum_{v=0}^{N-1} T(u,v) I_{u,v}. \qquad (4.37)$$

We can interpret this equation as a series expansion of the original image $G$ into a set of $N^2$ basis images $I_{u,v}$. The transform coefficients $T(u,v)$, $0 \leq u, v \leq N - 1$, become the coefficients of the expansion. Alternatively, the image $G$ is said to be a weighted sum of basis images. Note that, similar to the 1-D case, the coefficient or the weight $T(u,v)$ is a correlation measure between the image $G$ and the basis image $I_{u,v}$ (Wintz, 1972).

Note that basis images have nothing to do with the input image. Instead, it is completely defined by the transform itself. That is, basis images are the attribute of 2-D image transforms. Different transforms have different sets of basis images.

The motivation behind transform coding is that with a proper transform, hence, a proper set of basis images, the transform coefficients are more independent than the gray scales of the original input image. In the ideal case, the transform coefficients are statistically independent. We can then optimally encode the coefficients independently, which can make coding more efficient and simple. As pointed out in (Wintz, 1972), however, this is generally impossible because of the following two reasons. First, it requires the joint probability density function of the $N^2$ pixels, which have not been deduced from basic physical laws and cannot be measured. Second, even if the joint probability density functions were known, the problem of devising a reversible transform that can generate independent coefficients is unsolved. The optimum linear transform we can have results in uncorrelated coefficients. When Gaussian distribution is involved, we can have independent transform coefficients. In addition to the uncorrelatedness of coefficients, the variance of the coefficients varies widely. Insignificant coefficients can be ignored without introducing significant distortion in the reconstructed image. Significant coefficients can be allocated more bits in encoding. The coding efficiency is thus enhanced.

As shown in Figure 4.3, TC can be viewed as expanding the input image into a set of basis images, then quantizing and encoding the coefficients associated with the basis images separately. At the receiver the coefficients are reconstructed to produce a replica of the input image. This strategy is similar to that of subband coding, which is discussed in Chapter 8. From this point of view, transform coding can be considered a special case of subband coding, though transform coding was devised much earlier than subband coding.

It is worth mentioning an alternative way to define basis images. That is, a basis image with indexes $(u, v)$, $I_{u,v}$, of a transform can be constructed as the *outer product* of the $u$th basis vector, $\vec{b}_u$, and the $v$th basis vector, $\vec{b}_v$, of the transform. The basis vector, $\vec{b}_u$, is the $u$th column vector of the inverse transform matrix $I$ (Jayant and Noll, 1984). That is,

$$I_{u,v} = \vec{b}_u \vec{b}_v^T. \tag{4.38}$$

### 4.2.3 SUBIMAGE SIZE SELECTION

The selection of subimage (block) size, $N$, is important. Normally, the larger the size the more decorrelation the transform coding can achieve. It has been shown, however, that the correlation between image pixels becomes insignificant when the distance between pixels becomes large, e.g., it exceeds 20 pixels (Habibi, 1971a). On the other hand, a large size causes some problems. In adaptive transform coding, a large block cannot adapt to local statistics well. As will be seen later in this chapter, a transmission error in transform coding affects the whole associated subimage. Hence a large size implies a possibly severe effect of transmission error on reconstructed images. As will be shown in video coding (Section III and Section IV), transform coding is used together with motion-compensated coding. Consider that large block size is not used in motion estimation; subimage sizes of 4, 8, and 16 are used most often. In particular, $N = 8$ is adopted by the international still image coding standard, JPEG, as well as video coding standards H.261, H.263, MPEG 1, and MPEG 2.

## 4.3 TRANSFORMS OF PARTICULAR INTEREST

Several commonly used image transforms are discussed in this section. They include the discrete Fourier transform, the discrete Walsh transform, the discrete Hadamard transform, and the discrete Cosine and Sine transforms. All of these transforms are symmetric (hence, separable as well),

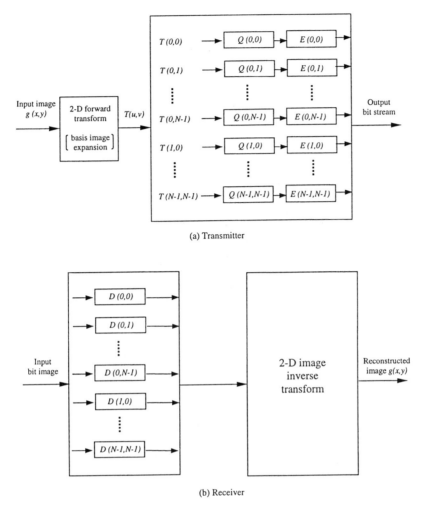

(a) Transmitter

(b) Receiver

**FIGURE 4.3**   Basis image interpretation of TC (Q: quantizer, E: encoder, D: decoder).

unitary, and reversible. For each transform, we define its transformation kernel and discuss its basis images.

### 4.3.1 DISCRETE FOURIER TRANSFORM (DFT)

The DFT is of great importance in the field of digital signal processing. Owing to the fast Fourier transform (FFT) based on the algorithm developed in (Cooley, 1965), the DFT is widely utilized for various tasks of digital signal processing. It has been discussed in many signal and image processing texts. Here we only define it by using the transformation kernel just introduced above. The forward and inverse transformation kernels of the DFT are

$$f(x,y,u,v) = \frac{1}{N}\exp\{-j2\pi(xu+yv)/N\} \tag{4.39}$$

and

$$i(x,y,u,v) = \frac{1}{N}\exp\{j2\pi(xu+yv)/N\} \tag{4.40}$$

Clearly, since complex quantities are involved in the DFT transformation kernels, the DFT is generally complex. Hence, we use the unitary matrix to handle the DFT (refer to Section 4.2.1.3). The basis vector of the DFT $\vec{b}_u$ is an $N \times 1$ column vector and is defined as

$$\vec{b}_u = \frac{1}{\sqrt{N}}\left[1, \exp\left(j2\pi\frac{u}{N}\right), \exp\left(j2\pi\frac{2u}{N}\right), \cdots, \exp\left(j2\pi\left(\frac{(N-1)u}{N}\right)\right)\right]^T \tag{4.41}$$

As mentioned, the basis image with index $(u,v)$, $I_{u,v}$, is equal to $\vec{b}_u\vec{b}_u^T$. A few basis images are listed below for $N = 4$.

$$I_{0,0} = \frac{1}{4}\begin{pmatrix} 1 & 1 & 1 & 1 \\ 1 & 1 & 1 & 1 \\ 1 & 1 & 1 & 1 \\ 1 & 1 & 1 & 1 \end{pmatrix} \tag{4.42}$$

$$I_{0,1} = \frac{1}{4}\begin{pmatrix} 1 & j & -1 & -j \\ 1 & j & -1 & -j \\ 1 & j & -1 & -j \\ 1 & j & -1 & -j \end{pmatrix} \tag{4.43}$$

$$I_{1,2} = \frac{1}{4}\begin{pmatrix} 1 & 1 & 1 & -j \\ j & -j & j & -j \\ -1 & 1 & -1 & 1 \\ -j & -j & -j & j \end{pmatrix} \tag{4.44}$$

$$I_{3,3} = \frac{1}{4}\begin{pmatrix} 1 & -j & -1 & j \\ -j & -1 & j & 1 \\ -1 & j & 1 & -j \\ j & 1 & -j & -1 \end{pmatrix} \tag{4.45}$$

### 4.3.2  DISCRETE WALSH TRANSFORM (DWT)

The transformation kernels of the DWT (Walsh, 1923) are defined as

$$f(x,y,u,v) = \frac{1}{N}\prod_{i=0}^{n-1}\left[(-1)^{p_i(x)p_{n-1-i}(u)}(-1)^{p_i(y)p_{n-1-i}(v)}\right] \tag{4.46}$$

and

$$i(x,y,u,v) = f(x,y,u,v). \tag{4.47}$$

where $n = \log_2 N$, $p_i(\text{arg})$ represents the $i$th bit in the natural binary representation of the arg, the $o$th bit corresponds to the least significant bit, and the $(n-1)$th bit corresponds to the most significant

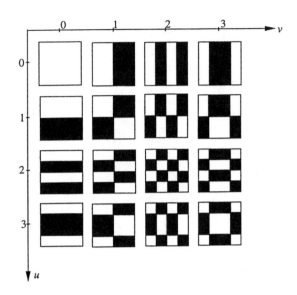

**FIGURE 4.4** When N = 4: a set of the 16 basis images of DWT.

bit. For instance, consider $N = 16$, then $n = 4$. The natural binary code of number 8 is 1000. Hence, $p_0(8) = p_1(8) = p_2(8) = 0$, and $p_3(8) = 1$. We see that if the factor $1/N$ is put aside then the forward transformation kernel is always an integer: either $+1$ or $-1$. In addition, the inverse transformation kernel is the same as the forward transformation kernel. Therefore, we conclude that the implementation of the DWT is simple.

When $N = 4$, the 16 basis images of the DWT are shown in Figure 4.4. Each corresponds to a specific pair of $(u, v)$ and is of resolution $4 \times 4$ in the $x$-$y$ coordinate system. They are binary images, where the bright represents $+1$, while the dark $-1$. The transform matrix of the DWT is shown below for $N = 4$.

$$F = \frac{1}{2}\begin{pmatrix} 1 & 1 & 1 & 1 \\ 1 & 1 & -1 & -1 \\ 1 & -1 & 1 & -1 \\ 1 & -1 & -1 & 1 \end{pmatrix} \tag{4.48}$$

### 4.3.3 DISCRETE HADAMARD TRANSFORM (DHT)

The DHT (Hadamard, 1893) is closely related to the DWT. This can be seen from the following definition of the transformation kernels.

$$f(x,y,u,v) = \frac{1}{N}\prod_{i=0}^{n}\left[(-1)^{p_i(x)p_i(u)}(-1)^{p_i(y)p_i(v)}\right] \tag{4.49}$$

and

$$i(x,y,u,v) = f(x,y,u,v) \tag{4.50}$$

where the definitions of $n$, $i$, and $p_i(\text{arg})$ are the same as in the DWT. For this reason, the term Walsh-Hadamard transform (DWHT) is frequently used to represent either of the two transforms.

When $N$ is a power of 2, the transform matrices of the DWT and DHT have the same row (or column) vectors except that the order of row (or column) vectors in the matrices are different. This is the only difference between the DWT and DHT under the circumstance $N = 2^n$. Because of this difference, while the DWT can be implemented by using the FFT algorithm with a straightforward modification, the DHT needs more work to use the FFT algorithm. On the other hand, the DHT possesses the following recursive feature, while the DWT does not:

$$F_2 = \begin{bmatrix} 1 & 1 \\ 1 & -1 \end{bmatrix} \qquad (4.51)$$

and

$$F_{2N} = \begin{bmatrix} F_N & F_N \\ F_N & -F_N \end{bmatrix} \qquad (4.52)$$

where the subscripts indicate the size of the transform matrices. It is obvious that the transform matrix of the DHT can be easily derived by using the recursion.

Note that the number of sign changes between consecutive entries in a row (or a column) of a transform matrix (from positive to negative and from negative to positive) is known as *sequency*. It is observed that the sequency does not monotonically increase as the order number of rows (or columns) increases in the DHT. Since sequency bears some similarity to frequency in the Fourier transform, sequency is desired as an increasing function of the order number of rows (or columns). This is realized by the *ordered* Hadamard transform (Gonzalez, 1992).

The transformation kernel of the ordered Hadamard transform is defined as

$$f(x, y, u, v) = \frac{1}{N} \prod_{i=0}^{N-1} \left[ (-1)^{p_i(x)d_i(u)} (-1)^{p_i(y)d_i(v)} \right] \qquad (4.53)$$

where the definitions of $i$, $p_i(\text{arg})$ are the same as defined above for the DWT and DHT. The $d_i(\text{arg})$ is defined as below.

$$d_0(\text{arg}) = b_{n-1}(\text{arg})$$

$$d_1(\text{arg}) = b_{n-1}(\text{arg}) + b_{n-2}(\text{arg}) \qquad (4.54)$$

$$d_{n-1}(\text{arg}) = b_1(\text{arg}) + b_0(\text{arg})$$

The 16 basis images of the ordered Hadamard transform are shown in Figure 4.5 for $N = 4$. It is observed that the variation of the binary basis images becomes more frequent monotonically when $u$ and $v$ increase. Also we see that the basis image expansion is similar to the frequency expansion of the Fourier transform in the sense that an image is decomposed into components with different variations. In transform coding, these components with different coefficients are treated differently.

### 4.3.4  DISCRETE COSINE TRANSFORM (DCT)

The DCT is the most commonly used transform for image and video coding.

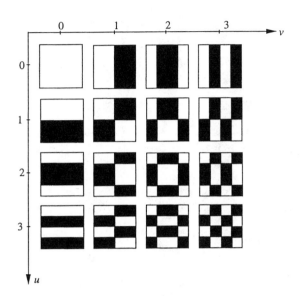

**FIGURE 4.5** When N = 4: a set of the 16 basis images of the ordered DHT.

### 4.3.4.1 Background

The DCT, which plays an extremely important role in image and video coding, was established by Ahmed et al. (1974). There, it was shown that the *basis member* $\cos[(2x + 1)u\pi/2N]$ is the $u$th Chebyshev polynomial $T_U(\xi)$ evaluated at the $x$th zero of $T_N(\xi)$. Recall that the Chebyshev polynomials are defined as

$$T_0(\xi) = 1/\sqrt{2} \tag{4.55}$$

$$T_K(\xi) = \cos\left[k\cos^{-1}(\xi)\right] \tag{4.56}$$

where $T_K(\xi)$ is the $k$th order Chebyshev polynomial and it has $k$ zeros, starting from the $I$st zero to the $k$th zero. Furthermore, it was demonstrated that the basis vectors of 1-D DCT provide a good approximation to the eigenvectors of the class of Toeplitz matrices defined as

$$
\begin{bmatrix}
1 & \rho & \rho^2 & \cdots & \rho^{N-1} \\
\rho & 1 & \rho & \cdots & \rho^{N-2} \\
\rho^2 & \rho & 1 & \cdots & \rho^{N-3} \\
\vdots & \vdots & \vdots & \cdots & \vdots \\
\rho^{N-1} & \rho^{N-2} & \rho^{N-3} & \cdots & 1
\end{bmatrix}, \tag{4.57}
$$

where $0 < \rho < 1$.

### 4.3.4.2 Transformation Kernel

The transformation kernel of the 2-D DCT can be extended straightforwardly from that of 1-D DCT as follows:

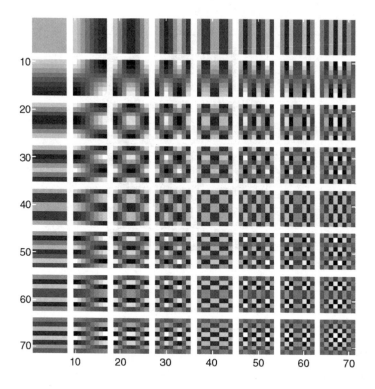

**FIGURE 4.6**   When N = 8: a set of the 64 basis images of the DCT.

$$f(x,y,u,v) = C(u)C(v)\cos\left(\frac{(2x+1)u\pi}{2N}\right)\cos\left(\frac{(2y+1)v\pi}{2N}\right) \qquad (4.58)$$

where

$$C(u) = \begin{cases} \sqrt{\dfrac{1}{N}} & for \quad u = 0 \\ \sqrt{\dfrac{2}{N}} & for \quad u = 1,2,\cdots,N-1 \end{cases} \qquad (4.59)$$

$$i(x,y,u,v) = f(x,y,u,v). \qquad (4.60)$$

Note that the $C(v)$ is defined the same way as in Equation 4.59. The 64 basis images of the DCT are shown in Figure 4.6 for $N = 8$.

### 4.3.4.3   Relationship with DFT

The DCT is closely related to the DFT. This can be examined from an alternative method of defining the DCT. It is known that applying the DFT to an $N$-point sequence $g_N(n)$, $n = 0,1,\cdots,N-1$, is equivalent to the following:

1. Repeating $g_N(n)$ every $N$ points, form a periodic sequence, $\tilde{g}_N(n)$, with a fundamental period $N$. That is,

$$\tilde{g}_N(n) = \sum_{i=-\infty}^{\infty} g_N(n - iN). \tag{4.61}$$

2. Determine the Fourier series expansion of the periodic sequence $\tilde{g}_N(n)$. That is, determine all the coefficients in the Fourier series which are known to be periodic with the same fundamental period $N$.
3. Truncate the sequence of the Fourier series coefficients so as to have the same support as that of the given sequence $g_N(n)$. That is, only keep the $N$ coefficients with indexes $0, 1, \cdots, N-1$ and set all the others to equal zero. These $N$ Fourier series coefficients form the DFT of the given $N$-point sequence $g_N(n)$.

An $N$-point sequence $g_N(n)$ and the periodic sequence $\tilde{g}_N(n)$, generated from $g_N(n)$, are shown in Figure 4.7(a) and (b), respectively. In summary, the DFT can be viewed as a correspondence between two periodic sequences. One is the periodic sequence $\tilde{g}_N(n)$, which is formed by periodically repeating $g_N(n)$. The other is the periodic sequence of Fourier series coefficients of $\tilde{g}_N(n)$.

The DCT of an $N$-point sequence is obtained via the following three steps:

1. Flip over the given sequence with respect to the end point of the sequence to form a $2N$-point sequence, $g_{2N}(n)$, as shown in Figure 4.7(c). Then form a periodic sequence $\tilde{g}_{2N}(n)$, shown in Figure 4.7(d), according to

$$\tilde{g}_{2N}(n) = \sum_{i=-\infty}^{\infty} g_{2N}(n - 2iN) \tag{4.62}$$

2. Find the Fourier series coefficients of the periodic sequences $\tilde{g}_{2N}(n)$.
3. Truncate the resultant periodic sequence of the Fourier series coefficients to have the support of the given finite sequence $g_N(n)$. That is, only keep the $N$ coefficients with indexes $0, 1, \cdots, N-1$ and set all the others to equal zero. These $N$ Fourier series coefficients form the DCT of the given $N$-point sequence $g_N(n)$.

A comparison between Figure 4.7(b) and (d) reveals that the periodic sequence $\tilde{g}_N(n)$ is not smooth. There usually exist discontinuities at the beginning and end of each period. These end-head discontinuities cause a high-frequency distribution in the corresponding DFT. On the contrary, the periodic sequence $\tilde{g}_{2N}(n)$ does not have this type of discontinuity due to flipping over the given finite sequence. As a result, there is no high-frequency component corresponding to the end-head discontinuities. Hence, the DCT possesses better energy compaction in the low frequencies than the DFT. By better energy compaction, we mean more energy is compacted in a fraction of transform coefficients. For instance, it is known that the most energy of an image is contained in a small region of low frequency in the DFT domain. Vivid examples can be found in (Gonzalez and Woods, 1992). In terms of energy compaction, when compared with the Karhunen-Loeve transform (the Hotelling transform is its discrete version), which is known as the optimal, the DCT is the best among the DFT, DWT, DHT, and discrete Haar transform.

Besides this advantage, the DCT can be implemented using the FFT. This can be seen from the above discussion. There, it has been shown that the DCT of an $N$-point sequence, $g_N(n)$, can be obtained from the DFT of the $2N$-point sequence $g_{2N}(n)$. Furthermore, the even symmetry

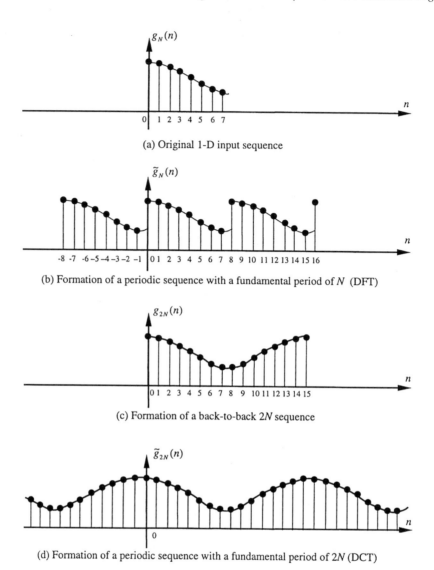

(a) Original 1-D input sequence

(b) Formation of a periodic sequence with a fundamental period of $N$ (DFT)

(c) Formation of a back-to-back $2N$ sequence

(d) Formation of a periodic sequence with a fundamental period of $2N$ (DCT)

**FIGURE 4.7**   An example to illustrate the differences and similarities between DFT and DCT.

in $\tilde{g}_{2N}(n)$ makes the computation required for the DCT of an $N$-point equal to that required for the DFT of the $N$-point sequence. Because of these two merits, the DCT is the most popular image transform used in image and video coding nowadays.

### 4.3.5   PERFORMANCE COMPARISON

In this subsection, we compare the performance of a few commonly used transforms in terms of energy compaction, mean square reconstruction error, and computational complexity.

#### 4.3.5.1   Energy Compaction

Since all the transforms we discussed are symmetric (hence separable) and unitary, the matrix form of the 2-D image transform can be expressed as $T = F^T G F$ as discussed in Section 4.2.1.3. In the 1-D case, the transform matrix $F$ is the counterpart of the matrix $\Phi$ discussed in the Hotelling transform. Using the $F$, one can transform a 1-D column vector $\vec{z}$ into another 1-D column vector $\vec{y}$. The components of the vector $\vec{y}$ are transform coefficients. The variances of these transform coefficients, and therefore the signal energy associated with the transform coefficients, can be arranged in a nondecreasing order. It can be shown that the total energy before and after the transform remains the same. Therefore, the more energy compacted in a fraction of total coefficients, the better energy compaction the transform has. One measure of energy compaction is the *transform coding gain* $G_{TC}$, which is defined as the ratio between the arithmetic mean and the geometric mean of the variances of all the components in the transformed vector (Jayant, 1984).

$$G_{TC} = \frac{\dfrac{1}{N}\displaystyle\sum_{i=0}^{N-1}\sigma_i^2}{\left(\displaystyle\prod_{i=0}^{N-1}\sigma_i^2\right)^{\frac{1}{N}}} \tag{4.63}$$

A larger $G_{TC}$ indicates higher energy compaction. The transform coding gains for a first-order autoregressive source with $\rho = 0.95$ achieved by using the DCT, DFT, and KLT was reported in (Zelinski and Noll, 1975; Jayant and Noll, 1984). The transform coding gain afforded by the DCT compares very closely to that of the optimum KLT.

### 4.3.5.2 Mean Square Reconstruction Error

The performance of the transforms can be compared in terms of the mean square reconstruction error as well. This was mentioned in Section 4.1.2 when we provided a statistical interpretation for transform coding. That is, after arranging all the $N$ transformed coefficients according to their variances in a nonincreasing order, if $L < N$ and we discard the last $(N - L)$ coefficients to reconstruct the original input signal $\vec{z}$ (similar to what we did with the Hotelling transform), then the mean square reconstruction error is

$$MSE_r = E\big[\|\vec{z} - \vec{z}'\|^2\big] = \sum_{i=L+1}^{N}\sigma_i^2, \tag{4.64}$$

where $\vec{z}'$ denotes the reconstructed vector. Note that in the above-defined mean square reconstruction error, the quantization error and transmission error have not been included. Hence, it is sometimes referred to as the mean square approximation error. Therefore it is desired to choose a transform so that the transformed coefficients are "more independent" and more energy is concentrated in the first $L$ coefficients. Then it is possible to discard the remaining coefficients to save coding bits without causing significant distortion in input signal reconstruction.

In terms of the mean square reconstruction error, the performance of the DCT, KLT, DFT, DWT, and discrete Haar transform for the 1-D case was reported in Ahmed et al. (1974). The variances of the 16 transform coefficients are shown in Figure 4.8 when $N = 16$, $\rho = 0.95$. Note that $N$ stands for the dimension of the 1-D vector, while the parameter $\rho$ is shown in the Toeplitz matrix (refer to Equation 4.57). We can see that the DCT compares most closely to the KLT, which is known to be optimum.

Note that the unequal variance distribution among transform coefficients has also found application in the field of pattern recognition. Similar results to those in Ahmed et al. (1974) for the DFT, DWT, and Haar transform were reported in (Andrews, 1971).

A similar analysis can be carried out for the 2-D case (Wintz, 1972). Recall that an image $g(x, y)$ can be expressed as a weighted sum of basis images $I_{u,v}$. That is,

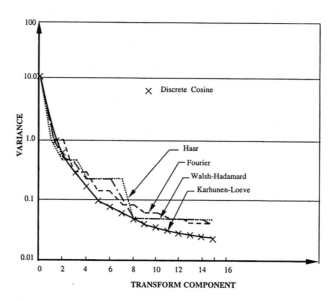

**FIGURE 4.8**  Transform coefficient variances when $N = 16$, $\rho = 0.95$. (From Ahmed, N. et al., *IEEE Trans. Comput.*, 90, 1974. With permission.)

$$G = \sum_{u=0}^{N-1} \sum_{v=0}^{N-1} T(u,v) I_{u,v} \tag{4.65}$$

where the weights are transform coefficients. We arrange the coefficients according to their variances in a nonincreasing order. For some choices of the transform (hence basis images), the coefficients become insignificant after the first $L$ terms, and the image can be approximated well by truncating the coefficients after $L$. That is,

$$G = \sum_{u=0}^{N-1} \sum_{v=0}^{N-1} T(u,v) I_{u,v} \approx \sum_{u=0}^{L} \sum_{v=0}^{L} T(u,v) I_{u,v} \tag{4.66}$$

The mean square reconstruction error is given by

$$MSE_r = \sum_{L}^{N-1} \sum^{N-1} \sigma_{u,v}^2 \tag{4.67}$$

A comparison among the KLT, DHT, and DFT in terms of the mean square reconstruction error for 2-D array of $16 \times 16$ (i.e., 256 transform coefficients) was reported in (Figure 5, Wintz, 1972). Note that the discrete KLT is image dependent. In the comparison, the KLT is calculated with respect to an image named "Cameraman." It shows that while the KLT achieves the best performance, the other transforms perform closely.

In essence, the criteria of mean square reconstruction error and energy compaction are closely related. It has been shown that the discrete Karhunen transform (KLT), also known as the Hotelling transform, is the optimum in terms of energy compaction and mean square reconstruction error. The DWT, DHT, DFT, and DCT are close to the optimum (Wintz, 1972; Ahmed et al., 1974); however, the DCT is the best among these several *suboptimum* transforms.

Note that the performance comparison among various transforms in terms of bit rate vs. distortion in the reconstructed image was reported in (Pearl et al., 1972; Ahmed et al., 1974). The same conclusion was drawn. That is, the KLT is optimum, while the DFT, DWT, DCT, and Haar transforms are close in performance. Among the suboptimum transforms, the DCT is the best.

### 4.3.5.3  Computational Complexity

Note that while the DWT, DHT, DFT, and DCT are input image independent, the discrete KLT (the Hotelling transform) is input dependent. More specifically, the row vectors of the Hotelling transform matrix are transposed eigenvectors of the covariance matrix of the input random vector. So far there is no fast transform algorithm available. This computational complexity prohibits the Hotelling transform from practical usage. It can be shown that the DWT, DFT, and DCT can be implemented using the FFT algorithm.

### 4.3.5.4  Summary

As pointed out above, the DCT is the best among the suboptimum transforms in terms of energy compaction. Moreover, the DCT can be implemented using the FFT. Even though a $2N$-point sequence is involved, the even symmetry makes the computation involved in the $N$-point DCT equivalent to that of the $N$-point FFT. For these two reasons, the DCT finds the widest application in image and video coding.

## 4.4  BIT ALLOCATION

As shown in Figure 4.2, in transform coding, an input image is first divided into blocks (subimages). Then a 2-D linear transform is applied to each block. The transformed blocks go through truncation, quantization, and codeword assignment. The last three functions: truncation, quantization, and codeword assignment, are combined and called bit allocation.

From the previous section, it is known that the applied transform decorrelates subimages. Moreover, it redistributes image energy in the transform domain in such a way that most of the energy is compacted into a small fraction of coefficients. Therefore, it is possible to discard the majority of transform coefficients without introducing significant distortion.

As a result, we see that in transform coding there are mainly three types of errors involved. One is due to truncation. That is, the majority of coefficients are truncated to zero. Others come from quantization. (Note that truncation can also be considered a special type of quantization). Transmission errors are the third type of error. Recall that the mean square reconstruction error discussed in Section 4.3.5.2 is in fact only related to truncation error. For this reason, it was referred to more precisely as a mean square approximation error. In general, the reconstruction error, i.e., the error between the original image signal and the reconstructed image at the receiver, includes three types of errors: truncation error, quantization error, and transmission error.

There are two different ways to truncate transform coefficients. One is called *zonal coding*, while the other is *threshold coding*. They are discussed below.

### 4.4.1  ZONAL CODING

In zonal coding, also known as *zonal sampling*, a zone in the transformed block is predefined according to a statistical average obtained from many blocks. All transform coefficients in the zone are retained, while all coefficients outside the zone are set to zero. As mentioned in Section 4.3.5.1, the total energy of the image remains the same after applying the transforms discussed there. Since it is known that the DC and low-frequency AC coefficients of the DCT occupy most of the energy, the zone is located in the top-left portion of the transformed block when the transform coordinate

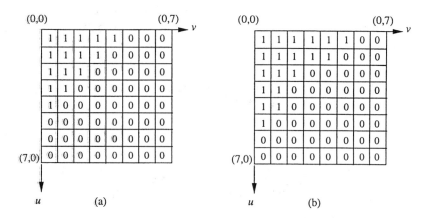

**FIGURE 4.9**   Two illustrations of zonal coding.

system is set conventionally. (Note that by DC we mean $u = v = 0$. By AC we mean $u$ and $v$ do not equal zero simultaneously.) That is, the origin is at the top-left corner of the transformed block. Two typical zones are shown in Figure 4.9. The simplest uniform quantization with natural binary coding can be used to quantize and encode the retained transform coefficients. With this simple technique, there is no overhead side information that needs to be sent to the receiver, since the structure of the zone, the scheme of the quantization, and encoding are known at both the transmitter and receiver.

The coding efficiency, however, may not be very high. This is because the zone is predefined based on average statistics. Therefore some coefficients outside the zone might be large in magnitude, while some coefficients inside the zone may be small in quantity. Uniform quantization and natural binary encoding are simple, but they are known not to be efficient enough.

For further improvement of coding efficiency, an adaptive scheme has to be used. There, a two-pass procedure is applied. In the first pass, the variances of transform coefficients are measured or estimated. Based on the statistics, the quantization and encoding schemes are determined. In the second pass, quantization and encoding are carried out (Habibi, 1971a; Chen and Smith, 1977).

### 4.4.2   THRESHOLD CODING

In threshold coding, also known as threshold sampling, there is not a predefined zone. Instead, each transform coefficient is compared with a threshold. If it is smaller than the threshold, then it is set to zero. If it is larger than the threshold, it will be retained for quantization and encoding. Compared with zonal coding, this scheme is adaptive in truncation in the sense that the coefficients with more energy are retained no matter where they are located. The addresses of these retained coefficients, however, have to be sent to the receiver as side information. Furthermore, the threshold is determined after an evaluation of all coefficients. Hence, it was usually a two-pass adaptive technique.

Chen and Pratt (1984) devised an efficient adaptive scheme to handle threshold coding. It is a one-pass adaptive scheme, in contrast to the two-pass adaptive schemes. Hence it is fast in implementation. With several effective techniques that will be addressed here, it achieved excellent results in transform coding. Specifically, it demonstrated a satisfactory quality of reconstructed frames at a bit rate of 0.4 bits per pixel for coding of color images, which corresponds to real-time color television transmission over a 1.5-Mb/sec channel. This scheme has been adopted by the international still coding standard JPEG. A block diagram of the threshold coding proposed by Chen and Pratt is shown in Figure 4.10. More details and modification made by JPEG will be described in Chapter 7.

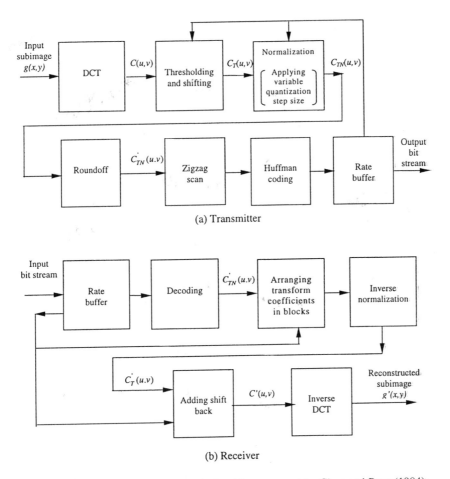

FIGURE 4.10 Block diagram of the algorithm proposed by Chen and Pratt (1984).

### 4.4.2.1 Thresholding and Shifting

The DCT is used in the scheme because of its superiority, described in Section 4.3. Here we use $C(u,v)$ to denote the DCT coefficients. The DC coefficient, $C(0,0)$, is processed differently. As mentioned in Chapter 3, the DC coefficients are encoded with a differential coding technique. For more details, refer to Chapter 7. For all the AC coefficients, the following thresholding and shifting are carried out:

$$C_T(u,v) = \begin{cases} C(u,v) - T & if \quad C(u,v) > T \\ 0 & if \quad C(u,v) \le T \end{cases} \tag{4.68}$$

where $T$ on the right-hand side is the threshold. Note that the above equation also implies a shifting of transform coefficients by $T$ when $C(u, v) > T$. The input-output characteristic of the thresholding and shifting is shown in Figure 4.11.

Figure 4.12 demonstrates that more than 60% of the DCT coefficients normally fall below a threshold value as low as 5. This indicates that with a properly selected threshold value it is possible to set most of the DCT coefficients equal to zero. The threshold value is adjusted by the feedback from the rate buffer, or by the desired bit rate.

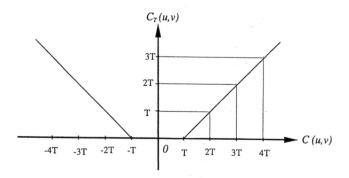

**FIGURE 4.11**   Input-output characteristic of thresholding and shifting.

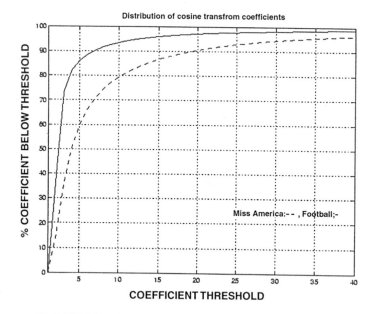

**FIGURE 4.12**   Amplitude distribution of the DCT coefficients.

### 4.4.2.2   Normalization and Roundoff

The threshold subtracted transform coefficients $C_T(u,v)$ are normalized before roundoff. The normalization is implemented as follows:

$$C_{TN}(u,v) = \frac{C_T(u,v)}{\Gamma_{u,v}} \tag{4.69}$$

where the normalization factor $\Gamma_{u,v}$ is controlled by the rate buffer. The roundoff process converts floating point to integer as follows.

$$R\left[C_{TN}(u,v)\right] = C_{TN}(u,v) = \begin{cases} \left\lfloor C_{TN}(u,v)+0.5 \right\rfloor & \text{if} \quad C_{TN}(u,v) \geq 0 \\ \left\lceil C_{TN}(u,v)-0.5 \right\rceil & \text{if} \quad C_{TN}(u,v) < 0 \end{cases} \tag{4.70}$$

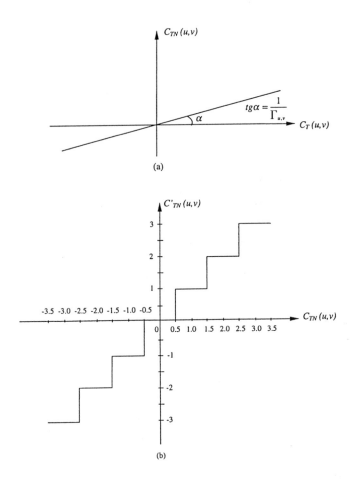

**FIGURE 4.13** Input-output characteristic of (a) normalization, (b) roundoff.

where the operator $\lfloor x \rfloor$ means the largest integer smaller than or equal to $x$, the operator $\lceil x \rceil$ means the smallest integer larger than or equal to $x$. The input-output characteristics of the normalization and roundoff are shown in Figure 4.13(a) and (b), respectively.

From these input-output characteristics, we can see that the roundoff is a uniform midtread quantizer with a unit quantization step. The combination of normalization and roundoff is equivalent to a uniform midtread quantizer with the quantization step size equal to the normalization factor $\Gamma_{u,v}$. Normalization is a scaling process, which makes the resultant uniform midtread quantizer adapt to the dynamic range of the associated transform coefficient. It is therefore possible for one quantizer design to be applied to various coefficients with different ranges. Obviously, by adjusting the parameter $\Gamma_{u,v}$ (quantization step size) a variable bit rate and mean square quantization error can be achieved. Hence, the selection of the normalization factors for different transform coefficients can take the statistical feature of the images and the characteristics of the human visual system (HVS) into consideration. In general, most image energy is contained in the DC and low-frequency AC transform coefficients. The HVS is more sensitive to a relatively uniform region than to a relatively detailed region, as discussed in Chapter 1. Chapter 1 also mentions that, with regard to the color image, the HVS is more sensitive to the luminance component than to the chrominance components.

These have been taken into consideration in JPEG. A matrix consisting of all the normalization factors is called a quantization table in JPEG. A luminance quantization table and a chrominance quantization table used in JPEG are shown in Figure 4.14. We observe that in general in both tables

| 16 | 11 | 10 | 16 | 24 | 40 | 51 | 61 |
|----|----|----|----|----|----|----|----|
| 12 | 12 | 14 | 19 | 26 | 58 | 60 | 55 |
| 14 | 13 | 16 | 24 | 40 | 57 | 69 | 56 |
| 14 | 17 | 22 | 29 | 51 | 87 | 80 | 62 |
| 18 | 22 | 37 | 56 | 68 | 109 | 103 | 77 |
| 24 | 35 | 55 | 64 | 81 | 104 | 113 | 92 |
| 49 | 64 | 78 | 87 | 103 | 121 | 120 | 101 |
| 72 | 92 | 95 | 98 | 112 | 100 | 103 | 99 |

| 17 | 18 | 24 | 47 | 99 | 99 | 99 | 99 |
|----|----|----|----|----|----|----|----|
| 18 | 21 | 26 | 66 | 99 | 99 | 99 | 99 |
| 24 | 26 | 56 | 99 | 99 | 99 | 99 | 99 |
| 47 | 66 | 99 | 99 | 99 | 99 | 99 | 99 |
| 99 | 99 | 99 | 99 | 99 | 99 | 99 | 99 |
| 99 | 99 | 99 | 99 | 99 | 99 | 99 | 99 |
| 99 | 99 | 99 | 99 | 99 | 99 | 99 | 99 |
| 99 | 99 | 99 | 99 | 99 | 99 | 99 | 99 |

(a) Luminance quantization table          (b) Chrominance quantization table

**FIGURE 4.14**    Quantization tables.

the small normalization factors are assigned to the DC and low-frequency AC coefficients. The large Γs are associated with the high-frequency transform coefficients. Compared with the luminance quantization table, the chrominance quantization table has larger quantization step sizes for the low- and middle-frequency coefficients and almost the same step sizes for the DC and high-frequency coefficients, indicating that the chrominance components are relatively coarsely quantized, compared with the luminance component.

### 4.4.2.3   Zigzag Scan

As mentioned at the beginning of this section, while threshold coding is adaptive to the local statistics and hence is more efficient in truncation, threshold coding needs to send the addresses of retained coefficients to the receiver as overhead side information. An efficient scheme, called the zigzag scan, was proposed by Chen and Pratt (1984) and is shown in Figure 4.14. As shown in Figure 4.12, a great majority of transform coefficients have magnitudes smaller than a threshold of 5. Consequently, most quantized coefficients are zero. Hence, in the 1-D sequence obtained by zigzag scanning, most of the numbers are zero. A code known as run-length code, discussed in Chapter 6, is very efficient under these circumstances to encode the address information of nonzero coefficients. Run-length of zero coefficients is understood as the number of consecutive zeros in the zigzag scan. Zigzag scanning minimizes the use of run-length codes in the block.

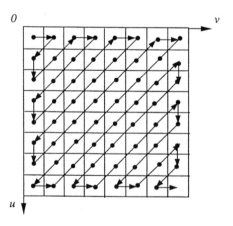

**FIGURE 4.15**    Zigzag scan of DCT coefficients within an 8 × 8 block.

### 4.4.2.4  Huffman Coding

Statistical studies of the magnitude of nonzero DCT coefficients and the run-length of zero DCT coefficients in zigzag scanning were conducted in (Chen and Pratt, 1984). The domination of the coefficients with small amplitudes and the short run-lengths was found and is shown in Figures 4.16 and 4.17. This justifies the application of the Huffman coding to the magnitude of nonzero transform coefficients and run-lengths of zeros.

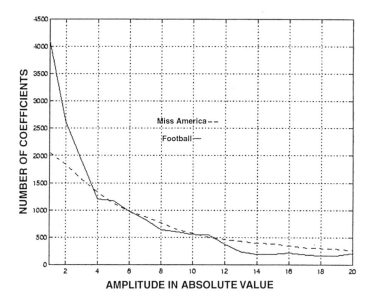

**FIGURE 4.16**  Histogram of DCT coefficients in absolute amplitude.

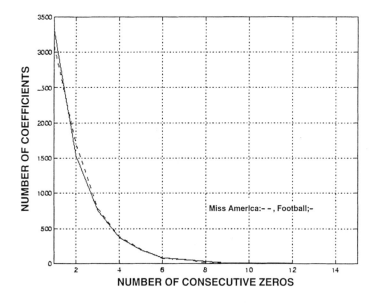

**FIGURE 4.17**  Histogram of zero run length.

#### 4.4.2.5   Special Codewords

Two special codewords were used by Chen and Pratt (1984). One is called *end of block* (EOB). Another is called *run-length prefix*. Once the last nonzero DCT coefficients in the zigzag is coded, EOB is appended, indicating the termination of coding the block. This further saves bits used in coding. A run-length prefix is used to discriminate the run-length codewords from the amplitude codewords.

#### 4.4.2.6   Rate Buffer Feedback and Equalization

As shown in Figure 4.10, a rate buffer accepts a variable-rate data input from the encoding process and provides a fixed-rate data output to the channel. The status of the rate buffer is monitored and fed back to control the threshold and the normalization factor. In this fashion a one-pass adaptation is achieved.

### 4.5   SOME ISSUES

#### 4.5.1   EFFECT OF TRANSMISSION ERRORS

In transform coding, each pixel in the reconstructed image relies on all transform coefficients in the subimage where the pixel is located. Hence, a bit reversal transmission error will spread. That is, an error in a transform coefficient will lead to errors in all the pixels within the subimage. As discussed in Section 4.2.3, this is one of the reasons the selected subimage size cannot be very large. Depending on which coefficient is in error, the effect caused by a bit reversal error on the reconstructed image varies. For instance, an error in the DC or a low-frequency AC coefficient may be objectionable, while an error in the high-frequency coefficient may be less noticeable.

#### 4.5.2   RECONSTRUCTION ERROR SOURCES

As discussed, three sources: truncation (discarding transform coefficients with small variances), quantization, and transmission contribute to the reconstruction error. It is noted that in TC the transform is applied block by block. Quantization and encoding of transform coefficients are also conducted blockwise. At the receiver, reconstructed blocks are put together to form the whole reconstructed image. In the process, block artifacts are produced. Sometimes, even though it may not severely affect an objective assessment of the reconstructed image quality, block artifacts can be annoying to the HVS, especially when the coding rate is low.

To alleviate the blocking effect, several techniques have been proposed. One is to overlap blocks in the source image. Another is to postfilter the reconstructed image along block boundaries. The selection of advanced transforms is an additional possible method (Lim, 1990).

In the block-overlapping method, when the blocks are finally organized to form the reconstructed image, each pixel in the overlapped regions takes an average value of all its reconstructed gray level values from multiple blocks. In this method, extra bits are used for those pixels involved in the overlapped regions. For this reason, the overlapped region is usually only one pixel wide.

Due to the sharp transition along block boundaries, block artifacts are of high frequency in nature. Hence, low-pass filtering is normally used in the postfiltering method. To avoid the blurring effect caused by low-pass filtering on the nonboundary image area, low-pass postfiltering is only applied to block boundaries. Unlike the block-overlapping method, the postfiltering method does not need extra bits. Moreover, it has been shown that the postfiltering method can achieve better results in combating block artifacts (Reeve and Lim, 1984; Ramamurthi and Gersho, 1986). For these two reasons, the postfiltering method has been adopted by the international coding standards.

### 4.5.3   COMPARISON BETWEEN DPCM AND TC

As mentioned at the beginning of the chapter, both differential coding and transform coding utilize interpixel correlation and are efficient coding techniques. Comparisons between these two techniques have been reported (Habibi, 1971b). Take a look at the techniques discussed in the previous chapter and in this chapter. We can see that differential coding is simpler than TC. This is because the linear prediction and differencing involved in differential coding are simpler than the 2-D transform involved in TC. In terms of the memory requirement and processing delay, differential coding such as DPCM is superior to TC. That is, DPCM needs less memory and has less processing delay than TC. The design of the DPCM system, however, is sensitive to image-to-image variation, and so is its performance. That is, an optimum DPCM design is matched to the statistics of a certain image. When the statistics change, the performance of the DPCM will be affected. On the contrary, TC is less sensitive to the variation in the image statistics. In general, the optimum DPCM coding system with a third or higher order predictor performs better than TC when the bit rate is about two to three bits per pixel for single images. When the bit rate is below two to three bits per pixel, TC is normally preferred. As a result, the international still image coding standard JPEG is based on TC, whereas, in JPEG, DPCM is used for coding the DC coefficients of DCT, and information-preserving differential coding is used for lossless still image coding.

### 4.5.4   HYBRID CODING

A method called hybrid transform/waveform coding, or simply hybrid coding, was devised in order to combine the merits of the two methods. By waveform coding, we mean coding techniques that code the waveform of a signal instead of the transformed signal. DPCM is a waveform coding technique. Hybrid coding combines TC and DPCM coding. That is, TC can be first applied rowwise, followed by DPCM coding columnwise, or vice versa. In this way, the two techniques complement each other. That is, the hybrid coding technique simultaneously has TC's small sensitivity to variable image statistics and DPCM's simplicity in implementation.

Worth mentioning is a successful hybrid coding scheme in interframe coding: predictive coding along the temporal domain. Specifically, it uses motion-compensated predictive coding. That is, the motion analyzed from successive frames is used to more accurately predict a frame. The prediction error (in the 2-D spatial domain) is transform coded. This hybrid coding scheme has been very efficient and was adopted by the international video coding standards H.261, H.263, and MPEG 1, 2, and 4.

## 4.6   SUMMARY

In transform coding, instead of the original image or some function of the original image in the spatial and/or temporal domain, the image in the transform domain is quantized and encoded. The main idea behind transform coding is that the transformed version of the image is less correlated. Moreover, the image energy is compacted into a small proper subset of transform coefficients.

The basis vector (1-D) and the basis image (2-D) provide a meaningful interpretation of transform coding. This type of interpretation considers the original image to be a weighted sum of basis vectors or basis images. The weights are the transform coefficients, each of which is essentially a correlation measure between the original image and the corresponding basis image. These weights are less correlated than the gray level values of pixels in the original image. Furthermore they have a great disparity in variance distribution. Some weights have large variances. They are retained and finely quantized. Some weights have small energy. They are retained and coarsely quantized. A vast majority of weights are insignificant and discarded. In this way, a high coding efficiency is achieved in transform coding. Because the quantized nonzero coefficients have a very nonuniform probability distribution, they can be encoded by using efficient variable-length codes. In summary, three factors: truncation (discarding a great majority of transform coefficients), adaptive quantization, and variable-length coding contribute mainly to a high coding efficiency of transform coding.

Several linear, reversible, unitary transforms have been studied and utilized in transform coding. They include the discrete Karhunen-Loeve transform (the Hotelling transform), the discrete Fourier transform, the Walsh transform, the Hadamard transform, and the discrete cosine transform. It is shown that the KLT is the optimum. The transform coefficients of the KLT are uncorrelated. The KLT can compact the most energy in the smallest fraction of transform coefficients. However, the KLT is image dependent. There is no fast algorithm to implement it. This prohibits the KLT from practical use in transform coding. While the rest of the transforms perform closely, the DCT appears to be the best. Its energy compaction is very close to the optimum KLT and it can be implemented using the fast Fourier transform. The DCT has been found to be efficient not only for still images coding but also for coding residual images (predictive error) in motion-compensated interframe predictive coding. These features make the DCT the most widely used transform in image and video coding.

There are two ways to truncate transform coefficients: zonal coding and threshold coding. In zonal coding, a zone is predefined based on average statistics. The transform coefficients within the zone are retained, while those outside the zone are discarded. In threshold coding, each transform coefficient is compared with a threshold. Those coefficients larger than the threshold are retained, while those smaller are discarded. Threshold coding is adaptive to local statistics. A two-pass procedure is usually taken. That is, the local statistics are measured or estimated in the first pass. The truncation takes place in the second pass. The addresses of the retained coefficients need to be sent to the receiver as overhead side information.

A one-step adaptive framework of transform coding has evolved as a result of the tremendous research efforts in image coding. It has become a base of the international still image coding standard JPEG. Its fundamental components include the DCT transform, thresholding and adaptive quantization of transform coefficients, zigzag scan, Huffman coding of the magnitude of the nonzero DCT coefficients and run-length of zeros in the zigzag scan, the codeword of EOB, and rate buffer feedback control.

The threshold and the normalization factor are controlled by rate buffer feedback. Since the threshold decides how many transform coefficients are retained and the normalization factor is actually the quantization step size, the rate buffer has direct impact on the bit rate of the transform coding system. Selection of quantization steps takes the energy compaction of the DCT and the characteristics of the HVS into consideration. That is, it uses not only statistical redundancy, but also psychovisual redundancy to enhance coding efficiency.

After thresholding, normalization and roundoff are applied to the DCT transform coefficients in a block; a great majority of transform coefficients are set to zero. A zigzag scan can convert the 2-D array of transform coefficients into a 1-D sequence. The number of consecutive zero-valued coefficients in the 1-D sequence is referred to as the run-length of zeros and is used to provide address information of nonzero DCT coefficients. Both the magnitude of nonzero coefficients and run-length information need to be coded. Statistical analysis has demonstrated that a small magnitude and short run-length are dominant. Therefore, efficient lossless entropy coding methods such as Huffman coding and arithmetic coding (the focus of the next chapter) can be applied to magnitude and run-length.

In a reconstructed subimage, there are three types of errors involved: truncation error (some transform coefficients have been set to zero), quantization error, and transmission error. In a broad sense, the truncation can be viewed as a part of the quantization. That is, these truncated coefficients are quantized to zero. The transmission error in terms of bit reversal will affect the whole reconstructed subimage. This is because, in the inverse transform (such as the inverse DCT), each transform coefficient makes a contribution.

In reconstructing the original image all the subimages are organized to form the whole image. Therefore the independent processing of individual subimages causes block artifacts. Though they may not severely affect the objective assessment of reconstructed image quality, block artifacts can be annoying, especially in low bit rate image coding. Block overlappling and postfiltering are two effective ways to alleviate block artifacts. In the former, neighboring blocks are purposely over-lapped by one pixel. In reconstructing the image, those pixels that have been coded more than once take an average of the multiple decoded values. Extra bits are used. In the latter technique, a low-pass filter is applied along boundaries of blocks. No extra bits are required in the process and the effect of combating block artifacts is better than with the former technique.

The selection of subimage size is an important issue in the implementation of transform coding. In general, a large size will remove more interpixel redundancy. But it has been shown that the pixel correlation becomes insignificant when the distance of pixels exceeds 20. On the other hand, a large size is not suitable for adaptation to local statistics, while adaptation is required in handling nonstationary images. A large size also makes the effect of a transmission error spread more widely. For these reasons, subimage size should not be large. In motion-compensated predictive interframe coding, motion estimation is normally carried out in sizes of $16 \times 16$ or $8 \times 8$. To be compatible, the subimage size in transform coding is normally chosen as $8 \times 8$.

Both predictive codings, say, DPCM and TC, utilize interpixel correlation and are efficient coding schemes. Compared with TC, DPCM is simpler in computation. It needs less storage and has less processing delay. But it is more sensitive to image-to-image variation. On the other hand, TC provides higher adaptation to statistical variation. TC is capable of removing more interpixel correlation, thus providing higher coding efficiency. Traditionally, predictive coding is preferred if the bit rate is in the range of two to three bits per pixel, while TC is preferred when bit rate is below two to three bits per pixel. However, the situation changes. TC becomes the core technology in image and video coding. Many special VLSI chips are designed and manufactured for reducing computational complexity. Consequently, predictive coding such as DPCM is only used in some very simple applications.

In the context of interframe coding, 3-D (two spatial dimensions and one temporal dimension) transform coding has not found wide application in practice due to the complexity in computation and storage. Hybrid transform/waveform coding has proven to be very efficient in interframe coding. There, motion-compensated predictive coding is used along the temporal dimension, while trans-form coding is used to code the prediction error in two spatial dimensions.

## 4.7  EXERCISES

**4-1.** Consider the following eight points in a 3-D coordinate system: $(0,0,0)^T$, $(1,0,0)^T$, $(0,1,0)^T$, $(0,0,1)^T$, $(0,1,1)^T$, $(1,0,1)^T$, $(1,1,0)^T$, $(1,1,1)^T$. Find the mean vector and covariance matrix using Equations 4.12 and 4.13.

**4-2.** For N = 4, find the basis images of the DFT, $I_{u,v}$ when (a) u = 0, v = 0; (b) u = 1, v = 0; (c) u = 2, v = 2; (d) u = 3, v = 2. Use both methods discussed in the text; i.e., the method with basis image and the method with basis vectors.

**4-3.** For N = 4, find the basis images of the ordered discrete Hadamard transform when (a) u = 0, v = 2; (b) u = 1, v = 3; (c) u = 2, v = 3; (d) u = 3, v = 3. Verify your results by comparing them with Figure 4.5.

**4-4.** Repeat the previous problem for the DWT, and verify your results by comparing them with Figure 4.4.

**4-5.** Repeat problem 4-3 for the DCT and N = 4.

**4-6.** When $N = 8$, draw the transform matrix $F$ for the DWT, DHT, the order DHT, DFT, and DCT.

**4-7.** The matrix form of forward and inverse 2-D symmetric image transforms are expressed in texts such as (Jayant and Noll, 1984) as $T = FGF^T$ and $G = ITI^T$, which are different from Equations 4.28 and 4.29. Can you explain this discrepancy?

**4-8.** Derive Equation 4.64. (Hint: use the concept of basis vectors and the orthogonality of basis vectors.)

**4-9.** Justify that the normalization factor is the quantization step.

**4-10.** The transform used in TC has two functions: decorrelation and energy compaction. Does decorrelation automatically lead to energy compaction? Comment.

**4-11.** Using your own words, explain the main idea behind transform coding.

**4-12.** Read the techniques by Chen and Pratt presented in Section 4.4.2. Compare them with JPEG discussed in Chapter 7. Comment on the similarity and dissimilarity between them.

**4-13.** How is one-pass adaptation to local statistics in the Chen and Pratt algorithm achieved?

**4-14.** Explain why the DCT is superior to the DFT in terms of energy compaction.

**4-15.** Why is the subimage size of $8 \times 8$ widely used?

## REFERENCES

Ahmed, N., T. Nararajan, and K. R. Rao, Discrete cosine transform, *IEEE Trans. Comput.*, 90-93, 1974.

Andrews, H. C. Multidimensional rotations in feature selection, *IEEE Trans. Comput.*, c-20, 1045-1051, 1971.

Chen, W. H. and C. H. Smith, Adaptive coding of monochrome and color images, *IEEE Trans. Commun.*, COM-25, 1285-1292, 1977.

Chen, W. H. and W. K. Pratt, Scene adaptive coder, *IEEE Trans. Commun.*, COM-32, 225-232, 1984.

Cooley, J. W. and J. W. Tukey, An algorithm for the machine calculation of complex Fourier series, *Math. Comput.*, 19, 297-301, 1965.

Gonzalez, R. C. and R. E. Woods, *Digital Image Processing*, Addison-Wesley, Reading, MA, 1992.

Hadamard, J. Resolution d'une question relative aux determinants, *Bull. Sci. Math.*, *Ser. 2*, 17, Part I, 240-246, 1893.

Habibi, A. and P. A. Wintz, Image coding by linear transformations and block quantization, *IEEE Trans. Commun. Technol.*, com-19, 50-60, 1971.

Habibi, A. Comparison of nth-order DPCM encoder with linear transformations and block quantization techniques, *IEEE Trans. Commun. Technol.*, com-19(6), 948-956, 1971.

Huang, J.-Y. and P. M. Schultheiss, Block quantization of correlated Gaussian random variables, *IEEE Trans. Commun. Syst.*, cs-11, 289-296, 1963.

Jayant, N. S. and P. Noll, *Digital Coding of Waveforms*, Prentice-Hall, Englewood Cliffs, NJ, 1984.

Karhunen, K., On Linear Methods in Probability Theory, Ann. Acad. Sci. Fennicae, Ser. A137. (Translated into English.) The Rand Corp., Santa Monica, CA, 1960.

Lim, J. S. *Two-Dimensional Signal and Image Processing*, Prentice-Hall, Englewood Cliffs, NJ, 1990.

Loève, M., Fonctions Aléatoires de Second Ordre, in P. Levy, *Processus Stochastique et Mouvement Brownien*, Hermann, Paris.

Pearl, J., H. C. Andrews, and W. K. Pratt, Performance measures for transform data coding, *IEEE Trans. Commun. Technol.*, com-20, 411-415, 1972.

Reeve, H. C. III and J. S. Lim, Reduction of blocking effects in image coding, *J. Opt. Eng.*, 23, 34-37, 1984.

Ramamurthi, B. and A. Gersho, Nonlinear space-variant postprocessing of block coded images, *IEEE Trans. Acoust. Speech Signal Process.*, 34, 1258-1267, 1986.

Strang, G., *Introduction to Linear Algebra*, Wellesley-Cambridge Press, 2nd ed., 1998.

Tasto, M. and P. A. Wintz, Image coding by adaptive block quantization, *IEEE Trans. Commun. Technol.*, com-19(6), 957-972, 1971.

Walsh, J. L. A closed set of normal orthogonal functions, *Am. J. Math.*, 45(1), 5-24, 1923.

Wintz, P. A. Transform picture coding, *Proc. IEEE*, 60(7), 809-820, 1972.

Zelinski, R. and P. Noll, Adaptive block quantization of speech signals, (in German), Tech. Rep. no. 181, Heinrich Hertz Institut, Berlin, 1975.

# 5 Variable-Length Coding: Information Theory Results (II)

Recall the block diagram of encoders shown in Figure 2.3. There are three stages that take place in an encoder: transformation, quantization, and codeword assignment. Quantization was discussed in Chapter 2. Differential coding and transform coding using two different transformation components were covered in Chapters 3 and 4, respectively. In differential coding it is the difference signal that is quantized and encoded, while in transform coding it is the transformed signal that is quantized and encoded. In this chapter and the next chapter, we discuss several codeword assignment (encoding) techniques. In this chapter we cover two types of variable-length coding: Huffman coding and arithmetic coding.

First we introduce some fundamental concepts of encoding. After that, the rules that must be obeyed by all optimum and instantaneous codes are discussed. Based on these rules, the Huffman coding algorithm is presented. A modified version of the Huffman coding algorithm is introduced as an efficient way to dramatically reduce codebook memory while keeping almost the same optimality.

The promising arithmetic coding algorithm, which is quite different from Huffman coding, is another focus of the chapter. While Huffman coding is a *block-oriented* coding technique, arithmetic coding is a *stream-oriented* coding technique. With improvements in implementation, arithmetic coding has gained increasing popularity. Both Huffman coding and arithmetic coding are included in the international still image coding standard JPEG (Joint Photographic [image] Experts Group coding). The adaptive arithmetic coding algorithms have been adopted by the international bilevel image coding standard JBIG (Joint Bi-level Image experts Group coding). Note that the material presented in this chapter can be viewed as a continuation of the information theory results presented in Chapter 1.

## 5.1 SOME FUNDAMENTAL RESULTS

Prior to presenting Huffman coding and arithmetic coding, we first provide some fundamental concepts and results as necessary background.

### 5.1.1 CODING AN INFORMATION SOURCE

Consider an information source, represented by a *source alphabet S*.

$$S = \{s_1, s_2, \cdots, s_m\} \tag{5.1}$$

where $s_i$, $i = 1, 2, \cdots, m$ are *source symbols*. Note that the terms source symbol and information message are used interchangeably in the literature. In this book, however, we would like to distinguish between them. That is, an information message can be a source symbol, or a combination of source symbols. We denote *code alphabet* by $A$ and

$$A = \{a_1, a_2, \cdots, a_r\} \tag{5.2}$$

where $a_j$, $j = 1, 2, \cdots, r$ are *code symbols*. A *message code* is a sequence of code symbols that represents a given information message. In the simplest case, a message consists of only a source symbol. Encoding is then a procedure to assign a *codeword* to the source symbol. Namely,

$$s_i \rightarrow A_i = \left( a_{i1}, a_{i2}, \cdots, a_{ik} \right) \qquad (5.3)$$

where the codeword $A_i$ is a string of $k$ code symbols assigned to the source symbol $s_i$. The term message ensemble is defined as the entire set of messages. A code, also known as an ensemble code, is defined as a mapping of all the possible sequences of symbols of $S$ (message ensemble) into the sequences of symbols in $A$.

Note that in binary coding, the number of code symbols $r$ is equal to 2, since there are only two code symbols available: the binary digits "0" and "1". Two examples are given below to illustrate the above concepts.

### Example 5.1

Consider an English article and the ASCII code. Refer to Table 5.1. In this context, the source alphabet consists of all the English letters in both lower and upper cases and all the punctuation marks. The code alphabet consists of the binary 1 and 0. There are a total of 128 7-bit binary codewords. From Table 5.1, we see that the codeword assigned to the capital letter A is 1000001. That is, A is a source symbol, while 1000001 is its codeword.

### Example 5.2

Table 5.2 lists what is known as the (5,2) code. It is a linear block code. In this example, the source alphabet consists of the four ($2^2$) source symbols listed in the left column of the table: 00, 01, 10, and 11. The code alphabet consists of the binary 1 and 0. There are four codewords listed in the right column of the table. From the table, we see that the code assigns a 5-bit codeword to each source symbol. Specifically, the codeword of the source symbol 00 is 00000. The source symbol 01 is encoded as 10100; 01111 is the codeword assigned to 10. The symbol 11 is mapped to 11011.

## 5.1.2 Some Desired Characteristics

To be practical in use, codes need to have some desirable characteristics (Abramson, 1963). Some of the characteristics are addressed in this subsection.

### 5.1.2.1 Block Code

A code is said to be a block code if it maps each source symbol in $S$ into a fixed codeword in A. Hence, the codes listed in the above two examples are block codes.

### 5.1.2.2 Uniquely Decodable Code

A code is uniquely decodable if it can be unambiguously decoded. Obviously, a code has to be uniquely decodable if it is to be of use.

### Example 5.3

Table 5.3 specifies a code. Obviously it is not uniquely decodable since if a binary string "00" is received we do not know which of the following two source symbols has been sent out: $s_1$ or $s_3$.

### Nonsingular Code

A block code is nonsingular if all the codewords are distinct (see Table 5.4).

### Example 5.4

Table 5.4 gives a nonsingular code since all four codewords are distinct. If a code is not a nonsingular code, i.e., at least two codewords are identical, then the code is not uniquely decodable. Notice, however, that a nonsingular code does not guarantee unique decodability. The code shown in Table 5.4 is such an example in that it is nonsingular while it is not uniquely decodable. It is not

## TABLE 5.1
## Seven-Bit American Standard Code for Information Interchange (ASCII)

| Bits | | | | 5 | 0 | 1 | 0 | 1 | 0 | 1 | 0 | 1 |
|---|---|---|---|---|---|---|---|---|---|---|---|---|
| | | | | 6 | 0 | 0 | 1 | 1 | 0 | 0 | 1 | 1 |
| 1 | 2 | 3 | 4 | 7 | 0 | 0 | 0 | 0 | 1 | 1 | 1 | 1 |
| 0 | 0 | 0 | 0 | NUL | DLE | SP | 0 | @ | P | ` | p |
| 1 | 0 | 0 | 0 | SOH | DC1 | ! | 1 | A | Q | a | q |
| 0 | 1 | 0 | 0 | STX | DC2 | " | 2 | B | R | b | r |
| 1 | 1 | 0 | 0 | ETX | DC3 | # | 3 | C | S | c | s |
| 0 | 0 | 1 | 0 | EOT | DC4 | $ | 4 | D | T | d | t |
| 1 | 0 | 1 | 0 | ENQ | NAK | % | 5 | E | U | e | u |
| 0 | 1 | 1 | 0 | ACK | SYN | & | 6 | F | V | f | v |
| 1 | 1 | 1 | 0 | BEL | ETB | ' | 7 | G | W | g | w |
| 0 | 0 | 0 | 1 | BS | CAN | ( | 8 | H | X | h | x |
| 1 | 0 | 0 | 1 | HT | EM | ) | 9 | I | Y | i | y |
| 0 | 1 | 0 | 1 | LF | SUB | * | : | J | Z | j | z |
| 1 | 1 | 0 | 1 | VT | ESC | + | ; | K | [ | k | { |
| 0 | 0 | 1 | 1 | FF | FS | , | < | L | \ | l | | |
| 1 | 0 | 1 | 1 | CR | GS | - | = | M | ] | m | } |
| 0 | 1 | 1 | 1 | SO | RS | . | > | N | ^ | n | ~ |
| 1 | 1 | 1 | 1 | SI | US | / | ? | O | — | o | DEL |

| | | | |
|---|---|---|---|
| NUL | Null, or all zeros | DC1 | Device control 1 |
| SOH | Start of heading | DC2 | Device control 2 |
| STX | Start of text | DC3 | Device control 3 |
| ETX | End of text | DC4 | Device control 4 |
| EOT | End of transmission | NAK | Negative acknowledgment |
| ENQ | Enquiry | SYN | Synchronous idle |
| ACK | Acknowledge | ETB | End of transmission block |
| BEL | Bell, or alarm | CAN | Cancel |
| BS | Backspace | EM | End of medium |
| HT | Horizontal tabulation | SUB | Substitution |
| LF | Line feed | ESC | Escape |
| VT | Vertical tabulation | FS | File separator |
| FF | Form feed | GS | Group separator |
| CR | Carriage return | RS | Record separator |
| SO | Shift out | US | Unit separator |
| SI | Shift in | SP | Space |
| DLE | Data link escape | DEL | Delete |

## TABLE 5.2
## A (5,2) Linear Block Code

| Source Symbol | Codeword |
|---|---|
| $S_1$ (0 0) | 0 0 0 0 0 |
| $S_2$ (0 1) | 1 0 1 0 0 |
| $S_3$ (1 0) | 0 1 1 1 1 |
| $S_4$ (1 1) | 1 1 0 1 1 |

---

**TABLE 5.3**
**A Not Uniquely Decodable Code**

| Source Symbol | Codeword |
|:---:|:---:|
| $S_1$ | 0 0 |
| $S_2$ | 1 0 |
| $S_3$ | 0 0 |
| $S_4$ | 1 1 |

---

**TABLE 5.4**
**A Nonsingular Code**

| Source Symbol | Codeword |
|:---:|:---:|
| $S_1$ | 1 |
| $S_2$ | 1 1 |
| $S_3$ | 0 0 |
| $S_4$ | 0 1 |

---

uniquely decodable because once the binary string "11" is received, we do not know if the source symbols transmitted are $s_1$ followed by $s_1$ or simply $s_2$.

**The *n*th Extension of a Block Code**

The *n*th extension of a block code, which maps the source symbol $s_i$ into the codeword $A_i$, is a block code that maps the sequences of source symbols $s_{i1}s_{i2}\cdots s_{in}$ into the sequences of codewords $A_{i1}A_{i2}\cdots A_{in}$.

**A Necessary and Sufficient Condition of a Block Code's Unique Decodability**

A block code is uniquely decodable if and only if the *n*th extension of the code is nonsingular for every finite *n*.

**Example 5.5**

The second extension of the nonsingular block code shown in Example 5.4 is listed in Table 5.5. Clearly, this second extension of the code is not a nonsingular code, since the entries $s_1s_2$ and $s_2s_1$ are the same. This confirms the nonunique decodability of the nonsingular code in Example 5.4.

---

**TABLE 5.5**
**The Second Extension of the Nonsingular Block Code in Example 5.4**

| Source Symbol | Codeword | Source Symbol | Codeword |
|:---:|:---:|:---:|:---:|
| $S_1\,S_1$ | 1 1 | $S_3\,S_1$ | 0 0 1 |
| $S_1\,S_2$ | 1 1 1 | $S_3\,S_2$ | 0 0 1 1 |
| $S_1\,S_3$ | 1 0 0 | $S_3\,S_3$ | 0 0 0 0 |
| $S_1\,S_4$ | 1 0 1 | $S_3\,S_4$ | 0 0 0 1 |
| $S_2\,S_1$ | 1 1 1 | $S_4\,S_1$ | 0 1 1 |
| $S_2\,S_2$ | 1 1 1 1 | $S_4\,S_2$ | 0 1 1 1 |
| $S_2\,S_3$ | 1 1 0 0 | $S_4\,S_3$ | 0 1 0 0 |
| $S_2\,S_4$ | 1 1 0 1 | $S_4\,S_4$ | 0 1 0 1 |

**TABLE 5.6**
**Three Uniquely Decodable Codes**

| Source Symbol | Code 1 | Code 2 | Code 3 |
|:---:|:---:|:---:|:---:|
| $S_1$ | 0 0 | 1 | 1 |
| $S_2$ | 0 1 | 0 1 | 10 |
| $S_3$ | 1 0 | 0 0 1 | 1 0 0 |
| $S_4$ | 1 1 | 0 0 0 1 | 1 0 0 0 |

### 5.1.2.3   Instantaneous Codes

**Definition of Instantaneous Codes**

A uniquely decodable code is said to be instantaneous if it is possible to decode each codeword in a code symbol sequence without knowing the succeeding codewords.

**Example 5.6**

Table 5.6 lists three uniquely decodable codes. The first one is in fact a two-bit natural binary code. In decoding, we can immediately tell which source symbols are transmitted since each codeword has the same length. In the second code, code symbol "1" functions like a comma. Whenever we see a "1", we know it is the end of the codeword. The third code is different from the previous two codes in that if we see a "10" string we are not sure if it corresponds to $s_2$ until we see a succeeding "1". Specifically, if the next code symbol is "0", we still cannot tell if it is $s_3$ since the next one may be "0" (hence $s_4$) or "1" (hence $s_3$). In this example, the next "1" belongs to the succeeding codeword. Therefore we see that code 3 is uniquely decodable. It is not instantaneous, however.

**Definition of the *j*th Prefix**

Assume a codeword $A_i = a_{i1}a_{i2}\cdots a_{ik}$. Then the sequences of code symbols $a_{i1}a_{i2}\cdots a_{ij}$ with $1 \leq j \leq k$ is the *j*th order prefix of the codeword $A_i$.

**Example 5.7**

If a codeword is 11001, it has the following five prefixes: 11001, 1100, 110, 11, 1. The first-order prefix is 1, while the fifth-order prefix is 11001.

**A Necessary and Sufficient Condition of Being an Instantaneous Code**

A code is instantaneous if and only if no codeword is a prefix of some other codeword. This condition is often referred to as the *prefix condition*. Hence, the instantaneous code is also called the prefix condition code or sometimes simply the prefix code. In many applications, we need a block code that is nonsingular, uniquely decodable, and instantaneous.

### 5.1.2.4   Compact Code

A uniquely decodable code is said to be compact if its average length is the minimum among all other uniquely decodable codes based on the same source alphabet $S$ and code alphabet $A$. A compact code is also referred to as a *minimum redundancy* code, or an *optimum* code.

Note that the average length of a code was defined in Chapter 1 and is restated below.

### 5.1.3   DISCRETE MEMORYLESS SOURCES

This is the simplest model of an information source. In this model, the symbols generated by the source are independent of each other. That is, the source is memoryless or it has a zero memory.

Consider the information source expressed in Equation 5.1 as a discrete memoryless source. The occurrence probabilities of the source symbols can be denoted by $p(s_1), p(s_2), \cdots, p(s_m)$. The lengths of the codewords can be denoted by $l_1$, $l_2$, $\cdots$, $l_m$. The average length of the code is then equal to

$$L_{avg} = \sum_{i=1}^{m} l_i p(s_i) \tag{5.4}$$

Recall Shannon's first theorem, i.e., the noiseless coding theorem described in Chapter 1. The average length of the code is bounded below by the entropy of the information source. The entropy of the source $S$ is defined as $H(S)$ and

$$H(S) = -\sum_{i=1}^{m} p(s_i) \log_2 p(s_i) \tag{5.5}$$

Recall that entropy is the average amount of information contained in a source symbol. In Chapter 1 the efficiency of a code, $\eta$, is defined as the ratio between the entropy and the average length of the code. That is, $\eta = H(S)/L_{avg}$. The redundancy of the code, $\zeta$, is defined as $\zeta = 1 - \eta$.

### 5.1.4  EXTENSIONS OF A DISCRETE MEMORYLESS SOURCE

Instead of coding each source symbol in a discrete source alphabet, it is often useful to code blocks of symbols. It is, therefore, necessary to define the $n$th extension of a discrete memoryless source.

#### 5.1.4.1  Definition

Consider the zero-memory source alphabet $S$ defined in Equation 5.1. That is, $S = \{s_1, s_2, \cdots, s_m\}$. If $n$ symbols are grouped into a block, then there is a total of $m^n$ blocks. Each block is considered as a new source symbol. These $m^n$ blocks thus form an information source alphabet, called the $n$th extension of the source $S$, which is denoted by $S^n$.

#### 5.1.4.2  Entropy

Let each block be denoted by $\beta_i$ and

$$\beta_i = (s_{i1}, s_{i2}, \cdots, s_{in}) \tag{5.6}$$

Then we have the following relation due to the memoryless assumption:

$$p(\beta_i) = \prod_{j=1}^{n} p(s_{ij}) \tag{5.7}$$

Hence, the relationship between the entropy of the source $S$ and the entropy of its $n$th extension is as follows:

$$H(S^n) = n \cdot H(S) \tag{5.8}$$

**Example 5.8**
Table 5.7 lists a source alphabet. Its second extension is listed in Table 5.8.

**TABLE 5.7**
**A Discrete Memoryless Source Alphabet**

| Source Symbol | Occurrence Probability |
|---|---|
| $S_1$ | 0.6 |
| $S_2$ | 0.4 |

**TABLE 5.8**
**The Second Extension of the Source Alphabet Shown in Table 5.7**

| Source Symbol | Occurrence Probability |
|---|---|
| $S_1 S_1$ | 0.36 |
| $S_2 S_2$ | 0.24 |
| $S_2 S_1$ | 0.24 |
| $S_2 S_2$ | 0.16 |

The entropy of the source and its second extension are calculated below.

$$H(S) = -0.6 \cdot \log_2(0.6) - 0.4 \cdot \log_2(0.4) \approx 0.97$$

$$H(S^2) = -0.36 \cdot \log_2(0.36) - 2 \cdot 0.24 \cdot \log_2(0.24) - 0.16 \cdot \log_2(0.16) \approx 1.94$$

It is seen that $H(S^2) = 2H(S)$.

### 5.1.4.3   Noiseless Source Coding Theorem

The noiseless source coding theorem, also known as Shannon's first theorem, defining the minimum average codeword length per source pixel, was presented in Chapter 1, but without a mathematical expression. Here, we provide some mathematical expressions in order to give more insight about the theorem.

For a discrete zero-memory information source $S$, the noiseless coding theorem can be expressed as

$$H(S) \le L_{avg} < H(S) + 1 \tag{5.9}$$

That is, there exists a variable-length code whose average length is bounded below by the entropy of the source (that is encoded) and bounded above by the entropy plus 1. Since the $n$th extension of the source alphabet, $S^n$, is itself a discrete memoryless source, we can apply the above result to it. That is,

$$H(S^n) \le L^n_{avg} < H(S^n) + 1 \tag{5.10}$$

where $L^n_{avg}$ is the average codeword length of a variable-length code for the $S^n$. Since $H(S^n) = nH(S)$ and $L^n_{avg} = nL^n$avg, we have

$$H(S) \le L_{avg} < H(S) + \frac{1}{n} \tag{5.11}$$

Therefore, when coding blocks of $n$ source symbols, the noiseless source coding theory states that for an arbitrary positive number $\varepsilon$, there is a variable-length code which satisfies the following:

$$H(S) \le L_{avg} < H(S) + \varepsilon \tag{5.12}$$

as $n$ is large enough. That is, the average number of bits used in coding per source symbol is bounded below by the entropy of the source and is bounded above by the sum of the entropy and an arbitrary positive number. To make $\varepsilon$ arbitrarily small, i.e., to make the average length of the code arbitrarily close to the entropy, we have to make the block size $n$ large enough. This version of the noiseless coding theorem suggests a way to make the average length of a variable-length code approach the source entropy. It is known, however, that the high coding complexity that occurs when $n$ approaches infinity makes implementation of the code impractical.

## 5.2  HUFFMAN CODES

Consider the source alphabet defined in Equation 5.1. The method of encoding source symbols according to their probabilities, suggested in (Shannon, 1948; Fano, 1949), is not optimum. It approaches the optimum, however, when the block size $n$ approaches infinity. This results in a large storage requirement and high computational complexity. In many cases, we need a direct encoding method that is optimum and instantaneous (hence uniquely decodable) for an information source with finite source symbols in source alphabet $S$. Huffman code is the first such optimum code (Huffman, 1952), and is the technique most frequently used at present. It can be used for r-ary encoding as $r > 2$. For notational brevity, however, we discuss only the Huffman coding used in the binary case presented here.

### 5.2.1  REQUIRED RULES FOR OPTIMUM INSTANTANEOUS CODES

Let us rewrite Equation 5.1 as follows:

$$S = \left(s_1, s_2, \cdots, s_m\right) \tag{5.13}$$

Without loss of generality, assume the occurrence probabilities of the source symbols are as follows:

$$p(s_1) \ge p(s_2) \ge \cdots \ge p(s_{m-1}) \ge p(s_m) \tag{5.14}$$

Since we are seeking the optimum code for $S$, the lengths of codewords assigned to the source symbols should be

$$l_1 \le l_2 \le \cdots \le l_{m-1} \le l_m. \tag{5.15}$$

Based on the requirements of the optimum and instantaneous code, Huffman derived the following rules (restrictions):

1.  $l_1 \le l_2 \le \cdots \le l_{m-1} = l_m$                                                         $(5.16)$

    Equations 5.14 and 5.16 imply that when the source symbol occurrence probabilities are arranged in a nonincreasing order, the length of the corresponding codewords should be in a nondecreasing order. In other words, the codeword length of a more probable source symbol should not be longer than that of a less probable source symbol. Furthermore, the length of the codewords assigned to the two least probable source symbols should be the same.

2. The codewords of the two least probable source symbols should be the same except for their last bits.

3. Each possible sequence of length $l_m - 1$ bits must be used either as a codeword or must have one of its prefixes used as a codeword.

*Rule 1* can be justified as follows. If the first part of the rule, i.e., $l_1 \leq l_2 \leq \cdots \leq l_{m-1}$ is violated, say, $l_1 > l_2$, then we can exchange the two codewords to shorten the average length of the code. This means the code is not optimum, which contradicts the assumption that the code is optimum. Hence it is impossible. That is, the first part of Rule 1 has to be the case. Now assume that the second part of the rule is violated, i.e., $l_{m-1} < l_m$. (Note that $l_{m-1} > l_m$ can be shown to be impossible by using the same reasoning we just used in proving the first part of the rule.) Since the code is instantaneous, codeword $A_{m-1}$ is not a prefix of codeword $A_m$. This implies that the last bit in the codeword $A_m$ is redundant. It can be removed to reduce the average length of the code, implying that the code is not optimum. This contradicts the assumption, thus proving Rule 1.

*Rule 2* can be justified as follows. As in the above, $A_{m-1}$ and $A_m$ are the codewords of the two least probable source symbols. Assume that they do not have the identical prefix of the order $l_m - 1$. Since the code is optimum and instantaneous, codewords $A_{m-1}$ and $A_m$ cannot have prefixes of any order that are identical to other codewords. This implies that we can drop the last bits of $A_{m-1}$ and $A_m$ to achieve a lower average length. This contradicts the optimum code assumption. It proves that Rule 2 has to be the case.

*Rule 3* can be justified using a similar strategy to that used above. If a possible sequence of length $l_m - 1$ has not been used as a codeword and any of its prefixes have not been used as codewords, then it can be used in place of the codeword of the $m$th source symbol, resulting in a reduction of the average length $L_{avg}$. This is a contradiction to the optimum code assumption and it justifies the rule.

### 5.2.2  HUFFMAN CODING ALGORITHM

Based on these three rules, we see that the two least probable source symbols have codewords of equal length. These two codewords are identical except for the last bits, the binary 0 and 1, respectively. Therefore, these two source symbols can be combined to form a single new symbol. Its occurrence probability is the sum of two source symbols, i.e., $p(s_{m-1}) + p(s_m)$. Its codeword is the common prefix of order $l_m - 1$ of the two codewords assigned to $s_m$ and $s_{m-1}$, respectively. The new set of source symbols thus generated is referred to as the first auxiliary source alphabet, which is one source symbol less than the original source alphabet. In the first auxiliary source alphabet, we can rearrange the source symbols according to a nonincreasing order of their occurrence probabilities. The same procedure can be applied to this newly created source alphabet. A binary 0 and a binary 1, respectively, are assigned to the last bits of the two least probable source symbols in the alphabet. The second auxiliary source alphabet will again have one source symbol less than the first auxiliary source alphabet. The procedure continues. In some step, the resultant source alphabet will have only two source symbols. At this time, we combine them to form a single source symbol with a probability of 1. The coding is then complete.

Let's go through the following example to illustrate the above Huffman algorithm.

### Example 5.9

Consider a source alphabet whose six source symbols and their occurrence probabilities are listed in Table 5.9. Figure 5.1 demonstrates the Huffman coding procedure applied. In the example, among the two least probable source symbols encountered at each step we assign binary 0 to the top symbol and binary 1 to the bottom symbol.

**TABLE 5.9**
**Source Alphabet and Huffman Codes in Example 5.9**

| Source Symbol | Occurrence Probability | Codeword Assigned | Length of Codeword |
|---|---|---|---|
| $S_1$ | 0.3 | 00 | 2 |
| $S_2$ | 0.1 | 101 | 3 |
| $S_3$ | 0.2 | 11 | 2 |
| $S_4$ | 0.05 | 1001 | 4 |
| $S_5$ | 0.1 | 1000 | 4 |
| $S_6$ | 0.25 | 01 | 2 |

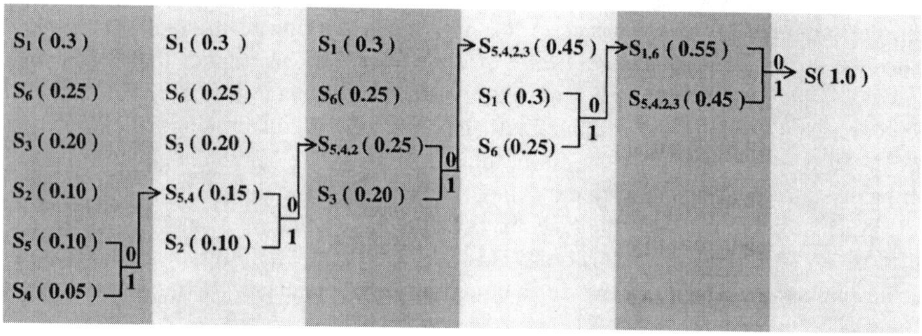

**FIGURE 5.1**   Huffman coding procedure in Example 5.9.

### 5.2.2.1   Procedures

In summary, the Huffman coding algorithm consists of the following steps.

1. Arrange all source symbols in such a way that their occurrence probabilities are in a nonincreasing order.
2. Combine the two least probable source symbols:
   - Form a new source symbol with a probability equal to the sum of the probabilities of the two least probable symbols.
   - Assign a binary 0 and a binary 1 to the two least probable symbols.
3. Repeat until the newly created auxiliary source alphabet contains only one source symbol.
4. Start from the source symbol in the last auxiliary source alphabet and trace back to each source symbol in the original source alphabet to find the corresponding codewords.

### 5.2.2.2   Comments

First, it is noted that the assignment of the binary 0 and 1 to the two least probable source symbols in the original source alphabet and each of the first $(u - 1)$ auxiliary source alphabets can be implemented in two different ways. Here $u$ denotes the total number of the auxiliary source symbols in the procedure. Hence, there is a total of $2^u$ possible Huffman codes. In Example 5.9, there are five auxiliary source alphabets, hence a total of $2^5 = 32$ different codes. Note that each is optimum: that is, each has the same average length.

Second, in sorting the source symbols, there may be more than one symbol having equal probabilities. This results in multiple arrangements of symbols, hence multiple Huffman codes. While all of these Huffman codes are optimum, they may have some other different properties.

For instance, some Huffman codes result in the minimum codeword length variance (Sayood, 1996). This property is desired for applications in which a constant bit rate is required.

Third, Huffman coding can be applied to r-ary encoding with $r > 2$. That is, code symbols are r-ary with $r > 2$.

### 5.2.2.3  Applications

As a systematic procedure to encode a finite discrete memoryless source, the Huffman code has found wide application in image and video coding. Recall that it has been used in differential coding and transform coding. In transform coding, as introduced in Chapter 4, the magnitude of the quantized transform coefficients and the run-length of zeros in the zigzag scan are encoded by using the Huffman code. This has been adopted by both still image and video coding standards.

## 5.3  MODIFIED HUFFMAN CODES

### 5.3.1  MOTIVATION

As a result of Huffman coding, a set of all the codewords, called a codebook, is created. It is an agreement between the transmitter and the receiver. Consider the case where the occurrence probabilities are skewed, i.e., some are large, while some are small. Under these circumstances, the improbable source symbols take a disproportionately large amount of memory space in the codebook. The size of the codebook will be very large if the number of the improbable source symbols is large. A large codebook requires a large memory space and increases the computational complexity. A modified Huffman procedure was therefore devised in order to reduce the memory requirement while keeping almost the same optimality (Hankamer, 1979).

**Example 5.10**
Consider a source alphabet consisting of 16 symbols, each being a 4-bit binary sequence. That is, $S = \{s_i, i = 1,2,\cdots,16\}$. The occurrence probabilities are

$$p(s_1) = p(s_2) = 1/4,$$

$$p(s_3) = p(s_4) = \cdots = p(s_{16}) = 1/28.$$

The source entropy can be calculated as follows:

$$H(S) = 2 \cdot \left(-\frac{1}{4}\log_2\frac{1}{4}\right) + 14 \cdot \left(-\frac{1}{28}\log_2\frac{1}{28}\right) \approx 3.404 \quad bits\ per\ symbol$$

Applying the Huffman coding algorithm, we find that the codeword lengths associated with the symbols are: $l_1 = l_2 = 2$, $l_3 = 4$, and $l_4 = l_5 = \cdots = l_{16} = 5$, where $l_i$ denotes the length of the $i$th codeword. The average length of Huffman code is

$$L_{avg} = \sum_{i=1}^{16} p(s_i)l_i = 3.464 \quad bits\ per\ symbol$$

We see that the average length of Huffman code is quite close to the lower entropy bound. It is noted, however, that the required codebook memory, $M$ (defined as the sum of the codeword lengths), is quite large:

$$M = \sum_{i=1}^{16} l_i = 73 \ \ bits$$

This number is obviously larger than the average codeword length multiplied by the number of codewords. This should not come as a surprise since the average here is in the statistical sense instead of in the arithmetic sense. When the total number of improbable symbols increases, the required codebook memory space will increase dramatically, resulting in a great demand on memory space.

### 5.3.2 ALGORITHM

Consider a source alphabet $S$ that consists of $2^v$ binary sequences, each of length $v$. In other words, each source symbol is a $v$-bit codeword in the natural binary code. The occurrence probabilities are highly skewed and there is a large number of improbable symbols in $S$. The modified Huffman coding algorithm is based on the following idea: lumping all the improbable source symbols into a category named ELSE (Weaver, 1978). The algorithm is described below.

1. Categorize the source alphabet $S$ into two disjoint groups, $S_1$ and $S_2$, such that

$$S_1 = \left\{ s_i \middle| p(s_i) > \frac{1}{2^v} \right\} \tag{5.17}$$

and

$$S_2 = \left\{ s_i \middle| p(s_i) \le \frac{1}{2^v} \right\} \tag{5.18}$$

2. Establish a source symbol ELSE with its occurrence probability equal to $p(S_2)$.
3. Apply the Huffman coding algorithm to the source alphabet $S_3$ with $S_3 = S_1 \cup ELSE$.
4. Convert the codebook of $S_3$ to that of $S$ as follows.
   - Keep the same codewords for those symbols in $S_1$.
   - Use the codeword assigned to ELSE as a prefix for those symbols in $S_2$.

### 5.3.3 CODEBOOK MEMORY REQUIREMENT

Codebook memory $M$ is the sum of the codeword lengths. The $M$ required by Huffman coding with respect to the original source alphabet $S$ is

$$M = \sum_{i \in S} l_i = \sum_{i \in S_1} l_i + \sum_{i \in S_2} l_i \tag{5.19}$$

where $l_i$ denotes the length of the $i$th codeword, as defined previously. In the case of the modified Huffman coding algorithm, the memory required $M_{mH}$ is

$$M_{mH} = \sum_{i \in S_3} l_i = \sum_{i \in S_1} l_i + l_{ELSE} \tag{5.20}$$

where $l_{ELSE}$ is the length of the codeword assigned to ELSE. The above equation reveals the big savings in memory requirement when the probability is skewed. The following example is used to illustrate the modified Huffman coding algorithm and the resulting dramatic memory savings.

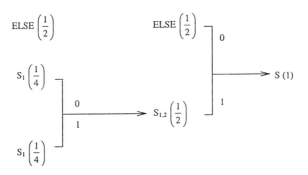

**FIGURE 5.2**  The modified Huffman coding procedure in Example 5.11.

**Example 5.11**

In this example, we apply the modified Huffman coding algorithm to the source alphabet presented in Example 5.10. We first lump the 14 symbols having the least occurrence probabilities together to form a new symbol ELSE. The probability of ELSE is the sum of the 14 probabilities. That is, $p(ELSE) = \frac{1}{28} \cdot 14 = \frac{1}{2}$.

Apply Huffman coding to the new source alphabet $S_3 = \{s_1, s_2, ELSE\}$, as shown in Figure 5.2. From Figure 5.2, it is seen that the codewords assigned to symbols $s_1$, $s_2$, and *ELSE*, respectively, are 10, 11, and 0. Hence, for every source symbol lumped into ELSE, its codeword is 0 followed by the original 4-bit binary sequence. Therefore, $M_{mH} = 2 + 2 + 1 = 5$ *bits*, i.e., the required codebook memory is only 5 bits. Compared with 73 bits required by Huffman coding (refer to Example 5.10), there is a savings of 68 bits in codebook memory space. Similar to the comment made in Example 5.10, the memory savings will be even larger if the probability distribution is skewed more severely and the number of improbable symbols is larger. The average length of the modified Huffman algorithm is $L_{avg,mH} = \frac{1}{4} \cdot 2 \cdot 2 + \frac{1}{28} \cdot 5 \cdot 14 = 3.5$ *bits per symbol*. This demonstrates that modified Huffman coding retains almost the same coding efficiency as that achieved by Huffman coding.

### 5.3.4  BOUNDS ON AVERAGE CODEWORD LENGTH

It has been shown that the average length of the modified Huffman codes satisfies the following condition:

$$H(S) \leq L_{avg} < H(S) + 1 - p \log_2 p \qquad (5.21)$$

where $p = \Sigma_{s_i \in s_2} p(s_i)$. It is seen that, compared with the noiseless source coding theorem, the upper bound of the code average length is increased by a quantity of $-p \log_2 p$. In Example 5.11 it is seen that the average length of the modified Huffman code is close to that achieved by the Huffman code. Hence the modified Huffman code is almost optimum.

## 5.4  ARITHMETIC CODES

Arithmetic coding, which is quite different from Huffman coding, is gaining increasing popularity. In this section, we will first analyze the limitations of Huffman coding. Then the principle of arithmetic coding will be introduced. Finally some implementation issues are discussed briefly.

### 5.4.1 LIMITATIONS OF HUFFMAN CODING

As seen in Section 5.2, Huffman coding is a systematic procedure for encoding a source alphabet, with each source symbol having an occurrence probability. Under these circumstances, Huffman coding is optimum in the sense that it produces a minimum coding redundancy. It has been shown that the average codeword length achieved by Huffman coding satisfies the following inequality (Gallagher, 1978).

$$H(S) \leq L_{avg} < H(S) + p_{\max} + 0.086 \tag{5.22}$$

where $H(S)$ is the entropy of the source alphabet, and $p_{\max}$ denotes the maximum occurrence probability in the set of the source symbols. This inequality implies that the upper bound of the average codeword length of Huffman code is determined by the entropy and the maximum occurrence probability of the source symbols being encoded.

In the case where the probability distribution among source symbols is skewed (some probabilities are small, while some are quite large), the upper bound may be large, implying that the coding redundancy may not be small. Imagine the following extreme situation. There are only two source symbols. One has a very small probability, while the other has a very large probability (very close to 1). The entropy of the source alphabet in this case is close to 0 since the uncertainty is very small. Using Huffman coding, however, we need two bits: one for each. That is, the average codeword length is 1, which means that the redundancy is very close to 1. This agrees with Equation 5.22. This inefficiency is due to the fact that Huffman coding always encodes a source symbol with an integer number of bits.

The noiseless coding theorem (reviewed in Section 5.1) indicates that the average codeword length of a block code can approach the source alphabet entropy when the block size approaches infinity. As the block size approaches infinity, the storage required, the codebook size, and the coding delay will approach infinity, however, and the complexity of the coding will be out of control.

The fundamental idea behind Huffman coding and Shannon-Fano coding (devised a little earlier than Huffman coding [Bell et al., 1990]) is block coding. That is, some codeword having an integral number of bits is assigned to a source symbol. A message may be encoded by cascading the relevant codewords. It is the *block-based* approach that is responsible for the limitations of Huffman codes.

Another limitation is that when encoding a message that consists of a sequence of source symbols, the $n$th extension Huffman coding needs to enumerate all possible sequences of source symbols having the same length, as discussed in coding the $n$th extended source alphabet. This is not computationally efficient.

Quite different from Huffman coding, arithmetic coding is *stream-based*. It overcomes the drawbacks of Huffman coding. A string of source symbols is encoded as a string of code symbols. Hence, it is free of the integral-bits-per-source symbol restriction and is more efficient. Arithmetic coding may reach the theoretical bound to coding efficiency specified in the noiseless source coding theorem for any information source. Below, we introduce the principle of arithmetic coding, from which we can see the stream-based nature of arithmetic coding.

### 5.4.2 PRINCIPLE OF ARITHMETIC CODING

To understand the different natures of Huffman coding and arithmetic coding, let us look at Example 5.12, where we use the same source alphabet and the associated occurrence probabilities used in Example 5.9. In this example, however, a string of source symbols $s_1 s_2 s_3 s_4 s_5 s_6$ is encoded. Note that we consider the terms *string* and *stream* to be slightly different. By stream, we mean a message, or possibly several messages, which may correspond to quite a long sequence of source symbols. Moreover, stream gives a dynamic "flavor." Later on we will see that arithmetic coding

**TABLE 5.10**
**Source Alphabet and Cumulative Probabilities in Example 5.12**

| Source Symbol | Occurrence Probability | Associated Subintervals | CP |
|---|---|---|---|
| $s_1$ | 0.3 | [0, 0.3) | 0 |
| $s_2$ | 0.1 | [0.3, 0.4) | 0.3 |
| $s_3$ | 0.2 | [0.4, 0.6) | 0.4 |
| $s_4$ | 0.05 | [0.6, 0.65) | 0.6 |
| $s_5$ | 0.1 | [0.65, 0.75) | 0.65 |
| $s_6$ | 0.25 | [0.75, 1.0) | 0.75 |

is implemented in an incremental manner. Hence stream is a suitable term to use for arithmetic coding. In this example, however, only six source symbols are involved. Hence we consider the term *string* to be suitable, aiming at distinguishing it from the term *block*.

**Example 5.12**
The set of six source symbols and their occurrence probabilities are listed in Table 5.10. In this example, the string to be encoded using arithmetic coding is $s_1 s_2 s_3 s_4 s_5 s_6$. In the following four subsections we will use this example to illustrate the principle of arithmetic coding and decoding.

### 5.4.2.1 Dividing Interval [0,1) into Subintervals

As pointed out by Elias, it is not necessary to sort out source symbols according to their occurrence probabilities. Therefore in Figure 5.3(a) the six symbols are arranged in their natural order, from symbols $s_1$, $s_2$, $\cdots$, up to $s_6$. The real interval between 0 and 1 is divided into six subintervals, each having a length of $p(s_i)$, $i = 1,2,\cdots,6$. Specifically, the interval denoted by [0,1) — where 0 is included in (the left end is closed) and 1 is excluded from (the right end is open) the interval — is divided into six subintervals. The first subinterval [0, 0.3) corresponds to $s_1$ and has a length of $P(s_1)$, i.e., 0.3. Similarly, the subinterval [0, 0.3) is said to be closed on the left and open on the right. The remaining five subintervals are similarly constructed. All six subintervals thus formed are disjoint and their union is equal to the interval [0, 1). This is because the sum of all the probabilities is equal to 1.

We list the sum of the preceding probabilities, known as *cumulative probability* (Langdon, 1984), in the right-most column of Table 5.10 as well. Note that the concept of cumulative probability (CP) is slightly different from that of cumulative distribution function (CDF) in probability theory. Recall that in the case of discrete random variables the CDF is defined as follows.

$$CDF(s_i) = \sum_{j=1}^{i} p(s_j) \tag{5.23}$$

The CP is defined as

$$CP(s_i) = \sum_{j=1}^{i-1} p(s_j) \tag{5.24}$$

where $CP(s_1) = 0$ is defined. Now we see each subinterval has its lower end point located at $CP(s_i)$. The width of each subinterval is equal to the probability of the corresponding source symbol. A subinterval can be completely defined by its lower end point and its width. Alternatively, it is

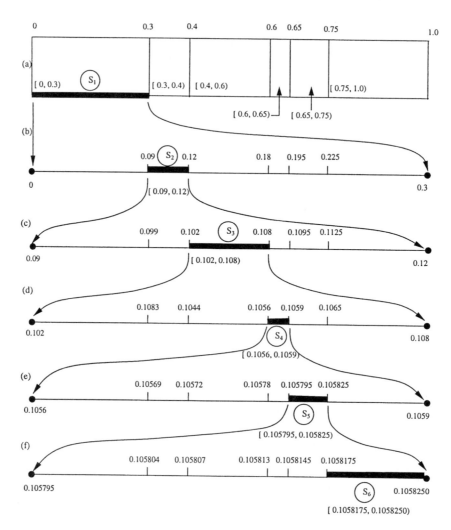

**FIGURE 5.3** Arithmetic coding working on the same source alphabet as that in Example 5.9. The encoded symbol string is $S_1 S_2 S_3 S_4 S_5 S_6$.

determined by its two end points: the lower and upper end points (sometimes also called the left and right end points).

Now we consider encoding the string of source symbols $s_1 s_2 s_3 s_4 s_5 s_6$ with the arithmetic coding method.

### 5.4.2.2 Encoding

**Encoding the First Source Symbol**
Refer to Figure 5.3(a). Since the first symbol is $s_1$, we pick up its subinterval [0, 0.3). Picking up the subinterval [0, 0.3) means that any real number in the subinterval, i.e., any real number equal to or greater than 0 and smaller than 0.3, can be a pointer to the subinterval, thus representing the source symbol $s_1$. This can be justified by considering that all the six subintervals are disjoint.

**Encoding the Second Source Symbol**
Refer to Figure 5.3(b). We use the same procedure as used in part (a) to divide the interval [0, 0.3) into six subintervals. Since the second symbol to be encoded is $s_2$, we pick up its subinterval [0.09, 0.12).

Notice that the subintervals are recursively generated from part (a) to part (b). It is known that an interval may be completely specified by its lower end point and width. Hence, the subinterval recursion in the arithmetic coding procedure is equivalent to the following two recursions: end point recursion and width recursion.

From interval [0, 0.3) derived in part (a) to interval [0.09, 0.12) obtained in part (b), we can conclude the following lower end point recursion:

$$L_{new} = L_{current} + W_{current} \cdot CP_{new} \tag{5.25}$$

where $L_{new}$, $L_{current}$ represent, respectively, the lower end points of the new and current recursions, and the $W_{current}$ and the $CP_{new}$ denote, respectively, the width of the interval in the current recursion and the cumulative probability in the new recursion. The width recursion is

$$W_{new} = W_{current} \cdot p(s_i) \tag{5.26}$$

where $W_{new}$, and $p(s_i)$ are, respectively, the width of the new subinterval and the probability of the source symbol $s_i$ that is being encoded. These two recursions, also called double recursion (Langdon, 1984), play a central role in arithmetic coding.

**Encoding the Third Source Symbol**
Refer to Figure 5.3(c). When the third source symbol is encoded, the subinterval generated above in part (b) is similarly divided into six subintervals. Since the third symbol to encode is $s_3$, its subinterval [0.102, 0.108) is picked up.

**Encoding the Fourth, Fifth, and Sixth Source Symbols**
Refer to Figure 5.3(d,e,f). The subinterval division is carried out according to Equations 5.25 and 5.26. The symbols $s_4$, $s_5$, and $s_6$ are encoded. The final subinterval generated is [0.1058175, 0.1058250).

That is, the resulting subinterval [0.1058175, 0.1058250) can represent the source symbol string $s_1 s_2 s_3 s_4 s_5 s_6$. Note that in this example decimal digits instead of binary digits are used. In binary arithmetic coding, the binary digits 0 and 1 are used.

### 5.4.2.3   Decoding

As seen in this example, for the encoder of arithmetic coding, the input is a source symbol string and the output is a subinterval. Let us call this the final subinterval or the resultant subinterval. Theoretically, any real numbers in the interval can be the code string for the input symbol string since all subintervals are disjoint. Often, however, the lower end of the final subinterval is used as the code string. Now let us examine how the decoding process is carried out with the lower end of the final subinterval.

Decoding sort of reverses what encoding has done. The decoder knows the encoding procedure and therefore has the information contained in Figure 5.3(a). It compares the lower end point of the final subinterval 0.1058175 with all the end points in (a). It is determined that $0 < 0.1058175 < 0.3$. That is, the lower end falls into the subinterval associated with the symbol $s_1$. Therefore, the symbol $s_1$ is first decoded.

Once the first symbol is decoded, the decoder may know the partition of subintervals shown in Figure 5.3(b). It is then determined that $0.09 < 0.1058175 < 0.12$. That is, the lower end is contained in the subinterval corresponding to the symbol $s_2$. As a result, $s_2$ is the second decoded symbol.

The procedure repeats itself until all six symbols are decoded. That is, based on Figure 5.3(c), it is found that $0.102 < 0.1058175 < 0.108$. The symbol $s_3$ is decoded. Then, the symbols $s_4$, $s_5$, $s_6$ are subsequently decoded because the following inequalities are determined:

$$0.1056 < 0.1058175 < 0.1059$$

$$0.105795 < 0.1058175 < 0.1058250$$

$$0.1058145 < 0.1058175 < 0.1058250$$

Note that a terminal symbol is necessary to inform the decoder to stop decoding.

The above procedure gives us an idea of how decoding works. The decoding process, however, does not need to construct parts (b), (c), (d), (e), and (f) of Figure 5.3. Instead, the decoder only needs the information contained in Figure 5.3(a). Decoding can be split into the following three steps: *comparison*, *readjustment* (subtraction), and *scaling* (Langdon, 1984).

As described above, through comparison we decode the first symbol $s_1$. From the way Figure 5.3(b) is constructed, we know the decoding of $s_2$ can be accomplished as follows. We subtract the lower end of the subinterval associated with $s_1$ in part (a), that is, 0 in this example, from the lower end of the final subinterval 0.1058175, resulting in 0.1058175. Then we divide this number by the width of the subinterval associated with $s_1$, i.e., the probability of $s_1$, 0.3, resulting in 0.352725. Looking at part (a) of Figure 5.3, it is found that $0.3 < 0.352725 < 0.4$. That is, the number is within the subinterval corresponding to $s_2$. Therefore the second decoded symbol is $s_2$. Note that these three decoding steps exactly "undo" what encoding did.

To decode the third symbol, we subtract the lower end of the subinterval with $s_2$, 0.3 from 0.352725, obtaining 0.052725. This number is divided by the probability of $s_2$, 0.1, resulting in 0.52725. The comparison of 0.52725 with end points in part (a) reveals that the third decoded symbol is $s_3$.

In decoding the fourth symbol, we first subtract the lower end of the $s_3$'s subinterval in part (a), 0.4 from 0.52725, getting 0.12725. Dividing 0.12725 by the probability of $s_3$, 0.2, results in 0.63625. Referring to part (a), we decode the fourth symbol as $s_4$ by comparison.

Subtraction of the lower end of the subinterval of $s_4$ in part (a), 0.6, from 0.63625 leads to 0.03625. Division of 0.03625 by the probability of $s_4$, 0.05, produces 0.725. The comparison between 0.725 and the end points in part (a) decodes the fifth symbol as $s_5$.

Subtracting 0.725 by the lower end of the subinterval associated with $s_5$ in part (a), 0.65, gives 0.075. Dividing 0.075 by the probability of $s_5$, 0.1, generates 0.75. The comparison indicates that the sixth decoded symbol is $s_6$.

In summary, considering the way in which parts (b), (c), (d), (e), and (f) of Figure 5.3 are constructed, we see that the three steps discussed in the decoding process: comparison, readjustment, and scaling, exactly "undo" what the encoding procedure has done.

### 5.4.2.4 Observations

Both encoding and decoding involve only arithmetic operations (addition and multiplication in encoding, subtraction and division in decoding). This explains the name *arithmetic coding*.

We see that an input source symbol string $s_1 s_2 s_3 s_4 s_5 s_6$, via encoding, corresponds to a subinterval [0.1058175, 0.1058250). Any number in this interval can be used to denote the string of the source symbols.

We also observe that arithmetic coding can be carried out in an *incremental* manner. That is, source symbols are fed into the encoder one by one and the final subinterval is refined continually, i.e., the code string is generated continually. Furthermore, it is done in a manner called *first in first out* (FIFO). That is, the source symbol encoded first is decoded first. This manner is superior to that of *last in first out* (LIFO). This is because FIFO is suitable for adaptation to the statistics of the symbol string.

It is obvious that the width of the final subinterval becomes smaller and smaller when the length of the source symbol string becomes larger and larger. This causes what is known as the precision problem. It is this problem that prohibited arithmetic coding from practical usage for quite a long period of time. Only after this problem was solved in the late 1970s, did arithmetic coding become an increasingly important coding technique.

It is necessary to have a termination symbol at the end of an input source symbol string. In this way, an arithmetic coding system is able to know when to terminate decoding.

Compared with Huffman coding, arithmetic coding is quite different. Basically, Huffman coding converts each source symbol into a fixed codeword with an integral number of bits, while arithmetic coding converts a source symbol string to a code symbol string. To encode the same source symbol string, Huffman coding can be implemented in two different ways. One way is shown in Example 5.9. We construct a fixed codeword for each source symbol. Since Huffman coding is instantaneous, we can cascade the corresponding codewords to form the output, a 17-bit code string 00.101.11.1001.1000.01, where, for easy reading, the five periods are used to indicate different codewords. As we see that for the same source symbol string, the final subinterval obtained by using arithmetic coding is [0.1058175, 0.1058250). It is noted that a decimal in binary number system, 0.000110111111111, which is of 15 bits, is equal to the decimal in decimal number system, 0.1058211962, which falls into the final subinterval representing the string $s_1 s_2 s_3 s_4 s_5 s_6$. This indicates that, for this example, arithmetic coding is more efficient than Huffamn coding.

Another way is to form a sixth extension of the source alphabet as discussed in Section 5.1.4: treat each group of six source symbols as a new source symbol; calculate its occurrence probability by multiplying the related six probabilities; then apply the Huffman coding algorithm to the sixth extension of the discrete memoryless source. This is called the sixth extension of Huffman block code (refer to Section 5.1.2.2). In other words, in order to encode the source string $s_1 s_2 s_3 s_4 s_5 s_6$, Huffman coding encodes all of the $6^6 = 46,656$ codewords in the sixth extension of the source alphabet. This implies a high complexity in implementation and a large codebook. It is therefore not efficient.

Note that we use the decimal fraction in this section. In binary arithmetic coding, we use the binary fraction. In (Langdon, 1984) both binary source and code alphabets are used in binary arithmetic coding.

Similar to the case of Huffman coding, arithmetic coding is also applicable to r-ary encoding with $r > 2$.

### 5.4.3   IMPLEMENTATION ISSUES

As mentioned, the final subinterval resulting from arithmetic encoding of a source symbol string becomes smaller and smaller as the length of the source symbol string increases. That is, the lower and upper bounds of the final subinterval become closer and closer. This causes a growing precision problem. It is this problem that prohibited arithmetic coding from practical usage for a long period of time. This problem has been resolved and the finite precision arithmetic is now used in arithmetic coding. This advance is due to the incremental implementation of arithmetic coding.

#### 5.4.3.1   Incremental Implementation

Recall Example 5.12. As source symbols come in one by one, the lower and upper ends of the final subinterval get closer and closer. In Figure 5.3, these lower and upper ends in Example 5.12 are listed. We observe that after the third symbol, $s_3$, is encoded, the resultant subinterval is [0.102, 0.108). That is, the two most significant decimal digits are the same and they remain the same in the encoding process. Hence, we can transmit these two digits without affecting the final code string. After the fourth symbol $s_4$ is encoded, the resultant subinterval is [0.1056, 0.1059). That is, one more digit, 5, can be transmitted. Or we say the cumulative output is now .105. After the sixth symbol is encoded, the final subinterval is [0.1058175, 0.1058250). The cumulative output is 0.1058. Refer to Table 5.11. This important observation reveals that we are able to incrementally transmit output (the code symbols) and receive input (the source symbols that need to be encoded).

**TABLE 5.11**
**Final Subintervals and Cumulative Output in Example 5.12**

| Source Symbol | Final Subinterval Lower End | Upper End | Cumulative Output |
|---|---|---|---|
| $S_1$ | 0 | 0.3 | — |
| $S_2$ | 0.09 | 0.12 | — |
| $S_3$ | 0.102 | 0.108 | 0.10 |
| $S_4$ | 0.1056 | 0.1059 | 0.105 |
| $S_5$ | 0.105795 | 0.105825 | 0.105 |
| $S_6$ | 0.1058175 | 0.1058250 | 0.1058 |

### 5.4.3.2  Finite Precision

With the incremental manner of transmission of encoded digits and reception of input source symbols, it is possible to use finite precision to represent the lower and upper bounds of the resultant subinterval, which gets closer and closer as the length of the source symbol string becomes long.

Instead of floating-point math, integer math is used. The potential problems known as underflow and overflow, however, need to be carefully monitored and controlled (Bell et al., 1990).

### 5.4.3.3  Other Issues

There are some other problems that need to be handled in implementation of binary arithmetic coding. Two of them are listed below (Langdon and Rissanen, 1981).

**Eliminating Multiplication**
The multiplication in the recursive division of subintervals is expensive in hardware as well as software. It can be avoided in binary arithmetic coding so as to simplify the implementation of binary arithmetic coding. The idea is to approximate the lower end of the interval by the closest binary fraction $2^{-Q}$, where Q is an integer. Consequently, the multiplication by $2^{-Q}$ becomes a right shift by Q bits. A simpler approximation to eliminate multiplication is used in the Skew Coder (Langdon and Rissanen, 1982) and the Q-Coder (Pennebaker et al., 1988).

**Carry-Over Problem**
Carry-over takes place in the addition required in the recursion updating the lower end of the resultant subintervals. A carry may *propagate* over q bits. If the q is larger than the number of bits in the fixed-length register utilized in finite precision arithmetic, the carry-over problem occurs. To block the carry-over problem, a technique known as "bit stuffing" is used, in which an additional buffer register is utilized.

For a detailed discussion on the various issues involved, readers are referred to (Langdon et al., 1981, 1982, 1984; Pennebaker et al., 1988, 1992). Some computer programs of arithmetic coding in C language can be found in (Bell et al., 1990; Nelson and Gailley, 1996).

### 5.4.4  History

The idea of encoding by using cumulative probability in some ordering, and decoding by comparison of magnitude of binary fraction, was introduced in Shannon's celebrated paper (Shannon, 1948). The recursive implementation of arithmetic coding was devised by Elias. This unpublished result was first introduced by Abramson as a note in his book on information theory and coding

(Abramson, 1963). The result was further developed by Jelinek in his book on information theory (Jelinek, 1968). The growing precision problem prevented arithmetic coding from attaining practical usage, however. The proposal of using finite precision arithmetic was made independently by Pasco (Pasco, 1976) and Rissanen (Rissanen, 1976). Practical arithmetic coding was developed by several independent groups (Rissanen and Langdon, 1979; Rubin, 1979; Guazzo, 1980). A well-known tutorial paper on arithmetic coding appeared in (Langdon, 1984). The tremendous efforts made in IBM led to a new form of adaptive binary arithmetic coding known as the Q-coder (Pennebaker et al., 1988). Based on the Q-coder, the activities of the international still image coding standards JPEG and JBIG combined the best features of the various existing arithmetic coders and developed the binary arithmetic coding procedure known as the QM-coder (Pennebaker and Mitchell, 1992).

### 5.4.5  APPLICATIONS

Arithmetic coding is becoming popular. Note that in text and bilevel image applications there are only two source symbols (black and white), and the occurrence probability is skewed. Therefore binary arithmetic coding achieves high coding efficiency. It has been successfully applied to bilevel image coding (Langdon and Rissanen, 1981) and adopted by the international standards for bilevel image compression, JBIG. It has also been adopted by the international still image coding standard, JPEG. More in this regard is covered in the next chapter when we introduce JBIG.

## 5.5  SUMMARY

So far in this chapter, not much has been explicitly discussed regarding the term variable-length codes. It is known that if source symbols in a source alphabet are equally probable, i.e., their occurrence probabilities are the same, then fixed-length codes such as the natural binary code are a reasonable choice. When the occurrence probabilities, however, are unequal, variable-length codes should be used in order to achieve high coding efficiency. This is one of the restrictions on the minimum redundancy codes imposed by Huffman. That is, the length of the codeword assigned to a probable source symbol should not be larger than that associated with a less probable source symbol. If the occurrence probabilities happen to be the integral powers of 1/2, then choosing the codeword length equal to $-\log_2 p(s_i)$ for a source symbol $s_i$ having the occurrence probability $p(s_i)$ results in minimum redundancy coding. In fact, the average length of the code thus generated is equal to the source entropy.

Huffman devised a systematic procedure to encode a source alphabet consisting of finitely many source symbols, each having an occurrence probability. It is based on some restrictions imposed on the optimum, instantaneous codes. By assigning codewords with variable lengths according to variable probabilities of source symbols, Huffman coding results in minimum redundancy codes, or optimum codes for short. These have found wide applications in image and video coding and have been adopted in the international still image coding standard JPEG and video coding standards H.261, H.263, and MPEG 1 and 2.

When some source symbols have small probabilities and their number is large, the size of the codebook of Huffman codes will require a large memory space. The modified Huffman coding technique employs a special symbol to lump all the symbols with small probabilities together. As a result, it can reduce the codebook memory space drastically while retaining almost the same coding efficiency as that achieved by the conventional Huffman coding technique.

On the one hand, Huffman coding is optimum as a block code for a fixed-source alphabet. On the other hand, compared with the source entropy (the lower bound of the average codeword length) it is not efficient when the probabilities of a source alphabet are skewed with the maximum probability being large. This is caused by the restriction that Huffman coding can only assign an integral number of bits to each codeword.

Another limitation of Huffman coding is that it has to enumerate and encode all the possible groups of *n* source symbols in the *n*th extension Huffman code, even though there may be only one such group that needs to be encoded.

Arithmetic coding can overcome the limitations of Huffman coding because it is stream-oriented rather than block-oriented. It translates a stream of source symbols into a stream of code symbols. It can work in an incremental manner. That is, the source symbols are fed into the coding system one by one and the code symbols are output continually. In this stream-oriented way, arithmetic coding is more efficient. It can approach the lower coding bounds set by the noiseless source coding theorem for various sources.

The recursive subinterval division (equivalently, the double recursion: the lower end recursion and width recursion) is the heart of arithmetic coding. Several measures have been taken in the implementation of arithmetic coding. They include the incremental manner, finite precision, and the elimination of multiplication. Due to its merits, binary arithmetic coding has been adopted by the international bilevel image coding standard, JBIG, and still image coding standard, JPEG. It is becoming an increasingly important coding technique.

## 5.6   EXERCISES

**5-1.** What does the noiseless source coding theorem state (using your own words)? Under what condition does the average code length approach the source entropy? Comment on the method suggested by the noiseless source coding theorem.

**5-2.** What characterizes a block code? Consider another definition of block code in (Blahut, 1986): a block code breaks the input data stream into blocks of fixed length n and encodes each block into a codeword of fixed length m. Are these two definitions (the one above and the one in Section 5.1, which comes from [Abramson, 1963]) essentially the same? Explain.

**5-3.** Is a uniquely decodable code necessarily a prefix condition code?

**5-4.** For text encoding, there are only two source symbols for black and white. It is said that Huffman coding is not efficient in this application. But it is known as the optimum code. Is there a contradiction? Explain.

**5-5.** A set of source symbols and their occurrence probabilities is listed in Table 5.12. Apply the Huffman coding algorithm to encode the alphabet.

**5-6.** Find the Huffman code for the source alphabet shown in Example 5.10.

**5-7.** Consider a source alphabet $S = \{s_i, i = 1,2,\cdots,32\}$ with $p(s_1) = 1/4$, $p(s_i) = 3/124$, $i = 2,3,\cdots,32$. Determine the source entropy and the average length of Huffman code if applied to the source alphabet. Then apply the modified Huffman coding algorithm. Calculate the average length of the modified Huffman code. Compare the codebook memory required by Huffman code and the modified Huffman code.

**5-8.** A source alphabet consists of the following four source symbols: $s_1$, $s_2$, $s_3$, and $s_4$, with their occurrence probabilities equal to 0.25, 0.375, 0.125, and 0.25, respectively. Applying arithmetic coding as shown in Example 5.12 to the source symbol string $s_2s_1s_3s_4$, determine the lower end of the final subinterval.

**5-9.** For the above problem, show step by step how we can decode the original source string from the lower end of the final subinterval.

**5-10.** In Problem 5.8, find the codeword of the symbol string $s_2s_1s_3s_4$ by using the fourth extension of the Huffman code. Compare the two methods.

**5-11.** Discuss how modern arithmetic coding overcame the growing precision problem.

**TABLE 5.12**
**Source Alphabet in Problem 5.5**

| Source Symbol | Occurrence Probability | Codeword Assigned |
|---|---|---|
| $S_1$ | 0.20 | |
| $S_2$ | 0.18 | |
| $S_3$ | 0.10 | |
| $S_4$ | 0.10 | |
| $S_5$ | 0.10 | |
| $S_6$ | 0.06 | |
| $S_7$ | 0.06 | |
| $S_8$ | 0.04 | |
| $S_9$ | 0.04 | |
| $S_{10}$ | 0.04 | |
| $S_{11}$ | 0.04 | |
| $S_{12}$ | 0.04 | |

# REFERENCES

Abramson, N. *Information Theory and Coding*, New York: McGraw-Hill, 1963.

Bell, T. C., J. G. Cleary, and I. H. Witten, *Text Compression*, Englewood, NJ: Prentice-Hall, 1990.

Blahut, R. E. *Principles and Practice of Information Theory*, Reading, MA: Addison-Wesley, 1986.

Fano, R. M. The transmission of information, Tech. Rep. 65, Research Laboratory of Electronics, MIT, Cambridge, MA, 1949.

Gallagher, R. G. Variations on a theme by Huffman, *IEEE Trans. Inf. Theory,* IT-24(6), 668-674, 1978.

Guazzo, M. A general minimum-redundacy source-coding algorithm, *IEEE Trans. Inf. Theory,* IT-26(1), 15-25, 1980.

Hankamer, M. A modified Huffman procedure with reduced memory requirement, *IEEE Trans. Commun.,* COM-27(6), 930-932, 1979.

Huffman, D. A. A method for the construction of minimum-redundancy codes, *Proc. IRE,* 40, 1098-1101, 1952.

Jelinek, F. *Probabilistic Information Theory*, New York: McGraw-Hill, 1968.

Langdon, G. G., Jr. and J. Rissanen, Compression of black-white images with arithmetic coding, *IEEE Trans. Commun.,* COM-29(6), 858-867, 1981.

Langdon, G. G., Jr. and J. Rissanen, A simple general binary source code, *IEEE Trans. Inf. Theory,* IT-28, 800, 1982.

Langdon, G. G., Jr., An introduction to arithmetic coding, *IBM J. Res. Dev.,* 28(2), 135-149, 1984.

Nelson, M. and J. Gailly, *The Data Compression Book*, 2nd ed., New York: M&T Books, 1996.

Pasco, R. Source Coding Algorithms for Fast Data Compression, Ph.D. dissertation, Stanford University, Stanford, CA, 1976.

Pennebaker, W. B., J. L. Mitchell, G. G. Langdon, Jr., and R. B. Arps, An overview of the basic principles of the Q-coder adaptive binary arithmetic Coder, *IBM J. Res. Dev.,* 32(6), 717-726, 1988.

Pennebaker, W. B. and J. L. Mitchell, *JPEG: Still Image Data Compression Standard*, New York: Van Nostrand Reinhold, 1992.

Rissanen, J. J. Generalized Kraft inequality and arithmetic coding, *IBM J. Res. Dev.,* 20, 198-203, 1976.

Rissanen, J. J. and G. G. Landon, Arithmetic coding, *IBM J. Res. Dev.,* 23(2), 149-162, 1979.

Rubin, F. Arithmetic stream coding using fixed precision registers, *IEEE Trans. Inf. Theory,* IT-25(6), 672-675, 1979.

Sayood, K. *Introduction to Data Compression*, San Francisco, CA: Morgan Kaufmann Publishers, 1996.

Shannon, C. E. A mathematical theory of communication, *Bell Syst. Tech. J.,* 27, 379-423, 1948; 623-656, 1948.

Weaver, C. S., Digital ECG data compression, in *Digital Encoding of Electrocardiograms,* H. K. Wolf, Ed., Springer-Verlag, Berlin/New York, 1979.

# 6 Run-Length and Dictionary Coding: Information Theory Results (III)

As mentioned at the beginning of Chapter 5, we are studying some codeword assignment (encoding) techniques in Chapters 5 and 6. In this chapter, we focus on run-length and dictionary-based coding techniques. We first introduce Markov models as a type of dependent source model in contrast to the memoryless source model discussed in Chapter 5. Based on the Markov model, run-length coding is suitable for facsimile encoding. Its principle and application to facsimile encoding are discussed, followed by an introduction to dictionary-based coding, which is quite different from Huffman and arithmetic coding techniques covered in Chapter 5. Two types of adaptive dictionary coding techniques, the LZ77 and LZ78 algorithms, are presented. Finally, a brief summary of and a performance comparison between international standard algorithms for lossless still image coding are presented.

Since the Markov source model, run-length, and dictionary-based coding are the core of this chapter, we consider this chapter as a third part of the information theory results presented in the book. It is noted, however, that the emphasis is placed on their applications to image and video compression.

## 6.1 MARKOV SOURCE MODEL

In the previous chapter we discussed the discrete memoryless source model, in which source symbols are assumed to be independent of each other. In other words, the source has zero memory, i.e., the previous status does not affect the present one at all. In reality, however, many sources are dependent in nature. Namely, the source has memory in the sense that the previous status has an influence on the present status. For instance, as mentioned in Chapter 1, there is an interpixel correlation in digital images. That is, pixels in a digital image are not independent of each other. As will be seen in this chapter, there is some dependence between characters in text. For instance, the letter $u$ often follows the letter $q$ in English. Therefore it is necessary to introduce models that can reflect this type of dependence. A Markov source model is often used in this regard.

### 6.1.1 DISCRETE MARKOV SOURCE

Here, as in the previous chapter, we denote a source alphabet by $S = \{s_1, s_2, \cdots, s_m\}$ and the occurrence probability by $p$. An $l$th order Markov source is characterized by the following equation of conditional probabilities.

$$p\left(s_j | s_{i1}, s_{i2}, \cdots, s_{il}, \cdots\right) = p\left(s_j | s_{i1}, s_{i2}, \cdots, s_{il}\right), \tag{6.1}$$

where $j, i1, i2, \cdots, il, \cdots \in \{1, 2, \cdots, m\}$, i.e., the symbols $s_j, s_{i1}, s_{i2}, \cdots, s_{il}, \cdots$ are chosen from the source alphabet $S$. This equation states that the source symbols are not independent of each other. The occurrence probability of a source symbol is determined by some of its previous symbols. Specifically, the probability of $s_j$ given its history being $s_{i1}, s_{i2}, \cdots, s_{il}, \cdots$ (also called the transition probability), is determined completely by the immediately previous $l$ symbols $s_{i1}, \cdots, s_{il}$. That is,

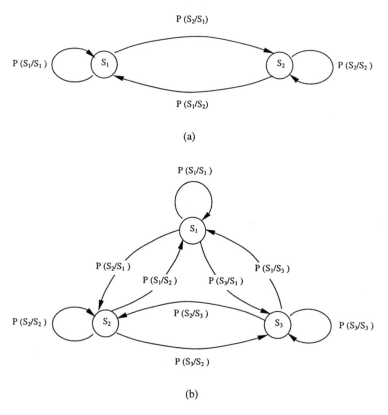

**FIGURE 6.1** State diagrams of the first-order Markov sources with their source alphabets having (a) two symbols and (b) three symbols.

the knowledge of the entire sequence of previous symbols is equivalent to that of the $l$ symbols immediately preceding the current symbol $s_j$.

An $l$th order Markov source can be described by what is called a *state diagram*. A state is a sequence of $(s_{i1}, s_{i2}, \cdots, s_{il})$ with $i1, i2, \cdots, il \in \{1,2,\cdots,m\}$. That is, any group of $l$ symbols from the $m$ symbols in the source alphabet $S$ forms a state. When $l = 1$, it is called a first-order Markov source. The state diagrams of the first-order Markov sources, with their source alphabets having two and three symbols, are shown in Figure 6.1(a) and (b), respectively. Obviously, an $l$th order Markov source with $m$ symbols in the source alphabet has a total of $m^l$ different states. Therefore, we conclude that a state diagram consists of all the $m^l$ states. In the diagram, all the transition probabilities together with appropriate arrows are used to indicate the state transitions.

The source entropy at a state $(s_{i1}, s_{i2}, \cdots, s_{il})$ is defined as

$$H\left(S\middle|s_{i1},s_{i2},\cdots,s_{il}\right) = -\sum_{j=1}^{m} p\left(s_j\middle|s_{i1},s_{i2},\cdots,s_{il}\right)\log_2 p\left(s_j\middle|s_{i1},s_{i2},\cdots,s_{il}\right) \tag{6.2}$$

The source entropy is defined as the statistical average of the entropy at all the states. That is,

$$H(S) = \sum_{(s_{i1},s_{i2},\cdots,s_{il})\in S^l} p\left(s_{i1},s_{i2},\cdots,s_{il}\right) H\left(S\middle|s_{i1},s_{i2},\cdots,s_{il}\right), \tag{6.3}$$

where, as defined in the previous chapter, $S^l$ denotes the $l$th extension of the source alphabet $S$. That is, the summation is carried out with respect to all $l$-tuples taking over the $S^l$. Extensions of a Markov source are defined below.

## 6.1.2　Extensions of a Discrete Markov Source

An extension of a Markov source can be defined in a similar way to that of an extension of a memoryless source in the previous chapter. The definition of extensions of a Markov source and the relation between the entropy of the original Markov source and the entropy of the $n$th extension of the Markov source are presented below without derivation. For the derivation, readers are referred to (Abramson, 1963).

### 6.1.2.1　Definition

Consider an $l$th order Markov source $S = \{s_1, s_2, \cdots, s_m\}$ and a set of conditional probabilities $p(s_j | s_{i1}, s_{i2}, \cdots, s_{il})$, where $j, i1, i2, \cdots, il \in \{1,2,\cdots,m\}$. Similar to the memoryless source discussed in Chapter 5, if $n$ symbols are grouped into a block, then there is a total of $m^n$ blocks. Each block can be viewed as a new source symbol. Hence, these $m^n$ blocks form a new information source alphabet, called the $n$th extension of the source $S$ and denoted by $S^n$. The $n$th extension of the $l$th-order Markov source is a $k$th-order Markov source, where $k$ is the smallest integer greater than or equal to the ratio between $l$ and $n$. That is,

$$k = \left\lceil \frac{l}{n} \right\rceil, \tag{6.4}$$

where the notation $\lceil a \rceil$ represents the operation of taking the smallest integer greater than or equal to the quantity $a$.

### 6.1.2.2　Entropy

Denote, respectively, the entropy of the $l$th order Markov source $S$ by $H(S)$, and the entropy of the $n$th extension of the $l$th order Markov source, $S^n$, by $H(S^n)$. The following relation between the two entropies can be shown:

$$H(S^n) = nH(S) \tag{6.5}$$

## 6.1.3　Autoregressive (AR) Model

The Markov source discussed above represents a kind of dependence between source symbols in terms of the transition probability. Concretely, in determining the transition probability of a present source symbol given all the previous symbols, only the set of finitely many immediately preceding symbols matters. The autoregressive model is another kind of dependent source model that has been used often in image coding. It is defined below.

$$s_j = \sum_{k=1}^{l} a_k s_{ik} + x_j, \tag{6.6}$$

where $s_j$ represents the currently observed source symbol, while $s_{ik}$ with $k = 1,2,\cdots,l$ denote the $l$ preceding observed symbols, $a_k$'s are coefficients, and $x_j$ is the current input to the model. If $l = 1$, the model defined in Equation 6.6 is referred to as the first-order AR model. Clearly, in this case, the current source symbol is a linear function of its preceding symbol.

## 6.2 RUN-LENGTH CODING (RLC)

The term *run* is used to indicate the repetition of a symbol, while the term *run-length* is used to represent the number of repeated symbols, in other words, the number of consecutive symbols of the same value. Instead of encoding the consecutive symbols, it is obvious that encoding the run-length and the value that these consecutive symbols commonly share may be more efficient. According to an excellent early review on binary image compression by Arps (1979), RLC has been in use since the earliest days of information theory (Shannon and Weaver, 1949; Laemmel, 1951).

From the discussion of the JPEG in Chapter 4 (with more details in Chapter 7), it is seen that most of the DCT coefficients within a block of $8 \times 8$ are zero after certain manipulations. The DCT coefficients are zigzag scanned. The nonzero DCT coefficients and their addresses in the $8 \times 8$ block need to be encoded and transmitted to the receiver side. There, the nonzero DCT values are referred to as labels. The position information about the nonzero DCT coefficients is represented by the run-length of zeros between the nonzero DCT coefficients in the zigzag scan. The labels and the run-length of zeros are then Huffman coded.

Many documents such as letters, forms, and drawings can be transmitted using facsimile machines over the general switched telephone network (GSTN). In digital facsimile techniques, these documents are quantized into binary levels: black and white. The resolution of these binary tone images is usually very high. In each scan line, there are many consecutive white and black pixels, i.e., many alternate white runs and black runs. Therefore it is not surprising to see that RLC has proven to be efficient in binary document transmission. RLC has been adopted in the international standards for facsimile coding: the CCITT Recommendations T.4 and T.6.

RLC using only the horizontal correlation between pixels on the same scan line is referred to as 1-D RLC. It is noted that the first-order Markov source model with two symbols in the source alphabet depicted in Figure 6.1(a) can be used to characterize 1-D RLC. To achieve higher coding efficiency, 2-D RLC utilizes both horizontal and vertical correlation between pixels. Both the 1-D and 2-D RLC algorithms are introduced below.

### 6.2.1 1-D RUN-LENGTH CODING

In this technique, each scan line is encoded independently. Each scan line can be considered as a sequence of alternating, independent white runs and black runs. As an agreement between encoder and decoder, the first run in each scan line is assumed to be a white run. If the first actual pixel is black, then the run-length of the first white run is set to be zero. At the end of each scan line, there is a special codeword called end-of-line (EOL). The decoder knows the end of a scan line when it encounters an EOL codeword.

Denote run-length by $r$, which is integer-valued. All of the possible run-lengths construct a source alphabet $R$, which is a random variable. That is,

$$R = \{r : r \in 0, 1, 2, \cdots\} \tag{6.7}$$

Measurements on typical binary documents have shown that the maximum compression ratio, $\zeta_{max}$, which is defined below, is about 25% higher when the white and black runs are encoded separately (Hunter and Robinson, 1980). The average white run-length, $\bar{r}_W$, can be expressed as

$$\bar{r}_W = \sum_{r=0}^{m} r \cdot P_W(r) \tag{6.8}$$

where $m$ is the maximum value of the run-length, and $P_W(r)$ denotes the occurrence probability of a white run with length $r$. The entropy of the white runs, $H_W$, is

$$H_W = -\sum_{r=0}^{m} P_W(r) \log_2 P_W(r) \tag{6.9}$$

For the black runs, the average run-length $\bar{r}_B$ and the entropy $H_B$ can be defined similarly. The maximum theoretical compression factor $\zeta_{max}$ is

$$\zeta_{max} = \frac{\bar{r}_W + \bar{r}_B}{H_W + H_B} \tag{6.10}$$

Huffman coding is then applied to two source alphabets. According to CCITT Recommendation T.4, A4 size ($210 \times 297$ mm) documents should be accepted by facsimile machines. In each scan line, there are 1728 pixels. This means that the maximum run-length for both white and black runs is 1728, i.e., $m = 1728$. Two source alphabets of such a large size imply the requirement of two large codebooks, hence the requirement of large storage space. Therefore, some modification was made, resulting in the "modified" Huffman (MH) code.

In the modified Huffman code, if the run-length is larger than 63, then the run-length is represented as

$$r = M \times 64 + T \quad as \quad r > 63, \tag{6.11}$$

where $M$ takes integer values from 1, 2 to 27, and $M \times 64$ is referred to as the makeup run-length; $T$ takes integer values from 0, 1 to 63, and is called the terminating run-length. That is, if $r \le 63$, the run-length is represented by a terminating codeword only. Otherwise, if $r > 63$, the run-length is represented by a makeup codeword and a terminating codeword. A portion of the modified Huffman code table (Hunter and Robinson, 1980) is shown in Table 6.1. In this way, the requirement of large storage space is alleviated. The idea is similar to that behind modified Huffman coding, discussed in Chapter 5.

### 6.2.2 2-D RUN-LENGTH CODING

The 1-D run-length coding discussed above only utilizes correlation between pixels within a scan line. In order to utilize correlation between pixels in neighboring scan lines to achieve higher coding efficiency, 2-D run-length coding was developed. In Recommendation T.4, the modified relative element address designate (READ) code, also known as the modified READ code or simply the MR code, was adopted.

The modified READ code operates in a line-by-line manner. In Figure 6.2, two lines are shown. The top line is called the reference line, which has been coded, while the bottom line is referred to as the coding line, which is being coded. There are a group of five changing pixels, $a_0$, $a_1$, $a_2$, $b_1$, $b_2$, in the two lines. Their relative positions decide which of the three coding modes is used. The starting changing pixel $a_0$ (hence, five changing points) moves from left to right and from top to bottom as 2-D run-length coding proceeds. The five changing pixels and the three coding modes are defined below.

#### 6.2.2.1 Five Changing Pixels

By a changing pixel, we mean the first pixel encountered in white or black runs when we scan an image line-by-line, from left to right, and from top to bottom. The five changing pixels are defined below.

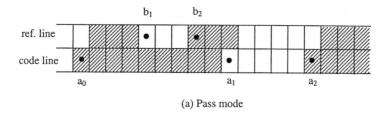

(a) Pass mode

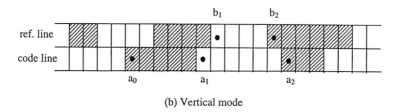

(b) Vertical mode

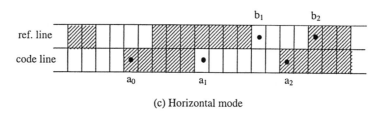

(c) Horizontal mode

**FIGURE 6.2**    2-D run-length coding.

$a_0$: The reference-changing pixel in the coding line. Its position is defined in the previous coding mode, whose meaning will be explained shortly. At the beginning of a coding line, $a_0$ is an imaginary white changing pixel located before the first actual pixel in the coding line.

$a_1$: The next changing pixel in the coding line. Because of the above-mentioned left-to-right and top-to-bottom scanning order, it is at the right-hand side of $a_0$. Since it is a changing pixel, it has an opposite "color" to that of $a_0$.

$a_2$: The next changing pixel after $a_1$ in the coding line. It is to the right of $a_1$ and has the same color as that of $a_0$.

$b_1$: The changing pixel in the reference line that is closest to $a_0$ from the right and has the same color as $a_1$.

$b_2$: The next changing pixel in the reference line after $b_1$.

### 6.2.2.2   Three Coding Modes

**Pass Coding Mode** — If the changing pixel $b_2$ is located to the left of the changing pixel $a_1$, it means that the run in the reference line starting from $b_1$ is not adjacent to the run in the coding line starting from $a_1$. Note that these two runs have the same color. This is called the pass coding mode. A special codeword, "0001", is sent out from the transmitter. The receiver then knows that the run starting from $a_0$ in the coding line does not end at the pixel below $b_2$. This pixel (below $b_2$ in the coding line) is identified as the reference-changing pixel $a_0$ of the new set of five changing pixels for the next coding mode.

**Vertical Coding Mode** — If the relative distance along the horizontal direction between the changing pixels $a_1$ and $b_1$ is not larger than three pixels, the coding is conducted in vertical coding

## TABLE 6.1
## Modified Huffman Code Table
## (Hunter and Robinson, 1980)

| Run-Length | White Runs | Black Runs |
|---|---|---|
| | **Terminating Codewords** | |
| 0 | 00110101 | 0000110111 |
| 1 | 000111 | 010 |
| 2 | 0111 | 11 |
| 3 | 1000 | 10 |
| 4 | 1011 | 011 |
| 5 | 1100 | 0011 |
| 6 | 1110 | 0010 |
| 7 | 1111 | 00011 |
| 8 | 10011 | 000101 |
| ⋮ | ⋮ | ⋮ |
| 60 | 01001011 | 000000101100 |
| 61 | 00110010 | 000001011010 |
| 62 | 00110011 | 000001100110 |
| 63 | 00110100 | 000001100111 |
| | **Makeup Codewords** | |
| 64 | 11011 | 0000001111 |
| 128 | 10010 | 000011001000 |
| 192 | 010111 | 000011001001 |
| 256 | 0110111 | 000001011011 |
| ⋮ | ⋮ | ⋮ |
| 1536 | 010011001 | 0000001011010 |
| 1600 | 010011010 | 0000001011011 |
| 1664 | 011000 | 0000001100100 |
| 1728 | 010011011 | 0000001100101 |
| EOL | 000000000001 | 000000000001 |

## TABLE 6.2
## 2-D Run-Length Coding Table

| Mode | Conditions | Output Codeword | Position of New $a_0$ |
|---|---|---|---|
| Pass coding mode | $b_2 a_1 < 0$ | 0001 | Under $b_2$ in coding line |
| Vertical coding mode | $a_1 b_1 = 0$ | 1 | $a_1$ |
| | $a_1 b_1 = 1$ | 011 | |
| | $a_1 b_1 = 2$ | 000011 | |
| | $a_1 b_1 = 3$ | 0000011 | |
| | $a_1 b_1 = -1$ | 010 | |
| | $a_1 b_1 = -2$ | 000010 | |
| | $a_1 b_1 = -3$ | 0000010 | |
| Horizontal coding mode | $|a_1 b_1| > 3$ | $001 + (a_0 a_1) + (a_1 a_2)$ | $a_2$ |

*Note:* $| x_i y_j |$: distance between $x_i$ and $y_j$, $x_i y_j > 0$: $x_i$ is right to $y_j$, $x_i y_j < 0$: $x_i$ is left to $y_j$.
$(x_i y_j)$: codeword of the run denoted by $x_i y_j$ taken from the modified Huffman code.

*Source:* From Hunter and Robinson (1980).

mode. That is, the position of $a_1$ is coded with reference to the position of $b_1$. Seven different codewords are assigned to seven different cases: the distance between $a_1$ and $b_1$ equals 0, $\pm 1$, $\pm 2$, $\pm 3$, where + means $a_1$ is to the right of $b_1$, while – means $a_1$ is to the left of $b_1$. The $a_1$ then becomes the reference changing pixel $a_0$ of the new set of five changing pixels for the next coding mode.

**Horizontal Coding Mode** — If the relative distance between the changing pixels $a_1$ and $b_1$ is larger than three pixels, the coding is conducted in horizontal coding mode. Here, 1-D run-length coding is applied. Specifically, the transmitter sends out a codeword consisting the following three parts: a flag "001"; a 1-D run-length codeword for the run from $a_0$ to $a_1$; a 1-D run-length codeword for the run from $a_1$ to $a_2$. The $a_2$ then becomes the reference changing pixel $a_0$ of the new set of five changing pixels for the next coding mode. Table 6.2 contains three coding modes and the corresponding output codewords. There, $(a_0 a_1)$ and $(a_1 a_2)$ represent 1-D run-length codewords of run-length $a_0 a_1$ and $a_1 a_2$, respectively.

### 6.2.3 EFFECT OF TRANSMISSION ERROR AND UNCOMPRESSED MODE

In this subsection, effect of transmission error in the 1-D and 2-D RLC cases and uncompressed mode are discussed.

#### 6.2.3.1 Error Effect in the 1-D RLC Case

As introduced above, the special codeword EOL is used to indicate the end of each scan line. With the EOL, 1-D run-length coding encodes each scan line independently. If a transmission error occurs in a scan line, there are two possibilities that the effect caused by the error is limited within the scan line. One possibility is that *resynchronization* is established after a few runs. One example is shown in Figure 6.3. There, the transmission error takes place in the second run from the left. Resynchronization is established in the fifth run in this example. Another possibility lies in the EOL, which forces resynchronization.

In summary, it is seen that the 1-D run-length coding will not propagate transmission error between scan lines. In other words, a transmission error will be restricted within a scan line. Although error detection and retransmission of data via an automatic repeat request (ARQ) system is supposed to be able to effectively handle the error susceptibility issue, the ARQ technique was not included in Recommendation T.4 due to the computational complexity and extra transmission time required.

Once the number of decoded pixels between two consecutive EOL codewords is not equal to 1728 (for an A4 size document), an error has been identified. Some *error concealment* techniques can be used to reconstruct the scan line (Hunter and Robinson, 1980). For instance, we can repeat

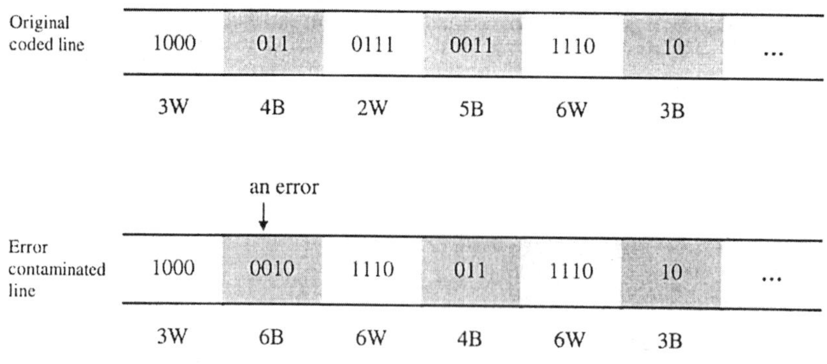

**FIGURE 6.3**  Establishment of resynchronization after a few runs.

the previous line, or replace the damaged line by a white line, or use a correlation technique to recover the line as much as possible.

### 6.2.3.2  Error Effect in the 2-D RLC Case

From the above discussion, we realize that 2-D RLC is more efficient than 1-D RLC on the one hand. The 2-D RLC is more susceptible to transmission errors than the 1-D RLC on the other hand. To prevent error propagation, there is a parameter used in 2-D RLC, known as the *K-factor*, which specifies the number of scan lines that are 2-D RLC coded.

Recommendation T.4 defined that no more than K-1 consecutive scan lines be 2-D RLC coded after a 1-D RLC coded line. For binary documents scanned at normal resolution, K = 2. For documents scanned at high resolution, K = 4.

According to Arps (1979), there are two different types of algorithms in binary image coding, *raster* algorithms and *area* algorithms. Raster algorithms only operate on data within one or two raster scan lines. They are hence mainly 1-D in nature. Area algorithms are truly 2-D in nature. They require that all, or a substantial portion, of the image is in random access memory. From our discussion above, we see that both 1-D and 2-D RLC defined in T.4 belong to the category of raster algorithms. Area algorithms require large memory space and are susceptible to transmission noise.

### 6.2.3.3  Uncompressed Mode

For some detailed binary document images, both 1-D and 2-D RLC may result in data expansion instead of data compression. Under these circumstances the number of coding bits is larger than the number of bilevel pixels. An uncompressed mode is created as an alternative way to avoid data expansion. Special codewords are assigned for the uncompressed mode.

For the performances of 1-D and 2-D RLC applied to eight CCITT test document images, and issues such as "fill bits" and "minimum scan line time (MSLT)," to name only a few, readers are referred to an excellent tutorial paper by Hunter and Robinson (1980).

## 6.3  DIGITAL FACSIMILE CODING STANDARDS

Facsimile transmission, an important means of communication in modern society, is often used as an example to demonstrate the mutual interaction between widely used applications and standardization activities. Active facsimile applications and the market brought on the necessity for international standardization in order to facilitate interoperability between facsimile machines worldwide. Successful international standardization, in turn, has stimulated wider use of facsimile transmission and, hence, a more demanding market. Facsimile has also been considered as a major application for binary image compression.

So far, facsimile machines are classified in four different groups. Facsimile apparatuses in groups 1 and 2 use analog techniques. They can transmit an A4 size (210 × 297 mm) document scanned at 3.85 lines/mm in 6 and 3 min, respectively, over the GSTN. International standards for these two groups of facsimile apparatus are CCITT (now ITU) Recommendations T.2 and T.3, respectively. Group 3 facsimile machines use digital techniques and hence achieve high coding efficiency. They can transmit the A4 size binary document scanned at a resolution of 3.85 lines/mm and sampled at 1728 pixels per line in about 1 min at a rate of 4800 b/sec over the GSTN. The corresponding international standard is CCITT Recommendations T.4. Group 4 facsimile apparatuses have the same transmission speed requirement as that for group 3 machines, but the coding technique is different. Specifically, the coding technique used for group 4 machines is based on 2-D run-length coding, discussed above, but modified to achieve higher coding efficiency. Hence it is referred to as the modified modified READ coding, abbreviated MMR. The corresponding standard is CCITT Recommendations T.6. Table 6.3 summarizes the above descriptions.

## TABLE 6.3 FACSIMILE CODING STANDARDS

| Group of Facsimile Apparatuses | Speed Requirement for A-4 Size Document | Analog or Digital Scheme | CCITT Recommendation | Compression Technique | | |
|---|---|---|---|---|---|---|
| | | | | Model | Basic Coder | Algorithm Acronym |
| $G_1$ | 6 min | Analog | T.2 | — | — | — |
| $G_2$ | 3 min | Analog | T.3 | — | — | — |
| $G_3$ | 1 min | Digital | T.4 | 1-D RLC | Modified Huffman | MH |
| | | | | 2-D RLC (optional) | | MR |
| $G_4$ | 1 min | Digital | T.6 | 2-D RLC | Modified Huffman | MMR |

## 6.4 DICTIONARY CODING

Dictionary coding, the focus of this section, is different from Huffman coding and arithmetic coding, discussed in the previous chapter. Both Huffman and arithmetic coding techniques are based on a statistical model, and the occurrence probabilities play a particular important role. Recall that in the Huffman coding the shorter codewords are assigned to more frequently occurring source symbols. In dictionary-based data compression techniques a symbol or a string of symbols generated from a source alphabet is represented by an index to a dictionary constructed from the source alphabet. A dictionary is a list of symbols and strings of symbols. There are many examples of this in our daily lives. For instance, the string "September" is sometimes represented by an index "9," while a social security number represents a person in the U.S.

Dictionary coding is widely used in text coding. Consider English text coding. The source alphabet includes 26 English letters in both lower and upper cases, numbers, various punctuation marks, and the space bar. Huffman or arithmetic coding treats each symbol based on its occurrence probability. That is, the source is modeled as a memoryless source. It is well known, however, that this is not true in many applications. In text coding, *structure* or *context* plays a significant role. As mentioned earlier, it is very likely that the letter *u* appears after the letter *q*. Likewise, it is likely that the word "concerned" will appear after "As far as the weather is." The strategy of the dictionary coding is to build a dictionary that contains frequently occurring symbols and string of symbols. When a symbol or a string is encountered and it is contained in the dictionary, it is encoded with an index to the dictionary. Otherwise, if not in the dictionary, the symbol or the string of symbols is encoded in a less efficient manner.

### 6.4.1 FORMULATION OF DICTIONARY CODING

To facilitate further discussion, we define dictionary coding in a precise manner (Bell et al., 1990). We denote a source alphabet by *S*. A dictionary consisting of two elements is defined as $D = (P, C)$, where *P* is a finite set of phrases generated from the *S*, and *C* is a coding function mapping *P* onto a set of codewords.

The set *P* is said to be complete if any input string can be represented by a series of phrases chosen from the *P*. The coding function *C* is said to obey the prefix property if there is no codeword that is a prefix of any other codeword. For practical usage, i.e., for reversible compression of any input text, the phrase set *P* must be complete and the coding function *C* must satisfy the prefix property.

### 6.4.2 CATEGORIZATION OF DICTIONARY-BASED CODING TECHNIQUES

The heart of dictionary coding is the formulation of the dictionary. A successfully built dictionary results in data compression; the opposite case may lead to data expansion. According to the ways in which dictionaries are constructed, dictionary coding techniques can be classified as static or adaptive.

### 6.4.2.1   Static Dictionary Coding

In some particular applications, the knowledge about the source alphabet and the related strings of symbols, also known as phrases, is sufficient for a fixed dictionary to be produced before the coding process. The dictionary is used at both the transmitting and receiving ends. This is referred to as static dictionary coding. The merit of the static approach is its simplicity. Its drawbacks lie in its relatively lower coding efficiency and less flexibility compared with adaptive dictionary techniques. By less flexibility, we mean that a dictionary built for a specific application is not normally suitable for utilization in other applications.

An example of static algorithms occurring is *digram* coding. In this simple and fast coding technique, the dictionary contains all source symbols and some frequently used pairs of symbols. In encoding, two symbols are checked at once to see if they are in the dictionary. If so, they are replaced by the index of the two symbols in the dictionary, and the next pair of symbols is encoded in the next step. If not, then the index of the first symbol is used to encode the first symbol. The second symbol is combined with the third symbol to form a new pair, which is encoded in the next step.

The digram can be straightforwardly extended to *n-gram*. In the extension, the size of the dictionary increases and so does its coding efficiency.

### 6.4.2.2   Adaptive Dictionary Coding

As opposed to the static approach, with the adaptive approach a completely defined dictionary does not exist prior to the encoding process and the dictionary is not fixed. At the beginning of coding, only an initial dictionary exists. It adapts itself to the input during the coding process. All the adaptive dictionary coding algorithms can be traced back to two different original works by Ziv and Lempel (1977, 1978). The algorithms based on Ziv and Lempel (1977) are referred to as the LZ77 algorithms, while those based on their 1978 work are the LZ78 algorithms. Prior to introducing the two landmark works, we will discuss the parsing strategy.

### 6.4.3   Parsing Strategy

Once we have a dictionary, we need to examine the input text and find a string of symbols that matches an item in the dictionary. Then the index of the item to the dictionary is encoded. This process of segmenting the input text into disjoint strings (whose union equals the input text) for coding is referred to as *parsing*. Obviously, the way to segment the input text into strings is not unique.

In terms of the highest coding efficiency, optimal parsing is essentially a shortest-path problem (Bell et al., 1990). In practice, however, a method called *greedy* parsing is used most often. In fact, it is used in all the LZ77 and LZ78 algorithms. With greedy parsing, the encoder searches for the longest string of symbols in the input that matches an item in the dictionary at each coding step. Greedy parsing may not be optimal, but it is simple in its implementation.

**Example 6.1**
Consider a dictionary, $D$, whose phrase set $P = \{a, b, ab, ba, bb, aab, bbb\}$. The codewords assigned to these strings are $C(a) = 10$, $C(b) = 11$, $C(ab) = 010$, $C(ba) = 0101$, $C(bb) = 01$, $C(abb) = 11$, and $C(bbb) = 0110$. Now the input text is *abbaab*.

Using greedy parsing, we then encode the text as $C(ab).C(ba).C(ab)$, which is a 10-bit string: 010.0101.010. In the above representations, the periods are used to indicate the division of segments in the parsing. This, however, is not an optimum solution. Obviously, the following parsing will be more efficient, i.e., $C(a).C(bb).C(aab)$, which is a 6-bit string: 10.01.11.

## 6.4.4  SLIDING WINDOW (LZ77) ALGORITHMS

As mentioned earlier, LZ77 algorithms are a group of adaptive dictionary coding algorithms rooted in the pioneering work of Ziv and Lempel (1977). Since they are adaptive, there is no complete and fixed dictionary before coding. Instead, the dictionary changes as the input text changes.

### 6.4.4.1  Introduction

In the LZ77 algorithms, the dictionary used is actually a portion of the input text, which has been recently encoded. The text that needs to be encoded is compared with the strings of symbols in the dictionary. The longest matched string in the dictionary is characterized by a *pointer* (sometimes called a *token*), which is represented by a triple of data items. Note that this triple functions as an index to the dictionary, as mentioned above. In this way, a variable-length string of symbols is mapped to a fixed-length pointer.

There is a sliding window in the LZ77 algorithms. The window consists of two parts: a search buffer and a look-ahead buffer. The search buffer contains the portion of the text stream that has recently been encoded which, as mentioned, is the dictionary; while the look-ahead buffer contains the text to be encoded next. The window slides through the input text stream from beginning to end during the entire encoding process. This explains the term *sliding* window. The size of the search buffer is much larger than that of the look-ahead buffer. This is expected because what is contained in the search buffer is in fact the adaptive dictionary. The sliding window is usually on the order of a few thousand symbols, whereas the look-ahead buffer is on the order of several tens to one hundred symbols.

### 6.4.4.2  Encoding and Decoding

Below we present more details about the sliding window dictionary coding technique, i.e., the LZ77 approach, via a simple illustrative example.

**Example 6.2**
Figure 6.4 shows a sliding window. The input text stream is *ikaccbadaccbaccbaccgikmoabc*. In part (a) of the figure, a search buffer of nine symbols and a look-ahead buffer of six symbols are shown. All the symbols in the search buffer, *accbadacc*, have just been encoded. All the symbols in the look-ahead buffer, *baccba*, are to be encoded. (It is understood that the symbols before the

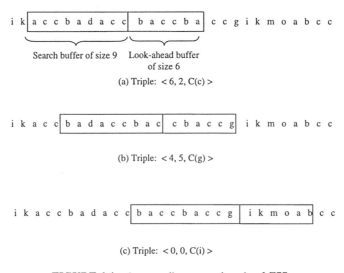

FIGURE 6.4   An encoding example using LZ77.

search buffer have been encoded and the symbols after the look-ahead buffer are to be encoded.) The strings of symbols, *ik* and *ccgikmoabcc*, are not covered by the sliding window at the moment.

At the moment, or in other words, in the first step of encoding, the symbol(s) to be encoded begin(s) with the symbol *b*. The pointer starts searching for the symbol *b* from the last symbol in the search buffer, *c*, which is immediately to the left of the first symbol *b* in the look-ahead buffer. It finds a match at the sixth position from *b*. It further determines that the longest string of the match is *ba*. That is, the maximum matching length is two. The pointer is then represented by a triple, <i,j,k>. The first item, "i", represents the distance between the first symbol in the look-ahead buffer and the position of the pointer (the position of the first symbol of the matched string). This distance is called *offset*. In this step, the offset is six. The second item in the triple, "j", indicates the length of the matched string. Here, the length of the matched string *ba* is two. The third item, "k", is the codeword assigned to the symbol immediately following the matched string in the look-ahead buffer. In this step, the third item is C(c), where C is used to represent a function to map symbol(s) to a codeword, as defined in Section 6.4.1. That is, the resulting triple after the first step is: <6, 2, C(c)>.

The reason to include the third item "k" into the triple is as follows. In the case where there is no match in the search buffer, both "i" and "j" will be zero. The third item at this moment is the codeword of the first symbol in the look-ahead buffer itself. This means that even in the case where we cannot find a match string, the sliding window still works. In the third step of the encoding process described below, we will see that the resulting triple is: <0, 0, C(i)>. The decoder hence understands that there is no matching, and the single symbol *i* is decoded.

The second step of the encoding is illustrated in part (b) of Figure 6.4. The sliding window has been shifted to the right by three positions. The first symbol to be encoded now is *c*, which is the left-most symbol in the look-ahead buffer. The search pointer moves towards the left from the symbol *c*. It first finds a match in the first position with a length of one. It then finds another match in the fourth position from the first symbol in the look-ahead buffer. Interestingly, the maximum matching can exceed the boundary between the search buffer and the look-ahead buffer and can enter the look-ahead buffer. Why this is possible will be seen shortly, when we discuss the decoding process. In this manner, it is found that the maximum length of matching is five. The last match is found at the fifth position. The length of the matched string, however, is only one. Since greedy parsing is used, the match with a length five is chosen. That is, the offset is four and the maximum match length is five. Consequently, the triple resulting from the second step is <4, 5, C(g)>.

The sliding window is then shifted to the right by six positions. The third step of the encoding is depicted in Part (c). Obviously, there is no matching of *i* in the search buffer. The resulting triple is hence <0, 0, C(i)>.

The encoding process can continue in this way. The possible cases we may encounter in the encoding, however, are described in the first three steps. Hence we end our discussion of the encoding process and discuss the decoding process. Compared with the encoding, the decoding is simpler because there is no need for matching, which involves many comparisons between the symbols in the look-ahead buffer and the symbols in the search buffer. The decoding process is illustrated in Figure 6.5.

In the above three steps, the resulting triples are <6, 2, C(c)>, <4, 5, C(g)>, and <0, 0, C(i)>. Now let us see how the decoder works. That is, how the decoder recovers the string *baccbaccgi* from these three triples.

In part (a) of Figure 6.5, the search buffer is the same as that in part (a) of Figure 6.4. That is, the string *accbadacc* stored in the search window is what was just decoded.

Once the first triple <6, 2, C(c)> is received, the decoder will move the decoding pointer from the first position in the look-ahead buffer to the left by six positions. That is, the pointer will point to the symbol *b*. The decoder then copies the two symbols starting from *b*, i.e., *ba*, into the look-ahead buffer. The symbol *c* will be copied right to *ba*. This is shown in part (b) of Figure 6.5. The window is then shifted to the right by three positions, as shown in part (c) of Figure 6.5.

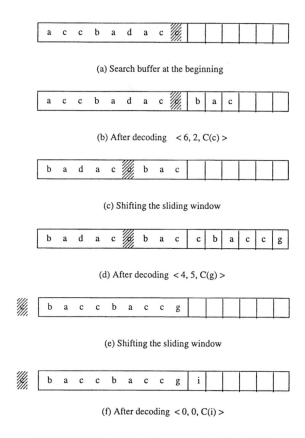

**FIGURE 6.5**    A decoding example using LZ77.

After the second triple <4, 5, C(g)> is received, the decoder moves the decoding pointer from the first position of the look-ahead buffer to the left by four positions. The pointer points to the symbol *c*. The decoder then copies five successive symbols starting from the symbol *c* pointed by the pointer. We see that at the beginning of this copying process there are only four symbols available for copying. Once the first symbol is copied, however, all five symbols are available. After copying, the symbol *g* is added to the end of the five copied symbols in the look-ahead buffer. The results are shown in part (c) of Figure 6.5. Part (d) then shows the window shifting to the right by six positions.

After receiving the triple <0, 0, C(i)>, the decoder knows that there is no match and a single symbol *i* is encoded. Hence, the decoder adds the symbol *i* following the symbol *g*. This is shown in part (f) of Figure 6.5.

In Figure 6.5, for each part, the last previously encoded symbol *c* prior to the receiving of the three triples is shaded. From part (f), we see that the string added after the symbol *c* due to the three triples is *baccbaccgi*. This agrees with the sequence mentioned at the beginning of our discussion about the decoding process. We thus conclude that the decoding process has correctly decoded the encoded sequence from the last encoded symbol and the received triples.

### 6.4.4.3    Summary of the LZ77 Approach

The sliding window consists of two parts: the search buffer and the look-ahead buffer. The most recently encoded portion of the input text stream is contained in the search buffer, while the portion of the text that needs to be encoded immediately is in the look-ahead buffer. The first symbol in the look-ahead buffer, located to the right of the boundary between the two buffers, is the symbol or the beginning of a string of symbols to be encoded at the moment. Let us call it the symbol $s$. The size of the search buffer is usually much larger than that of the look-ahead buffer.

In encoding, the search pointer moves to the left, away from the symbol $s$, to find a match of the symbol $s$ in the search buffer. Once a match is found, the encoding process will further determine the length of the matched string. When there are multiple matches, the match that produces the longest matched string is chosen. The match is denoted by a triple <i, j, k>. The first item in the triple, "i", is the offset, which is the distance between the pointer pointing to the symbol giving the maximum match and the symbol $s$. The second item, "j", is the length of the matched string. The third item, "k", is the codeword of the symbol following the matched string in the look-ahead buffer. The sliding window is then shifted to the right by j+1 positions before the next coding step takes place.

When there is no matching in the search buffer, the triple is represented by <0, 0, C(s)>, where C(s) is the codeword assigned to the symbol $s$. The sliding window is then shifted to the right by one position.

The sliding window is shifted along the input text stream during the encoding process. The symbol $s$ moves from the beginning symbol to the ending symbol of the input text stream.

At the very beginning, the content of the search buffer can be arbitrarily selected. For instance, the symbols in the search buffer may all be the space symbol.

Let us denote the size of the search buffer by $SB$, the size of the look-ahead buffer by $L$, and the size of the source alphabet by $A$. Assume that the natural binary code is used. Then we see that the LZ77 approach encodes variable-length strings of symbols with fixed-length codewords. Specifically, the offset "i" is of coding length $\lceil \log_2 SB \rceil$, the length of matched string "j" is of coding length $\lceil \log_2 (SB + L) \rceil$, and the codeword "k" is of coding length $\lceil \log_2 (A) \rceil$, where the sign $\lceil a \rceil$ denotes the smallest integer larger than $a$.

The length of the matched string is equal to $\lceil \log_2 (SB + L) \rceil$ because the search for the maximum matching can enter into the look-ahead buffer as shown in Example 6.2.

The decoding process is simpler than the encoding process since there are no comparisons involved in the decoding.

The most recently encoded symbols in the search buffer serve as the dictionary used in the LZ77 approach. The merit of doing so is that the dictionary is well adapted to the input text. The limitation of the approach is that if the distance between the repeated patterns in the input text stream is larger than the size of the search buffer, then the approach cannot utilize the structure to compress the text. A vivid example can be found in (Sayood, 1996).

A window with a moderate size, say, $SB + L \leq 8192$, can compress a variety of texts well. Several reasons have been analyzed by Bell et al. (1990).

Many variations have been made to improve coding efficiency of the LZ77 approach. The LZ77 produces a triple in each encoding step; i.e., the offset (position of the matched string), the length of the matched string, and the codeword of the symbol following the matched string. The transmission of the third item in each coding step is not efficient. This is true especially at the beginning of coding. A variant of the LZ77, referred to as the LZSS algorithm, improves this inefficiency.

## 6.4.5 LZ78 Algorithms

### 6.4.5.1 Introduction

As mentioned above, the LZ77 algorithms use a sliding window of fixed size, and both the search buffer and the look-ahead buffer have a fixed size. This means that if the distance between two repeated patterns is larger than the size of the search buffer, the LZ77 algorithms cannot work efficiently. The fixed size of both the buffers implies that the matched string cannot be longer than the sum of the sizes of the two buffers, placing another limitation on coding efficiency. Increasing the sizes of the search buffer and the look-ahead buffer seemingly will resolve the problem. A close look, however, reveals that it also leads to increases in the number of bits required to encode the offset and matched string length, as well as an increase in processing complexity.

The LZ78 algorithms (Ziv and Lempel, 1978) eliminate the use of the sliding window. Instead, these algorithms use the encoded text as a dictionary which, potentially, does not have a fixed size. Each time a pointer (token) is issued, the encoded string is included in the dictionary. Theoretically, the LZ78 algorithms reach optimal performance as the encoded text stream approaches infinity. In practice, however, as mentioned above with respect to the LZ77, a very large dictionary will affect coding efficiency negatively. Therefore, once a preset limit to the dictionary size has been reached, either the dictionary is fixed for the future (if the coding efficiency is good), or it is reset to zero, i.e., it must be restarted.

Instead of the triples used in the LZ77, only pairs are used in the LZ78. Specifically, only the position of the pointer to the matched string and the symbol following the matched string need to be encoded. The length of the matched string does not need to be encoded since both the encoder and the decoder have exactly the same dictionary, i.e., the decoder knows the length of the matched string.

### 6.4.5.2   Encoding and Decoding

Like the discussion of the LZ77 algorithms, we will go through an example to describe the LZ78 algorithms.

**Example 6.3**

Consider the text stream: *baccbaccacbcabccbbacc*. Table 6.4 shows the coding process. We see that for the first three symbols there is no match between the individual input symbols and the entries in the dictionary. Therefore, the doubles are, respectively, <0, C(b)>, <0, C(a)>, and <0, C(c)>, where 0 means no match, and C(b), C(a), and C(c) represent the codewords of *b*, *a*, and *c*, respectively. After symbols b, a, c, comes *c*, which finds a match in the dictionary (the third entry). Therefore, the next symbol *b* is combined to be considered. Since the string *cb* did not appear before, it is encoded as a double and it is appended as a new entry into the dictionary. The first item in the double is the index of the matched entry *c*, 3, the second item is the index/codeword of the symbol following the match *b*, 1. That is, the double is <3, 1>. The following input symbol is *a*, which appeared in the dictionary. Hence, the next symbol *c* is taken into consideration. Since the string *ac* is not an entry of the dictionary, it is encoded with a double. The first item in the double is the index of symbol *a*, 2; the second item is the index of symbol *c*, 3, i.e., <2, 3>. The encoding proceeds in this way. Take a look at Table 6.4. In general, as the encoding proceeds, the entries in the dictionary become longer and longer. First, entries with single symbols come out, but later, more and more entries with two symbols show up. After that, more and more entries with three symbols appear. This means that coding efficiency is increasing.

Now consider the decoding process. Since the decoder knows the rule applied in the encoding, it can reconstruct the dictionary and decode the input text stream from the received doubles. When the first double <0, C(b)> is received, the decoder knows that there is no match. Hence, the first entry in the dictionary is *b*. So is the first decoded symbol. From the second double <0, C(a)>, symbol *a* is known as the second entry in the dictionary as well as the second decoded symbol. Similarly, the next entry in the dictionary and the next decoded symbol are known as *c*. When the following double <3, 1> is received. The decoder knows from two items, 3 and 1, that the next two symbols are the third and the first entries in the dictionary. This indicates that the symbols *c* and *b* are decoded, and the string *cb* becomes the fourth entry in the dictionary.

We omit the next two doubles and take a look at the double <4, 3>, which is associated with Index 7 in Table 6.4. Since the first item in the double is 4, it means that the maximum matched string is *cb*, which is associated with Index 4 in Table 6.4. The second item in the double, 3, implies that the symbol following the match is the third entry *c*. Therefore the decoder decodes a string *cbc*. Also the string *cbc* becomes the seventh entry in the reconstructed dictionary. In this way, the

**TABLE 6.4**
**An Encoding Example Using the LZ78 Algorithm**

| Index | Doubles | Encoded Symbols |
|-------|---------|-----------------|
| 1 | < 0, C(b) > | b |
| 2 | < 0, C(a) > | a |
| 3 | < 0, C(c) > | c |
| 4 | < 3, 1 > | cb |
| 5 | < 2, 3 > | ac |
| 6 | < 3, 2 > | ca |
| 7 | < 4, 3 > | cbc |
| 8 | < 2, 1 > | ab |
| 9 | < 3, 3 > | cc |
| 10 | < 1, 1 > | bb |
| 11 | < 5, 3 > | acc |

decoder can reconstruct the exact same dictionary as that established by the encoder and decode the input text stream from the received doubles.

### 6.4.5.3   LZW Algorithm

Both the LZ77 and LZ78 approaches, when published in 1977 and 1978, respectively, were theory oriented. The effective and practical improvement over the LZ78 by Welch (1984) brought much attention to the LZ dictionary coding techniques. The resulting algorithm is referred to the LZW algorithm. It removed the second item in the double (the index of the symbol following the longest matched string) and, hence, it enhanced coding efficiency. In other words, the LZW only sends the indexes of the dictionary to the decoder. For the purpose, the LZW first forms an initial dictionary, which consists of all the individual source symbols contained in the source alphabet. Then, the encoder examines the input symbol. Since the input symbol matches to an entry in the dictionary, its succeeding symbol is cascaded to form a string. The cascaded string does not find a match in the initial dictionary. Hence, the index of the matched symbol is encoded and the enlarged string (the matched symbol followed by the cascaded symbol) is listed as a new entry in the dictionary. The encoding process continues in this manner.

For the encoding and decoding processes, let us go through an example to see how the LZW algorithm can encode only the indexes and the decoder can still decode the input text string.

**Example 6.4**

Consider the following input text stream: *accbadaccbaccbacc*. We see that the source alphabet is $S = \{a,b,c,d,\}$. The top portion of Table 6.5 (with indexes 1,2,3,4) gives a possible initial dictionary used in the LZW. When the first symbol *a* is input, the encoder finds that it has a match in the dictionary. Therefore the next symbol *c* is taken to form a string *ac*. Because the string *ac* is not in the dictionary, it is listed as a new entry in the dictionary and is given an index, 5. The index of the matched symbol *a*, 1, is encoded. When the second symbol, *c*, is input the encoder takes the following symbol *c* into consideration because there is a match to the second input symbol *c* in the dictionary. Since the string *cc* does not match any existing entry, it becomes a new entry in the dictionary with an index, 6. The index of the matched symbol (the second input symbol), *c*, is encoded. Now consider the third input symbol *c*, which appeared in the dictionary. Hence, the following symbol *b* is cascaded to form a string *cb*. Since the string *cb* is not in the dictionary, it becomes a new entry in the dictionary and is given an index, 7. The index of matched symbol *c*, 3, is encoded. The process proceeds in this fashion.

**TABLE 6.5**
**An Example of the Dictionary Coding**
**Using the LZW Algorithm**

| Index | Entry | Input Symbols | Encoded Index |
|-------|-------|---------------|---------------|
| 1 | a | ⎫ | |
| 2 | b | ⎬ Initial dictionary | |
| 3 | c | ⎪ | |
| 4 | d | ⎭ | |
| 5 | ac | a | 1 |
| 6 | cc | c | 3 |
| 7 | cb | c | 3 |
| 8 | ba | b | 2 |
| 9 | ad | a | 1 |
| 10 | da | d | 4 |
| 11 | acc | a,c | 5 |
| 12 | cba | c,b | 7 |
| 13 | accb | a,c,c | 11 |
| 14 | bac | b,a, | 8 |
| 15 | cc... | c,c,... | |

Take a look at entry 11 in the dictionary shown in Table 6.5. The input symbol at this point is *a*. Since it has a match in the previous entries, its next symbol *c* is considered. Since the string *ac* appeared in entry 5, the succeeding symbol *c* is combined. Now the new enlarged string becomes *acc* and it does not have a match in the previous entries. It is thus added to the dictionary. And a new index, 11, is given to the string *acc*. The index of the matched string *ac*, 5, is encoded and transmitted. The final sequence of encoded indexes is 1, 3, 3, 2, 1, 4, 5, 7, 11, 8. Like the LZ78, the entries in the dictionary become longer and longer in the LZW algorithm. This implies high coding efficiency since long strings can be represented by indexes.

Now let us take a look at the decoding process to see how the decoder can decode the input text stream from the received index. Initially, the decoder has the same dictionary (the top four rows in Table 6.5) as that in the encoder. Once the first index 1 comes, the decoder decodes a symbol *a*. The second index is 3, which indicates that the next symbol is *c*. From the rule applied in encoding, the decoder knows further that a new entry *ac* has been added to the dictionary with an index 5. The next index is 3. It is known that the next symbol is also *c*. It is also known that the string *cc* has been added into the dictionary as the sixth entry. In this way, the decoder reconstructs the dictionary and decodes the input text stream.

### 6.4.5.4  Summary

The LZW algorithm, as a representative of the LZ78 approach, is summarized below.

The initial dictionary contains the indexes for all the individual source symbols. At the beginning of encoding, when a symbol is input, since it has a match in the initial dictionary, the next symbol is cascaded to form a two-symbol string. Since the two-symbol string cannot find a match in the initial dictionary, the index of the former symbol is encoded and transmitted, and the two-symbol string is added to the dictionary with a new, incremented index. The next encoding step starts with the latter symbol of the two.

In the middle, the encoding process starts with the last symbol of the latest added dictionary entry. Since it has a match in the previous entries, its succeeding symbol is cascaded after the symbol to form a string. If this string appeared before in the dictionary (i.e., the string finds a match), the next symbol is cascaded as well. This process continues until such an enlarged string cannot find a match in the dictionary. At this moment, the index of the last matched string (the longest match) is encoded and transmitted, and the enlarged and unmatched string is added into the dictionary as a new entry with a new, incremented index.

Decoding is a process of transforming the index string back to the corresponding symbol string. In order to do so, however, the dictionary must be reconstructed in exactly the same way as that established in the encoding process. That is, the initial dictionary is constructed first in the same way as that in the encoding. When decoding the index string, the decoder reconstructs the same dictionary as that in the encoder according to the rule used in the encoding.

Specifically, at the beginning of the decoding, after receiving an index, a corresponding single symbol can be decoded. Via the next received index, another symbol can be decoded. From the rule used in the encoding, the decoder knows that the two symbols should be cascaded to form a new entry added into the dictionary with an incremented index. The next step in the decoding will start from the latter symbol among the two symbols.

Now consider the middle of the decoding process. The presently received index is used to decode a corresponding string of input symbols according to the reconstructed dictionary at the moment. (Note that this string is said to be with the present index.) It is known from the encoding rule that the symbols in the string associated with the next index should be considered. (Note that this string is said to be with the next index.) That is, the first symbol in the string with the next index should be appended to the last symbol in the string with the present index. The resultant combination, i.e., the string with the present index followed by the first symbol in the string with the next index, cannot find a match to an entry in the dictionary. Therefore, the combination should be added to the dictionary with an incremented index. At this moment, the next index becomes the new present index, and the index following the next index becomes the new next index. The decoding process then proceeds in the same fashion in a new decoding step.

Compared with the LZ78 algorithm, the LZW algorithm eliminates the necessity of having the second item in the double, an index/codeword of the symbol following a matched string. That is, the encoder only needs to encode and transmit the first item in the double. This greatly enhances the coding efficiency and reduces the complexity of the LZ algorithm.

### 6.4.5.5   Applications

The CCITT Recommendation V.42 bis is a data compression standard used in modems that connect computers with remote users via the GSTN. In the compressed mode, the LZW algorithm is recommended for data compression.

In image compression, the LZW finds its application as well. Specifically, it is utilized in the graphic interchange format (GIF) which was created to encode graphical images. GIF is now also used to encode natural images, though it is not very efficient in this regard. For more information, readers are referred to Sayood (1996). The LZW algorithm is also used in the UNIX Compress command.

## 6.5   INTERNATIONAL STANDARDS FOR LOSSLESS STILL IMAGE COMPRESSION

In the previous chapter, we studied Huffman and arithmetic coding techniques. We also briefly discussed the international standard for bilevel image compression, known as the JBIG. In this chapter, so far we have discussed another two coding techniques: the run-length and dictionary coding techniques. We also introduced the international standards for facsimile compression, in which the techniques known as the MH, MR, and MMR were recommended. All of these techniques involve lossless compression. In the next chapter, the international still image coding standard JPEG will be introduced. As we will see, the JPEG has four different modes. They can be divided into two compression categories: lossy and lossless. Hence, we can talk about the lossless JPEG. Before leaving this chapter, however, we briefly discuss, compare, and summarize various techniques used in the international standards for lossless still image compression. For more details, readers are referred to an excellent survey paper by Arps and Truong (1994).

### 6.5.1 Lossless Bilevel Still Image Compression

#### 6.5.1.1 Algorithms

As mentioned above, there are four different international standard algorithms falling into this category.

**MH (Modified Huffman coding)** — This algorithm is defined in CCITT Recommendation T.4 for facsimile coding. It uses the 1-D run-length coding technique followed by the "modified" Huffman coding technique.

**MR (Modified READ [Relative Element Address Designate] coding)** — Defined in CCITT Recommendation T.4 for facsimile coding. It uses the 2-D run-length coding technique followed by the "modified" Huffman coding technique.

**MMR (Modified Modified READ coding)** — Defined in CCITT Recommendation T.6. It is based on MR, but is modified to maximize compression.

**JBIG (Joint Bilevel Image experts Group coding)** — Defined in CCITT Recommendation T.82. It uses an adaptive 2-D coding model, followed by an adaptive arithmetic coding technique.

#### 6.5.1.2 Performance Comparison

The JBIG test image set was used to compare the performance of the above-mentioned algorithms. The set contains scanned business documents with different densities, graphic images, digital halftones, and mixed (document and halftone) images.

Note that digital halftones, also named (digital) halftone images, are generated by using only binary devices. Some small black units are imposed on a white background. The units may assume different shapes: a circle, a square, and so on. The more dense the black units in a spot of an image, the darker the spot appears. The digital halftoning method has been used for printing gray-level images in newspapers and books. Digital halftoning through character overstriking, used to generate digital images in the early days for the experimental work associated with courses on digital image processing, has been described by Gonzalez and Woods (1992).

The following two observations on the performance comparison were made after the application of the several techniques to the JBIG test image set.

1. For bilevel images excluding digital halftones, the compression ratio achieved by these techniques ranges from 3 to 100. The compression ratio increases monotonically in the order of the following standard algorithms: MH, MR, MMR, JBIG.
2. For digital halftones, MH, MR, and MMR result in data expansion, while JBIG achieves compression ratios in the range of 5 to 20. This demonstrates that among the techniques, JBIG is the only one suitable for the compression of digital halftones.

### 6.5.2 Lossless Multilevel Still Image Compression

#### 6.5.2.1 Algorithms

There are two international standards for multilevel still image compression:

**JBIG (Joint Bilevel Image experts Group coding)** — Defined in CCITT Recommendation T.82. It uses an adaptive arithmetic coding technique. To encode multilevel images, the JIBG decomposes multilevel images into bit-planes, then compresses these bit-planes using its bilevel image compression technique. To further enhance the compression ratio, it uses Gary coding to represent pixel amplitudes instead of weighted binary coding.

**JPEG (Joint Photographic (image) Experts Group coding)** — Defined in CCITT Recommendation T.81. For lossless coding, it uses the differential coding technique. The predictive error is encoded using either Huffman coding or adaptive arithmetic coding techniques.

### 6.5.2.2   Performance Comparison

A set of color test images from the JPEG standards committee was used for performance comparison. The luminance component (Y) is of resolution $720 \times 576$ pixels, while the chrominance components (U and V) are of $360 \times 576$ pixels. The compression ratios calculated are the combined results for all the three components. The following observations have been reported.

1. When quantized in 8 bits per pixel, the compression ratios vary much less for multilevel images than for bilevel images, and are roughly equal to 2.
2. When quantized with 5 bits per pixel down to 2 bits per pixel, compared with the lossless JPEG the JBIG achieves an increasingly higher compression ratio, up to a maximum of 29%.
3. When quantized with 6 bits per pixel, JBIG and lossless JPEG achieve similar compression ratios.
4. When quantized with 7 bits per pixel to 8 bits per pixel, the lossless JPEG achieves a 2.4 to 2.6% higher compression ratio than JBIG.

## 6.6   SUMMARY

Both Huffman coding and arithmetic coding, discussed in the previous chapter, are referred to as variable-length coding techniques, since the lengths of codewords assigned to different entries in a source alphabet are different. In general, a codeword of a shorter length is assigned to an entry with higher occurrence probabilities. They are also classified as fixed-length to variable-length coding techniques (Arps, 1979), since the entries in a source alphabet have the same fixed length. Run-length coding (RLC) and dictionary coding, the focus of this chapter, are opposite, and are referred to as variable-length to fixed-length coding techniques. This is because the runs in the RLC and the string in the dictionary coding are variable and are encoded with codewords of the same fixed length.

Based on RLC, the international standard algorithms for facsimile coding, MH, MR, and MMR have worked successfully except for dealing with digital halftones. That is, these algorithms result in data expansion when applied to digital halftones. The JBIG, based on an adaptive arithmetic coding technique, not only achieves a higher coding efficiency than MH, MR, and MMR for facsimile coding, but also compresses the digital halftones effectively.

Note that 1-D RLC utilizes the correlation between pixels within a scan line, whereas 2-D RLC utilizes the correlation between pixels within a few scan lines. As a result, 2-D RLC can obtain higher coding efficiency than 1-D RLC. On the other hand, 2-D RLC is more susceptible to transmission errors than 1-D RLC.

In text compression, the dictionary-based techniques have proven to be efficient. All the adaptive dictionary-based algorithms can be classified into two groups. One is based on a work by Ziv and Lempel in 1977, and another is based on their pioneering work in 1978. They are called the LZ77 and LZ78 algorithms, respectively. With the LZ77 algorithms, a fixed-size window slides through the input text stream. The sliding window consists of two parts: the search buffer and the look-ahead buffer. The search buffer contains the most recently encoded portion of the input text, while the look-ahead buffer contains the portion of the input text to be encoded immediately. For the symbols to be encoded, the LZ77 algorithms search for the longest match in the search buffer. The information about the match: the distance between the matched string in the search buffer and that in the look-ahead buffer, the length of the matched string, and the codeword of the symbol following the matched string in the look-ahead buffer are encoded. Many improvements have been made in the LZ77 algorithms.

The performance of the LZ77 algorithms is limited by the sizes of the search buffer and the look-ahead buffer. With a finite size for the search buffer, the LZ77 algorithms will not work well in the case where repeated patterns are apart from each other by a distance longer than the size of the search buffer. With a finite size for the sliding window, the LZ77 algorithms will not work well in the case where matching strings are longer than the window. In order to be efficient, however, these sizes cannot be very large.

In order to overcome the problem, the LZ78 algorithms work in a different way. They do not use the sliding window at all. Instead of using the most recently encoded portion of the input text as a dictionary, the LZ78 algorithms use the index of the longest matched string as an entry of the dictionary. That is, each matched string cascaded with its immediate next symbol is compared with the existing entries of the dictionary. If this combination (a new string) does not find a match in the dictionary constructed at the moment, the combination will be included as an entry in the dictionary. Otherwise, the next symbol in the input text will be appended to the combination and the enlarged new combination will be checked with the dictionary. The process continues until the new combination cannot find a match in the dictionary. Among the several variants of the LZ78 algorithms, the LZW algorithm is perhaps the most important one. It only needs to encode the indexes of the longest matched strings to the dictionary. It can be shown that the decoder can decode the input text stream from the given index stream. In doing so, the same dictionary as that established in the encoder needs to be reconstructed at the decoder, and this can be implemented since the same rule used in the encoding is known in the decoder.

The size of the dictionary cannot be infinitely large because, as mentioned above, the coding efficiency will not be high. The common practice of the LZ78 algorithms is to keep the dictionary fixed once a certain size has been reached and the performance of the encoding is satisfactory. Otherwise, the dictionary will be set to empty and will be reconstructed from scratch.

Considering the fact that there are several international standards concerning still image coding (for both bilevel and multilevel images), a brief summary of them and a performance comparison have been presented in this chapter. At the beginning of this chapter, a description of the discrete Markov source and its $n$th extensions was provided. The Markov source and the autoregressive model serve as important models for the dependent information sources.

## 6.7   EXERCISES

**6-1.** Draw the state diagram of a second-order Markov source with two symbols in the source alphabet. That is, $S = \{s_1, s_2\}$. It is assumed that the conditional probabilities are

$$p\left(s_1 \middle| s_1 s_1\right) = p\left(s_2 \middle| s_2 s_2\right) = 0.7,$$

$$p\left(s_2 \middle| s_1 s_1\right) = p\left(s_1 \middle| s_2 s_2\right) = 0.3, \text{ and}$$

$$p\left(s_1 \middle| s_1 s_2\right) = p\left(s_1 \middle| s_2 s_1\right) = p\left(s_2 \middle| s_1 s_2\right) = p\left(s_2 \middle| s_2 s_1\right) = 0.5.$$

**6-2.** What are the definitions of raster algorithm and area algorithm in binary image coding? To which category does 1-D RLC belong? To which category does 2-D RLC belong?

**6-3.** What effect does a transmission error have on 1-D RLC and 2-D RLC, respectively? What is the function of the codeword EOL?

**6-4.** Make a convincing argument that the "modified" Huffman (MH) algorithm reduces the requirement of large storage space.

**6-5.** Which three different modes does 2-D RLC have? How do you view the vertical mode?

**6-6.** Using your own words, describe the encoding and decoding processes of the LZ77 algorithms. Go through Example 6.2.

**6-7.** Using your own words, describe the encoding and decoding processes of the LZW algorithm. Go through Example 6.3.

**6-8.** Read the reference paper (Arps and Truong, 1994), which is an excellent survey on the international standards for lossless still image compression. Pay particular attention to all the figures and to Table 1.

## REFERENCES

Abramson, N. *Information Theory and Coding*, New York: McGraw-Hill, 1963.

Arps, R. B. Binary Image Compression, in *Image Transmission Techniques*, W. K. Pratt (Ed.), New York: Academic Press, 1979.

Arps, R. B. and T. K. Truong, Comparison of international standards for lossless still image compression, *Proc. IEEE*, 82(6), 889-899, 1994.

Bell, T. C., J. G. Cleary, and I. H. Witten, *Text Compression*, Englewood Cliffs, NJ: Prentice-Hall, 1990.

Gonzalez, R. C. and R. E. Woods, *Digital Image Processing*, Reading, MA: Addison-Wesley, 1992.

Hunter, R. and A. H. Robinson, International digital facsimile coding standards, *Proc. IEEE*, 68(7), 854-867, 1980.

Laemmel, A. E. Coding Processes for Bandwidth Reduction in Picture Transmission, Rep. R-246-51, PIB-187, Microwave Res. Inst., Polytechnic Institute of Brooklyn, New York.

Nelson, M. and J.-L. Gailly, *The Data Compression Book*, 2nd ed., New York: M&T Books, 1995.

Sayood, K. *Introduction to Data Compression*, San Francisco, CA: Morgan Kaufmann Publishers, 1996.

Shannon, C. E. and W. Weaver, *The Mathematical Theory of Communication*, Urbana, IL: University of Illinois Press, 1949.

Welch, T. A technique for high-performance data compression, *IEEE Trans. Comput.*, 17(6), 8-19, 1984.

Ziv, J. and A. Lempel, A universal algorithm for sequential data compression, *IEEE Trans. Inf. Theory*, 23(3), 337-343, 1977.

Ziv, J. and A. Lempel, Compression of individual sequences via variable-rate coding, *IEEE Trans. Inf. Theory*, 24(5), 530-536, 1978.

# Section II

## Still Image Compression

# 7 Still Image Coding Standard: JPEG

In this chapter, the JPEG standard is introduced. This standard allows for lossy and lossless encoding of still images and four distinct modes of operation are supported: sequential DCT-based mode, progressive DCT-based mode, lossless mode and hierarchical mode.

## 7.1  INTRODUCTION

Still image coding is an important application of data compression. When an analog image or picture is digitized, each pixel is represented by a fixed number of bits, which correspond to a certain number of gray levels. In this uncompressed format, the digitized image requires a large number of bits to be stored or transmitted. As a result, compression become necessary due to the limited communication bandwidth or storage size. Since the mid-1980s, the ITU and ISO have been working together to develop a joint international standard for the compression of still images. Officially, JPEG [jpeg] is the ISO/IEC international standard 10918-1; digital compression and coding of continuous-tone still images, or the ITU-T Recommendation T.81. JPEG became an international standard in 1992. The JPEG standard allows for both lossy and lossless encoding of still images. The algorithm for lossy coding is a DCT-based coding scheme. This is the baseline of JPEG and is sufficient for many applications. However, to meet the needs of applications that cannot tolerate loss, e.g., compression of medical images, a lossless coding scheme is also provided and is based on a predictive coding scheme. From the algorithmic point of view, JPEG includes four distinct modes of operation, namely, sequential DCT-based mode, progressive DCT-based mode, lossless mode, and hierarchical mode. In the following sections, an overview of these modes is provided. Further technical details can be found in the books by Pennelbaker and Mitchell (1992) and Symes (1998).

In the sequential DCT-based mode, an image is first partitioned into blocks of 8 × 8 pixels. The blocks are processed from left to right and top to bottom. The 8 × 8 two-dimensional Forward DCT is applied to each block and the 8 × 8 DCT coefficients are quantized. Finally, the quantized DCT coefficients are entropy encoded and output as part of the compressed image data.

In the progressive DCT-based mode, the process of block partitioning and Forward DCT transform is the same as in the sequential DCT-based mode. However, in the progressive mode, the quantized DCT coefficients are first stored in a buffer before the encoding is performed. The DCT coefficients in the buffer are then encoded by a multiple scanning process. In each scan, the quantized DCT coefficients are partially encoded by either spectral selection or successive approximation. In the method of spectral selection, the quantized DCT coefficients are divided into multiple spectral bands according to a zigzag order. In each scan, a specified band is encoded. In the method of successive approximation, a specified number of most significant bits of the quantized coefficients are first encoded and the least significant bits are then encoded in subsequent scans.

The difference between sequential coding and progressive coding is shown in Figure 7.1. In the sequential coding an image is encoded part by part according to the scanning order, while in the progressive coding the image is encoded by a multiscanning process and in each scan the full image is encoded to a certain quality level.

As mentioned earlier, lossless coding is achieved by a predictive coding scheme. In this scheme, three neighboring pixels are used to predict the current pixel to be coded. The prediction difference

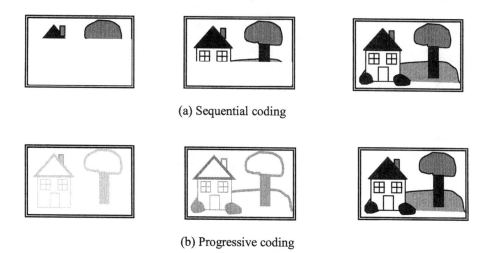

(a) Sequential coding

(b) Progressive coding

**FIGURE 7.1**    (a) Sequential coding, (b) progressive coding.

is entropy coded using either Huffman or arithmetic coding. Since the prediction is not quantized, the coding is lossless.

Finally, in the hierarchical mode, an image is first spatially down-sampled to a multilayered pyramid, resulting in a sequence of frames as shown in Figure 7.2. This sequence of frames is encoded by a predictive coding scheme. Except for the first frame, the predictive coding process is applied to the differential frames, i.e., the differences between the frame to be coded and the predictive reference frame. It is important to note that the reference frame is equivalent to the previous frame that would be reconstructed in the decoder. The coding method for the difference frame may use the DCT-based coding method, the lossless coding method, or the DCT-based processes with a final lossless process. Down-sampling and up-sampling filters are used in the hierarchical mode. The hierarchical coding mode provides a progressive presentation similar to the progressive DCT-based mode, but is also useful in the applications that have multiresolution requirements. The hierarchical coding mode also provides the capability of progressive coding to a final lossless stage.

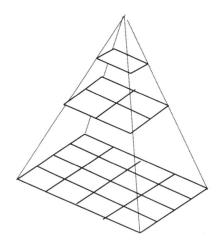

**FIGURE 7.2**    Hierarchical multiresolution encoding.

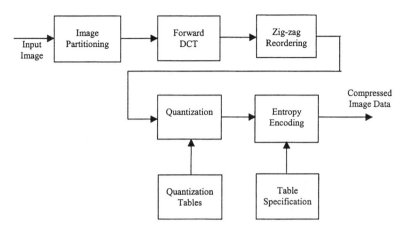

**FIGURE 7.3**   Block diagram of a sequential DCT-based encoding process.

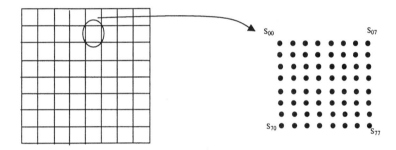

**FIGURE 7.4**   Partitioning to 8 × 8 blocks.

## 7.2   SEQUENTIAL DCT-BASED ENCODING ALGORITHM

The sequential DCT-based coding algorithm is the baseline algorithm of the JPEG coding standard. A block diagram of the encoding process is shown in Figure 7.3. As shown in Figure 7.4, the digitized image data are first partitioned into blocks of 8 × 8 pixels. The two-dimensional forward DCT is applied to each 8 × 8 block. The two-dimensional forward and inverse DCT of 8 × 8 block are defined as follows:

$$\text{FDCT:}\quad S_{uv} = \frac{1}{4} C_u C_v \sum_{i=0}^{7} \sum_{j=0}^{7} s_{ij} \cos\frac{(2i+1)u\pi}{16} \cos\frac{(2j+1)v\pi}{16}$$

$$\text{IDCT:}\quad s_{ij} = \frac{1}{4} \sum_{u=0}^{7} \sum_{v=0}^{7} C_u C_v S_{uv} \cos\frac{(2i+1)u\pi}{16} \cos\frac{(2j+1)v\pi}{16} \tag{7.1}$$

$$C_u C_v = \begin{cases} \dfrac{1}{\sqrt{2}} & \text{for } u, v = 0 \\ 1 & \text{otherwise} \end{cases}$$

where $s_{ij}$ is the value of the pixel at position $(i,j)$ in the block, and $S_{uv}$ is the transformed $(u,v)$ DCT coefficient.

**TABLE 7.1**

**Two Examples of Quantization Tables Used by JPEG**

| 16 | 11 | 10 | 16 | 24 | 40 | 51 | 61 |
|----|----|----|----|----|----|----|----|
| 12 | 12 | 14 | 19 | 26 | 58 | 60 | 55 |
| 14 | 13 | 16 | 24 | 40 | 57 | 69 | 56 |
| 14 | 17 | 22 | 29 | 51 | 87 | 80 | 62 |
| 18 | 22 | 37 | 56 | 68 | 109 | 103 | 77 |
| 24 | 35 | 55 | 64 | 81 | 104 | 113 | 92 |
| 49 | 64 | 78 | 87 | 103 | 121 | 120 | 101 |
| 72 | 92 | 95 | 98 | 112 | 100 | 103 | 99 |

Luminance quantization table

| 17 | 18 | 24 | 47 | 99 | 99 | 99 | 99 |
|----|----|----|----|----|----|----|----|
| 18 | 21 | 26 | 66 | 99 | 99 | 99 | 99 |
| 24 | 26 | 56 | 99 | 99 | 99 | 99 | 99 |
| 47 | 66 | 99 | 99 | 99 | 99 | 99 | 99 |
| 99 | 99 | 99 | 99 | 99 | 99 | 99 | 99 |
| 99 | 99 | 99 | 99 | 99 | 99 | 99 | 99 |
| 99 | 99 | 99 | 99 | 99 | 99 | 99 | 99 |
| 99 | 99 | 99 | 99 | 99 | 99 | 99 | 99 |

Chrominance quantization table

After the forward DCT, quantization of the transformed DCT coefficients is performed. Each of the 64 DCT coefficients is quantized by a uniform quantizer:

$$S_{quv} = round\left(\frac{S_{uv}}{Q_{uv}}\right) \qquad (7.2)$$

where the $S_{quv}$ is the quantized value of the DCT coefficient, $S_{uv}$, and $Q_{uv}$ is the quantization step obtained from the quantization table. There are four quantization tables that may be used by the encoder, but there is no default quantization table specified by the standard. Two particular quantization tables are shown in Table 7.1.

At the decoder, the dequantization is performed as follows:

$$R_{quv} = S_{quv} \times Q_{uv} \qquad (7.3)$$

where $R_{quv}$ is the value of the dequantized DCT coefficient. After quantization, the DC coefficient, $S_{q00}$, is treated separately from the other 63 AC coefficients. The DC coefficients are encoded by a predictive coding scheme. The encoded value is the difference (*DIFF*) between the quantized DC coefficient of the current block ($S_{q00}$) and that of the previous block of the same component (*PRED*):

$$DIFF = S_{q00} - PRED \qquad (7.4)$$

The value of *DIFF* is entropy coded with Huffman tables. More specifically, the two's complement of the possible *DIFF* magnitudes are grouped into 12 categories, "SSSS". The Huffman codes for these 12 difference categories and additional bits are shown in the Table 7.2.

For each nonzero category, additional bits are added to the codeword to uniquely identify which difference within the category actually occurred. The number of additional bits is defined by "SSSS" and the additional bits are appended to the least significant bit of the Huffman code (most significant bit first) according to the following rule. If the difference value is positive, the "SSSS" low-order bits of *DIFF* are appended; if the difference value is negative, then the "SSSS" low-order bits of *DIFF-1* are appended. As an example, the Huffman tables used for coding the luminance and chrominance DC coefficients are shown in Tables 7.3 and 7.4, respectively. These two tables have been developed from the average statistics of a large set of images with 8-bit precision.

**TABLE 7.2**
**Huffman Coding of DC Coefficients**

| SSSS | DIFF Values | Additional Bits |
|------|-------------|-----------------|
| 0 | 0 | – |
| 1 | –1,1 | 0,1 |
| 2 | –3,–2,2,3 | 00,01,10,11 |
| 3 | –7,...,–4,4,...,7 | 000,...,011,100,.,111 |
| 4 | –15,...,–8,8,...,15 | 0000,.,0111,1000,...,1111 |
| 5 | –31,...,–16,16,...,31 | 00000,...,01111,10000,...,11111 |
| 6 | –63,...,–32,32,...63 | ....,... |
| 7 | –127,...,–64,64,...,127 | ....,... |
| 8 | –255,...,–128,128,...,255 | ....,... |
| 9 | –511,...,–256,256,...,511 | ....,... |
| 10 | –1023,...,–512,512,...,1023 | ....,... |
| 11 | –2047,...,–1024,1024,...,2047 | ....,... |

**TABLE 7.3**
**Huffman Table for Luminance**
**DC Coefficient Differences**

| Category | Code Length | Codeword |
|----------|-------------|----------|
| 0 | 2 | 00 |
| 1 | 3 | 010 |
| 2 | 3 | 011 |
| 3 | 3 | 100 |
| 4 | 3 | 101 |
| 5 | 3 | 110 |
| 6 | 4 | 1110 |
| 7 | 5 | 11110 |
| 8 | 6 | 111110 |
| 9 | 7 | 1111110 |
| 10 | 8 | 11111110 |
| 11 | 9 | 111111110 |

In contrast to the coding of DC coefficients, the quantized AC coefficients are arranged to a zigzag order before being entropy coded. This scan order is shown in Figure 7.5.

According to the zigzag scanning order, the quantized coefficients can be represented as:

$$ZZ(0) = S_{q00}, ZZ(1) = S_{q01}, ZZ(2) = S_{q10}, ...., ZZ(63) = S_{q77}. \tag{7.5}$$

Since many of the quantized AC coefficients become zero, they can be very efficiently encoded by exploiting the run of zeros. The run-length of zeros are identified by the nonzero coefficients. An 8-bit code 'RRRRSSSS' is used to represent the nonzero coefficient. The four least significant bits, 'SSSS', define a category for the value of the next nonzero coefficient in the zigzag sequence, which ends the zero run. The four most significant bits, 'RRRR', define the run-length of zeros in the zigzag sequence or the position of the nonzero coefficient in the zigzag sequence. The composite value, RRRRSSSS, is shown in Figure 7.6. The value 'RRRRSSSS' = '11110000' is defined as ZRL, "RRRR" = "1111" represents a run-length of 16 zeros and "SSSS" = "0000" represents a zero amplitude. Therefore, ZRL is used to represent a run-length of 16 zero coefficients followed

**TABLE 7.4**
**Huffman table for chrominance**
**DC coefficient differences**

| Category | Code Length | Codeword |
|----------|-------------|----------|
| 0 | 2 | 00 |
| 1 | 2 | 01 |
| 2 | 2 | 10 |
| 3 | 3 | 110 |
| 4 | 4 | 1110 |
| 5 | 5 | 11110 |
| 6 | 6 | 111110 |
| 7 | 7 | 1111110 |
| 8 | 8 | 11111110 |
| 9 | 9 | 111111110 |
| 10 | 10 | 1111111110 |
| 11 | 11 | 11111111110 |

DC

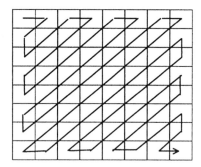

**FIGURE 7.5**   Zigzag scanning order of DCT coefficients.

SSSS

| RRRR | | 0 | 1 | 2 | | 9 | 10 |
|------|---|-----|---|---|---|---|----|
| | 0 | EOB | | | | | |
| | . | N/A | | | | | |
| | . | N/A | Composite values | | | | |
| | | N/A | | | | | |
| | 15 | ZRL | | | | | |

**FIGURE 7.6**   Two-dimensional value array for Huffman coding.

by a zero-amplitude coefficient, it is not an *abbreviation*. In the case of a run-length of zero coefficients that exceeds 15, multiple symbols will be used. A special value 'RRRRSSSS' = '00000000' is used to code the end-of-block (EOB). An EOB occurs when the remaining coefficients in the block are zeros. The entries marked "N/A" are undefined.

---

**TABLE 7.5**
**Huffman Coding for AC Coefficients**

| Category (SSSS) | AC Coefficient Range |
|---|---|
| 1 | −1,1 |
| 2 | −3,−2,2,3 |
| 3 | −7,...,−4,4,...,7 |
| 4 | −15,...,−8,8,...,15 |
| 5 | −31,...,−16,16,...,31 |
| 6 | −63,...,−32,32,...,63 |
| 7 | −127,...,−64,.64,...,127 |
| 8 | −255,...,−128,128,...,255 |
| 9 | −511,...,−256,256,...,511 |
| 10 | −1023,.,−512,512,...,1023 |
| 11 | −2047,...,−1024,1024,...,2047 |

---

The composite value, RRRRSSSS, is then Huffman coded. SSSS is actually the number to indicate "category" in the Huffman code table. The coefficient values for each category are shown in Table 7.5.

Each Huffman code is followed by additional bits that specify the sign and exact amplitude of the coefficients. As with the DC code tables, the AC code tables have also been developed from the average statistics of a large set of images with 8-bit precision. Each composite value is represented by a Huffman code in the AC code table. The format for the additional bits is the same as in the coding of DC coefficients. The value of SSSS gives the number of additional bits required to specify the sign and precise amplitude of the coefficient. The additional bits are either the low-order SSSS bits of $ZZ(k)$ when $ZZ(k)$ is positive, or the low-order SSSS bits of $ZZ(k)-1$ when $ZZ(k)$ is negative. Here, $ZZ(k)$ is the *kth* coefficient in the zigzag scanning order of coefficients being coded. The Huffman tables for AC coefficients can be found in Annex K of the JPEG standard (jpeg) and are not listed here due to space limitations.

As described above, Huffman coding is used as the means of entropy coding. However, an adaptive arithmetic coding procedure can also be used. As with the Huffman coding technique, the binary arithmetic coding technique is also lossless. It is possible to transcode between two systems without either of the FDCT or IDCT processes. Since this transcoding is a lossless process, it does not affect the picture quality of the reconstructed image. The arithmetic encoder encodes a series of binary symbols, zeros or ones, where each symbol represents the possible result of a binary decision. The binary decisions include the choice between positive and negative signs, a magnitude being zero or nonzero, or a particular bit in a sequence of binary digits being zero or one. There are four steps in the arithmetic coding: initializing the statistical area, initializing the encoder, terminating the code string, and adding restart markers.

## 7.3 PROGRESSIVE DCT-BASED ENCODING ALGORITHM

In progressive DCT-based coding, the input image is first partitioned to blocks of $8 \times 8$ pixels. The two-dimensional $8 \times 8$ DCT is then applied to each block. The transformed DCT-coefficient data are then encoded with multiple scans. At each scan, a portion of the transformed DCT coefficient data is encoded. This partially encoded data can be reconstructed to obtain a full image size with lower picture quality. The coded data of each additional scan will enhance the reconstructed image quality until the full quality has been achieved at the completion of all scans. Two methods have been used in the JPEG standard to perform the DCT-based progressive coding. These include spectral selection and successive approximation.

In the method of spectral selection, the transformed DCT coefficients are first reordered as a zigzag sequence and then divided into several bands. A frequency band is defined in the scan header by specifying the starting and ending indexes in the zigzag sequence. The band containing the DC coefficient is encoded at the first scan. In the following scan, it is not necessary for the coding procedure to follow the zigzag ordering.

In the method of the successive approximation, the DCT coefficients are first reduced in precision by the point transform. The point transform of the DCT coefficients is an arithmetic shift right by a specified number of bits, or division by a power of 2 (near zero, there is slight difference in truncation of precision between an arithmetic shift and division by 2, see annex K10 of [jpeg]). This specified number is the successive approximation of bit position. To encode using successive approximations, the significant bits of the DCT coefficient are encoded in the first scan, and each successive scan that follows progressively improves the precision of the coefficient by one bit. This continues until full precision is reached.

The principles of spectral selection and successive approximation are shown in Figure 7.7. For both methods, the quantized coefficients are coded with either Huffman or arithmetic codes at each scan. In spectral selection and the first scan of successive approximation for an image, the AC coefficient coding model is similar to that used in the sequential DCT-based coding mode. However, the Huffman code tables are extended to include coding of runs of end-of-bands (EOBs). For distinguishing the end-of-band and end-of-block, a number, n, which is used to indicate the range of run length, is added to the end-of-band (EOBn). The EOBn code sequence is defined as follows. Each EOBn is followed by an extension field, which has the minimum number of bits required to specify the run length. The end-of-band run structure allows efficient coding of blocks which have only zero coefficients. For example, an EOB run of length 5 means that the current block and the next 4 blocks have an end-of-band with no intervening nonzero coefficients. The Huffman coding structure of the subsequent scans of successive approximation for a given image is similar to the coding structure of the first scan of that image. Each nonzero quantized coefficient is described by a composite 8-bit run length-magnitude value of the form: RRRRSSSS. The four most significant bits, RRRR, indicate the number of zero coefficients between the current coefficient and the previously coded coefficient. The four least significant bits, SSSS, give the magnitude category of the nonzero coefficient. The run length-magnitude composite value is Huffman coded. Each Huffman code is followed by additional bits: one bit is used to code the sign of the nonzero coefficient and another bit is used to code the correction, where "0" means no correction and "1" means add one to the decoded magnitude of the coefficient. Although the above technique has been described using Huffman coding, it should be noted that arithmetic encoding can also be used in its place.

## 7.4 LOSSLESS CODING MODE

In the lossless coding mode, the coding method is spatially based coding instead of DCT-based coding. However, the coding method is extended from the method for coding the DC coefficients in the sequential DCT-based coding mode. Each pixel is coded with a predictive coding method, where the predicted value is obtained from one of three one-dimensional or one of four two-dimensional predictors, which are shown in Figure 7.8.

In Figure 7.8, the pixel to be coded is denoted by x, and the three causal neighbors are denoted by a, b, and c. The predictive value of x, Px, is obtained from three neighbors, a, b, and c in the one of seven ways as listed in Table 7.6.

In Table 7.6, the selection value 0 is only used for differential coding in the hierarchical coding mode. Selections 1, 2, and 3 are one-dimensional predictions and 4, 5, 6, and 7 are two-dimensional predictions. Each prediction is performed with full integer precision, and without clamping of either the underflow or overflow beyond the input bounds. In order to achieve lossless coding, the prediction differences are coded with either Huffman coding or arithmetic coding. The prediction

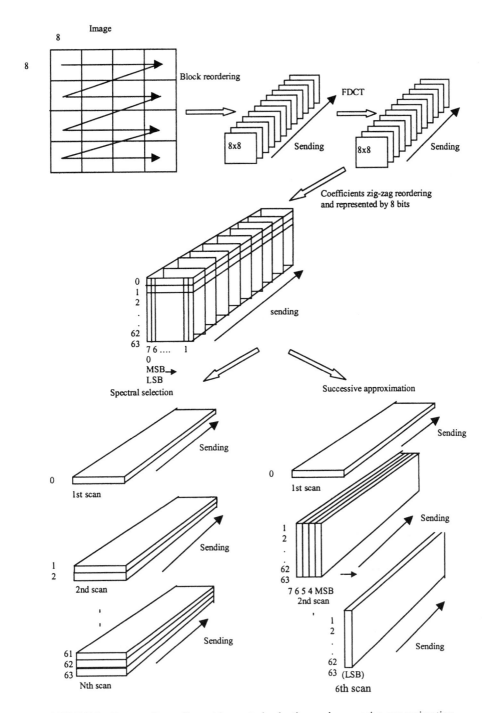

**FIGURE 7.7** Progressive coding with spectral selection and successive approximation.

difference values can be from 0 to $2^{16}$ for 8-bit pixels. The Huffman tables developed for coding DC coefficients in the sequential DCT-based coding mode are used with one additional entry to code the prediction differences. For arithmetic coding, the statistical model defined for the DC coefficients in the sequential DCT-based coding mode is generalized to a two-dimensional form in which differences are conditioned on the pixel to the left and the line above.

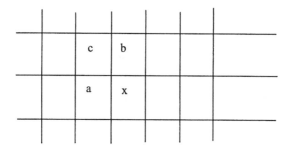

**FIGURE 7.8**   Spatial relationship between the pixel to be coded and three decoded neighbors.

**TABLE 7.6**
**Predictors for Lossless Coding**

| Selection-Value | Prediction |
|---|---|
| 0 | No prediction (hierarchical mode) |
| 1 | Px = a |
| 2 | Px = b |
| 3 | Px = c |
| 4 | Px = a+b-c |
| 5 | Px = a + ((b-c)/2)[a] |
| 6 | Px = b + ((a-c)/2)[a] |
| 7 | Px = (a+b)/2 |

[a] Shift right arithmetic operation.

## 7.5   HIERARCHICAL CODING MODE

The hierarchical coding mode provides a progressive coding similar to the progressive DCT-based coding mode, but it offers more functionality. This functionality addresses applications with multi-resolution requirements. In the hierarchical coding mode, an input image frame is first decomposed to a sequence of frames, such as the pyramid shown in Figure 7.2. Each frame is obtained through a down-sampling process, i.e., low-pass filtering followed by subsampling. The first frame (the lowest resolution) is encoded as a nondifferential frame. The following frames are encoded as differential frames, where the differential is with respect to the previously coded frame. Note that an up-sampled version that would be reconstructed in the decoder is used. The first frame can be encoded by the methods of sequential DCT-based coding, spectral selection, method of progressive coding, or lossless coding with either Huffman code or arithmetic code. However, within an image, the differential frames are either coded by the DCT-based coding method, the lossless coding method, or the DCT-based process with a final lossless coding. All frames within the image must use the same entropy coding, either Huffman or arithmetic, with the exception that nondifferential frames coded with the baseline coding may occur in the same image with frames coded with arithmetic coding methods. The differential frames are coded with the same method used for the nondifferential frames except the final frame. The final differential frame for each image may use a differential lossless coding method.

In the hierarchical coding mode, resolution changes in frames may occur. These resolution changes occur if down-sampling filters are used to reduce the spatial resolution of some or all frames of an image. When the resolution of a reference frame does not match the resolution of the frame to be coded, a up-sampling filter is used to increase the resolution of the reference frame. The block diagram of coding of a differential frame is shown in Figure 7.9.

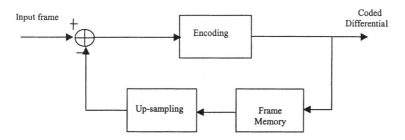

**FIGURE 7.9**   Hierarchical coding of a differential frame.

The up-sampling filter increases the spatial resolution by a factor of two in both horizontal and vertical directions by using bilinear interpolation of two neighboring pixels. The up-sampling with bilinear interpolation is consistent with the down-sampling filter that is used for the generation of down-sampled frames. It should be noted that the hierarchical coding mode allows one to improve the quality of the reconstructed frames at a given spatial resolution.

## 7.6   SUMMARY

In this chapter, the still image coding standard, JPEG, has been introduced. The JPEG coding standard includes four coding modes: sequential DCT-based coding mode, progressive DCT-based coding mode, lossless coding mode, and hierarchical coding mode. The DCT-based coding method is probably the one that most of us are familiar with; however, the lossless coding modes in JPEG which use a spatial domain predictive coding process have many interesting applications as well. For each coding mode, entropy coding can be implemented with either Huffman coding or arithmetic coding. JPEG has been widely adopted for many applications.

## 7.7   EXERCISES

**7-1.** What is the difference between sequential coding and progressive coding in JPEG? Conduct a project to encode an image with sequence coding and progressive coding, respectively.

**7-2.** Use the JPEG lossless mode to code several images and explain why different bit rates are obtained.

**7-3.** Generate a Huffman code table using a set of images with 8-bit precision (aproximately 2~3) using the method presented in Annex C of the JPEG specification. This set of images is called the training set. Use this table to code an image within the training set and an image which is not in the training set, and explain the results.

**7-4.** Design a three-layer progressive JPEG coder using (a) spectral selection, and (b) progressive approximation (0.3 bits per pixel at the first layer, 0.2 bits per pixel at the second layer, and 0.1 bits per pixel at the third layer).

## REFERENCES

Digital compression and coding of continuous-tone still images. Requirements and Guidelines, ISO-/IEC International Standard 10918-1, CCITT T.81, September, 1992.

Pennelbaker, W. B. and J. L. Mitchell, *JPEG: Still Image Data Compression Standard*, Van Nostrand Reinhold, New York, 1992.

Symes, P. *Compression: Fundamental Compression Techniques and an Overview of the JPEG and MPEG Compression Systems*, McGraw-Hill, New York, 1998.

# 8  Wavelet Transform for Image Coding

During the last decade, a number of signal processing applications have emerged using wavelet theory. Among those applications, the most widespread developments have occurred in the area of data compression. Wavelet techniques have demonstrated the ability to provide not only high coding efficiency, but also spatial and quality scalability features. In this chapter, we focus on the utility of the wavelet transform for image data compression applications.

## 8.1  REVIEW OF THE WAVELET TRANSFORM

### 8.1.1  DEFINITION AND COMPARISON WITH SHORT-TIME FOURIER TRANSFORM

The wavelet transform, as a specialized research field, started over a decade ago (Grossman and Morlet, 1984). To better understand the theory of wavelets, we first give a very short review of the Short-Time Fourier Transform (STFT) since there are some similarities between the STFT and the wavelet transform. As we know, the STFT uses sinusoidal waves as its orthogonal basis and is defined as:

$$F(\omega,\tau) = \int_{-\infty}^{+\infty} f(t)w(t-\tau)e^{-j\omega t}dt \tag{8.1}$$

where $w(t)$ is a time-domain windowing function, the simplest of which is a rectangular window that has a unit value over a time interval and has zero elsewhere. The value $\tau$ is the starting position of the window. Thus, the STFT maps a function $f(t)$ into a two-dimensional plane $(\omega,\tau)$. The STFT is also referred to as Gabor transform (Cohen, 1989). Similar to the STFT, the wavelet transform also maps a time or spatial function into a two-dimensional function in $a$ and $\tau$ ($\omega$ and $\tau$ for STFT). The wavelet transform is defined as follows. Let $f(t)$ be any square integrable function, i.e., it satisfies:

$$\int_{-\infty}^{+\infty} |f(t)|^2 dt < \infty \tag{8.2}$$

The continuous-time wavelet transform of $f(t)$ with respect to a wavelet $\psi(t)$ is defined as:

$$W(a,\tau) = \int_{-\infty}^{+\infty} f(t)\frac{1}{\sqrt{|a|}}\psi^*\left(\frac{t-\tau}{a}\right)dt \tag{8.3}$$

where $a$ and $\tau$ are real variables and * denotes complex conjugation. The wavelet is defined as:

$$\psi_{a\tau}(t) = |a|^{-1/2}\psi\left(\frac{t-\tau}{a}\right) \tag{8.4}$$

The above equation represents a set of functions that are generated from a single function, $\psi(t)$, by dilations and translations. The variable $\tau$ represents the time shift and the variable $a$ corresponds to the amount of time-scaling or dilation. If $a > 1$, there is an expansion of $\psi(t)$, while if $0 < a < 1$, there is a contraction of $\psi(t)$. For negative values of $a$, the wavelet experiences a time reversal in combination with a dilation. The function, $\psi(t)$, is referred to as the mother wavelet and it must satisfy two conditions:

1. The function integrates to zero:

$$\int_{-\infty}^{+\infty} \psi(t)dt = 0 \tag{8.5}$$

2. The function is square integrable, or has finite energy:

$$\int_{-\infty}^{+\infty} \psi|(t)|^2 dt < \infty \tag{8.6}$$

The continuous-time wavelet transform can now be rewritten as:

$$W(a,\tau) = \int_{-\infty}^{+\infty} f(t)\psi_{a\tau}^*(t)dt \tag{8.7}$$

In the following, we give two well-known examples of $\psi(t)$ and their Fourier transforms. The first example is the Morlet (modulated Gaussian) wavelet (Daubechies, 1990),

$$\Psi(\omega) = \sqrt{2\pi}\, e^{\frac{(\omega-\omega_0)^2}{2}} \tag{8.8}$$

and the second example is the Haar wavelet:

$$\psi = \begin{cases} 1 & 0 \leq t \leq 1/2 \\ -1 & 1/2 \leq t \leq 1 \\ 0 & otherwise \end{cases} \tag{8.9}$$

$$\psi(\omega) = je^{-j\frac{\omega}{2}}\frac{sm^2(\omega/4)}{\omega/4}$$

From the above definition and examples, we can find that the wavelets have zero DC value. This is clear from Equation 8.5. In order to have good time localization, the wavelets are usually bandpass signals and they decay rapidly towards zero with time. We can also find several other important properties of the wavelet transform and several differences between STFT and the wavelet transform.

The STFT uses a sinusoidal wave as its basis function. These basis functions keep the same frequency over the entire time interval. In contrast, the wavelet transform uses a particular wavelet as its basis function. Hence, wavelets vary in both position and frequency over the time interval. Examples of two basis functions for the sinusoidal wave and wavelet are shown in Figure 8.1(a) and (b), respectively.

The STFT uses a single analysis window. In contrast, the wavelet transform uses a short time window at high frequencies and a long time window at low frequencies. This is referred to as constant Q-factor filtering or relative constant bandwidth frequency analysis. A comparison of the

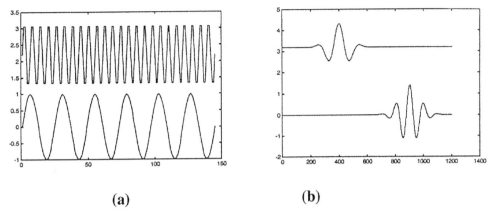

**(a)**                                          **(b)**

**FIGURE 8.1**    (a) Two sinusoidal waves, and (b) two wavelets.

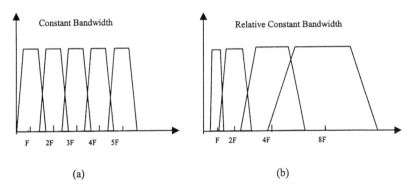

(a)                                          (b)

**FIGURE 8.2**    (a) Constant bandwidth analysis (for Fourier transform), and (b) relative constant bandwidth analysis (for wavelet transform).

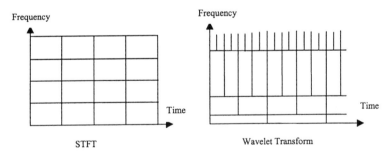

**FIGURE 8.3**    Comparison of the STFT and the wavelet transform in the time-frequency plane.

constant bandwidth analysis of the STFT and the relative constant bandwidth wavelet transform is shown in Figure 8.2(a) and (b), respectively.

This feature can be further explained with the concept of a time-frequency plane, which is shown in Figure 8.3.

As shown in Figure 8.3, the window size of the STFT in the time domain is always chosen to be constant. The corresponding frequency bandwidth is also constant. In the wavelet transform, the window size in the time domain varies with the frequency. A longer time window is used for a lower frequency and a shorter time window is used for a higher frequency. This property is very important for image data compression. For image data, the concept of a time-frequency plane becomes a spatial-frequency plane. The spatial resolution of a digital image is measured with pixels, as described in Chapter 15. To overcome the limitations of DCT-based coding, the wavelet transform allows the spatial resolution and frequency bandwidth to vary in the spatial-frequency plane. With this variation, better bit allocation for active and smooth areas can be achieved.

The continuous-time wavelet transform can be considered as a correlation. For fixed $a$, it is clear from Equation 8.3 that $W(a, \tau)$ is the cross-correlation of functions $f(t)$ with related wavelet conjugate dilated to scale factor $a$ at time lag $\tau$. This is an important property of the wavelet transform for multiresolution analysis of image data. Since the convolution can be seen as a filtering operation, the integral wavelet transform can be seen as a bank of linear filters acting upon $f(t)$. This implies that the image data can be decomposed by a bank of filters defined by the wavelet transform.

The continuous-time wavelet transform can be seen as an operator. First, it has the property of linearity. If we rewrite $W(a, \tau)$ as $W_{a\tau}[f(t)]$, then we have

$$W_{a\tau}[\alpha f(t) + \beta g(t)] = \alpha W_{a\tau}[f(t)] + \beta \, W_{a\tau}[g(t)] \qquad (8.10)$$

where $\alpha$ and $\beta$ are constant scalars. Second, it has the property of translation:

$$W_{a\tau}[f(t - \lambda)] = W(a, \tau - \lambda) \qquad (8.11)$$

where $\lambda$ is a time lag.

Finally, it has the property of scaling

$$W_{a\tau}[f(t/\alpha)] = W(a/\alpha, \tau/\alpha) \qquad (8.12)$$

### 8.1.2 DISCRETE WAVELET TRANSFORM

In the continuous-time wavelet transform, the function $f(t)$ is transformed to a function $W(a, \tau)$ using the wavelet $\psi(t)$ as a basis function. Recall that the two variables $a$ and $\tau$ are the dilation and translation, respectively. Now let us to find a means of obtaining the inverse transform, i.e., given $W(a, b)$, find $f(t)$. If we know how to get the inverse transform, we can then represent any arbitrary function $f(t)$ as a summation of wavelets, such as in the Fourier transform and DCT that provide a set of coefficients for reconstructing the original function using sine and cosine as the basis functions. In fact, this is possible if the mother wavelet satisfies the admissibility condition:

$$C = \int_{-\infty}^{+\infty} \frac{|\Psi(\omega)|^2}{|\omega|} d\omega \qquad (8.13)$$

where $C$ is a finite constant and $\Psi(\omega)$ is the Fourier transform of the mother wavelet function $\psi(t)$. Then, the inverse wavelet transform is

$$f(t) = \frac{1}{C} \int_{-\infty}^{+\infty} \int_{-\infty}^{+\infty} \frac{1}{|a|^2} W(a, \tau) \psi_{a\tau}(t) da d\tau \qquad (8.14)$$

The above results can be extended for two-dimensional signals. If $f(x,y)$ is a two-dimensional function, its continuous-time wavelet transform is defined as:

$$W\left(a,\tau_x,\tau_y\right)=\int_{-\infty}^{+\infty}\int_{-\infty}^{+\infty}f(x,y)\psi^*_{a\tau_x\tau_y}(x,y)dxdy \tag{8.15}$$

where $\tau_x$ and $\tau_y$ specify the transform in two dimensions. The inverse two-dimensional continuous-time wavelet transform is then defined as:

$$f(x,y)=\frac{1}{C}\int_{-\infty}^{+\infty}\int_{-\infty}^{+\infty}\int_{-\infty}^{+\infty}\frac{1}{|a|^3}W\left(a,\tau_x,\tau_y\right)\psi_{a\tau_x\tau_y}(x,y)dad\tau_xd\tau_y \tag{8.16}$$

where the $C$ is defined as in Equation 8.13 and $\psi(x,y)$ is a two-dimensional wavelet

$$\psi_{a\tau_x\tau_y}(x,y)=\frac{1}{|a|}\psi\left(\frac{x-\tau_x}{a},\frac{y-\tau_y}{a}\right) \tag{8.17}$$

For image coding, the wavelet is used to decompose the image data into wavelets. As indicated in the third property of the wavelet transform, the wavelet transform can be viewed as the cross-correlation of the function $f(t)$ and the wavelets $\psi_{a\tau}(t)$. Therefore, the wavelet transform is equivalent to finding the output of a bank of bandpass filters specified by the wavelets of $\psi_{a\tau}(t)$ as shown in Figure 8.4. This process decomposes the input signal into several subbands. Since each subband can be further partitioned, the filter bank implementation of the wavelet transform can be used for multiresolution analysis (MRA). Intuitively, when the analysis is viewed as a filter bank, the time resolution must increase with the central frequency of the analysis filters. This can be exactly obtained by the scaling property of the wavelet transform, where the center frequencies of the bandpass filters increase as the bandwidth becomes wider. Again, the bandwidth becomes wider by reducing the dilation parameter $a$. It should be noted that such a multiresolution analysis is consistent with the constant Q-factor property of the wavelet transform. Furthermore, the resolution limitation of the STFT does not exist in the wavelet transform since the time-frequency resolutions in the wavelet transform vary, as shown in Figure 8.2(b).

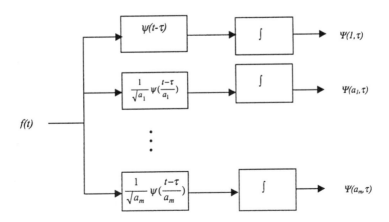

**FIGURE 8.4** The wavelet transform implemented with a bank of filters.

For digital image compression, it is preferred to represent $f(t)$ as a discrete superposition sum rather than an integral. With this move to the discrete space, the dilation parameter $a$ in Equation 8.10 takes the values $a = 2^k$ and the translation parameter $\tau$ takes the values $\tau = 2^k l$, where both $k$ and $l$ are integers. From Equation 8.4, the discrete version of $\psi_{a\tau}(t)$ becomes:

$$\psi_{kl}(t) = 2^{-\frac{k}{2}} \psi\left(2^{-k}t - l\right) \tag{8.18}$$

Its corresponding wavelet transform can be rewritten as:

$$W(k,l) = \int_{-\infty}^{+\infty} f(t)\psi_{kl}^*(t)dt \tag{8.19}$$

and the inverse transform becomes:

$$f(t) = \sum_{k=-\infty}^{+\infty}\sum_{l=-\infty}^{+\infty} d(k,l)2^{-\frac{k}{2}}\psi\left(2^{-k}t - l\right) \tag{8.20}$$

The values of the wavelet transform at those $a$ and $\tau$ are represented by $d(k,l)$:

$$d(k,l) = W(k,l)/C \tag{8.21}$$

The $d(k,l)$ coefficients are referred to as the discrete wavelet transform of the function $f(t)$ (Daubechies, 1992; Vetterli and Kovacevic, 1995). It is noted that the discretization so far is only applied to the parameters $a$ and $\tau$; $d(k,l)$ is still a continuous-time function. If the discretization is further applied to the time domain by letting $t = mT$, where $m$ is an integer and $T$ is the sampling interval (without loss of generality, we assume T = 1), then the discrete-time wavelet transform is defined as:

$$W_d(k,l) = \sum_{m=-\infty}^{+\infty} f(m)\psi_{kl}^*(m) \tag{8.22}$$

Of course, the sampling interval has to be chosen according to the Nyquist sampling theorem so that no information is lost in the process of sampling. The inverse discrete-time wavelet transform is then

$$f(m) = \sum_{m=-\infty}^{+\infty}\sum_{l=-\infty}^{+\infty} d(k,l)2^{-\frac{k}{2}}\psi\left(2^{-k}m - l\right) \tag{8.23}$$

## 8.2 DIGITAL WAVELET TRANSFORM FOR IMAGE COMPRESSION

### 8.2.1 BASIC CONCEPT OF IMAGE WAVELET TRANSFORM CODING

From the previous section, we have learned that the wavelet transform has several features that are different from traditional transforms. It is noted from Figure 8.2 that each transform coefficient in the STFT represents a constant interval of time regardless of which band the coefficient belongs to, whereas for the wavelet transform, the coefficients at the course level represent a larger time

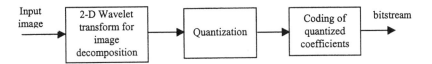

**FIGURE 8.5**  Block diagram of the image coding with the wavelet transform coding.

interval but a narrower band of frequencies. This feature of the wavelet transform is very important for image coding. In traditional image transform coding, which makes use of the Fourier transform or discrete cosine transform (DCT), one difficult problem is to choose the block size or window width so that statistics computed within that block provide good models of the image signal behavior. The choice of the block size has to be compromised so that it can handle both active and smooth areas. In the active areas, the image data are more localized in the spatial domain, while in the smooth areas the image data are more localized in the frequency domain. With traditional transform coding, it is very hard to reach a good compromise. The main contribution of wavelet transform theory is that it provides an elegant framework in which both statistical behaviors of image data can be analyzed with equal importance. This is because that wavelets can provide a signal representation in which some of the coefficients represent long data lags corresponding to a narrow band or low frequency range, and some of the coefficients represent short data lags corresponding to a wide band or high frequency range. Therefore, it is possible to obtain a good trade-off between spatial and frequency domain with the wavelet representation of image data.

To use the wavelet transform for image coding applications, an encoding process is needed which includes three major steps: image data decomposition, quantization of the transformed coefficients, and coding of the quantized transformed coefficients. A simplified block diagram of this process is shown in Figure 8.5. The image decomposition is usually a lossless process which converts the image data from the spatial domain to frequency domain, where the transformed coefficients are decorrelated. The information loss happens in the quantization step and the compression is achieved in the coding step. To begin the decomposition, the image data are first partitioned into four subbands labeled as $LL_1$, $HL_1$, $LH_1$, and $HH_1$, as shown in Figure 8.6(a). Each coefficient represents a spatial area corresponding to one-quarter of the original image size. The low frequencies represent a bandwidth corresponding to $0 < |\omega| < \pi/2$, while the high frequencies represent the band $\pi/2 < |\omega| < \pi$. To obtain the next level of decomposition, the $LL_1$ subband is further decomposed into the next level of four subbands, as shown in Figure 8.6(b). The low frequencies of the second level decomposition correspond to $0 < |\omega| < \pi/4$, while the high frequencies at the second level correspond to $\pi/4 < |\omega| < \pi/2$. This decomposition can be continued

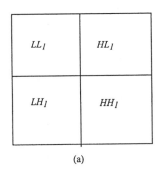

(a)

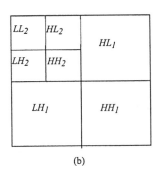

(b)

**FIGURE 8.6**  Two-dimensional wavelet transform. (a) First-level decomposition, and (b) second-level decomposition. (*L* denotes a low band, *H* denotes a high band, and the subscript denotes the number of the level. For example, $LL_1$ denotes the low-low band at level 1.)

to as many levels as needed. The filters used to compute the discrete wavelet transform are generally the symmetric quadrature mirror filters (QMF), as described by Woods (1991). A QMF-pyramid subband decomposition is illustrated in Figure 8.6(b).

During quantization, each subband is quantized differently depending on its importance, which is usually based on its energy or variance (Jayant and Noll, 1984). To reach the predetermined bit rate or compression ratio, coarse quantizers or large quantization steps would be used to quantize the low-energy subbands while the finer quantizers or small quantization steps would be used to quantize the large-energy subbands. This results in fewer bits allocated to those low-energy subbands and more bits for large-energy subbands.

### 8.2.2 Embedded Image Wavelet Transform Coding Algorithms

As with other transform coding schemes, most wavelet coefficients in the high-frequency bands have very low energy. After quantization, many of these high-frequency wavelet coefficients are quantized to zero. Based on the statistical property of the quantized wavelet coefficients, Huffman coding tables can be designed. Generally, most of the energy in an image is contained in the low-frequency bands. The data structure of the wavelet-transformed coefficients is suitable to exploit this statistical property.

Consider a multilevel decomposition of an image with the discrete wavelet transform, where the lowest levels of decomposition would correspond to the highest-frequency subbands and the finest spatial resolution, and the highest level of decomposition would correspond to the lowest-frequency subband and the coarsest spatial resolution. Arranging the subbands from lowest to highest frequency, we expect a decrease in energy. Also, we expect that if the wavelet-transformed coefficients at a particular level have lower energy, then coefficients at the lower levels or high-frequency subbands, which correspond to the same spatial location, would have smaller energy.

Another feature of the wavelet coefficient data structure is spatial self-similarity across subbands. Several algorithms that have been developed to exploit this and the above-mentioned properties for image coding. Among them, one of the first was proposed by Shapiro (1993) and used an embedded zerotree technique referred to as EZW. Another algorithm is the so-called set partitioning in hierarchical trees (SPIHT) developed by Said and Pearlman (1996). This algorithm also produces an embedded bitstream. The advantage of the embedded coding schemes allows an encoding process to terminate at any point so that a target bit rate or distortion metric can be met exactly. Intuitively, for a given bit rate or distortion requirement a nonembedded code should be more efficient than an embedded code since it has no constraints imposed by embedding requirements. However, embedded wavelet transform coding algorithms are currently the best. The additional constraints do not seem to have deleterious effect. In the following, we introduce the two embedded coding algorithms: the zerotree coding and the set partitioning in hierarchical tree coding.

As with DCT-based coding, an important aspect of wavelet-based coding is to code the positions of those coefficients that will be transmitted as nonzero values. After quantization the probability of the zero symbol must be extremely high for the very low bit rate case. A large portion of the bit budget will then be spent on encoding the significance map, or the binary decision map that indicates whether a transformed coefficient has a zero or nonzero quantized value. Therefore, the ability to efficiently encode the significance map becomes a key issue for coding images at very low bit rates. A new data structure, the zerotree, has been proposed for this purpose (Shapiro, 1993). To describe zerotree, we first must define insignificance. A wavelet coefficient is insignificant with respect to a given threshold value if the absolute value of this coefficient is smaller than this threshold. From the nature of the wavelet transform we can assume that every wavelet transformed at a given scale can be strongly related to a set of coefficients at the next finer scale of similar orientation. More specially, we can further assume that if a wavelet coefficient at a coarse scale is insignificant with respect to the preset threshold, then all wavelet coefficients at finer scales are likely to be insignificant with respect to this threshold. Therefore, we can build a tree with these

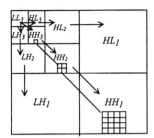 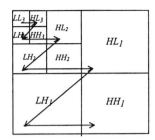

**FIGURE 8.7**   (Left) Parent-children dependencies of subbands; the arrow points from the subband of the parents to the subband of the children. At top left is the lowest-frequency band. (Right) The scanning order of the subbands for encoding a significance map.

parent-child relationships, such that coefficients at a coarse scale are called parents, and all coefficients corresponding to the same spatial location at the next finer scale of similar orientation are called children. Furthermore, for a parent, the set of all coefficients at all finer scales of similar orientation corresponding to the same spatial location are called descendants. For a QMF-pyramid decomposition the parent-children dependencies are shown in Figure 8.7(a). For a multiscale wavelet transform, the scan of the coefficients begins at the lowest frequency subband and then takes the order of *LL*, *HL*, *LH*, and *HH* from the lower scale to the next higher scale, as shown in Figure 8.7(b).

The zerotree is defined such that if a coefficient itself and all of its descendants are insignificant with respect to a threshold, then this coefficient is considered an element of a zerotree. An element of a zerotree is considered as a zerotree root if this element is not the descendant of a previous zerotree root with respect to the same threshold value. The significance map can then be efficiently represented by a string with three symbols: zerotree root, isolated zero, and significant. The isolated zero means that the coefficient is insignificant, but it has some significant descendant. At the finest scale, only two symbols are needed since all coefficients have no children, thus the symbol for zerotree root is not used. The symbol string is then entropy encoded. Zerotree coding efficiently reduces the cost for encoding the significance map by using self-similarity of the coefficients at different scales. Additionally, it is different from the traditional run-length coding that is used in DCT-based coding schemes. Each symbol in a zerotree is a single terminating symbol, which can be applied to all depths of the zerotree, similar to the end-of-block (EOB) symbol in the JPEG and MPEG video coding standards. The difference between the zerotree and EOB is that the zerotree represents the insignificance information at a given orientation across different scale layers. Therefore, the zerotree can efficiently exploit the self-similarity of the coefficients at the different scales corresponding to the same spatial location. The EOB only represents the insignificance information over the spatial area at the same scale.

In summary, the zerotree-coding scheme tries to reduce the number of bits to encode the significance map, which is used to encode the insignificant coefficients. Therefore, more bits can be allocated to encode the important significant coefficients. It should be emphasized that this zerotree coding scheme of wavelet coefficients is an embedded coder, which means that an encoder can terminate the encoding at any point according to a given target bit rate or target distortion metric. Similarly, a decoder which receives this embedded stream can terminate at any point to reconstruct an image that has been scaled in quality.

Another embedded wavelet coding method is the SPIHT-based algorithm (Said and Pearlman, 1996). This algorithm includes two major core techniques: the set partitioning sorting algorithm and the spatial orientation tree. The set partitioning sorting algorithm is the algorithm that hierarchically divides coefficients into significant and insignificant, from the most significant bit to the least significant bit, by decreasing the threshold value at each hierarchical step for constructing a significance map. At each threshold value, the coding process consists of two passes: the sorting

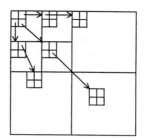

**FIGURE 8.8**   Relationship between pixels in
the spatial orientation tree.

pass and the refinement pass — except for the first threshold that has only the sorting pass. Let
$c(i,j)$ represent the wavelet-transformed coefficients and $m$ is an integer. The sorting pass involves
selecting the coefficients such that $2^m \le |c(i,j)| \le 2^{m+1}$, with $m$ being decreased at each pass. This
process divides the coefficients into subsets and then tests each of these subsets for significant
coefficients. The significance map constructed in the procedure is tree-encoded. The significant
information is store in three ordered lists: list of insignificant pixels (LIP), list of significant pixels
(LSP), and list of insignificant sets (LIS). At the end of each sorting pass, the LSP contains the
coordinates of all significant coefficients with respect to the threshold at that step. The entries in
the LIS can be one of two types: type A represents all its descendants, type B represents all its
descendants from its grandchildren onward. The refinement pass involves transmitting the $m$th-
most significant bit of all the coefficients with respect to the threshold, $2^{m+1}$.

The idea of a spatial orientation tree is based on the following observation. Normally, among
the transformed coefficients most of the energy is concentrated in the low frequencies. For the
wavelet transform, when we move from the highest to the lowest levels of the subband pyramid
the energy usually decreases. It is also observed that there exists strong spatial self-similarity
between subbands in the same spatial location such as in the zerotree case. Therefore, a spatial
orientation tree structure has been proposed for the SPIHT algorithm. The spatial orientation tree
naturally defines the spatial relationship on the hierarchical pyramid as shown in Figure 8.8.

During the coding, the wavelet-transformed coefficients are first organized into spatial orientation
trees as in Figure 8.8. In the spatial orientation tree, each pixel $(i,j)$ from the former set of subbands
is seen as a root for the pixels $(2i, 2j)$, $(2i+1, 2j)$, $(2i, 2j+1)$, and $(2i+1, 2j+1)$ in the subbands of the
current level. For a given $n$-level decomposition, this structure is used to link pixels of the adjacent
subbands from level $n$ until to level $1$. In the highest-level $n$, the pixels in the low-pass subband are
linked to the pixels in the three high-pass subbands at the same level. In the subsequent levels, all
the pixels of a subband are involved in the tree-forming process. Each pixel is linked to the pixels
of the adjacent subband at the next lower level. The tree stops at the lowest level.

The implementation of the SPIHT algorithm consists of four steps: initialization, sorting pass,
refinement pass, and quantization scale update. In the initialization step, we find an integer $m = \lfloor \log_2(\max_{(i,j)}\{|c(i,j)|\}) \rfloor$. Here $\lfloor \ \rfloor$ represent an operation of obtaining the largest integer less than
$|c(i,j)|$. The value of $m$ is used for testing the significance of coefficients and constructing the
significance map. The LIP is set as an empty list. The LIS is initialized to contain all the coefficients
in the low-pass subbands that have descendants. These coefficients can be used as roots of spatial
trees. All these coefficients are assigned to be of type A. The LIP is initialized to contain all the
coefficients in the low-pass subbands.

In the sorting pass, each entry of the LIP is tested for significance with respect to the threshold
value $2^m$. The significance map is transmitted in the following way. If it is significant, a "1" is
transmitted, a sign bit of the coefficient is transmitted, and the coefficient coordinates are moved
to the LSP. Otherwise, a "0" is transmitted. Then, each entry of the LIS is tested for finding the
significant descendants. If there are none, a "0" is transmitted. If the entry has at least one significant
descendant, then a "1" is transmitted and each of the immediate descendants are tested for significance. The significance map for the immediate descendants is transmitted in such a way that if it

is significant, a "1" plus a sign bit are transmitted and the coefficient coordinates are appended to the LSP. If it is not significant, a "0" is transmitted and the coefficient coordinates are appended to the LIP. If the coefficient has more descendants, then it is moved to the end of the LIS as an entry of type B. If an entry in the LIS is of type B, then its descendants are tested for significance. If at least one of them is significant, then this entry is removed from the list, and its immediate descendants are appended to the end of the list of type A. For the refinement pass, the $m$th-most significant bit of the magnitude of each entry of the LSP is transmitted except those in the current sorting pass. For the quantization scale update step, $m$ is decreased by 1 and the procedure is repeated from the sorting pass.

## 8.3   WAVELET TRANSFORM FOR JPEG-2000

### 8.3.1   INTRODUCTION TO JPEG-2000

Most image coding standards so far have exploited the DCT as their core technique for image decomposition. However, recently there has been a noticeable change. The wavelet transform has been adopted by MPEG-4 for still image coding (mpeg4). Also, JPEG-2000 is considering using the wavelet transform as its core technique for the next generation of the still image coding standard (jpeg2000 vm). This is because the wavelet transform can provide not only excellent coding efficiency, but also good spatial and quality scalable functionality. JPEG-2000 is a new type of image compression system under development by Joint Photographic Experts Group for still image coding. This standard is intended to meet a need for image compression with great flexibility and efficient interchangeability. JPEG-2000 is also intended to offer unprecedented access into the image while still in compressed domain. Thus, images can be accessed, manipulated, edited, transmitted, and stored in a compressed form. As a new coding standard, the detailed requirements of JPEG-2000 include:

*Low bit-rate compression performance*: JPEG-2000 is required to offer excellent coding performance at bit rates lower than 0.25 bits per pixel for highly detailed gray-bits per level images since the current JPEG (10918-1) cannot provide satisfactory results at this range of bit rates. This is the primary feature of JPEG-2000.

*Lossless and lossy compression*: it is desired to provide lossless compression naturally in the course of progressive decoding. This feature is especially important for medical image coding where the loss is not always allowed. Also, other applications such as high-quality image archival systems and network applications desire to have the functionality of lossless reconstruction.

*Large images*: currently, the JPEG image compression algorithm does not allow for images greater than 64K by 64K without tiling.

*Single decomposition architecture*: the current JPEG standard has 44 modes; many of these modes are for specific applications and not used by the majority of JPEG decoders. It is desired to have a single decomposition architecture that can encompass the interchange between applications.

*Transmission in noisy environments*: it is desirable to consider error robustness while designing the coding algorithm. This is important for the application of wireless communication. The current JPEG has provision for restart intervals, but image quality suffers dramatically when bit errors are encountered.

*Computer-generated imagery*: the current JPEG is optimized for natural imagery and does not perform well on computer-generated imagery or computer graphics.

*Compound documents*: the new coding standard is desired to be capable of compressing both continuous-tone and bilevel images. The coding scheme can compress and decompress images from 1 bit to 16 bits for each color component. The current JPEG standard does not work well for bilevel images.

*Progressive transmission by pixel accuracy and resolution*: progressive transmission that allows images to be transmitted with increasing pixel accuracy or spatial resolution is important for many applications. The image can be reconstructed with different resolutions and pixel accuracy as needed for different target devices such as in World Wide Web applications and image archiving.

*Real-time encoding and decoding*: for real-time applications, the coding scheme should be capable of compressing and decompressing with a single sequential pass. Of course, optimal performance cannot be guaranteed in this case.

*Fixed rate, fixed size, and limited workspace memory*: the requirement of fixed bit rate allows the decoder to run in real time through channels with limited bandwidth. The limited memory space is required by the hardware implementation of decoding.

There are also some other requirements such as backwards compatibility with JPEG, open architecture for optimizing the system for different image types and applications, interface with MPEG-4, and so on. All these requirements are seriously being considered during the development of JPEG-2000. However, it is still too early to comment whether all targets can be reached at this moment. There is no doubt, though, that the basic requirement on the coding performance at very low bit rate for still image coding will be achieved by using wavelet-based coding as the core technique instead of DCT-based coding.

## 8.3.2 VERIFICATION MODEL OF JPEG-2000

Since JPEG-2000 is still awaiting finalization, we introduce the techniques that are very likely to be adopted by the new standard. As in other standards such as MPEG-2 and MPEG-4, the verification model (VM) plays an important role during the development of standards. This is because the VM or TM (test model for MPEG-2) is a platform for verifying and testing the new techniques before they are adopted as standards. The VM is updated by completing a set of core experiments from one meeting to another. Experience has shown that the decoding part of the final version of VM is very close to the final standard. Therefore, in order to give an overview of the related wavelet transform parts of the JPEG-2000, we start to introduce the newest version of JPEG-2000 VM (jpeg2000 vm). The VM of JPEG-2000 describes the encoding process, decoding process, and the bitstream syntax, which eventually completely defines the functionality of the existing JPEG-2000 compression system.

The newest version of the JPEG-2000 verification model, currently VM 4.0, was revised on April 22, 1999. In this VM, the final convergence has not been reached, but several candidates have been introduced. These techniques include a DCT-based coding mode, which is currently the baseline JPEG, and a wavelet-based coding mode. In the wavelet-based coding mode, several algorithms have been proposed: overlapped spatial segmented wavelet transform (SSWT), non-overlapped SSWT, and the embedded block-based coding with optimized truncation (EBCOT). Among these techniques, and according to current consensus, EBCOT is a very likely candidate for adoption into the final JPEG-2000 standard.

The basic idea of EBCOT is the combination of block coding with wavelet transform. First, the image is decomposed into subbands using the wavelet transform. The wavelet transform is not restricted to any particular decomposition. However, the Mallat wavelet provides the best compression performance, on average, for natural images; therefore, the current bitstream syntax is restricted to the standard Mallat wavelet transform in VM 4.0. After decomposition, each subband is divided into 64 × 64 blocks, except at image boundaries where some blocks may have smaller sizes. Every block is then coded independently. For each block, a separate bitstream is generated without utilizing any information from other blocks. The key techniques used for coding include an embedded quadtree algorithm and fractional bit-plane coding.

| $B_i^1$ | $B_i^2$ | $B_i^3$ | $B_i^4$ |
|---|---|---|---|
| $B_i^5$ | $B_i^6$ | $B_i^7$ | $B_i^8$ |
| $B_i^9$ | $B_i^{10}$ | $B_i^{11}$ | $B_i^{12}$ |
| $B_i^{13}$ | $B_i^{14}$ | $B_i^{15}$ | $B_i^{16}$ |

**FIGURE 8.9** Example of sub-block partitioning for a block of $64 \times 64$.

The idea of an embedded quad-tree algorithm is that it uses a single bit to represent whether or not each leading bit-plane contains any significant samples. The quad-tree is formed in the following way. The subband is partitioned into a basic block. The basic block size is $64 \times 64$. Each basic block is further partitioned into $16 \times 16$ sub-blocks, as shown in Figure 8.9. Let $\sigma^j(B_i^k)$ denote the significance of sub-block, $B_i^k$($k$ is the $k$th sub-block as shown in Figure 8.9), in $j$th bit plane of $i$th block. If one or more samples in the sub-block have the magnitude greater than $2^j$, then $\sigma^j(B_i^k) = 1$; otherwise, $\sigma^j(B_i^k) = 0$. For each bit-plane, the information concerning the significant sub-blocks is first encoded. All other sub-blocks can then be bypassed in the remaining coding procedure for that bit-plane. To specify the exact coding sequence, we define a two-level quad-tree for the block size of $64 \times 64$ and sub-block size of $16 \times 16$. The level-1 quads, $Q_i^1[k]$, consist of four sub-blocks, $B_i^1$, $B_i^2$, $B_i^3$, $B_i^4$ from Figure 8.9. In the same way, we define level-2 quads, $Q_i^2[k]$, to be $2 \times 2$ groupings of level-1 quads. Let $\sigma^j(Q_i^1[k])$ denote the significance of the level-1 quad, $Q_i^1[k]$, in $j$th bit-plane. If at least one member sub-block is significant in the $j$th bit-plane, then $\sigma^j(Q_i^1[k]) = 1$; otherwise, $\sigma^j(Q_i^1[k]) = 0$. At each bit-plane, the quad-tree coder visits the level-2 quad first, followed by level-1 quads. When visiting a particular quad, $Q_i^L[k]$(L = 1 or 2, it is the number of the level), the coder sends the significance of each of the four child quads, $\sigma^j(Q_i^L[k])$, or sub-blocks, $\sigma^j(B_i^k)$, as appropriate, except if the significance value can be deduced from the decoder. Under the following three cases, the significance may be deduced by the decoder: (1) the relevant quad or sub-block was significant in the previous bit-plane; (2) the entire sub-block is insignificant; or (3) this is the last child or sub-block visited in $Q_i^L[k]$ and all previous quads or sub-blocks are insignificant.

The idea of bit-plane coding is to entropy code the most significant bit first for all samples in the sub-blocks and to send the resulting bits. Then, the next most significant bit will be coded and sent, this process will be continued until all bit-planes have been coded and sent. This kind of bitstream structure can be used for robust transmission. If the bitstream is truncated due to a transmission error or some other reason, then some or all the samples in the block may lose one or more least significant bits. This will be equivalent to having used a coarser quantizer for the relevant samples and we can still obtain a reduced-quality reconstructed image. The idea of fractional bit-plane coding is to code each bit-plane with four passes: a forward significance propagation pass, a backward significance propagation pass, a magnitude refinement pass, and a normalization pass. For the technical details of fractional bit-plane coding, the interested readers can refer to the VM of JPEG-2000 (jpeg2000 vm).

Finally, we briefly describe the optimization issue of EBCOT. The encoding optimization algorithm is not a part of the standard, since the decoder does not need to know how the encoder generates the bitstream. From the viewpoint of the standard, the only requirement from the decoder to the encoder is that the bitstream must be compliant with the syntax of the standard. However, from the other side, the bitstream syntax could always be defined to favor certain coding algorithms for generating optimized bitstreams. The optimization algorithm described here is justified only if the distortion measure adopted for the code blocks is additive. That is, the final distortion, $D$, of the whole reconstructed image should satisfy

$$D = \sum_i D_i^{Ti} \qquad (8.24)$$

where $D_i$ is the distortion for block $B_i$ and $T_i$ is the truncation point for $B_i$. Let $R$ be the total number of bits for coding all blocks of the image for a set of truncation point $T_i$, then

$$R = \sum_i R_i^{Ti} \qquad (8.25)$$

where $R_i^{Ti}$ are the bits for coding block $B_i$. The optimization process wishes to find the suitable set of $T_i$ values, which minimizes $D$ subject to the constraint $R \leq R_{max}$. $R_{max}$ is the maximum number of bits assigned for coding the image. The solution is obtained by the method of Lagrange multipliers:

$$L = \sum_i \left( R_i^{Ti} - \lambda D_i^{Ti} \right) \qquad (8.26)$$

where the value $\lambda$ must be adjusted until the rate obtained by the truncation points, which minimize the value of $L$, satisfies $R = R_{max}$. From Equation 8.26, we have a separate trivial optimization problem for each individual block. Specially, for each block, $B_i$, we find the truncation point, $T_i$, which minimizes the value $(R_i^{Ri} - \lambda D_i^{Ti})$. This can be achieved by finding the slope turning points of rate distortion curves. In the VM, the set of truncation points and the slopes of rate distortion curves are computed immediately after each block is coded, and we only store enough information to later determine the truncation points which correspond to the slope turning points of rate distortion curves. This information is generally much smaller than the bitstream which is stored for the block itself. Also, the search for the optimal $\lambda$ is extremely fast and occupies a negligible portion of the overall computation time.

## 8.4   SUMMARY

In this chapter, image coding using the wavelet transform has been introduced. First, an overview of wavelet theory was given, and second, the principles of image coding using wavelet transform have been presented. Additionally, two particular embedded image coding algorithms have been explained, namely, the embedded zerotree and set partitioning in hierarchical trees. Finally, the new standard for still image coding, JPEG-2000, which may adopt the wavelet as its core technique, has been described.

## 8.5   EXERCISES

**8-1.** For a given function, the Mexican hat wavelet,

$$f(t) = \begin{cases} 1, & for \ |t| \leq 1, \\ 0, & otherwise \end{cases}$$

Use Equations 8.3 and 8.4 to derive a closed-form expression for the continuous wavelet transform, $\psi_{ab}(t)$.

**8-2.** Consider the dilation equation

$$\varphi(t) = \sqrt{2} \sum_k h(k) \varphi(2t - k)$$

How does $\varphi(t)$ change if $h(k)$ is shifted? Specifically, let $g(k) = h(n-l)$.

$$u(t) = \sqrt{2} \sum_k g(k) u(2t - k)$$

How does $u(t)$ relate to $\varphi(t)$?

**8-3.** Let $\varphi_a(t)$ and $\varphi_b(t)$ be two scaling functions generated by the two scaling filters $h_a(k)$ and $h_b(k)$. Show that the convolution $j_a(t)* \ j_b(t)$ satisfies a dilation equation with $h_a(k)* \ h_b(k)/\sqrt{2}$.

**8-4.** In the applications of denoising and image enhancement, how can the wavelet transform improve the results?

**8-5.** For a given function

$$f(t) = \begin{cases} 0 & t < 0 \\ t & 0 \le t < 1 \\ 1 & t \ge 1 \end{cases}$$

show that the wavelet transform of $f(t)$ will be

$$W(a,b) = \text{sgn} \frac{2f\left(b + \dfrac{a}{2}\right) - f(b) - f(b + a)}{\sqrt{|a|}}$$

where sgn(x) is the signum function defined as

$$\text{sgn}(x) = \begin{cases} -1 & t < 0 \\ 1 & t > 0 \\ 0 & t = 0 \end{cases}$$

# REFERENCES

Cohen, L. Time-Frequency Distributions — A Review, *Proc. IEEE,* Vol. 77, No. 7, July 1989, pp. 941-981.

Daubechies, I. *Ten Lectures on Wavelets,* CBMS Series, Philadelphia, SIAM, 1992.

Grossman, A. and J. Morlet, Decompositions of hard functions into square integrable wavelets of constant shape, *SIAM J. Math. Anal.,* 15(4), 723-736, 1984.

Jayant, N. S. and P. Noll, *Digital Coding of Waveforms,* Englewood Cliffs, NJ: Prentice-Hall, 1984.

jpeg2000 vm, JPEG-2000 Verification Model 4.0 (Tech. description), sc29wg01 N1282, April 22, 1999.

mpeg4, ISO/IEC 14496-2, Coding of Audio-Visual Objects, Nov. 1998.

Said, A. and W. A. Pearlman, A new fast and efficient image codec based on set partitioning in hierarchical trees, *IEEE Trans. Circuits Syst. Video Technol.,* 243-250, 1996.

Shapiro, J. Embedded image coding using zerotrees of wavelet coefficients, *IEEE Trans. Signal Process.,* 3445-3462, Dec. 1993.

Vetterli, M. and J. Kovacevic, *Wavelets and Subband Coding,* Englewood Cliffs, NJ: Prentice-Hall, 1995.

Woods, J., Ed., *Subband Image Coding,* Kluwer Academic Publishers, 1991.

# 9 Nonstandard Image Coding

In this chapter, we introduce three nonstandard image coding techniques: vector quantization (VQ) (Nasrabadi and King, 1988), fractal coding (Barnsley and Hurd, 1993; Fisher, 1994; Jacquin, 1993), and model-based coding (Li et al., 1994).

## 9.1 INTRODUCTION

The VQ, fractal coding, and model-based coding techniques have not yet been adopted as an image coding standard. However, due to their unique features these techniques may find some special applications. Vector quantization is an effective technique for performing data compression. Theoretically, vector quantization is always better than scalar quantization because it fully exploits the correlation between components within the vector. The optimal coding performance will be obtained when the dimension of the vector approaches infinity, and then the correlation between all components is exploited for compression. Another very attractive feature of image vector quantization is that its decoding procedure is very simple since it only consists of table look-ups. However, there are two major problems with image VQ techniques. The first is that the complexity of vector quantization exponentially increases with the increasing dimensionality of vectors. Therefore, for vector quantization it is important to solve the problem of how to design a practical coding system which can provide a reasonable performance under a given complexity constraint. The second major problem of image VQ is the need for a codebook, which causes several problems in practical application such as generating a universal codebook for a large number of images, scaling the codebook to fit the bit rate requirement, and so on. Recently, the lattice VQ schemes have been proposed to address these problems (Li, 1997).

Fractal theory has a long history. Fractal-based techniques have been used in several areas of digital image processing such as image segmentation, image synthesis, and computer graphics, but only in recent years have they been extended to the applications of image compression (Jacquin, 1993). A fractal is a geometric form which has the unique feature of having extremely high visual self-similar irregular details while containing very low information content. Several methods for image compression have been developed based on different characteristics of fractals. One method is based on Iterated Function Systems (*IFS*) proposed by Barnsley (1988). This method uses the self-similar and self-affine property of fractals. Such a system consists of sets of transformations including translation, rotation, and scaling. On the encoder side of a fractal image coding system, a set of fractals is generated from the input image. These fractals can be used to reconstruct the image at the decoder side. Since these fractals are represented by very compact fractal transformations, they require very small amounts of data to be expressed and stored as formulas. Therefore, the information needed to be transmitted is very small. The second fractal image coding method is based on the fractal dimension (Lu, 1993; Jang and Rajala, 1990). Fractal dimension is a good representation of the roughness of image surfaces. In this method, the image is first segmented using the fractal dimension and then the resultant uniform segments can be efficiently coded using the properties of the human visual system. Another fractal image coding scheme is based on fractal geometry, which is used to measure the length of a curve with a yardstick (Walach, 1989). The details of these coding methods will be discussed in Section 9.3.

The basic idea of model-based coding is to reconstruct an image with a set of model parameters. The model parameters are then encoded and transmitted to the decoder. At the decoder the decoded

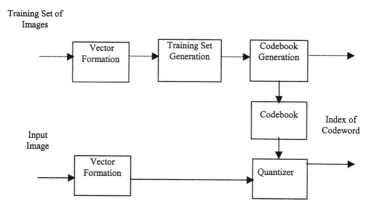

**FIGURE 9.1**  Principle of image vector quantization. The dashed lines correspond to training set generation, codebook generation, and transmission (if it is necessary).

model parameters are used to reconstruct the image with the same model used at the encoder. Therefore, the key techniques in the model-based coding are image modeling, image analysis, and image synthesis.

## 9.2  VECTOR QUANTIZATION

### 9.2.1  BASIC PRINCIPLE OF VECTOR QUANTIZATION

An N-level vector quantizer, $Q$, is mapping from a $K$-dimensional vector set $\{V\}$, into a finite codebook, $W = \{w_1, w_2, ..., w_N\}$:

$$Q: V \rightarrow W \tag{9.1}$$

In other words, it assigns an input vector, $v$, to a representative vector (codeword), $w$ from a codebook, $W$. The vector quantizer, $Q$, is completely described by the codebook, $W = \{w_1, w_2, ..., w_N\}$, together with the disjoint partition, $R = \{r_1, r_2, ..., r_N\}$, where

$$r_i = \{v: Q(v) = w_i\} \tag{9.2}$$

and $w$ and $v$ are $K$-dimensional vectors. The partition should identically minimize the quantization error (Gersho, 1982). A block diagram of the various steps involved in image vector quantization is depicted in Figure 9.1.

The first step in image vector quantization is the image formation. The image data are first partitioned into a set of vectors. A large number of vectors from various images are then used to form a training set. The training set is used to generate a codebook, normally using an iterative clustering algorithm. The quantization or coding step involves searching each input vector for the closest codeword in the codebook. Then the corresponding index of the selected codeword is coded and transmitted to the decoder. At the decoder, the index is decoded and converted to the corresponding vector with the same codebook as at the encoder by look-up table. Thus, the design decisions in implementing image vector quantization include (1) vector formation; (2) training set generation; (3) codebook generation; and (4) quantization.

#### 9.2.1.1  Vector Formation

The first step of vector quantization is vector formation; that is, the decomposition of the images into a set of vectors. Many different decompositions have been proposed; examples include the intensity values of a spatially contiguous block of pixels (Gersho and Ramamuthi, 1982; Baker and Gray, 1983); these same intensity values, but now normalized by the mean and variance of the block (Murakami et al., 1982); the transformed coefficients of the block pixels (Li and Zhang, 1995); and the adaptive linear predictive coding coefficients for a block of pixels (Sun, 1984). Basically, the approaches of vector formation can be classified into two categories: direct spatial or temporal, and feature extraction. Direct spatial or temporal is a simple approach to forming vectors from the intensity values of a spatial or temporal contiguous block of pixels in an image or an image sequence. A number of image vector quantizaton schemes have been investigated with this method. The other method is feature extraction. An image feature is a distinguishing primitive characteristic. Some features are natural in the sense that they are defined by the visual appearance of an image, while the other so-called artificial features result from specific manipulations or measurements of images or image sequences. In vector formation, it is well known that the image data in a spatial domain can be converted to a different domain so that subsequent quantization and joint entropy encoding can be more efficient. For this purpose, some features of image data, such as transformed coefficients and block means can be extracted and vector quantized. The practical significance of feature extraction is that it can result in the reduction of vector size, consequently reducing the complexity of coding procedure.

### 9.2.1.2  Training Set Generation

An optimal vector quantizer should ideally match the statistics of the input vector source. However, if the statistics of an input vector source are unknown, a training set representative of the expected input vector source can be used to design the vector quantizer. If the expected vector source has a large variance, then a large training set is needed. To alleviate the implementation complexity caused by a large training set, the input vector source can be divided into subsets. For example, in (Gersho, 1982) the single input source is divided into "edge" and "shade" vectors, and then the separate training sets are used to generate the separate codebooks. Those separate codebooks are then concatenated into a final codebook. In other methods, small local input sources corresponding to portions of the image are used as the training sets, thus the codebook can better match the local statistics. However, the codebook needs to be updated to track the changes in local statistics of the input sources. This may increase the complexity and reduce the coding efficiency. Practically, in most coding systems a set of typical images is selected as the training set and used to generate the codebook. The coding performance can then be insured for the images with the training set, or for those not in the training set but with statistics similar to those in the training set.

### 9.2.1.3  Codebook Generation

The key step in conventional image vector quantization is the development of a good codebook. The optimal codebook, using the mean squared error (MSE) criterion, must satisfy two necessary conditions (Gersho, 1982). First, the input vector source is partitioned into a predecided number of regions with the minimum distance rule. The number of regions is decided by the requirement of the bit rate, or compression ratio and coding performance. Second, the codeword or the representative vector of this region is the mean value, or the statistical center, of the vectors within the region. Under these two conditions, a generalized Lloyd clustering algorithm proposed by Linde, Buzo, and Gray (1980) — the so-called LBG algorithm — has been extensively used to generate the codebook. The clustering algorithm is an iterative process, minimizing a performance index calculated from the distances between the sample vectors and their cluster centers. The LBG clustering algorithm can only generate a codebook with a local optimum, which depends on the initial cluster seeds. Two basic procedures have been used to obtain the initial codebook or cluster seeds. In the first approach, the starting point involves finding a small codebook with only two codewords, and then recursively splitting the codebook until the required number of codewords is obtained. This approach is referred to as binary splitting. The second procedure starts with initial seeds for the required number of codewords, these seeds being generated by preprocessing the training sets. To address the problem of a local optimum, Equitz (1989) proposed a new clustering algorithm, the pairwise nearest neighbor (PNN) algorithm. The PNN algorithm begins with a separate cluster for each vector in the training set and merges together two clusters at a time until the desired codebook size is obtained. At the beginning of the clustering process, each cluster contains only one vector. In the following process the two closest vectors in the training set are merged to their statistical mean value, in such a way the error incurred by replacing these two vectors with a single codeword is minimized. The PNN algorithm significantly reduces computational complexity without sacrificing performance. This algorithm can also be used as an initial codebook generator for the LBG algorithm.

### 9.2.1.4  Quantization

Quantization in the context of a vector quantization involves selecting a codeword in the codebook for each input vector. The optimal quantization, in turn, implies that for each input vector, $v$, the closest codeword, $w_i$, is found as shown in Figure 9.2. The measurement criterion could be mean squared error, absolute error, or other distortion measures.

A full-search quantization is an exhaustive search process over the entire codebook for finding the closest codeword, as shown in Figure 9.3(a). It is optimal for the given codebook, but the computation is more expensive. An alternative approach is a tree-search quantization, where the search is carried out based on a hierarchical partition. A binary tree search is shown in Figure 9.3(b). A tree search is much faster than a full search, but it is clear that the tree search is suboptimal for the given codebook and requires more memory for the codebook.

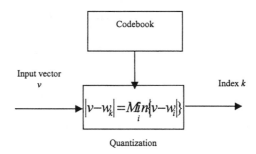

**FIGURE 9.2**  Principle of vector quantization.

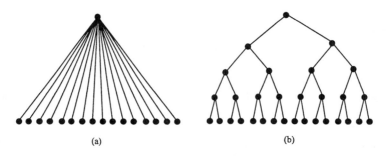

**FIGURE 9.3**  (a) Full search quantization; (b) binary tree search quantization.

## 9.2.2 SEVERAL IMAGE CODING SCHEMES WITH VECTOR QUANTIZATION

In this section, we are going to present several image coding schemes using vector quantization which include residual vector quantization, classified vector quantization, transform domain vector quantization, predictive vector quantization, and block truncation coding (BTC) which can be seen as a binary vector quantization.

### 9.2.2.1 Residual VQ

In the conventional image vector quantization, the vectors are formed by spatially partitioning the image data into blocks of 8 × 8 or 4 × 4 pixels. In the original spatial domain the statistics of vectors may be widely spread in the multidimensional vector space. This causes difficulty in generating the codebook with a finite size and limits the coding performance. Residual VQ is proposed to alleviate this problem. In residual VQ, the mean of the block is extracted and coded separately. The vectors are formed by subtracting the block mean from the original pixel values. This scheme can be further modified by considering the variance of the blocks. The original blocks are converted to the vectors with zero mean and unit standard deviation with the following conversion formula (Murakami et al., 1982):

$$m_i = \frac{1}{K} \sum_{j=0}^{k-1} s_j \tag{9.3}$$

$$x_j = \frac{\left(s_j - m_i\right)}{\sigma_i} \tag{9.4}$$

$$\sigma_i = \left[ \frac{1}{K} \sum_{j=0}^{K-1} \left(s_j - m_i\right)^2 \right]^{\frac{1}{2}} \tag{9.5}$$

where $m_i$ is the mean value of $i$th block, $\sigma_i$ is the variance of $i$th block, $s_j$ is the pixel value of pixel $j$ ($j = 0, ..., K-1$) in the $i$th block, $K$ is the total number of pixels in the block, and $x_j$ is the normalized value of pixel $j$. The new vector $X_i$ is now formed by $x_j$ ($j = 0, 1, ..., k-1$):

$$X_i = [x_0, x_1, ..., x_K]_i \tag{9.6}$$

With the above normalization the probability function $P(X)$ of input vector $X$ is approximately similar for image data from different scenes. Therefore, it is easy to generate a codebook for the new vector set. The problem with this method is that the mean and variance values of blocks have to be coded separately. This increases the overhead and limits the coding efficiency. Several methods have been proposed to improve the coding efficiency. One of these methods is to use predictive coding to code the block mean values. The mean value of the current block can be predicted by one of the previously coded neighbors. In such a way, the coding efficiency increases as the use of interblock correlation.

### 9.2.2.2 Classified VQ

In image vector quantization, the codebook is usually generated using training set under constraint of minimizing the mean squared error. This implies that the codeword is the statistical mean of the region. During quantization, each input vector is replaced by its closest codeword. Therefore, the coded images usually suffer from edge distortion at very low bit rates, since edges are smoothed by the operation of averaging with the small-sized codebook. To overcome this problem, we can classify the training vector set into edge vectors and shade vectors (Gersho, 1982). Two separate codebooks can then be generated with the two types of training sets. Each input vector can be coded by the appropriate codeword in the codebook. However, the edge vectors can be further classified into many types according to their location and angular orientation. The classified VQ can be extended into a system which contains many sub-codebooks, each representing a type of edge. However, this would increase the complexity of the system and would be hard to implement in practical applications.

### 9.2.2.3 Transform Domain VQ

Vector quantization can be performed in the transform domain. A spatial block of $4 \times 4$ or $8 \times 8$ pixels is first transformed to the $4 \times 4$ or $8 \times 8$ transformed coefficients. There are several ways to form vectors with transformed coefficients. In the first method, a number of high-order coefficients can be discarded since most of the energy is usually contained in the low-order coefficients for most blocks. This reduces the VQ computational complexity at the expense of a small increase in distortion. However, for some active blocks, the edge information is contained in the high frequencies, or high-order coefficients. Serious subjective distortion will be caused by discarding high frequencies. In the second method, the transformed coefficients are divided into several bands and each band is used to form its corresponding vector set. This method is equivalent to the classified VQ in spatial domain. An adaptive scheme is then developed by using two kinds of vector formation methods. The first method is used for the blocks containing the moderate intensity variation and the second method is used for the blocks with high spatial activities. However, the complexity increases as more codebooks are needed in this kind of adaptive coding system.

### 9.2.2.4 Predictive VQ

The vectors are usually formed by the spatially consecutive blocks. The consecutive vectors are then highly statistically dependent. Therefore, better coding performance can be achieved if the correlation between vectors is exploited. Several predictive VQ schemes have been proposed to address this problem. One kind of predictive VQ is finite state VQ (Foster et al., 1985). The finite-state VQ is similar to a trellis coder. In the finite state VQ, the codebook consists of a set of sub-codebooks. A state variable is then used to specify which sub-codebook should be selected for coding the input vector. The information about the state variable must be inferred from the received sequence of state symbols and initial state such as in a trellis coder. Therefore, no side information or no overhead need be transmitted to the decoder. The new encoder state is a function of the previous encoder state and the selected sub-codebook. This permits the decoder to track the encoder state if the initial condition is known. The finite-state VQ needs additional memory to store the previous state, but it takes advantage of correlation between successive input vectors by choosing the appropriate codebook for the given past history. It should be noted that the minimum distortion selection rule of conventional VQ is not necessary optimum for finite-state VQ for a given decoder since a low-distortion codeword may lead to a bad state and hence to poor long-term behavior. Therefore, the key design issue of finite-state VQ is to find a good next-state function.

Another predictive VQ was proposed by Hang and Woods (1985). In this system, the input vector is formed in such a way that the current pixel is as the first element of the vector and the previous inputs as the remaining elements in the vector. The system is like a mapping or a recursive filter which is used to predict the next pixel. The mapping is implemented by a vector quantizer look-up table and provides the predictive errors.

### 9.2.2.5 Block Truncation Coding

In the block truncation code (BTC) (Delp and Mitchell, 1979), an image is first divided into $4 \times 4$ blocks. Each block is then coded individually. The pixels in each block are first converted into two-level signals by using the first two moments of the block:

$$a = m + \sigma \sqrt{\frac{q}{N - q}}$$

$$b = m - \sigma \sqrt{\frac{N - 1}{q}}$$

(9.7)

where $m$ is the mean value of the block, $\sigma$ is the standard deviation of the block, $N$ is the number of total pixels in the block, and $q$ is the number of pixels which are greater in value than $m$. Therefore, each block can be described by the values of block mean, variance, and a binary-bit plane which indicates whether the pixels have values above or below the block mean. The binary-bit plane can be seen as a binary vector quantizer. If the mean and variance of the block are quantized to 8 bits, then 2 bits per pixel is achieved for blocks of $4 \times 4$ pixels. The conventional BTC scheme can be modified to increase the coding efficiency. For example, the block mean can be coded by a DPCM coder which exploits the interblock correlation. The bit plane can be coded with an entropy coder on the patterns (Udpikar and Raina, 1987).

### 9.2.3 LATTICE VQ FOR IMAGE CODING

In conventional image vector quantization schemes, there are several issues, which cause some difficulties for the practical application of image vector quantization. The first problem is the limitation of vector dimension. It has been indicated that the coding performance of vector quantization increases as the vector dimension while the coding complexity exponentially increases at the same time as the increasing vector dimension. Therefore, in practice only a small vector dimension is possible under the complexity constraint. Another important issue in VQ is the need for a codebook. Much research effort has gone into finding how to generate a codebook. However, in practical applications there is another problem of how to scale the codebook for various rate-distortion requirements. The codebook generated by LBG-like algorithms with a training set is usually only suitable for a specified bit rate and does not have the flexibility of codebook scalability. For example, a codebook generated for an image with small resolution may not be suitable for images with high resolution. Even for the same spatial resolution, different bit rates would require different codebooks. Additionally, the VQ needs a table to specify the codebook and, consequently, the complexity of storing and searching is too high to have a very large table. This further limits the coding performance of image VQ.

These problems become major obstacles for implementing image VQ. Recently, an algorithm of lattice VQ has been proposed to address these problems (Li et al., 1997). Lattice VQ does not have the above problems. The codebook for lattice VQ is simply a collection of lattice points uniformly distributed over the vector space. Scalability can be achieved by scaling the cell size associated with every lattice point just like in the scalar quantizer by scaling the quantization step. The basic concept of the lattice can be found in (Conway and Slone, 1991). A typical lattice VQ scheme is shown in Figure 9.4. There are two steps involved in the image lattice VQ. The first step is to find the closest lattice point for the input vector. The second step is to label the lattice point, i.e., mapping a lattice point to an index. Since lattice VQ does need a codebook, the index assignment is based on a lattice labeling algorithm instead of a look-up table such as in conventional VQ. Therefore, the key issue of lattice VQ is to develop an efficient lattice-labeling algorithm. With this

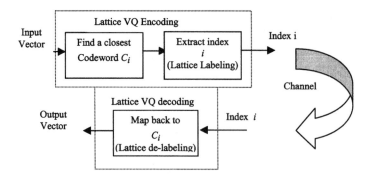

**FIGURE 9.4** Block diagram of lattice VQ.

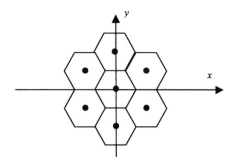

**FIGURE 9.5** Labeling a two-dimensional lattice.

algorithm the closest lattice point and its corresponding index within a finite boundary can be obtained by a calculation at the encoder for each input vector.

At the decoder, the index is converted to the lattice point by the same labeling algorithm. The vector is then reconstructed with the lattice point. The efficiency of a labeling algorithm for lattice VQ is measured by how many bits are needed to represent the indices of the lattice points within a finite boundary. We use a two-dimensional lattice to explain the lattice labeling efficiency. A two-dimensional lattice is shown in Figure 9.5.

In Figure 9.5, there are seven lattice points. One method used to label these seven 2-D lattice points is to use their coordinates $(x,y)$ to label each point. If we label $x$ and $y$ separately, we need two bits to label three values of $x$ and three bits to label a possible five values of $y$, and need a total of five bits. It is clear that three bits are sufficient to label seven lattice points. Therefore, different labeling algorithms may have different labeling efficiency. Several algorithms have been developed for multidimensional lattice labeling. In (Conway, 1983), the labeling method assigns an index to every lattice point within a Voronoi boundary where the shape of the boundary is the same as the shape of Voronoi cells. Apparently, for different dimension, the boundaries have different shapes. In the algorithm proposed in (Laroia, 1993), the same method is used to assign an index to each lattice point. Since the boundaries are defined by the labeling algorithm, this algorithm might not achieve a 100% labeling efficiency for a prespecified boundary such as a pyramid boundary. The algorithm proposed by Fischer (1986) can assign an index to every lattice point within a prespecified pyramid boundary and achieves a 100% labeling efficiency, but this algorithm can only be used for the $Z^n$ lattice. In a recently proposed algorithm (Wang et al., 1998), the technical breakthrough was obtained. In this algorithm a labeling method was developed for Construction-A and Construction-B lattices (Conway, 1983), which is very useful for VQ with proper vector dimensions, such as 16, and achieves 100% efficiency. Additionally, these algorithms are used for labeling lattice points with 16 dimensions and provide minimum distortion. These algorithms were developed based on the relationship between lattices and linear block codes. Construction-A and Construction-B are the two simplest ways to construct a lattice from a binary linear block code C = (n, k, d), where n, k, and d are the length, the dimension, and the minimum distance of the code, respectively.

A Construction-A lattice is defined as:

$$\Lambda_n = C + 2Z^n \tag{9.8}$$

where $Z^n$ is the $n$-dimensional cubic lattice and $C$ is a binary linear block code. There are two steps involved for labeling a Construction-A lattice. The first is to order the lattice points according to the binary linear block code $C$, and then to order the lattice points associated with a particular nonzero binary codeword. For the lattice points associated with a nonzero binary codeword, two sub-lattices are considered separately. One sub-lattice consists of all the dimensions that have a "0" component in the binary codeword and the other consists of all the dimensions that have a "1" component in the binary codeword. The first sub-lattice is considered as a 2Z lattice while the second is considered as a translated 2Z lattice. Therefore, the labeling problem is reduced to labeling the Z lattice at the final stage.

A Construction-B lattice is defined as:

$$\Lambda_n = C + 2D_n \tag{9.9}$$

where $D_n$ is an $n$-dimensional Construction-A lattice with the definition as:

$$D_n = (n, n-1, 2) + 2Z^n \tag{9.10}$$

and $C$ is a binary doubly even linear block code. When $n$ is equal to 16, the binary even linear block code associated with $\Lambda_{16}$ is $C = (16, 5, 8)$. The method for labeling a Construction-B lattice is similar to the method for labeling a Construction-A lattice with two minor differences. The first difference is that for any vector $y = c + 2x$, $x \in Z^n$, if $y$ is a Construction-A lattice point; and $x \in D_n$, if $y$ is a Construction-B lattice point. The second difference is that $C$ is a binary doubly even linear block code for Construction-B lattices while it is not necessarily doubly even for Construction-A lattices. In the implementation of these lattice point labeling algorithms, the encoding and decoding functions for lattice VQ have been developed in (Li et al., 1997). For a given input vector, an index representing the closest lattice point will be found by the encoding function, and for an input index the reconstructed vector will be generated by the decoding function. In summary, the idea of lattice VQ for image coding is an important achievement in eliminating the need for a codebook for image VQ. The development of efficient algorithms for lattice point labeling makes lattice VQ feasible for image coding.

## 9.3 FRACTAL IMAGE CODING

### 9.3.1 MATHEMATICAL FOUNDATION

A fractal is a geometric form whose irregular details can be represented by some objects with different scale and angle, which can be described by a set of transformations such as affine transformations. Additionally, the objects used to represent the image's irregular details have some form of self-similarity and these objects can be used to represent an image in a simple recursive way. An example of fractals is the Von Koch curve as shown in Figure 9.6. The fractals can be used to generate an image. The fractal image coding that is based on iterated function systems

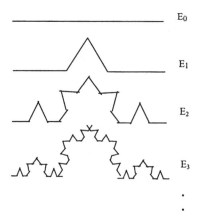

**FIGURE 9.6**   Construction of the Von Koch curve.

(*IFS*) is the inverse process of image generation with fractals. Therefore, the key technology of fractal image coding is the generation of fractals with an *IFS*.

To explain *IFS*, we start from the contractive affine transformation. A two-dimensional affine transformation A is defined as follows:

$$A\begin{bmatrix} x \\ y \end{bmatrix} = \begin{bmatrix} a & b \\ c & d \end{bmatrix}\begin{bmatrix} x \\ y \end{bmatrix} + \begin{bmatrix} e \\ f \end{bmatrix} \tag{9.11}$$

This is a transformation which consists of a linear transformation followed by a shift or translation, and maps points in the Euclidean plane into new points in the another Euclidean plane. We define that a transformation is contractive if the distance between two points $P_1$ and $P_2$ in the new plane is smaller than their distance in the original plane, i.e.,

$$d\big(A(P_1), A(P_2)\big) < s\, d(P_1, P_2) \tag{9.12}$$

where $s$ is a constant and $0 < s < 1$. The contractive transformations have the property that when the contractive transformations are repeatedly applied to the points in a plane, these points will converge to a fixed point. *An iterated function system (IFS) is defined as a collection of contractive affine transformations.* A well-known example of *IFS* contains four following transformations:

$$A_i\begin{bmatrix} x \\ y \end{bmatrix} = \begin{bmatrix} a & b \\ c & d \end{bmatrix}\begin{bmatrix} x \\ y \end{bmatrix} + \begin{bmatrix} e \\ f \end{bmatrix} \quad i = 1, 2, 3, 4. \tag{9.13}$$

This is the *IFS* of a fern leaf, whose parameters are shown in Table 9.1.

The transformation $A_1$ is used to generate the stalk, the transformation $A_2$ is used to generate the right leaf, the transformation $A_3$ is used to generate the left leaf, and the transformation $A_4$ is used to generate main fern. A fundamental theorem of fractal geometry is that each *IFS* defines a unique fractal image. This image is referred to as the attractor of the *IFS*. In other words, an image corresponds to the attractor of an *IFS*. Now let us explain how to generate the image using the *IFS*. Let us suppose that an *IFS* contains N affine transformations, $A_1, A_2, \ldots A_N$, and each transformation has an associated probability, $p_1, p_2, \ldots, p_N$, respectively. Suppose that this is a complete set and the sum of the probability equals to 1, i.e.,

**TABLE 9.1**
**The Parameters of the *IFS* of a Fern Leaf**

|       | a     | b     | c     | d    | e | f   |
|-------|-------|-------|-------|------|---|-----|
| $A_1$ | 0     | 0     | 0     | 0.16 | 0 | 0.2 |
| $A_2$ | 0.2   | −0.26 | 0.23  | 0.22 | 0 | 0.2 |
| $A_3$ | −0.15 | 0.28  | 0.26  | 0.24 | 0 | 0.2 |
| $A_4$ | 0.85  | 0.04  | −0.04 | 0.85 | 0 | 0.2 |

$$p_1 + p_2 + ... + p_N = 1 \text{ and } p_i > 0 \text{ for } i = 0, 1, ..., N. \tag{9.14}$$

The procedure for generating an attractor is as follows. For any given point $(x_0, y_0)$ in a Euclidean plane, one transformation in the *IFS* according to its probability is selected and applied to this point to generate a new point $(x_1, y_1)$. Then another transformation is selected according to its probability and applied to the point $(x_1, y_1)$ to obtain a new point $(x_2, y_2)$. This process is repeated over and over again to obtain a long sequence of points: $(x_0, y_0)$, $(x_1, y_1)$, ..., $(x_n, y_n)$, .... According to the theory of iterated function systems, these points will converge to an image that is the attractor of the given *IFS*. The above-described procedure is shown in the flowchart of Figure 9.7. With the above algorithm and the parameters in Table 9.1, initially the point can be anywhere within the large square, but after several iterations it will converge onto the fern. The 2-D affine transformations are extended to 3-D transformations, which can be used to create fractal surfaces with the iterated function systems. This fractal surface can be considered as the gray level or brightness of a 2-D image.

### 9.3.2 *IFS*-BASED FRACTAL IMAGE CODING

As described in the last section, an *IFS* can be used to generate a unique image, which is referred to as an attractor of the *IFS*. In other words, this image can be simply represented by the parameters of the *IFS*. Therefore, if we can use an inverse procedure to generate a set of transformations, i.e.,

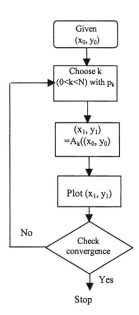

**FIGURE 9.7**   Flowchart of image generation with an *IFS*.

an *IFS* from an image, then these transformations or the *IFS* can be used to represent the approximation of the image. The image coding system can use the parameters of the transformations in the *IFS* instead of the original image data for storage or transmission. Since the *IFS* contains only very limited data such as transformation parameters, this image coding method may result in a very high compression ratio. For example, the fern image is represented by 24 integers or 192 bits (if each integer is represented by 8 bits). This number is much smaller than the number needed to represent the fern image pixel by pixel. Now the key issue of the *IFS*-based fractal image coding is to generate the *IFS* for the given input image. Three methods have been proposed to obtain the *IFS* (Lu, 1993). One is the direct method, that directly finds a set of contractive affine transformations from the image based on the self-similarity of the image. The second method is to partition an image into the smaller objects whose *IFS*s are known. These *IFS*s are used to form a library. The encoding procedure is to look for an *IFS* from the library for each small object. The third method is called partitioned *IFS* (*PIFS*). In this method, the image is first divided into smaller blocks and then the *IFS* for each block is found by mapping a larger block into a small block.

In the direct approach, the image is first partitioned into nonoverlapped blocks in such a way that each block is similar to the whole image and a transformation can map the whole image to the block. The transformation for each individual block may be different. The combination of these transformations can be taken as the *IFS* of the given image. Then much fewer data are required to represent the *IFS* or the transformations than to transmit or store the given image in the pixel by pixel way. For the second approach, the key issue is how to partition the given image into objects whose *IFS*s are known. The image processing techniques such as color separation, edge detection, spectrum analysis, and texture variation analysis can be used for image partitioning. However, for natural images or arbitrary images, it may be impossible or very difficult to find an *IFS* whose attractor perfectly covers the original image. Therefore, for most natural images the partitioned *IFS* method has been proposed (Lu, 1993). In this method, the transformations do not map the whole image into small block. For encoding an image, the whole image is first partitioned into a number of larger blocks that are referred to as domain blocks. The domain blocks can be overlapped. Then the image is partitioned into a number of smaller blocks that are called as range blocks. The range blocks do not overlap and the sum total of the range blocks covers the whole image. In the third step, a set of contractive transformations is chosen. Each range block is mapped into a domain block with a searching method and a matching criterion. The combination of the transformations is used to form a partitioned *IFS* (*PIFS*). The parameters of *PIFS* are transmitted to the decoder. It is noted that no domain blocks are transmitted. The decoding starts with a flat background. The iterated process is then applied with the set of transformations. The reconstructed image is then obtained after the process converges. From the above discussion, it is found that there are three main design issues involved in the block fractal image coding system. First are partitioning techniques which include range block partitioning and domain block partitioning. As mentioned earlier, the domain block is larger than the range block. Dividing the image into square blocks is the simplest partitioning approach. The second issue is the choice of distortion measurement and a searching method. The common distortion measurement in the block fractal image coding is the root mean square (*RMS*) error. The closest match between the range block and transformed domain block is found by the *RMS* distortion measurement. The third method is the selection of a set of contractive transformations defined consistently with a partition.

It is noted that the partitioned *IFS* (*PIFS*)-based fractal image coding has several similar features with image vector quantization. Both coding schemes are block-based coding schemes and need a codebook for encoding. For *PIFS*-based fractal image coding the domain blocks can be seen as forming a virtual codebook. One difference is that the fractal image coding does not need to transmit the codebook data (domain blocks) to the decoder while VQ does. The second difference is the block size. For VQ, block size for the code vector and input vector is the same while in *PIFS* fractal coding the size of the domain block is different from the size of the range blocks. Another difference is that in fractal image coding the image itself serves as the codebook, while this is not true for VQ image coding.

### 9.3.3 OTHER FRACTAL IMAGE CODING METHODS

Besides the *IFS*-based fractal image coding, there are several other fractal image coding methods. One is the segmentation-based coding scheme using fractal dimensions. In this method, the image is segmented into regions based on the properties of the human visual system (*HVS*). The image is segmented into the regions, each of these regions is homogeneous in the sense of having similar features by visual perception. This is different from the traditional image segmentation techniques that try to segment an image into regions of constant intensity. For a complicated image, good representation of an image needs a large number of small segmentations. However, in order to obtain a high compression ratio, the number of segmentations is limited. The trade-off between image quality and bit rate has to be considered. A parameter, fractal dimension, is used as a measure to control the trade-off. Fractal dimension is a characteristic of a fractal. It is related to a metric property such as the length of a curve and the area of a surface. The fractal dimension can provide a good measurement of the perceptual roughness of the curve and surface. For example, if we use many segments of straight lines to approximate a curve, by increasing the length of the straight lines perceptually rougher curves are represented.

## 9.4  MODEL-BASED CODING

### 9.4.1  BASIC CONCEPT

In the model-based coding, an image model that can be a 2-D model for still images or a 3-D model for video sequence is first constructed. At the encoder, the model is used to analyze the input image. The model parameters are then transmitted to the decoder. At the decoder the reconstructed image is synthesized by the model parameters, with the same image model used at the encoder. This basic idea of model-based coding is shown in the Figure 9.8. Therefore, the basic techniques in the model-based coding are the image modeling, image analysis, and image synthesis techniques. Both image analysis and synthesis are based on the image model. The image modeling techniques used for image coding can normally be divided into two classes: structure modeling and motion modeling. Motion modeling is usually used for video sequences and moving pictures, while structure modeling is usually used for still image coding. The structure model is used for reconstruction of a 2-D or 3-D scene model.

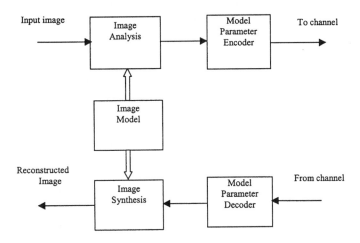

**FIGURE 9.8**  Basic principle of model-based coding.

### 9.4.2 IMAGE MODELING

The geometric model is usually used for image structure description. The geometric model can be classified into a surface-based description and volume-based description. The major advantage of surface description is that such description is easily converted into a surface representation that can be encoded and transmitted. In these models the surface is approximated by planar polygonal patches such as triangle patches. The surface shape is represented by a set of points that represent the vertices of these triangle meshes. The size of these triangle patches can be adjusted according to the surface complexity. In other words, for more complicated areas, more triangle meshes are needed to approximate the surface while for smoothing areas, the mesh sizes can be larger or less vertices of the triangle meshes are needed to represent the surface. The volume-based description is a natural approach for modeling most solid world objects. Most existing research work on volume-based description focuses on the parametric volume description. The volume-based description is used for 3-D objects or video sequences.

However, model-based coding is successfully applicable only to certain kinds of images since it is very hard to find general image models suitable for most natural scenes. The few successful examples of image models include the human face, head, and body. These models are developed for the analysis and synthesis of moving images. The face animation has been adopted for the MPEG-4 visual coding. The body animation is under consideration for version 2 of MPEG-4 visual coding.

## 9.5 SUMMARY

In this chapter three kinds of image coding techniques, vector quantization, fractal image coding, and model-based coding, which are not used in the current standards, have been presented. All three techniques have several important features such as very high compression ratios for certain kinds of images and very simple decoding procedures (especially for VQ). However, due to some limitations these techniques have not been adopted by industry standards. It should be noted that recently the facial model face animation technique has been adopted for the MPEG-4 visual standard (mpeg4 visual).

## 9.6 EXERCISES

**9-1.** In the modified residual VQ described in Equation 9.5, with a 4 × 4 block and 8 bits for each pixel of original image, we use 8 bits for coding block mean and block variance. We want to obtain the final bit rate of 2 bits per pixel. What codebook size do we have to use for the coding residual, assuming that we use fixed-length coding to code vector indices?

**9-2.** In the block truncation coding described in Equation 9.7, what is the bit rate for a block size of 4 × 4 if the mean and variance are both encoded with 8 bits? Do you have any suggestions for reducing the bit rate without seriously affecting the reconstruction quality?

**9-3.** Is the codebook generated with the LBG algorithm local optimum? List the several important factors that will affect the quality of codebook generation.

**9-4.** In image coding using VQ, what kind of problems will be caused by using the codebook in practical applications (hint: changing bit rate).

**9-5.** What is the most important improvement of the lattice VQ over traditional VQ in practical application. What is the key issue for lattice VQ for image coding application?

**9-6.** Write a subroutine to generate a fern leaf (using C).

# REFERENCES

Baker, R. L. and R. M. Gray, Image compression using nonadaptive spatial vector quantization, ISCAS'83, 1983, 55-61.

Barnsley, M.F. and A.E. Jacquin, Application of recurrent iterated function systems, SPIE, vol. 1001, *Visual Communications and Image Processing*, 1988, 122-131.

Barnsley, M. and L. P. Hurd, *Fractal Image Compression*, A.K. Peters, Wellesley, MA, 1993.

Conway, J. H. and N. J. A. Slone, A fast encoding method for lattice codes and quantizers, *IEEE Trans. Inform. Theory*, vol. IT-29, 820-824, 1983.

Conway, J. H. and N. J. A. Slone, *Sphere Packings, Lattices and Groups*, New York: Springer-Verlag, 1991.

Delp, E. J. and D. R. Mitchell, Image compression using block truncation coding, *IEEE Trans. Commun.*, COM-27, 1979.

Dunham, M. and R. Gray, An algorithm for the design of labelled-transition finite-state vector quantizer, *IEEE Trans. Commun.*, COM-33, 83-89, 1985.

Equits, W. H. A new vector quantization clustering algorithm, *IEEE Trans. Acoust. Speech Signal Process.*, 37, 1568-1575, 1989.

Fischer, T. R., A paramid vector quantization, *IEEE Trans. Inform. Theory*, vol. IT-32, 568-583, 1986.

Fisher, Y. *Fractal Image Compression — Theory and Application,* New York: Springer-Verlag, 1994.

Foster, J., R. M. Gray, and M. O. Dunham, Finite-state vector quantization for waveform coding, *IEEE Trans. Inf. Theory*, IT-31, 348-359, 1985.

Gersho, A. and B. Ramamurthi, Image coding using vector quantization, ICASSP'82, Paris, May 1982, 428-431.

Gersho, A. On the structure of vector quantizer, *IEEE Trans. Inf. Theory*, IT-28, 157-166, 1982.

Hang, H. M. and J. W. Woods, Predictive vector quantization of images, *IEEE Trans. Commun.*, COM-33, 1208-1219, 1985.

ISO/IEC 14496-2, Coding of Audio-Visual Objects, Part 2, Dec. 18, 1998.

Jacquin, A. E. Fractal Image Coding: A Review, *Proc. IEEE,* 81(10), 1451-1465, 1993.

Jang, J. and S. A. Rajala, Segmentation-based image coding using fractals and the human visual system, IEEE Int. Conf. Acoust. Speech Signal Processing, 1990, pp. 1957-1960.

Laroia, R. and N. Farvardin, A structured fixed rate vector quantizer derived from a variable length scaler quantizer: I & II, *IEEE Trans. Inform. Theory*, vol. 39, 851-876, 1993.

Li, H., A. Lundmark, and R. Forchheimer, Image Sequence Coding at Very Low Bitrates: A Review, *IEEE Trans. Image Process.*, 3(5), 1994.

Li, W. and Y. Zhang, Vector-based signal processing and quantization for image and video compression, *Proc. IEEE,* Volume 83(2), 317-335, 1995.

Li, W. et al., A video coding algorithm using vector-based tehnique, *IEEE Trans. Circuits Syst. Video Technol.*, 7(1), 146-157, 1997.

Linde, Y. A. Buzo, and R. M. Gray, An algorithm for vector quantizer design, *IEEE Trans. Commun.*, 28, 84-95, 1980.

Lu, G. Fractal image compression, *Signal Process. Image Commun.*, 5, 327-343, 1993.

Murakami, T., K. Asai, and E. Yamazaki, Vector quantization of video signals, *Electron. Lett.*, 7, 1005-1006, 1982.

Nasrabadi, N. M. and R. A. King, Image Coding using Vector Quantization: A Review, *IEEE Trans. Commun.*, COM-36(8), 957-971, 1988.

Stewart, L. C., R. M. Gray and Y. Linde, The design of trellis waveform coders, *IEEE Trans. Commun.*, COM-30, 702-710, 1982.

Sun, H. and M. Goldberg, Image coding using LPC with vector quantization, *IEEE Proc. Int. Conf. Digital Signal Processing*, Florence, Italy, Sept. 1984, 508-512.

Udpikar, V. R. and J. P. Raina, BTC image coding using vector quantization, *IEEE Trans. Commun.*, COM-35, 352-356, 1987.

Walach, E. and E. Karnin, A fractal-based approach to image compression, *ICASSP 1986*, 529-532.

Wang, C., H. Q. Cao, W. Li, and K. K. Tzeng, Lattice Labeling Algorithm for Vector Quantization, *IEEE Trans. Circuits Syst. Video Technol.*, 8(2), 206-220, 1998.

# Section III

## Motion Estimation and Compression

# 10 Motion Analysis and Motion Compensation

Up to this point, what we have discussed in the previous chapters were basic techniques in image coding, specifically, techniques utilized in still image coding. From here on, we are going to address the issue of video sequence compression. To fulfill the task, we will first define the concepts of image and video sequences. Then we address the issue of interframe correlation between successive frames. Two techniques in exploitation of interframe correlation, frame replenishment and motion-compensated coding, will then be discussed. The rest of the chapter covers the concepts of motion analysis and motion compensation in general.

## 10.1 IMAGE SEQUENCES

In this section the concept of various image sequences is defined in a theoretical and systematic manner. The relationship between image sequences and video sequences is also discussed.

It is well known that in the 1960s the advent of the semiconductor computer and the space program swiftly brought the field of digital image processing into public focus. Since then the field has experienced rapid growth and has entered every aspect of modern technology. Since the early 1980s, digital image sequence processing has been an attractive research area (Huang, 1981a, 1983). This is not surprising, because an image sequence, as a collection of images, may provide more information than a single image frame. The increased computational complexity and memory space associated with image sequence processing are becoming more affordable due to more advanced, achievable computational capability. With the tremendous advancements continuously made in VLSI computer and information processing, image and video sequences are evermore indispensable elements of modern life. While the pace and the future of this development cannot be predicted, one thing is certain: this process is going to drastically change all aspects of our world in the next several decades.

As far as image sequence processing is concerned, it is noted that in addition to temporal image sequences, stereo image pair and stereo image sequences also received attention in the middle of the 1980s (Waxman and Duncan, 1986). The concepts of temporal and spatial image sequences, and the imaging space (which may be considered as a next-higher-level unification of temporal and spatial image sequences) may be illustrated as follows.

Consider a sensor located in a specific position in the three-dimensional (3-D) world space. It generates images about the scene, one after another. As time goes by, the images form a sequence. The set of these images can be represented with a brightness function g(x,y,t), where x and y are coordinates on the image planes. This is referred to as a *temporal image sequence*. This is the basic outline about the brightness function g(x,y,t) dealt with by researchers in both computer vision, e.g., Horn and Schunck (1980) and signal processing fields, e.g., Pratt (1979).

Now consider a generalization of the above basic outline. A sensor, as a solid article, can be translated (in three free dimensions) and rotated (in two free dimensions). It is noted that here the rotation of a sensor about its optical axis is not counted, since the images generated will remain unchanged when this type of rotation takes place. So, we can obtain a variety of images when a sensor is translated to different coordinates and rotated to different angles in the 3-D world space. Equivalently, we can imagine that there is an infinite number of sensors in the 3-D world space that occupies all possible spatial coordinates and assumes all possible orientations at each coordinate; i.e., they are located on all possible positions. At one specific moment, all of these images form a set, which can be referred to as a *spatial image sequence*. When time varies, these sets of images form a much larger set of images, called an *imaging space*.

Clearly, it is impossible to describe such a set of images by using the above-mentioned g(x,y,t). Instead, it should be described by a more general brightness function,

$$g(x, y, t, \bar{s}),\qquad(10.1)$$

where $\bar{s}$ indicates the sensor's position in the 3-D world space; i.e., the coordinates of the sensor center and the orientation of the optical axis of the sensor. Hence $\bar{s}$ is a 5-D vector. That is,

$$\bar{s} = (\tilde{x}, \tilde{y}, \tilde{z}, \beta, \gamma),\qquad(10.2)$$

where $\tilde{x}$, $\tilde{y}$, and $\tilde{z}$ represent the coordinates of the optical center of the sensor in the 3-D world space; and $\beta$ and $\gamma$ represent the orientation of the optical axis of the sensor in the 3-D world space. More specifically, each sensor in the 3-D world space may be considered associated with a 3-D Cartesian coordinate system such that its optical center is located on the origin and its optical axis is aligned with the $OZ$ axis. In the 3-D world space we choose a 3-D Cartesian coordinate system as the reference coordinate system. Hence, a sensor with its Cartesian coordinate system coincident with the reference coordinate system has its position in the 3-D world space denoted by $\bar{s}$ = (0,0,0,0,0). An arbitrary sensor position denoted by $\bar{s}$ = ($\tilde{x}$, $\tilde{y}$, $\tilde{z}$, $\hat{a}$, $\tilde{a}$) can be described as follows. The sensor's associated Cartesian coordinate system is first shifted from the reference coordinate system in the 3-D world space with its origin settled at ($\tilde{x}$, $\tilde{y}$, $\tilde{z}$) in the reference coordinate system. Then it is rotated with the rotation angles $\beta$ and $\gamma$ being the same as Euler angles (Shu and Shi, 1991; Shi et al., 1994). Figure 10.1 shows the reference coordinate system and an arbitrary Cartesian coordinate system (indicating an arbitrary sensor position). There, $oxy$ and $o'x'y'$ represent, respectively, the related image planes.

Assume now a world point $P$ in the 3-D space that is projected onto the image plane as a pixel with the coordinates $x_P$ and $y_P$. Then, $x_P$ and $y_P$ are also dependent on t and $\bar{s}$. That is, the coordinates of the pixel can be denoted by $x_P = x_P (t, \bar{s})$ and $y_P = y_P (t, \bar{s})$. So generally speaking, we have

$$g = g\left(x_p(t, \bar{s}), y_p(t, \bar{s}), t, \bar{s}\right).\qquad(10.3)$$

As far as temporal image sequences are concerned, let us take a look at the framework of Pratt (1979), and Horn and Schunck (1980). There, $g = g (x_P (t), y_P (t), t)$ is actually a special case of Equation 10.3, i.e., $g = g(x_P(t, \bar{s} = \text{constant vector}), y_P(t, \bar{s} = \text{constant vector}), (t, \bar{s} = \text{constant vector})$. In other words, the variation of $\bar{s}$ is restricted to be zero, i.e., $\Delta\bar{s} = 0$. This means the sensor is fixed in a certain position in the 3-D world space.

Obviously, an alternative is to define the imaging space as a set of all temporal image sequences; i.e., those taken by sensors located at all possible positions in the 3-D world space. Stereo image sequences can thus be viewed as a proper subset of the imaging space, just like a stereo pair of images can be considered as a proper subset of a spatial image sequence.

In summary, the imaging space is a collection of all possible forms assumed by the general brightness function g (x, y, t, $\bar{s}$). Each picture taken by a sensor located on a particular position at a specific moment is merely a special cross section of this imaging space. Both temporal and spatial image sequences are special proper subsets of the imaging space. They are in the middle level, between the imaging space and the individual images. This hierarchical structure is depicted in Figure 10.2.

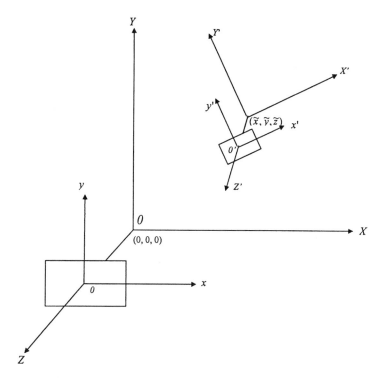

**FIGURE 10.1**  Two sensor positions $\bar{s} = (0,0,0,0,0)$ and $\bar{s} = (\tilde{x}, \tilde{y}, \tilde{z}, \hat{a}, \hat{a})$.

Before we conclude this section, we should discuss the relationship between image sequences and video sequences. It is noted that the term *video* is used very often nowadays in addition to the terms *image frames* and *image sequence*. It is necessary to pause for a while to discuss the relationship between these terms. Image frames and image sequence have been defined clearly above with the introduction of the concept of the imaging space. Video can mean an individual video frame or video sequences. It refers, however, to those frames and sequences that are associated with the visible frequency band in the electromagnetic spectrum. For image frames and image sequences, there is no such restriction. For instance, infrared image frames and sequences correspond to a band outside the visible band in the spectrum. From this point of view, the scope of image frames and sequences is wider than that of video frames and sequences. When the visible band is concerned, the terms *image frame and sequence* are interchangeable with that *video frame and sequence*.

Another point we would like to bring to the reader's attention is as follows. Though video is referred to as visual information, which includes both a single frame and frame sequences, in practice it is often used to mean sequences exclusively. Such an example can be found in *Digital Video Processing* (Tekalp, 1995).

In this book, we use *image compression* to indicate still image compression, and *video compression* to indicate video sequence compression. Readers should keep in mind, however, that (1) video can mean a single frame or sequences of frames; and (2) the scope of image is wider than that of video, and video is more pertinent to multimedia engineering.

## 10.2  INTERFRAME CORRELATION

As far as video compression is concerned, all the techniques discussed in the previous chapters are applicable. By this we mean two classes of techniques. The first class, which is also the most straightforward way to handle video compression, is to code each frame separately. That is,

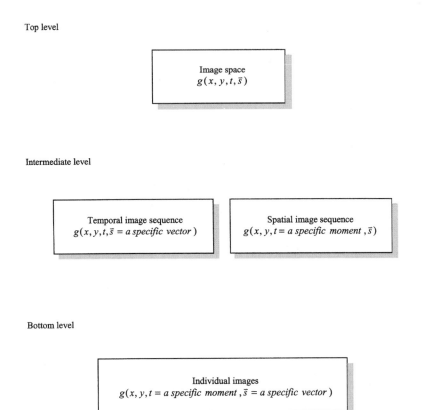

Top level

Image space
$g(x, y, t, \bar{s})$

Intermediate level

Temporal image sequence
$g(x, y, t, \bar{s} = a\ specific\ vector\ )$

Spatial image sequence
$g(x, y, t = a\ specific\ moment\ , \bar{s})$

Bottom level

Individual images
$g(x, y, t = a\ specific\ moment\ , \bar{s} = a\ specific\ vector\ )$

**FIGURE 10.2**   A hierarchical structure.

individual frames are coded independently on each other. For instance, using a JPEG compression algorithm to code each frame in a video sequence results in *motion JPEG* (Westwater and Furht, 1997). In the second class, methods utilized for still image coding can be generalized for video compression. For instance, (DCT) transform coding can be generalized and applied to video coding by extending 2-D DCT to 3-D DCT. That is, instead of 2-D DCT, say, $8 \times 8$, applied to a single image frame, we can apply 3-D DCT, say, $8 \times 8 \times 8$, to a video sequence. Refer to Figure 10.3. That is, 8 blocks of $8 \times 8$ each located, respectively, at the same position in one of the 8 successive frames from a video sequence are coded together with the 3-D DCT. It was reported that this 3-D DCT technique is quite efficient (Lim, 1990; Westwater and Furht, 1997). In addition, the DPCM technique and the hybrid technique can be generalized and applied to video compression in a similar fashion (Jain, 1989; Lim, 1990). It is noted that in the second class of techniques several successive frames are grouped and coded together, while in the first class each frame is coded independently.

   Video compression has its own characteristics, however, that make it quite different from still image compression. The major difference lies in the exploitation of interframe correlation that exists between successive frames in video sequences, in addition to the intraframe correlation that exists within each frame. As mentioned in Chapter 1, the interframe correlation is also referred to as *temporal redundancy*, while the intraframe correlation is referred to as *spatial redundancy*. In order to achieve coding efficiency, we need to remove these redundancies for video compression. To do so we must first understand these redundancies.

   Consider a video sequence taken in a videophone service. There, the camera is static most of the time. A typical scene is a head-and-shoulder view of a person imposed on a background. In this type of video sequence the background is usually static. Only the speaker is experiencing motion, which is not severe. Therefore, there is a strong similarity between successive frames, that

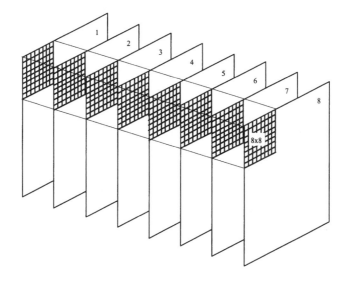

**FIGURE 10.3**   3-D DCT of $8 \times 8 \times 8$.

is, a strong adjacent-frame correlation. In other words, there is a strong interframe correlation. It was reported by Mounts (1969) that when using videophone-like signals with moderate motion in the scene, on average, less than one-tenth of the elements change between frames by an amount which exceeds 1% of the peak signal. Here, a 1% change is regarded as significant. Our experiment on the first 40 frames of the Miss America sequence supports this observation. Two successive frames of the sequence, frames 24 and 25, are shown in Figure 10.4.

Now, consider a video sequence generated in a television broadcast. It is well known that television signals are generated with a scene scanned in a particular manner in order to maintain a steady picture for a human being to view, regardless of whether there is a scenery change or not. That is, even if there is no change from one frame to the next, the scene is still scanned constantly. Hence there is a great deal of frame-to-frame correlation (Haskell et al., 1972b; Netravali and Robbins, 1979). In TV broadcasts, the camera is most likely not static, and it may be panned, tilted, and zoomed. Furthermore, more movement is involved in the scene. As long as the TV frames are taken

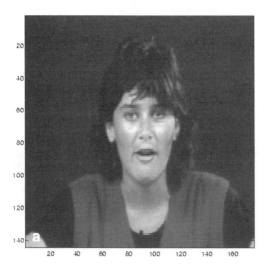 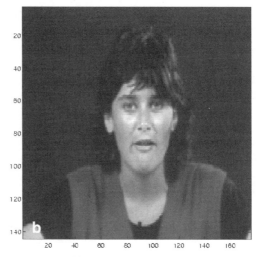

**FIGURE 10.4**   Two frames of the Miss America sequence: (a) frame 24, (b) frame 25.

densely enough, then most of the time we think the changes between successive frames are due mainly to the apparent motion of the objects in the scene that takes place during the frame intervals. This implies that there is also a high correlation between sequential frames. In other words, there is an interframe redundancy (interpixel redundancy between pixels in successive frames). There is more correlation between television picture elements along the frame-to-frame temporal dimension than there is between adjacent elements in a single frame along the spatial dimension. That is, there is generally more interframe correlation than intraframe correlation. Taking advantage of the interframe correlation, i.e., eliminating or decreasing the uncertainty of successive frames, leads to video data compression. This is analogous to the case of still image coding with the DPCM technique, where we can predict part of an image by knowing the other part. Now the knowledge of the previous frames can remove the uncertainty of the next frame. In both cases, knowledge of the past removes the uncertainty of the future, leaving less actual information to be transmitted (Kretzmer, 1952). In Chapter 16, we will see that the words "past" and "future" used here are not necessary. They can be changed, respectively, to "some frames" and "some other frames" in advanced video coding techniques such as MPEG. There, a frame might be predicted from both its previous frames and its future frames.

At this point, it becomes clear that the second class of techniques (mentioned at the beginning of this section), which generalizes techniques originally developed for still image coding and applies them to video coding, exploits interframe correlation. For instance, in the case of the 3-D DCT technique, a strong temporal correlation causes an energy compaction within the low temporal frequency region. The 3-D DCT technique drops transform coefficients associated with high temporal frequency, thus achieving data compression.

The two techniques specifically developed to exploit interframe redundancy, i.e., frame replenishment and motion-compensated coding, are introduced below. The former is the early work, while the latter is the more popular recent work.

## 10.3   FRAME REPLENISHMENT

As mentioned in Chapter 3, frame-to-frame redundancy has long been recognized in TV signal compression. The first few experiments of a frame sequence coder exploiting interframe redundancy may be traced back to the 1960s (Seyler, 1962, 1965; Mounts, 1969). In (Mounts, 1969) the first real demonstration was presented and was termed *conditional replenishment*. This frame replenishment technique can be briefly described as follows. Each pixel in a frame is classified into *changing* or *unchanging* areas depending on whether or not the intensity difference between its present value and its previous one (the intensity value at the same position on the previous frame) exceeds a threshold. If the difference does exceed the threshold, i.e., a *significant* change has been identified, the address and intensity of this pixel are coded and stored in a buffer and then transmitted to the receiver to replenish intensity. For those unchanging pixels, nothing is coded and transmitted. Their previous intensities are repeated in the receiver. It is noted that the buffer is utilized to make the information presented to the transmission channel occur at a smooth bit rate. The threshold is to make the average replenishment rate match the channel capacity.

Since the replenishment technique only encodes those pixels whose intensity value has changed significantly between successive frames, its coding efficiency is much higher than the coding techniques which encode every pixel of every frame, say, the DPCM technique applied to each single frame. In other words, utilizing interframe correlation, the replenishment technique achieves a lower bit rate, while keeping the equivalent reconstructed image quality.

Much effort had been made to further improve this type of simple replenishment algorithm. As mentioned in the discussion of 3-D DPCM in Chapter 3, for instance, it was soon realized that intensity values of pixels in a changing area need not be transmitted independently of one another. Instead, using both spatial and temporal neighbors' intensity values to predict the intensity value

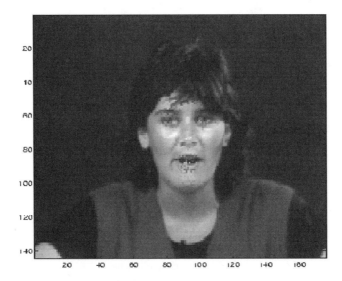

**FIGURE 10.5**  Dirty window effect.

of a changing pixel leads to a *frame-difference* predictive coding technique. There, the differential signal is coded instead of the original intensity values, thus achieving a lower bit rate. For more detail, readers are referred to Section 3.5.2. Another example of the improvements is that measures have been taken to distinguish the intensity difference caused by noise from those associated with changing to avoid the dirty window effect, whose meaning is given in the next paragraph. For more detailed information on these improvements over the simple frame replenishment technique, readers are referred to two excellent reviews by Haskell et al. (1972b, 1979).

The main drawback associated with the frame replenishment technique is that it is difficult to handle frame sequences containing more rapid changes. When there are more rapid changes, the number of pixels whose intensity values need to be updated increases. In order to maintain the transmission bit-rate at a steady and proper level the threshold has to be raised, thus causing many slow changes that cannot show up in the receiver. This poorer reconstruction in the receiver is somewhat analogous to viewing a scene through a dirty window. This is referred to as the dirty window effect. The result of one experiment on the dirty window effect is displayed in Figure 10.5. From frame 22 to frame 25 of the Miss America sequence, there are 2166 pixels (less than 10% of the total pixels) that change their gray level values by more than 1% of the peak signal. When we only update the gray level values for 25% (randomly chosen) of these changing pixels, we can clearly see the dirty window effect. When rapid scene changes exceed a certain level, buffer saturation will result, causing picture breakup (Mounts, 1969). Motion-compensated coding, which is discussed below, has been proved to be able to provide better performance than the replenishment technique in situations with rapid changes.

## 10.4  MOTION-COMPENSATED CODING

In addition to the frame-difference predictive coding technique (a variant of the frame replenishment technique discussed above), another technique: displacement-based predictive coding, was developed at almost the same time (Rocca, 1969; Haskell and Limb, 1972a). In this technique, a motion model is assumed. That is, the changes between successive frames are considered due to the translation of moving objects in the image planes. Displacement vectors of objects are first estimated. Differential signals between the intensity value of the picture elements in the moving areas and those of their counterparts in the previous frame, which are translated by the estimated

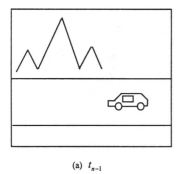
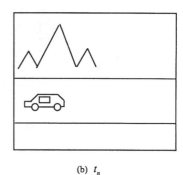

(a) $t_{n-1}$                                                     (b) $t_n$

**FIGURE 10.6**   Two consecutive frames of a video sequence.

displacement, are encoded. This approach, which takes motion into account to compress video sequences, is referred to as motion-compensated predictive coding. It has been found to be much more efficient than the frame-difference prediction technique.

To understand the above statement, let us look at the diagram shown in Figure 10.6. Assume a car translating from the right side to the left side in the image planes at a uniform speed during the time interval between the two consecutive image frames. Other than this, there are no movements or changes in the frames. Under this circumstance, if we know the displacement vector of the car on the image planes during the time interval between two consecutive frames, we can then predict the position of the car in the latter frame from its position in the former frame. One may think that if the translation vector is estimated well, then so is the prediction of the car position. This is true. In reality, however, estimation errors occurring in determination of the motion vector, which may be caused by various noises existing in the frames, may cause the predicted position of the car in the latter frame to differ from the actual position of the car in the latter frame.

The above translational model is a very simple one; it cannot accommodate motions other than translation, say, rotation, and camera zooming. Occlusion and disocclusion of objects make the situation even more complicated since in the occlusion case some portions of the images may disappear, while in the disocclusion case some newly exposed areas may appear. Therefore, the prediction error is almost inevitable. In order to have good-quality frames in the receiver, we can find the prediction error by subtracting the predicted version of the latter frame from the actual version of latter frame. If we encode both the displacement vectors and the prediction error, and transmit these data to the receiver, we may be able to obtain high-quality reconstructed images in the receiver. This is because in the receiving end, using the displacement vectors transmitted from the transmitter and the reconstructed former frame, we can predict the latter frame. Adding the transmitted prediction error to the predicted frame, we may reconstruct the latter frame with satisfactory quality. Furthermore, if manipulating the procedure properly, we are able to achieve data compression.

The displacement vectors are referred to as side or overhead information to indicate their auxiliary nature. It is noted that motion estimation drastically increases the computational complexity of the coding algorithm. In other words, the higher coding efficiency is obtained in motion-compensated coding, but with a higher computational burden. As we pointed out in Section 10.1, this is both technically feasible and economically desired since the cost of digital signal processing decreases much faster than that of transmission (Dubois et al., 1981).

Motion-compensated video compression has become a major development in coding. For more information, readers should refer to several excellent survey papers (Musmann et al., 1985; Zhang et al., 1995; Kunt, 1995).

The common practice of motion-compensated coding in video compression can be split into the following three stages. First, the *motion analysis* stage; that is, displacement vectors for either

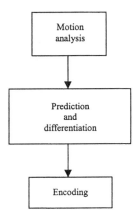

**FIGURE 10.7** Block diagram of motion-compensated coding.

every pixel or a set of pixels in image planes from sequential images are estimated. Second, the present frame is predicted by using estimated motion vectors and the previous frame. The prediction error is then calculated. This stage is called *prediction and differentiation*. The third stage is *encoding*. The prediction error (difference between the present and the predicted present frames) and the motion vectors are encoded. Through an appropriate manipulation, the total amount of data for both the motion vectors and prediction error is expected to be much less than the raw data existing in the image frames, thus resulting in data compression. A block diagram of motion-compensated coding is shown in Figure 10.7.

Before leaving this section, we compare the frame replenishment technique with the motion-compensated coding technique. Qualitatively speaking, we see from the above discussion that the replenishment technique is also a kind of predictive coding in nature. This is particularly true if we consider the frame-difference predictive technique used in frame replenishment. There, it uses a pixel's intensity value in the previous frame as an estimator of its intensity value in the present frame.

Now let's look at motion-compensated coding. Consider a pixel on the present frame. Through motion analysis, the motion-compensated technique finds its counterpart in the previous frame. That is, a pixel in the previous frame is identified such that it is supposed to translate to the position on the present frame of the pixel under consideration during the time interval between successive frames. This counterpart's intensity value is used as an estimator of that of the pixel under consideration. We can see that the model used for motion-compensated coding is much more advanced than that used for frame replenishment, therefore, it achieves a much higher coding efficiency. A motion-compensated coding technique that utilized the first pel-recursive algorithm for motion estimation (Netravali and Robbins, 1979) was reported to achieve a bit rate 22 to 50% lower than that obtained by simple frame-difference prediction, a version of frame replenishment.

The more advanced model utilized in motion-compensated coding, on the other hand, leads to higher computational complexity. Consequently, both the coding efficiency and the computational complexity in motion-compensated coding are higher than that in frame replenishment.

## 10.5 MOTION ANALYSIS

As discussed above, we usually conduct motion analysis in video sequence compression. There, 2-D displacement vectors of a pixel or a group of pixels on image planes are estimated from given image frames. Motion analysis can be viewed from a much broader point of view. It is well known that the vision systems of both humans and animals observe the outside world to ascertain motion and to navigate themselves in the 3-D world space. Two groups of scientists study vision. Scientists in the first group, including psychophysicists, physicians, and neurophysiologists study human and animal vision. Their goal is to understand biological vision systems — their operation, features, and limitations. Computer scientists and electrical engineers form the second group. As pointed out by Aggarwal and Nandhakumar (1988), their ultimate goal is to develop computer vision systems with the ability to navigate, recognize, and track objects, and estimate their speed and direction. Each group benefits from the research results of the other group. The knowledge and results of research in psychophysics, physiology, and neurophysiology have influenced the design of computer vision systems. Simultaneously, the research results achieved in computer vision have provided a framework in modeling biological vision systems and have helped in remedying faults in biological vision systems. This process will continue to advance research in both groups, hence benefiting society.

### 10.5.1  BIOLOGICAL VISION PERSPECTIVE

In the field of biological vision, most scientists consider motion perception as a two-step process, even though there is no ample biological evidence to support this view (Singh, 1991). The two steps are measurement and interpretation. The first step measures the 2-D motion projected on the imaging surfaces. The second step interprets the 2-D motion to induce the 3-D motion and structure on the scene.

### 10.5.2  COMPUTER VISION PERSPECTIVE

In the field of computer vision, motion analysis from image sequences is traditionally split into two steps. In the first step, intermediate variables are derived. By *intermediate variables*, we mean 2-D motion parameters in image planes. In the second step, 3-D motion variables, say, speed, displacement, position, and direction, are determined.

Depending on the different intermediate results, all approaches to motion analysis can be basically classified into two categories: feature correspondence and optical flow. In the former category, a few distinct features are first extracted from image frames. For instance, consider an image sequence containing an aircraft. Two consecutive frames are shown in Figure 10.8. The head and tail of the aircraft, and the tips of its wings may be chosen as features. The correspondence of these features on successive image frames needs to be established. In the second step, 3-D motion can then be analyzed from the extracted features and their correspondence in successive frames. In the latter category of approaches, the intermediate variables are optical flow. An optical flow vector is defined as a velocity vector of a pixel on an image frame. An optical flow field is referred to as the collection of the velocity vectors of all the pixels on the frame. In the first step, optical flow vectors are determined from image sequences as the intermediate variables. In the second step, 3-D motion is estimated from optical flow. It is noted that optical flow vectors are closely

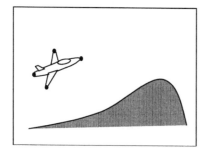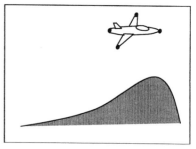

**FIGURE 10.8**  Feature extraction and correspondence from two consecutive frames in a temporal image sequence.

related to displacement vectors in that a velocity vector multiplying by the time interval between two consecutive frames results in the corresponding displacement vector. Optical flow and its determination will be discussed in detail in Chapter 13.

It is noted that there is a so-called direct method in motion analysis. Contrary to the above optical flow approach, instead of determining 2-D motion variables, (i.e., the intermediate variables), prior to 3-D motion estimation, the direct method attempts to estimate 3-D motion without explicitly solving for the intermediate variables. In (Huang and Tsai, 1981b) the equation characterizing displacement vectors in the 2-D image plane and the equation characterizing motion parameters in 3-D world space are combined so that the motion parameters in 3-D world space can be directly derived. This method has been utilized to recover structure (object surfaces) in 3-D world space as well (Negahdaripour and Horn, 1987; Horn and Weldon, 1988; Shu and Shi, 1993). The direct method has certain limitations. That is, if the geometry of object surfaces is not known in advance, then the method fails.

The feature correspondence approach is sometimes referred to as the discrete approach, while the optical flow approach is sometimes referred to as the continuous approach. This is because the correspondence approach concerns only a set of relatively sparse but highly discriminatory 2-D features on image planes. The optical flow approach is concerned with a dense field of motion vectors.

It has been found that both feature extraction and correspondence establishment are not trivial tasks. Occlusion and disocclusion which, respectively, cause some features to disappear and some features to reappear, make feature correspondence even more difficult. The development of robust techniques to solve the correspondence problem is an active research area and is still in its infancy. So far, only partial solutions suitable for simplistic situations have been developed (Aggarwal and Nandhakumar, 1988). Hence the feature correspondence approach is rarely used in video compression. Because of this, we will not discuss this approach any further.

Motion analysis (sometimes referred to as motion estimation or motion interpretation) from image sequences is necessary in automated navigation. It has played a central role in the field of computer vision since the late 1970s and early 1980s. A great deal of the papers presented at the International Conference on Computer Vision cover this and related topics. Many workshops, symposiums, and special sessions are organized around this subject (Thompson, 1989).

### 10.5.3 SIGNAL PROCESSING PERSPECTIVE

In the field of signal processing, motion analysis is mainly considered in the context of bandwidth reduction and/or data compression in the transmission of visual signals. Therefore, instead of the motion in 3-D world space, only the 2-D motion in the image plane is concerned.

Because of the real-time nature in visual transmission, the motion model cannot be very complicated. So far, the 2-D translational model is most frequently assumed in the field. In the 2-D translational model it is assumed that the change between a frame and its previous one is due to the motion of objects in the frame plane during the time interval between two consecutive frames. In many cases, as long as frames are taken densely enough, this assumption is valid. By *motion analysis* we mean the estimation of translational motion — either the displacement vectors or velocity vectors. With this kind of motion analysis, one can apply the motion-compensated coding discussed above, making coding more efficient.

Basically there are three techniques in 2-D motion analysis: correlation, and recursive and differential techniques. Philosophically speaking, the first two techniques belong to the same group: region matching.

Refer to Figure 10.6, where the moving car is the object under investigation. By *motion analysis* we mean finding the displacement vector, i.e., a vector representing the relative positions of the car in the two consecutive frames. With region matching, one may consider the car (or a portion of the car) as a region of interest, and seek the best match between the two regions in the two frames: specifically, the region in the present frame and the region in the previous frame. For identifying the best match, two techniques, the correlation and the recursive methods, differ in methodology. The correlation technique finds the best match by searching the maximum correlation between the two regions in a predefined search range, while the recursive technique estimates the best match by recursively minimizing a nonlinear measurement of the dissimilarities between the two regions.

A couple of comments are in order. First, it is noted that the most frequently used technique in motion analysis is called block matching, which is a type of the correlation technique. There, a video frame is divided into nonoverlapped rectangular blocks with each block having the same size, usually 16 × 16. Each block thus generated is assumed to move as one, i.e., all pixels in a block share the same displacement vector. For each block, we find its best match in the previous frame with correlation. That is, the block in the previous frame, which gives the maximum correlation, is identified. The relative position of these two best matched blocks produces a displacement vector. This block matching technique is simple and very efficient, and will be discussed in detail in Chapter 11. Second, as multimedia finds more and more applications, the regions occupied by arbitrarily-shaped objects (no longer always rectangular blocks) become increasingly important in content-based video retrieval and manipulation. Motion analysis in this case is discussed in Chapter 18. Third, although the recursive technique is categorized as a region matching technique, it may be used for finding displacement vectors for individual pixels. In fact the recursive technique was originally developed for determining displacement vectors of pixels and, hence, it is called pel-recursive. This technique is discussed in Chapter 12. Fourth, both correlation and recursive techniques can be utilized for determining optical flow vectors. Optical flow is discussed in Chapter 13.

The third technique in 2-D motion analysis is the differential technique. This is one of the main techniques utilized in determining optical flow vectors. It is named after the term of differentials because it uses partial differentiation of an intensity function with respect to the spatial coordinates x and y, as well as the temporal coordinate $t$. This technique is also discussed in Chapter 13.

## 10.6  MOTION COMPENSATION FOR IMAGE SEQUENCE PROCESSING

Motion analysis has long been considered a key issue in image sequence processing (Huang, 1981a; Shi, 1997). Obviously, in an area like automated navigation, motion analysis plays a central role. From the discussion in this chapter, we see that motion analysis also plays a key role in video data compression. Specifically, we have discussed the concept of motion-compensated video coding in Section 10.4. In this section we would like to consider motion compensation for image sequence processing, in general. Let us first consider motion-compensated interpolation. Then, we will discuss motion-compensated enhancement, restoration, and down-conversion.

### 10.6.1  MOTION-COMPENSATED INTERPOLATION

Interpolation is a simple yet efficient and important method in image and video compression. In image compression, we may only transmit, say, every other row. We then try to interpolate these missing rows from the other half of the transmitted rows in the receiver. In this way, we compress the data to half. Since the interpolation is carried out within a frame, it is referred to as *spatial* interpolation. In video compression, for instance, in videophone service, instead of transmitting 30 frames per second, we may choose a lower frame rate, say, 10 frames per second. In the receiver, we may try to interpolate the dropped frames from the transmitted frames. This strategy immediately drops the transmitted data to one third. Another example is the conversion of a motion picture into an NTSC (National Television System Commission) TV signal. There, every first frame in the

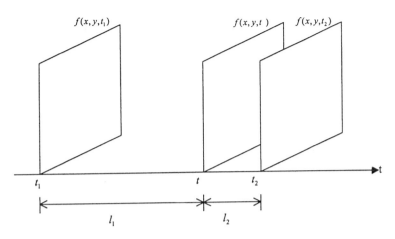

**FIGURE 10.9**  Weighted linear interpolation.

motion picture is repeated three times and the next frame twice, thus converting a 24-frame-per-second motion picture to a 60-field-per-second NTSC signal. This is commonly referred to as 3:2 pulldown. In these two examples concerning video, interpolation is along the temporal dimension, which is referred to as *temporal* interpolation.

For basic concepts of zero-order interpolation, bilinear interpolation, and polynomial interpolation, readers are referred to signal processing texts, for instance (Lim, 1990). In temporal interpolation, the zero-order interpolation means creation of a frame by copying its nearest frame along the time dimension. The conversion of a 24-frame-per-second motion picture to a 60-field-per-second NTSC signal can be classified into this type of interpolation. Weighted linear interpolation can be illustrated with Figure 10.9.

There, the weights are determined according to the lengths of time intervals, which is similar to the bilinear interpolation widely used in spatial interpolation, except that here only one index (along the time axes) is used, while two indexes (along two spatial axes) are used in spatial bilinear interpolation. That is,

$$f(x,y,t) = \frac{l_2}{l_1 + l_2} f(x,y,t_1) + \frac{l_1}{l_1 + l_2} f(x,y,t_2)$$

(10.4)

If there are one or multiple moving objects existing in successive frames, however, the weighted linear interpolation will blur the interpolated frames. Taking motion into account in the interpolation results in motion-compensated interpolation. In Figure 10.10, we still use the three frames shown in Figure 10.9 to illustrate the concept of motion-compensated interpolation. First, motion between two given frames is estimated. That is, the displacement vectors for each pixel are determined. Second, we choose a frame that is nearer to the frame we want to interpolate. Third, the displacement vectors determined in the first step are proportionally converted to the frame to be created. Each pixel in this frame is projected via the determined motion trajectory to the frame chosen in step 2. In the process of motion-compensated interpolation, spatial interpolation in the frame chosen in step 2 usually is needed.

## 10.6.2  MOTION-COMPENSATED ENHANCEMENT

It is well known that when an image is corrupted by additive white Gaussian noise (AWGN) or burst noise, linear low-pass filtering, such as simple averaging or nonlinear low-pass filtering, such as a median filter, performs well in removing the noise. When an image sequence is concerned,

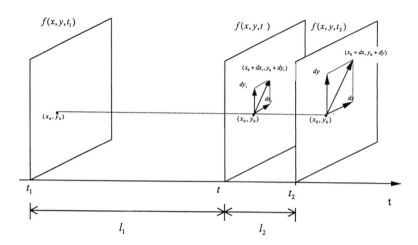

**FIGURE 10.10**  Motion-compensated interpolation.

we may apply such types of filtering along the temporal dimension to remove noise. This is called temporal filtering. These types of low-pass filtering may blur images, an effect that may become quite serious when motion exists in image planes. The enhancement, which takes motion into account, is referred to as motion-compensated enhancement, and has been found very efficient in temporal filtering (Huang and Hsu, 1981c).

   To facilitate the discussion, we consider simple averaging as a means for noise filtering in what follows. It is understood that other filtering techniques are possible, and that everything discussed here is applicable there. Instead of simply averaging $n$ successive image frames in a video sequence, motion-compensated temporal filtering will first analyze the motion existing in these frames. That is, we estimate the motion of pixels in successive frames first. Then averaging will be conducted only on those pixels along the same motion trajectory. In Figure 10.11, three successive frames are shown and denoted by f (x, y, $t_1$), f (x, y, $t_2$), and f (x, y, $t_3$), respectively. Assume that three pixels, denoted by $(x_1, y_1)$, $(x_2, y_2)$, and $(x_3, y_3)$, respectively, are identified to be perspective projections of the same object point in the 3-D world space on the three frames. The averaging is then applied to these three pixels. It is noted that the number of successive frames, $n$, may not necessarily have to be three. Motion analysis can use any one of the several techniques discussed in Section 10.5.

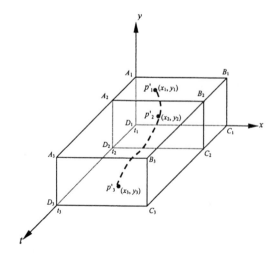

**FIGURE 10.11**  Motion-compensated temporal filtering.

Motion-compensated temporal filtering is not necessarily implemented pixelwise; it can also be used objectwise, or regionwise.

### 10.6.3 MOTION-COMPENSATED RESTORATION

Extensive attention has been paid to the restoration of full-length feature films. There, typical artifacts are due to dirt and sparkle. Early work in the detection of these artifacts ignored motion information completely. Late motion estimation has been utilized to detect these artifacts based on the assumption that the artifacts occur occasionally along the temporal dimension. Once the artifacts have been found, motion-compensated temporal filtering and/or interpolation will be used to remove the artifacts. One successful algorithm for the detection and removal of anomalies in digitized animation film can be found in (Tom et al., 1998).

### 10.6.4 MOTION-COMPENSATED DOWN-CONVERSION

Here we present one more example in which motion compensation finds application in digital video processing.

It is believed that there will be a need to down-convert a high definition television (HDTV) image sequence for display onto an NTSC monitor during the upcoming transition to digital television broadcast. The most straightforward approach is to fully decode the image sequence first, then apply a prefiltering and subsampling process to each field of the interlaced sequence. This is referred to as a full-resolution decoder (FRD). The merit of this approach is the high quality achieved, while the drawback is a high cost in terms of the large amount of memory required to store the reference frames. To reduce the required memory space, another approach is considered. In this approach, the down-conversion is conducted within the decoding loop and is referred to as a low-resolution decoder (LRD). It can significantly reduce the required memory and still achieve a reasonably good picture quality.

The prediction drift is a major type of artifact existing in the down-conversion. It is defined as the successive blurring of forward-predicted frames with a group of pictures. It is caused mainly by non-ideal interpolation of sub-pixel intensities and the loss of high-frequency data within the block. An optimal set of filters to perform low-resolution motion compensation has been derived to effectively minimize the drift. For details on an algorithm in the down-conversion utilizing an optimal motion compensation scheme, readers are referred to Vetro and Sun (1998).

## 10.7  SUMMARY

After Section II, still image compression, we shift our attention to video compression. Prior to Section IV, where we discuss various video compression algorithms and standards, however, we first address the issue of motion analysis and motion compensation in this chapter that starts Section III, motion estimation and compensation. This is because video compression has its own characteristics, which are different from those of still image compression. The main difference lies in interframe correlation.

In this chapter, the concept of various image sequences is discussed in a broad scope. In doing so, a single image, temporal image sequences, and spatial image sequences are all unified under the concept of imaging space. The redundancy between pixels in successive frames is analyzed for both videoconferencing and TV broadcast cases. In these applications, there is more interframe correlation than intraframe correlation, in general. Therefore, the utilization of interframe correlation becomes a key issue in video compression.

There are two major techniques in exploitation of interframe correlation: frame replenishment and motion compensation. In the conditional replenishment technique, only those pixel gray level values, whose variation from their counterparts in the previous frame exceeds a threshold, are encoded and transmitted to the receiver. These pixels are called changing pixels. For pixels other than the changing pixels, their gray values are just repeated in the receiver. This simplest frame replenishment technique achieves higher coding efficiency than coding each pixel in each frame due to the utilization of interframe redundancy. In the more advanced frame replenishment techniques, say, the frame-difference predictive coding technique, both temporal and spatial neighboring gray values of the pixels are used to predict that of a changing pixel. Instead of the intensity values of the changing pixels, the prediction error is encoded and transmitted. Because the variance of the prediction error is smaller than that of the intensity values, this more advanced frame replenishment technique is more efficient than the conditional replenishment technique.

The main drawback of frame replenishment techniques is associated with rapid motion and/or intensity variation occurring on the image planes. Under these circumstances, frame replenishment will suffer from the dirty window effect, and even buffer saturation.

In motion-compensated coding, the motion of pixels is first analyzed. Based on the previous frame and the estimated motion, the current frame is predicted. The prediction error together with motion vectors are encoded and transmitted to the receiver. Due to more accurate prediction based on a motion model, motion-compensated coding achieves higher coding efficiency compared with frame replenishment. This is conceivable because frame replenishment basically uses the intensity value of a pixel in the previous frame to predict that of the pixel in the same location in the present frame, while the prediction in motion-compensated coding uses motion trajectory. This implies that higher coding efficiency is obtained in motion compensation at the cost of higher computational complexity. This is technically feasible and economically desired since the cost of digital signal processing decreases much faster than that of transmission.

Because of the real-time requirement in video coding, only a simple 2-D translational model is used. There are mainly three types of motion analysis techniques used in motion-compensated coding. They are block matching, pel-recursion, and optical flow. By far, block matching is used most frequently. These three techniques are discussed in detail in the following three chapters.

Motion compensation is also widely utilized in other tasks of digital video sequence processing. Examples include motion-compensated interpolation, motion-compensated enhancement, motion-compensated restoration, and motion-compensated down-conversion.

## 10.8  EXERCISES

**10-1.** Explain the analogy between a stereo image sequence vs. the imaging space, and a stereo image pair vs. the spatial image sequence to which the stereo image pair belongs.

**10-2.** Explain why the imaging space can be considered as a unification of image frames, spatial image sequences, and temporal image sequences.

**10-3.** Give the definitions of the following several concepts: image, image sequence, and video. Discuss the relationship between them.

**10-4.** What feature causes video compression to be quite different from still image compression?

**10-5.** Describe the conditional replenishment technique. Why can it achieve higher coding efficiency in video coding than those techniques encoding each pixel in each frame?

**10-6.** Describe the frame-difference predictive coding technique. You may want to refer to Section 3.5.2.

**10-7.** What is the main drawback of frame replenishment?

**10-8.** Both the frame-difference predictive coding and motion-compensated coding are predictive codings in nature.

(a) What is the main difference between the two?

(b) Explain why motion-compensated coding is usually more efficient.

(c) What is the price paid for higher coding efficiency with motion-compensated coding?

**10-9.** Motion analysis is an important task encountered in both computer vision and video coding. What is the major different requirement for motion analysis in these two fields?

**10-10.** Work on the first 40 frames of a video sequence other than the Miss America sequence. Determine, on an average basis, what percentage of the total pixels change their gray-level values by more than 1% of the peak signal between two consecutive frames.

**10-11.** Similar to the experiment associated with Figure 10.5, do your own experiment to observe the dirty window effect. That is, work on two successive frames of a video sequence chosen by yourself, and only update a part of those changing pixels.

**10-12.** Take two frames from the Miss America sequence or from another sequence of your own choice in which a relatively large amount of motion is involved.

    (a) Using the weighted linear interpolation defined in Equation 10.4, create an interpolated frame, which is located in the 1/3 of the time interval from the second frame (i.e., $l_2 = \frac{1}{3}(l_1 + l_2)$ according to Figure 10.9).

    (b) Using motion-compensated interpolation, create an interpolated frame at the same position along the temporal dimension.

    (c) Compare the two interpolated frames and make your comments.

# REFERENCES

Aggarwal, J. K. and N. Nandhakumar, On the computation of motion from sequences of images — a review, *Proc. IEEE,* 76(8), 917-935, 1988.

Dubois, E., B. Prasada and M. S. Sabri, Image Sequence Coding, in *Image Sequence Analysis*, T. S. Huang, Ed., Springer-Verlag, New York, 1981, chap. 3.

Haskell, B. G. and J. O. Limb, Predictive Video Encoding Using Measured Subject Velocity, U.S. Patent 3,632, 865, January 1972.

Haskell, B. G., F. W. Mounts and J. C. Candy, Interframe coding of videotelephone pictures, *Proc. IEEE,* 60(7), 792-800, July 1972.

Haskell, B. G. Frame Replenishment Coding of Television, in *Image Transmission Techniques*, W. K. Pratt, Ed., Academic Press, New York, 1979, chap. 6.

Horn, B. K. P. and B. G. Schunck, Determining optical flow, *Artificial Intelligence*, 17, 185-203, 1981.

Horn, B. K. P. and E. J. Weldon, Jr., Direct methods for recovering motion, *Int. J. Comput. Vision*, 2, 51-76, 1988.

Huang, T. S. Ed., *Image Sequence Analysis*, Springer-Verlag, New York, 1981.

Huang, T. S. and R. Y. Tsai, Image Sequence Analysis: Motion Estimation, in *Image Sequence Analysis*, T. S. Huang, Ed., Springer-Verlag, New York, 1981.

Huang, T. S. and Y. P. Hsu, Image Sequence Enhancement, in *Image Sequence Analysis*, T. S. Huang, Ed., Springer-Verlag, New York, 1981.

Huang, T. S., Ed., *Image Sequence Processing and Dynamic Scene Analysis*, Springer-Verlag, New York, 1983.

Jain, A. K. *Fundamentals of Digital Image Processing*, Prentice-Hall, Englewood Cliffs, NJ, 1989.

Kretzmer, E. R. Statistics of television signal, *Bell Syst. Tech. J.,* 31(4), 751-763, 1952.

Kunt, M., Ed., Special Issue on Digital Television. Part I: Technologies, *Proc. IEEE*, 83(6), 1995.

Lim, J. S. *Two-Dimensional Signal and Image Processing*, Prentice-Hall, Englewood Cliffs, NJ, 1990.

Mounts, F. W. A video encoding system with conditional picture-element replenishment, *Bell Syst. Tech. J.,* 48(7), 2545-1554, 1969.

Musmann, H. G., P. Pirsch, and H. J. Grallert, Advances in picture coding, *Proc. IEEE,* 73(4), 523-548, 1985.

Negahdaripour, S. and B. K. P. Horn, Direct passive navigation, *IEEE Trans. Pattern Anal. Machine Intell.,* PAMI-9(1), 168-176, 1987.

Netravali, A. N. and J. D. Robbins, Motion-compensated television coding: Part I, *Bell Syst. Tech. J.,* 58(3), 631-670, 1979.

Pratt, W. K., Ed., *Image Transmission Techniques*, Academic Press, New York, 1979.

Rocca, F. Television bandwidth compression utilizing frame-to-frame correlation and movement compensation, *Symposium on Picture Bandwidth Compression*, Gordon and Breach, Newark, NJ, 1972.

Seyler, A. J. The coding of visual signals to reduce channel-capacity requirements, *IEEE Monogr.* 533E, July 1962.

Seyler, A. J. Probability distributions of television frame difference, *Proc. IREE (Australia)*, 26, 335, 1965.

Singh, A. *Optical Flow Computation: A Unified Perspective*, IEEE Press, New York, 1991.

Shi, Y. Q., C. Q. Shu, and J. N. Pan, Unified optical flow field approach to motion analysis from a sequence of stereo images, *Pattern Recognition*, 27(12), 1577-1590, 1994.

Shi, Y. Q. Editorial introduction to special issue on image sequence processing, *Int. J. Imaging Syst. Technol.*, 9(4), 189-191, 1998.

Shu, C. Q. and Y. Q. Shi, On unified optical flow field, *Pattern Recognition*, 24(6), 579-586, 1991.

Shu, C. Q. and Y. Q. Shi, Direct recovering of Nth order surface structure using unified optical flow field, *Pattern Recognition*, 26(8), 1137-1148, 1993.

Tekalp, A. M. *Digital Video Processing*, Prentice-Hall, Englewood Cliffs, NJ, 1995.

Thompson, W. B. Introduction to special issue on visual motion, *IEEE Trans. Pattern Anal. Machine Intell.*, 11(5), 449-450, 1989.

Tom, B. C., M. G. Kang, M. C. Hong and A. K. Katsaggelos, Detection and removal of anomalies in digitized animation film, *Int. J. Imaging Syst. Technol.*, 9(4), 283-293, 1998.

Vetro, A. and H. Sun, Frequency domain down-conversion of HDTV using an optimal motion compensation scheme, *Int. J. Imaging Syst. Technol.*, 9(4), 274-282, 1998.

Waxman, A. M. and J. H. Duncan, Binocular image flow: steps towards stereo-motion fusion, *IEEE Trans. Pattern Anal. Machine Intell.*, PAMI-8(6), 715-729, 1986.

Westwater, R. and B. Furht, *Real-Time Video Compression*, Kluwer Academic, Norwall, MA, 1997.

Zhang, Y.-Q., W. Li, and M. L. Liou, Ed., Special Issue on Advances in Image and Video Compression, *Proc. IEEE*, 83(2), 1995.

# 11  Block Matching

As mentioned in the previous chapter, displacement vector measurement and its usage in motion compensation in interframe coding for a TV signal can be traced back to the 1970s. Netravali and Robbins (1979) developed a pel-recursive technique, which estimates the displacement vector for each pixel recursively from its neighboring pixels using an optimization method. Limb and Murphy (1975), Rocca and Zanoletti (1972), Cafforio and Rocca (1976), and Brofferio and Rocca (1977) developed techniques for the estimation of displacement vectors of a block of pixels. In the latter approach, an image is first segmented into areas with each having an approximately uniform translation. Then the motion vector is estimated for each area. The segmentation and motion estimation associated with these arbitrarily shaped blocks are very difficult. When there are multiple moving areas in images, the situation becomes more challenging. In addition to motion vectors, the shape information of these areas needs to be coded. Hence, when moving areas have various complicated shapes, both computational complexity and coding load will increase remarkably.

In contrast, the block matching technique, which is the focus of this chapter, is simple, straightforward, and yet very efficient. It has been by far the most popularly utilized motion estimation technique in video coding. In fact, it has been adopted by all the international video coding standards: ISO, MPEG-1 and MPEG-2, and ITU H.261, and H.263. These standards will be introduced in detail in Chapters 16, 17, and 19, respectively.

It is interesting to note that even nowadays, with the tremendous advancements in multimedia engineering, object-based and/or content-based manipulation of audiovisual information is still very demanding, particularly in audiovisual data storage, retrieval, and distribution. The applications include digital library, video on demand, audiovisual databases, and so on. Therefore, the coding of arbitrarily shaped objects has attracted great research attention these days. It has been included in the MPEG-4 activities (Brailean, 1997), and will be discussed in Chapter 18.

In this chapter various aspects of block matching are addressed. They include the concept and algorithm, matching criteria, searching strategies, limitations, and new improvements.

## 11.1  NONOVERLAPPED, EQUALLY SPACED, FIXED SIZE, SMALL RECTANGULAR BLOCK MATCHING

To avoid the kind of difficulties encountered in motion estimation and motion compensation with arbitrarily shaped blocks, the block matching technique was proposed by Jain and Jain (1981) based on the following simple motion model.

An image is partitioned into a set of nonoverlapped, equally spaced, fixed size, small rectangular blocks; and the translation motion within each block is assumed to be uniform. Although this simple model considers translation motion only, other types of motions, such as rotation and zooming of large objects, may be closely approximated by the piecewise translation of these small blocks provided that these blocks are small enough. This observation, originally made by Jain and Jain, has been confirmed again and again since then.

Displacement vectors for these blocks are estimated by finding their best matched counterparts in the previous frame. In this manner, motion estimation is significantly easier than that for arbitrarily shaped blocks. Since the motion of each block is described by only one displacement vector, the side information on motion vectors decreases. Furthermore, the rectangular shape

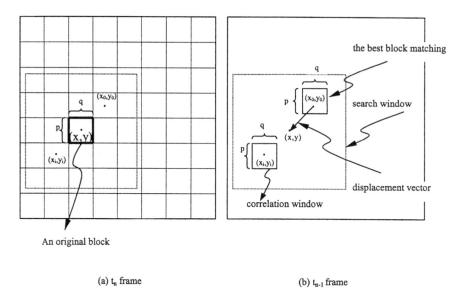

**FIGURE 11.1**   Block matching.

information is known to both the encoder and the decoder, and hence does not need to be encoded, which saves both computation load and side information.

The block size needs to be chosen properly. In general, the smaller the block size, the more accurate is the approximation. It is apparent, however, that the smaller block size leads to more motion vectors being estimated and encoded, which means an increase in both computation and side information. As a compromise, a size of $16 \times 16$ is considered to be a good choice. (This has been specified in international video coding standards such as H.261, H.263, and MPEG-1 and MPEG-2.) Note that for finer estimation a block size of $8 \times 8$ is sometimes used.

Figure 11.1 is utilized to illustrate the block matching technique. In Figure 11.1(a) an image frame at moment $t_n$ is segmented into nonoverlapped $p \times q$ rectangular blocks. As mentioned above, in common practice, square blocks of $p = q = 16$ are used most often. Consider one of the blocks centered at (x, y). It is assumed that the block is translated as a whole. Consequently, only one displacement vector needs to be estimated for this block. Figure 11.1(b) shows the previous frame: the frame at moment $t_{n-1}$. In order to estimate the displacement vector, a rectangular search window is opened in the frame $t_{n-1}$ and centered at the pixel (x, y). Consider a pixel in the search window, a rectangular correlation window of the same size $p \times q$ is opened with the pixel located in its center. A certain type of similarity measure (correlation) is calculated. After this matching process has been completed for all candidate pixels in the search window, the correlation window corresponding to the largest similarity becomes the best match of the block under consideration in frame $t_n$. The relative position between these two blocks (the block and its best match) gives the displacement vector. This is shown in Figure 11.1(b).

The size of the search window is determined by the size of the correlation window and the maximum possible displacement along four directions: upward, downward, rightward, and leftward. In Figure 11.2 these four quantities are assumed to be the same and are denoted by $d$. Note that $d$ is estimated from *a priori* knowledge about the translation motion, which includes the largest possible motion speed and the temporal interval between two consecutive frames, i.e., $t_n - t_{n-1}$.

## 11.2   MATCHING CRITERIA

Block matching belongs to image matching and can be viewed from a wider perspective. In many image processing tasks, we need to examine two images or two portions of images on a pixel-by-pixel

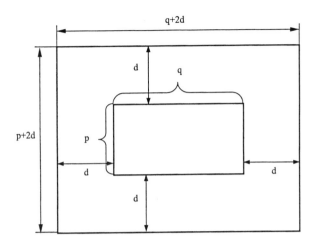

**FIGURE 11.2** Search window and correlation window.

basis. These two images or two image regions can be selected from a spatial image sequence, i.e., from two frames taken at the same time with two different sensors aiming at the same object, or from a temporal image sequence, i.e., from two frames taken at two different moments by the same sensor. The purpose of the examination is to determine the similarity between the two images or two portions of images. Examples of this type of application include image registration (Pratt, 1974) and template matching (Jain, 1989). The former deals with spatial registration of images, while the latter extracts and/or recognizes an object in an image by matching the object template and a certain area of the image.

The similarity measure, or correlation measure, is a key element in the matching process. The basic correlation measure between two images $t_n$ and $t_{n-1}$, C (s, t), is defined as follows (Anuta, 1969).

$$C(s,t) = \frac{\sum_{j=1}^{p} \sum_{k=1}^{q} f_n(j,k) f_{n-1}(j+s,k+t)}{\sqrt{\sum_{j=1}^{p} \sum_{k=1}^{q} f_n(j,k)^2} \sqrt{\sum_{j=1}^{p} \sum_{k=1}^{q} f_{n-1}(j+s,k+t)^2}}. \tag{11.1}$$

This is also referred to as a normalized two-dimensional cross-correlation function (Musmann et al., 1985).

Instead of finding the maximum similarity or correlation, an equivalent but yet more computationally efficient way of block matching is to find the minimum dissimilarity, or matching error. The dissimilarity (sometimes referred to as the error, distortion, or distance) between two images $t_n$ and $t_{n-1}$, D (s, t) is defined as follows.

$$D(s,t) = \frac{1}{lm} \sum_{j=1}^{p} \sum_{k=1}^{q} M\big(f_n(j,k), f_{n-1}(j+s,k+t)\big), \tag{11.2}$$

where M(u,v) is a metric that measures the dissimilarity between the two arguments u and v. The D (s, t) is also referred to as the matching criterion or the D values.

In the literature there are several types of matching criteria, among which the mean square error (MSE) (Jain and Jain, 1981) and mean absolute difference (MAD) (Koga et al., 1981) are used most often. It is noted that the sum of the squared difference (SSD) (Anandan, 1987) or the sum of the squared error (SSE) (Chan et al., 1990) is essentially the same as MSE. The mean absolute difference is sometimes referred to as the mean absolute error (MAE) in the literature (Nogaki and Ohta, 1972).

In the MSE matching criterion, the dissimilarity metric M (u, v) is defined as

$$M(u,v) = (u-v)^2. \tag{11.3}$$

In the MAD,

$$M(u,v) = |u-v|. \tag{11.4}$$

Obviously, both criteria are simpler than the normalized two-dimensional cross-correlation measure defined in Equation 11.1.

Before proceeding to the next section, a comment on the selection of the dissimilarity measure is due. A study based on experimental works reported that the matching criterion does not significantly affect the search (Srinivasan, 1984). Hence, the MAD is preferred due to its simplicity in implementation (Musmann et al., 1985).

## 11.3   SEARCHING PROCEDURES

The searching strategy is another important issue to deal with in block matching. Several searching strategies are discuused below.

### 11.3.1   FULL SEARCH

Figure 11.2 shows a search window, a correlation window, and their sizes. In searching for the best match, the correlation window is moved to each candidate position within the search window. That is, there are a total (2 d+1) × (2 d+1) positions that need to be examined. The minimum dissimilarity gives the best match. Apparently, this full search procedure is brute force in nature. While the full search delivers good accuracy in searching for the best match (thus, good accuracy in motion estimation), a large amount of computation is involved.

In order to lower computational complexity, several fast searching procedures have been developed. They are introduced below.

### 11.3.2   2-D LOGARITHMIC SEARCH

Jain and Jain (1981) developed a 2-D logarithmic searching procedure. Based on a 1-D logarithmic search procedure (Knuth, 1973), the 2-D procedure successively reduces the search area, thus reducing the computational burden. The first steps computes the matching criteria for five points in the search window. These five points are as follows: the central point of the search window and the four points surrounding it, with each being a midpoint between the central point and one of the four boundaries of the window. Among these five points, the one corresponding to the minimum dissimilarity is picked as the winner. In the next step, surrounding this winner, another set of five points are selected in a similar fashion to that in the first step, with the distances between the five points remaining unchanged. The exception takes place when either a central point of a set of five points or a boundary point of the search window gives a minimum D value. In these circumstances, the distances between the five points need to be reduced. The procedure continues until the final step, in which a set of candidate points are located in a 3 × 3 2-D grid. Figure 11.3 demonstrates two cases of the procedure. Figure 11.3(a) shows that the minimum D value takes place on a boundary, while Figure 11.3(b) shows the minimum D value in the central position.

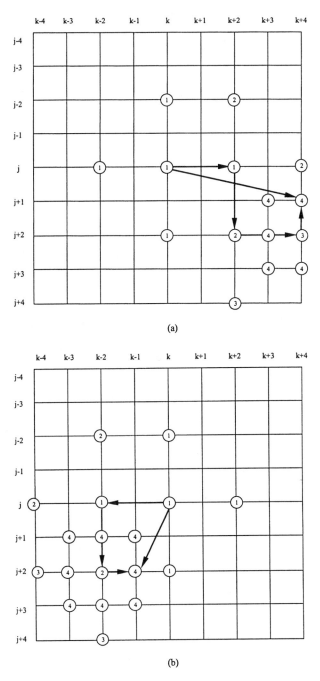

**FIGURE 11.3**  (a) A 2-D logarithmic search procedure. Points at (j, k+2), (j+2, k+2), (j+2, k+4), and (j+1, k+4) are found to give the minimum dissimilarity in steps 1, 2, 3, and 4, respectively. (b) A 2-D logarithmic search procedure. Points at (j, k-2), (j+2, k-2), and (j+2, k-1) are found to give the minimum dissimilarity in steps 1, 2, 3, and 4, respectively.

A convergence proof of the procedure is presented by Jain and Jain (1981), under the assumption that the dissimilarity monotonically increases as the search point moves away from the point corresponding to the minimum dissimilarity.

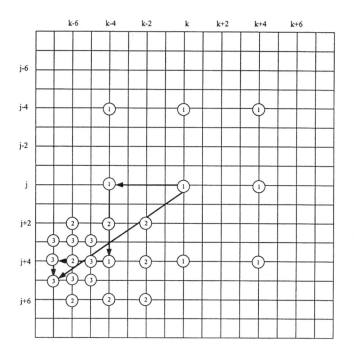

**FIGURE 11.4**   Three-step search procedure. Points (j+4, k-4), (j+4, k-6), and (j+5, k-7) give the minimum dissimilarity in steps 1, 2, and 3, respectively.

### 11.3.3   COARSE-FINE THREE-STEP SEARCH

Another important work on the block matching technique was completed at almost the same time by Koga et al. (1981). A coarse-fine three-step procedure was developed for fast searching.

The three-step search is very similar to the 2-D logarithm search. There are, however, three main differences between the two procedures. First, each step in the three-step search compares a set of nine points that form a $3 \times 3$ 2-D grid structure. Second, the distances between the points in the $3 \times 3$ 2-D grid structure in the three-step search decrease monotonically in steps 2 and 3. Third, a total of only three steps are carried out. Obviously, these three items are different from the 2-D logarithmic search described in Section 11.3.2. An illustrative example of the three-step search is shown in Figure 11.4.

### 11.3.4   CONJUGATE DIRECTION SEARCH

The conjugate direction search is another fast search algorithm that was developed by Srinivasan and Rao (1984). In principle, the procedure consists of two parts. In the first part, it finds the minimum dissimilarity along the horizontal direction with the vertical coordinate fixed at an initial position. In the second part, it finds the minimum D value along the vertical direction with the horizontal coordinate fixed in the position determined in the first part. Starting with the vertical direction followed by the horizontal direction is, of course, functionally equivalent. It was reported that this search procedure works quite efficiently (Srinivasan and Rao, 1984).

Figure 11.5 illustrates the principle of the conjugate direction search. In this example, each step involves a comparison between three testing points. If a point assumes the minimum D value compared with both of its two immediate neighbors (in one direction), then it is considered to be the best match along this direction, and the search along another direction is started. Specifically, the procedure starts to compare the D values for three points (j, k–1), (j, k), and (j, k+1). If the D value of point (j, k–1) appears to be the minimum among the three, then points (j, k–2), (j, k–1),

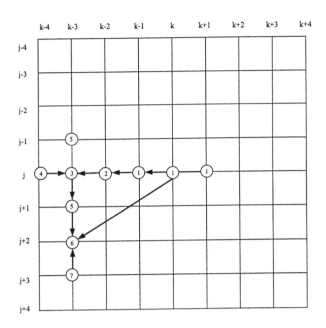

**FIGURE 11.5**  Conjugate direction search.

and (j, k) are examined. The procedure continues, finding point (j, k–3) as the best match along the horizontal direction since its D value is smaller than that of points (j, k–4) and (j, k–2). The procedure is then conducted along the vertical direction. In this example the best matching is finally found at point (j+2, k–3).

## 11.3.5  SUBSAMPLING IN THE CORRELATION WINDOW

In the evaluation of the matching criterion, either MAD or MSE, all pixels within a correlation window at the $t_{n-1}$ frame and an original block at the $t_n$ frame are involved in the computation. Note that the correlation window and the original block are the same size (refer to Figure 11.1). In order to further reduce the computational effort, a subsampling inside the window and the block is performed (Bierling, 1988). Aliasing effects can be avoided by using low-pass filtering. For instance, only every second pixel, both horizontally and vertically inside the window and the block, is taken into account for the evaluation of the matching criterion. Obviously, by using this subsampling technique, the computational burden is reduced by a factor of 4. Since 3/4 of the pixels within the window and the block are not involved in the matching computation, however, the use of such a subsampling procedure may affect the accuracy of the estimated motion vectors, especially in the case of small-size blocks. Therefore, the subsampling technique is recommended only for those cases with a large enough block size so that the matching accuracy will not be seriously affected. Figure 11.6 shows an example of $2 \times 2$ subsampling applied to both an original block of $16 \times 16$ at the $t_n$ frame and a correlation window of the same size at the $t_{n-1}$ frame.

## 11.3.6  MULTIRESOLUTION BLOCK MATCHING

It is well known that a multiresolution structure, also known as a pyramid structure, is a very powerful computational configuration for various image processing tasks. To save computation in block matching, it is natural to resort to the pyramid structure. In fact, the multiresolution technique has been regarded as one of the most efficient methods in block matching (Tzovaras et al., 1994). In a named top-down multiresolution technique, a typical Gaussian pyramid is formed first.

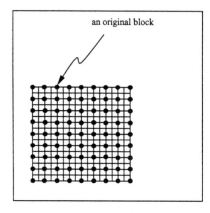
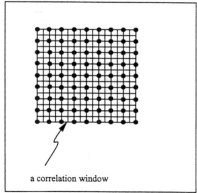

(a) An original block of 16×16 in frame at $t_n$          (b) A correlation window of 16×16 in frame at t

**FIGURE 11.6**   An example of $2 \times 2$ subsampling in the original block and correlation window for a fast search.

Before diving into further description, let us pause here to give those readers who have not been exposed to the Gaussian pyramid a short introduction to the concept. For those who know the concept, this paragraph can be skipped. Briefly speaking, a Gaussian pyramid can be understood as a set of images with different resolutions related to an original image in a certain way. The original image has the highest resolution and is considered as the lowest level, sometimes called the bottom level, in the set. From the bottom level to the top level, the resolution decreases monotonically. Specifically, between two consecutive levels, the upper level is half as large as the lower level in both horizontal and vertical directions. The upper level is generated by applying a low-pass filter (which has a group of weights) to the lower level, followed by a $2 \times 2$ subsampling. That is, each pixel in the upper level is a weighted average of some pixels in the lower level. In general, this iterative procedure of generating a level in the set is equivalent to convolving a specific weight function with the original image at the bottom level followed by an appropriate subsampling. Under certain conditions, these weight functions can closely approximate the Gaussian probability density function, which is why the pyramid is named after Gauss. (For a detailed discussion, readers are referred to Burt and Adelson [1983, 1984].) A Gaussian pyramid structure is depicted in Figure 11.7. Note that the Gaussian pyramid depicted in Figure 11.7 resembles a so-called quad-tree structure in which each node has four children nodes. In the simplest quad-tree pyramid, each pixel in an upper level is assigned an average value of its corresponding four pixels in the next lower level.

Now let's return to our discussion on the top-down multiresolution technique. After a Gaussian pyramid has been constructed, motion search ranges are allocated among the different pyramid levels. Block matching is initiated at the lowest resolution level to obtain an initial estimation of motion vectors. These computed motion vectors are then propagated to the next higher resolution level, where they are corrected and then propagated to the next level. This procedure continues until the highest resolution level is reached. As a result, a large amount of computation can be saved. Tzovaras et al. (1994) showed that a two-level Gaussian pyramid outperforms a three-level pyramid. Compared with full search block matching, the top-down multiresolution block search saves up to 67% of computations without seriously affecting the quality of the reconstructed images.

In conclusion, it has been demonstrated that multiresolution is indeed an efficient computational structure in block matching. This once again confirms the high computational efficiency of the multiresolution structure.

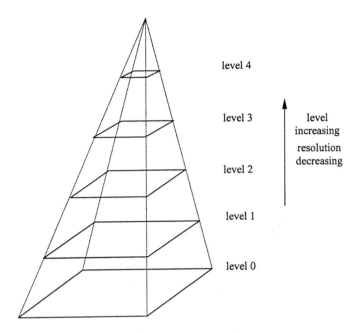

**FIGURE 11.7**   Gaussian pyramid structure.

### 11.3.7   THRESHOLDING MULTIRESOLUTION BLOCK MATCHING

With the multiresolution technique discussed above, the computed motion vectors at any intermediate pyramid level are projected to the next higher resolution level. In reality, some computed
motion vectors at the lower resolution levels may be inaccurate and have to be further refined,
while others may be relatively accurate and able to provide satisfactory motion compensation for
the corresponding block. From a computation-saving point of view, for the latter class it may not
be worth propagating the motion vectors to the next higher resolution level for further processing.

Motivated by the above observation, a new multiresolution block matching method with a
thresholding technique was developed by Shi and Xia (1997). The thresholding technique prevents
those blocks, whose estimated motion vectors provide satisfactory motion compensation, from
further processing, thus saving a lot of computation. In what follows, this technique is presented
in detail so as to provide readers with an insight to both multiresolution block matching and
thresholding multiresolution block matching techniques.

**Algorithm** — Let $f_n(x, y)$ be the frame of an image sequence at current moment $n$. First, two
Gaussian pyramids are formed, pyramids $n$ and $n - 1$, from image frames $f_n(x, y)$ and $f_{n-1}(x, y)$,
respectively. Let the levels of the pyramids be denoted by $l$, $l = 0, 1, \ldots, L$, where 0 is the lowest
resolution level (top level), $L$ is the full resolution level (bottom level), and $L+1$ is the total number
of layers in the pyramids. If $(i, j)$ are the coordinates of the upper-left corner of a block at level $l$
of pyramid $n$, the block is referred to as block $(i, j)_n^l$. The horizontal and vertical dimensions of a
block at level $l$ are denoted by $b_x^l$ and $b_y^l$, respectively. Like the variable block size method (refer
to Method 1 in Tzovaras et al. [1994]), the size of the block in this work varies with the pyramid
levels. That is, if the size of a block at level $l$ is $b_x^l$, then the size of the block at level $l + 1$ becomes
$2b_x^l \times 2b_y^l$. The variable block size method is used because it gives more efficient motion estimation
than the fixed block size method. Here, the matching criterion used for motion estimation is the
MAD because it does not require multiplication and performs similar to the MSE. The MAD
between block $(i, j)^l b_n^l$ of the current frame and block $(i + v_x, j + v_y)^l b_{n-1}^l$ of the previous frame at
level $l$ can be calculated as

$$MAD_{(i,j)^l_n}\left(v^l_x, v^l_y\right) = \frac{1}{b^l_x \times b^l_y} \sum_{k=0}^{b^l_x-1} \sum_{m=0}^{b^l_y-1} \left|f^l_n(i+k, j+m) - f^l_{n-1}(i+k+v^l_x, j+m+v^l_y)\right| \quad (11.5)$$

where $V^1 = (v^1_x, v^1_y)$ is one of the candidates of the motion vector of block $(i, j)^l_n$, $v^l_x, v^l_y$ are the two components of the motion vector along the x and y directions, respectively.

A block diagram of the algorithm is shown in Figure 11.8. The threshold in terms of MAD needs to be determined in advance according to the accuracy requirement of the motion estimation. Determining the threshold is discussed below in Part B of this subsection. Gaussian pyramids are formed for two consecutive frames of an image sequence from which motion estimation is desired. Block matching is then performed at the top level with the full-search scheme. The estimated motion vectors are checked to see if they provide satisfactory motion compensation. If the accuracy requirement is met, then the motion vectors will be directly transformed to the bottom level of the pyramid. Otherwise, the motion vectors will be propagated to the next higher resolution level for further refinement. This thresholding process is discussed below in Part C of this subsection. The algorithm continues in this fashion until either the threshold has been satisfied or the bottom level has been reached. The skipping of some intermediate-level calculations provides for computational saving. Experimental work with quite different motion complexities demonstrates that the proposed algorithm reduces the processing time from 14 to 20%, while maintaining almost the same quality in the reconstructed image compared with the fastest existing multiresolution block matching algorithm (Tzovaras et al., 1994).

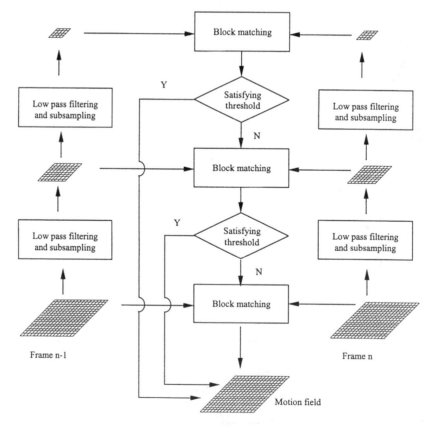

**FIGURE 11.8**   Block diagram for a three-level threshold multiresolution block matching.

**TABLE 11.1**
**Parameters Used in the Experiments**

| Parameters at Level | Low Resolution Level | Full Resolution Level |
|---|---|---|
| **Miss America** | | |
| Search range | $3 \times 3$ | $1 \times 1$ |
| Block size | $4 \times 4$ | $8 \times 8$ |
| Thresholding value | 2 | None (not applicable) |
| **Train** | | |
| Search range | $4 \times 4$ | $1 \times 1$ |
| Block size | $4 \times 4$ | $8 \times 8$ |
| Thresholding value | 3 | None (not applicable) |
| **Football** | | |
| Search range | $4 \times 4$ | $1 \times 1$ |
| Block size | $4 \times 4$ | $8 \times 8$ |
| Thresholding value | 4 | None (not applicable) |

**Threshold Determination** — The MAD accuracy criterion is used in this work for the sake of saving computations. The threshold value has a direct impact on the performance of the proposed algorithm. A small threshold value can improve the reconstructed image quality at the expense of increased computational effort. On the other hand, a large threshold value can reduce the computational complexity, but the quality of the reconstructed image may be degraded. One possible way to determine a threshold value, which was used in many experiments by Shi and Xia (1997), is as follows.

The peak signal-to-noise ratio (PSNR) is commonly used as a measure of the quality of the reconstructed image. As introduced in Chapter 1, it is defined as

$$PSNR = 10\log_{10}\frac{255^2}{MSE} \qquad (11.6)$$

From the given required PSNR, one can find the necessary MSE value. A square root of this MSE value can be chosen as a threshold value, which is applied to the first two images from the sequence. If the resulting PSNR and required processing time are satisfactory, it is then used for the rest of the sequence. Otherwise, the threshold can be slightly adjusted accordingly and applied to the second and third images to check the PSNR and processing time. It was reported in numerous experiments that this adjusted threshold value was accurate enough, and that there was no need for further adjustment. As shown in Table 11.1, the threshold values used for the "Miss America," "Train," and "Football" sequences (three sequences having quite different motion complexities) are 2, 3, and 4, respectively. They are all determined in this fashion and give satisfactory performance, as shown in the three rows marked "New Method (TH=2)," "New Method (TH=3)" and "New Method (TH=4)," respectively, in Table 11.2. That is, the PSNR experiences only about 0.1 dB loss and the processing time decreases drastically. In the experiments, the threshold value of 3, i.e., the average value of 2, 3, and 4, was also tried. Refer to the three rows marked "New Method (TH=3)" in Table 11.2. It is noted that this average threshold value 3 has already given satisfactory performance for all three sequences. Specifically, for the "Miss America" sequence, since the criterion increases from 2 to 3, the PSNR loss increases from 0.12 to 0.48 dB, and the reduction in processing time increases from 20 to 38%. For the "Football" sequence, since the criterion decreases from 4 to 3, the PSNR loss decreases from 0.08 to 0.05 dB, and the reduction in processing time decreases from 14 to 9%. Obviously, for the "Train" sequence, the criterion, as well as the performance, remains the same. One can therefore conclude that the threshold determination may not require much computation at all.

**Thresholding** — Motion vectors estimated at each pyramid level will be checked to see if they provide satisfactory motion compensation. Assume $V^l(i, j) = (v_x^l, v_y^l)$ is the estimated motion vector for block $(i, j)_n^l$ at level $l$ of pyramid $n$. For thresholding, $V^l(i, j)$ should be directly projected to the bottom level $L$. The corresponding motion vector for the same block at the bottom level of pyramid $n$ will be $V^L(2^{(L-l)}i, 2^{(L-l)}j)$, and is given as

$$V^L\left(2^{(L-l)}i,\ 2^{(L-l)}j\right) = 2^{(L-l)}V^l(i, j) \tag{11.7}$$

The MAD between the block at the bottom pyramid level of the current frame and its counterpart in the previous frame can be determined according to Equation 11.5, where the motion vector is $V^L = V^L(2^{(L-l)}i, 2^{(L-l)}j)$. This computed MAD value can be compared with the predefined threshold. If this MAD value is less than the threshold, the computed motion vector $V^L(2^{(L-l)}i, 2^{(L-l)}j)$ will be assigned to block $(2^{(L-l)}i, 2^{(L-l)}j)_n^L$ at level L in the current frame and motion estimation for this block will be stopped. If not, the estimated motion vector $V^l(i, j)$ at level $l$ will be propagated to level $l + 1$ for further refinement. Figure 11.9 gives an illustration of the above thresholding process.

**Experiments** — To verify the effectiveness of the proposed algorithm, extensive experiments have been conducted. The performance of the new algorithm is evaluated and compared with that of Method 1, one of the most efficient multiresolution block matching methods (Tzovaras et al., 1994) in terms of PSNR, error image entropy, motion vector entropy, the number of blocks stopped at the top level vs. the total number of blocks, and processing time. The number of blocks stopped at the top level is the number of blocks withheld from further processing, while the total number of blocks is the number of blocks existing at the top level. It is noted that the total number of blocks is the same for each level in the pyramid. The processing time is the sum of the total number of additions involved in the evaluation of the MAD and the thresholding operation.

In the experiments, two-level pyramids are used since they give better performance for motion estimation purposes (Tzovaras et al., 1994). The algorithms are tested on three video sequences with different motion complexities, i.e., the "Miss America," "Train," and "Football." The "Miss America" sequence has a speaker imposed on a static background and contains less motion. The "Train" sequence has more detail and contains a fast-moving object (train). The 20th frame of the sequence is shown in Figure 11.10. The "Football" sequence contains the most complicated motion

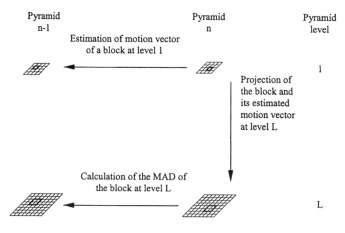

**FIGURE 11.9** The thresholding process.

**FIGURE 11.10**   The 20th frame of the "Train" sequence.

**FIGURE 11.11**   The 20th frame in the "Football" sequence.

compared with the other two sequences. The 20th frame is shown in Figure 11.11. Table 11.1 is the list of implementing parameters used in the experiments. Tables 11.2 and 11.3 give the performance of the proposed algorithm compared with Method 1. In all three cases, the motion estimation has a half-pixel accuracy, the meaning of which will be explained in the next section. All performance measures listed there are averaged for the first 25 frames of the testing sequences.

Each frame of the "Miss America" sequence is of $360 \times 288$ pixels. For convenience, only the central portion, $320 \times 256$ pixels, is processed. Using the operational parameters listed in Table 11.1 (with a criterion value of 2), 38% of the total blocks at the top level satisfy the predefined criterion and are not propagated to the bottom level. The processing time needed by the proposed algorithm is 20% less than Method 1, while the PSNR, the error image entropy, and the vector entropy are almost the same. Compared with Method 1, an extra amount of computation (around $0.16 \times 10^6$

**TABLE 11.2**
**Experimental Results (I)**

| | PSNR (dB) | Error Image Entropy (bits per pixel) | Vector Entropy (bits/vector) | Block Stopped at Top Level/Total Block | Processing Times (No. of Additions, $10^6$) |
|---|---|---|---|---|---|
| | | | **Miss America Sequence** | | |
| Method 1 (Tzovaras et al., 1994) | 38.91 | 3.311 | 6.02 | 0/1280 | 10.02 |
| New method (TH=2) | 38.79 | 3.319 | 5.65 | 487/1280 | 8.02 |
| New method (TH=3) | 38.43 | 3.340 | 5.45 | 679/1280 | 6.17 |
| | | | **Train Sequence** | | |
| Method 1 (Tzovaras et al., 1994) | 27.37 | 4.692 | 6.04 | 0/2560 | 22.58 |
| New method (TH=3) | 27.27 | 4.788 | 5.65 | 1333/2560 | 18.68 |
| | | | **Football Sequence** | | |
| Method 1 (Tzovaras et al., 1994) | 24.26 | 5.379 | 7.68 | 0/3840 | 30.06 |
| New method (TH=4) | 24.18 | 5.483 | 7.58 | 1464/3840 | 25.90 |
| New method (TH=3) | 24.21 | 5.483 | 7.57 | 1128/3840 | 27.10 |

additions) is conducted on the thresholding operation, but a large computational savings (around $2.16 \times 10^6$ additions) is achieved by withholding from further processing those blocks whose MAD values at the full resolution level are less than the predefined accuracy criterion.

The frames of the "Train" sequence are $720 \times 288$ pixels, and only the central portion, $640 \times 256$ pixels, is processed. Using the operational parameters listed in Table 11.1 (with a criterion value of 3), about 52% of the total blocks are stopped at the top level. The processing time is reduced about 17% by the new algorithm, compared with Method 1. The PSNR, the error image entropy, and the vector entropy are almost the same.

The frames of the "Football" sequence are $720 \times 480$ pixels, and only the central portion, $640 \times 384$ pixels, is processed. Using the operational parameters listed in Table 11.1 (with a criterion value of 4), about 38% of the total blocks are stopped at the top level. The processing time is about 14% less than that required by Method 1, while the PSNR, the error image entropy, and the vector entropy are almost the same.

As discussed, the experiments with a single accuracy criterion of 3 also produce similarly good performance for the three different image sequences.

In summary, it is clear that with the three different testing sequences, the thresholding multi-resolution block matching algorithm works faster than the fastest existing top-down multiresolution block matching algorithm while achieving almost the same quality of the reconstructed image.

## 11.4   MATCHING ACCURACY

Apparently, the two components of the displacement vectors obtained using the technique described above are an integer multiple of pixels. This is referred to as one-pixel accuracy. If a higher accuracy is desired, i.e., the components of the displacement vectors may be a non-integer multiple of pixels, then spatial interpolation is required. Not only will more computation be involved, but also more bits will be required to represent motion vectors. The gain is a more accurate motion estimation, hence less prediction error. In practice, half-pixel or quarter-pixel accuracy are two widely utilized accuracies other than one-pixel accuracy.

## 11.5 LIMITATIONS WITH BLOCK MATCHING TECHNIQUES

Although very simple, straightforward, and efficient, hence, utilized most widely in video coding, the block matching motion compensation technique has its drawbacks. First, it has an unreliable motion vector field with respect to the true motion in 3-D world space. In particular, it has unsatisfactory motion estimation and compensation along moving boundaries. Second, it causes block artifacts. Third, it needs to handle side information. That is, it needs to encode and transmit motion vectors as an overhead to the receiving end, thus making it difficult to use smaller block size to achieve higher accuracy in motion estimation.

All these drawbacks are due to its simple model: each block is assumed to experience a uniform translation and the motion vectors of partitioned blocks are estimated independently of each other. Unreliable motion estimation, particularly along moving boundaries, causes more prediction error, hence reduced coding efficiency.

The block artifacts do not cause severe perceptual degradation to the human visual system (HVS) when the available coding bit rate is adequately high. This is because, with a high bit rate, a sufficient amount of the motion-compensated prediction error can be transmitted to the receiving end, hence improving the subjective visual effect to such an extent that the block artifacts do not appear to be annoying. However, when the available bit rate is low, particularly lower than 64 kbps, the artifacts become visually unpleasant. In Figure 11.12, a reconstructed frame of the "Miss America" sequence at a low bit rate is shown. Obviously, block artifacts are very annoying,

**FIGURE 11.12** The 21st reconstructed frame of the "Miss America" sequence using a codec following H.263.

especially where the mouth and hair are involved. The sequence was coded and decoded by using a codec following ITU-T Recommendations H.263, an international standard in which block matching is utilized for motion estimation.

The assumption that motion within each block is uniform requires a small block size such as $16 \times 16$ and $8 \times 8$. A small block size leads to a large number of motion vectors, however, resulting in a large overhead of side information. A study by Chan et al. (1990) indicated that $8 \times 8$ block matching performs much better than $16 \times 16$ in terms of decoded image quality due to better motion estimation and compensation. The bits used for encoding motion vectors, however, increase significantly (about four times), which may be prohibitive for very low bit rate coding since the total bit rate needed for both prediction error and motion vectors may exceed the available bit rate. It is noted that when the coding bit rate is quite low, say, on the order of 20 kbps, the *side* information becomes compatible with the *main* information (prediction error) (Lin et al., 1997).

Tremendous research efforts have been made to overcome the limitations of block-matching techniques. Some improvements have been achieved and are discussed next. It should be kept in mind, however, that block matching is still by far the most popular and efficient motion estimation and compensation technique utilized for video coding, and it has been adopted for use by various international coding standards. In other words, block matching is the most appropriate technique in the framework of first-generation video coding (Dufaux and Moscheni, 1995).

## 11.6   NEW IMPROVEMENTS

### 11.6.1   HIERARCHICAL BLOCK MATCHING

Bierling (1988) developed the hierarchical search based on the following two observations. On the one hand, for a relatively large displacement, accurate block matching requires a relatively large block size. This is conceivable if one considers its opposite case: a large displacement with a small correlation window. Under this circumstance, the search range is large. Therefore the probability of finding multiple matches is high, resulting in unreliable motion estimation. On the other hand, a large block size may violate the assumption that all pixels in the block share the same displacement vector. Hence a relatively small block size is required in order to meet the assumption. These observations shed light on the problem of using a fixed block size, which may lead to unreliable motion estimation.

To satisfy these two contradicting requirements simultaneously, in a hierarchical search procedure a set of different sizes of blocks and correlation windows is utilized. To facilitate the discussion, consider a three-level hierarchical block-matching algorithm, in which three block-matching procedures are conducted, each with its own parameters. Block matching is first conducted with respect to the largest size of blocks and correlation windows. Using the estimated displacement vector as an initial vector at the second level, a new search is carried out with respect to the second largest size of blocks and correlation windows. The third search procedure is carried out similarly, based on the results of the second search. An example with three correlation windows is illustrated in Figure 11.13. It is noted that the resultant displacement vector is the sum of the three displacement vectors determined by three searches.

The parameters in these three levels are listed in Table 11.4. The algorithm is described below with an explanation of the various parameters in Table 11.4. Prior to each block matching, a separate low-pass filter is applied to the whole image in order to achieve reliable block matching. The low-pass filtering used is simply a local averaging. That is, the gray value of every pixel is replaced by the mean value of the gray values of all pixels within a square area centered at the pixel to which the mean value is assigned. In calculating the matching criterion D value, a subsampling is applied to the original block and the correlation window in order to save computation, which was discussed in Section 11.3.5.

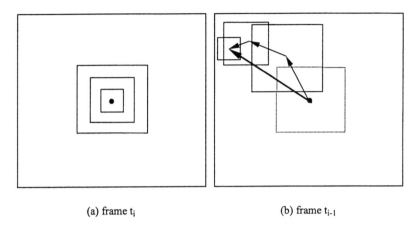

(a) frame $t_i$                                         (b) frame $t_{i-1}$

**FIGURE 11.13**   Hierarchical block matching.

**TABLE 11.3**
**Experimental Results (II)**

| Frames Tested | Total Blocks Stopped at Top Level (%) | Saved Processing Time Compared with Method 1 in Tzovaras et al. (1994) (%) |
|---|---|---|
| "Miss America" sequence (TH = 2) | 38 | 20 |
| "Train" sequence (TH = 3) | 52 | 17 |
| "Football" sequence (TH = 4) | 38 | 14 |

In the first level, for every 8th pixel horizontally and vertically (a step size of $8 \times 8$), block matching is conducted with the maximum displacement being $\pm 7$ pixels, a correlation window size of $64 \times 64$, and a subsampling factor of $4 \times 4$. A $5 \times 5$ averaging low-pass filter is applied prior to first level block matching. Second-level block matching is conducted with respect to every 4th pixel horizontally and vertically (a step size of $4 \times 4$). Note that for a pixel whose displacement vector estimate has not been determined in first-level block matching, an average of the four nearest neighboring estimates will be taken as its estimate. All the parameters for the second level are listed in Table 11.4. One thing that needs to be emphasized is that in block matching at this level the search window should be displaced by the estimated displacement vector obtained in the first level. Third-level block matching is dealt with accordingly for every 2nd pixel horizontally and vertically (a step size of $2 \times 2$). The different parameters are listed in Table 11.4. In each of the three levels, the three-step search discussed in Section 11.3.3 is utilized.

Experimental work has demonstrated a more reliable motion estimation due to the usage of a set of different sizes for both the original block and the correlation window. The first level with a large window size and a large displacement range determines a major portion of the displacement vector reliably. The successive levels with smaller window sizes and smaller displacement ranges are capable of adaptively estimating motion vectors more locally.

Figure 11.14 shows a portion of an image with pixels processed in the three levels, respectively. It is noted that it is possible to apply one more interpolation after these three levels so that a motion vector field of full resolution is available. Such a full-resolution motion vector field is useful in

**TABLE 11.4**

**Parameters Used in a Three-Level Hierarchical Block Matching**

| Hierarchical Level | Maximum Displacement | Correlation Window Size | Step Size | LPF Window Size | Subsampling |
|---|---|---|---|---|---|
| 1 | ±7 pel | 64 × 64 | 8 | 5 × 5 | 4 × 4 |
| 2 | ±3 pel | 28 × 28 | 4 | 5 × 5 | 4 × 4 |
| 3 | ±1 pel | 12 × 12 | 2 | 3 × 3 | 2 × 2 |

*Source:* Data from Bierling (1988).

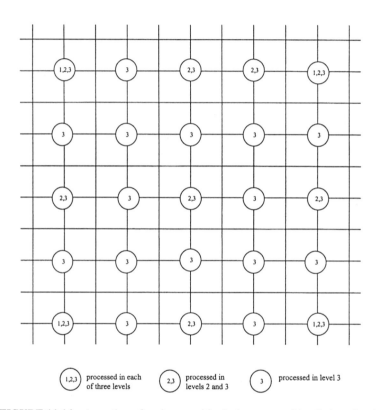

**FIGURE 11.14**   A portion of an image with pixels processed in all three levels.

such applications as motion-compensated interpolation in the context of videophony. There, in order to maintain a low bit rate some frames are skipped for transmission. At the receiving end these skipped frames need to be interpolated. As discussed in Chapter 10, motion-compensated interpolation is able to produce better frame quality than that achievable by using weighted linear interpolation.

## 11.6.2   MULTIGRID BLOCK MATCHING

Multigrid theory was developed originally in mathematics (Hackbusch and Trottenberg, 1982). It is a useful computational structure in image processing besides the multiresolution one described in Section 11.3.6. A diagram with three different levels used to illustrate a multigrid structure is shown in Figure 11.15. Although it is also a hierarchical structure, each level within the hierarchy

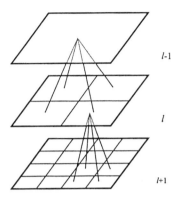

**FIGURE 11.15**   Illustration of a three-level hierarchical structure.

is of the same resolution. A few algorithms based on multigrid structure have been developed in order to improve the block-matching technique. Two advanced methods are introduced below.

**Thresholding Multigrid Block Matching** — Realizing that the simple block-based motion model (assuming a uniform motion within a fixed-size block) in the block matching technique causes several drawbacks, Chan et al. (1990) proposed a variable size block matching technique. The main idea is using a split-and-merge strategy with a multigrid structure in order to segment an image into a set of variable size blocks, each of which has an approximately uniform motion. A binary tree (also known as bin-tree) structure is used to record the relationship between these blocks of different sizes.

Specifically, an image frame is initially split into a set of square blocks by cutting the image alternately horizontally and vertically. With respect to each block thus generated, a block matching is performed in conjunction with its previous frame. Then the matching accuracy in terms of the sum squared error is compared with a preset threshold. If it is smaller than or equal to the threshold, the block remains unchanged in the whole process and the estimated motion vector is final. Otherwise, the block will be split into two blocks, and a new run of block matching is conducted for each of these two children blocks. The process continues until either the estimated vector satisfies a preset accuracy requirement or the block size has reached a predefined minimum. At this point, a merge process is proposed by Chan et al. Neighboring blocks under the same intermediate nodes in the bin-tree are checked to see if they can be merged, i.e., if the merged block can be approximated with adequate accuracy by a block in the reconstructed previous frame. It is noted that the merge operation may be optional depending on the specific application.

A block diagram of multigrid block matching is shown in Figure 11.16. Note that it is similar to that shown in Figure 11.8 for the thresholding multiresolution block matching discussed in Section 11.3.6. This observation reflects the similarities between multigrid and multiresolution structures: both are hierarchical in nature and the splitting and merging can be easily performed. An example of an image decomposition and its corresponding bin-tree are shown in Figure 11.17.

It was reported by Chan et al. (1990) that, with respect to a picture of a computer mouse and a coin, the proposed variable size block matching achieves up to a 6-dB improvement in SNR and about 30% reduction in required bits compared with fixed-size ($16 \times 16$) block matching. For several typical videoconferencing sequences, the proposed algorithm constantly performs better than the fixed-size block matching technique in terms of improved SNR of reconstructed frames with the same bit rate.

A similar algorithm was reported by Xia and Shi (1996) where a quad-tree-based segmentation is used. The thresholding technique is similar to that used by Shi and Xia (1997) and the emphasis is placed on the reduction of computational complexity. It was found that for the head-shoulder type of videophony sequences the thresholding multigrid block matching algorithm performs better

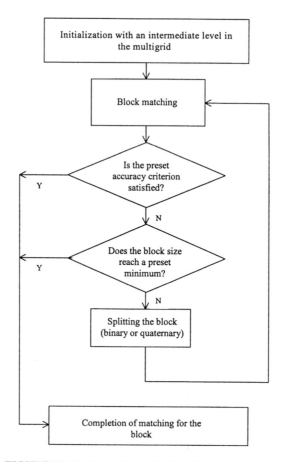

**FIGURE 11.16**  Flow chart of multigrid block matching.

than the thresholding multiresolution block matching algorithm. For video sequences that contain more complicated details and motion, however, the performance comparison turns out to be reversed.

A few remarks can be made as a conclusion for the thresholding technique. Although it needs to encode and transmit the bin-tree or quad-tree as a portion of side information, and it has to resolve the preset threshold issue, overall, the proposed algorithms achieve better performance compared with fixed-size block matching. With the flexibility provided through the variable-size methodology, the proposed approach is capable of making the model of the uniform motion within each block more accurate than fixed-size block matching can do.

**Optimal Multigrid Block Matching** — As pointed out in Chapter 10, the ultimate goal of motion estimation and motion compensation in the context of video coding is to provide a high code efficiency in real time. In other words, accurate true motion estimation is not the final goal, although accurate motion estimation is certainly desired. This point was presented by Bierling (1988) as well. There, the different requirements with respect to motion-compensated coding and motion-compensated interpolation were discussed. While the former requires motion vector estimation leading to minimum prediction error and at the same time a low amount of motion vector information, the latter requires accurate estimation of true vectors and a high resolution of the motion vector field.

This point was very much emphasized by Dufaux and Moscheni (1995). They clearly stated that in the context of video coding, estimation of true motion in 3-D world space is not the ultimate

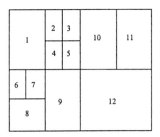

(a) An example of a decomposition

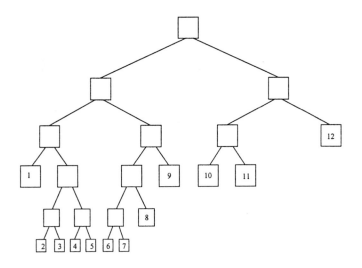

(b) The corresponding bin-tree

**FIGURE 11.17**   Thresholding multigrid block matching.

goal. Instead, motion estimation should be able to provide good temporal prediction and at the same time require low overhead information. In a word, the total amount of information that needs to be encoded should be minimized. Based on this observation, a multigrid block matching technique with an advanced entropy criterion was proposed.

Since it belongs to the category of thresholding multigrid block matching, it shares many similarities with those of Chan et al. (1990) and Xia and Shi (1996). It also bears some resemblance to thresholding multiresolution block matching (Shi and Xia, 1997). What really distinguishes this approach from other algorithms is its segmentation decision rule. Instead of a preset threshold, the algorithm works with an adaptive entropy criterion, which aims at controlling the segmentation in order to achieve an optimal solution in such a way that the total number of bits needed for representing both the prediction error and motion overhead is minimized. The decision of splitting a block is made only when the extra motion overhead involved in the splitting is lower than the gain obtained from less prediction error due to more accurate motion estimation. Not only is it optimal in the sense of bit saving, but it also eliminates the need for setting a threshold.

The number of bits needed for encoding motion information can be estimated in a straightforward manner. As far as the prediction error is concerned, the bits required can be represented by a total entropy of the prediction error, which can be estimated by using an analytical expression presented by Dufaux (1994) and Moscheni et al. (1993). Note that the coding cost for quad-tree segmentation information is negligible compared with that used for encoding prediction error and motion vectors and, hence, is omitted in determining the criterion.

**FIGURE 11.18**  The 20th frame of the "Flower Garden" sequence.

In addition to this entropy criterion, a more advanced procedure is adopted in the algorithm for down-projecting the motion vectors between two consecutive grids in the coarse-to-fine iterative refinement process.

Both qualitative and quantitative assessments in experiments demonstrate its good performance. It was reported that, when the PSNR is fixed, the bit rate saving for the sequence "Flower Garden" is from 10 to 20%, for "Mobile Calendar" from 6 to 12%, and for "Table Tennis" up to 8%. This can be translated into a gain in the PSNR ranging from 0.5 to 1.5 dB. Subjectively, the visual quality is improved greatly. In particular, moving edges become much sharper. Figures 11.18, 11.19, and 11.20 show a frame from "Flower Garden," "Mobile Calendar," and "Table Tennis" sequences, respectively.

### 11.6.3  PREDICTIVE MOTION FIELD SEGMENTATION

As pointed at the beginning of Section 11.5, the block-based model, which assumes constant motion within each block, leads to unreliable motion estimation and compensation. This block effect becomes more obvious and severe for motion-discontinuous areas in image frames. This is because there are two or more regions in a block in the areas, each having a different motion. Using one motion vector to represent and compensate for the whole block results in a significant prediction error increase.

Orchard (1993) proposed a predictive motion field segmentation technique to improve motion estimation and compensation along boundaries of moving objects. Significant improvement in the accuracy of the motion-compensated frame was achieved through relaxing the restrictive block-based model along moving boundaries. That is, for those blocks involving moving boundaries, the motion field assumes pixel resolution instead of block resolution.

Two key issues have to be resolved in order to realize the idea. One is the segmentation issue. It is known that the segmentation information is needed at the receiving end for motion compensation. This gives rise to a large increase in side information. To maintain almost the same amount of coding cost as the conventional block matching technique, the motion field segmentation was proposed to be conducted based on previously decoded frames. This scheme is based on the following observation: the shape of a moving object does not change from frame to frame.

**FIGURE 11.19**   The 20th frame of the "Mobile and Calendar" sequence.

**FIGURE 11.20**   The 20th frame of the "Table Tennis" sequence.

This segmentation is similar to the pel recursive technique (which will be discussed in detail in the next chapter) in the sense that both techniques operate *backwards*: based on previously decoded frames. The segmentation is different from the pel recursive method in that it only uses previously decoded frames to predict the shape of discontinuity in the motion field; not the whole motion field itself. Motion vectors are still estimated using the current frame at the encoder.

Consequently, this scheme is capable of achieving high accuracy in motion estimation, and at the same time it does not cause a large increase in side information due to the motion field segmentation.

Another key issue is how to achieve a reconstructed motion field with pixel resolution along moving boundaries. In order to avoid extra motion vectors that need to be encoded and transmitted, the motion vectors applied to these segmented regions in the areas of motion discontinuity are selected from a set of neighboring motion vectors. As a result, the proposed technique is capable of reconstructing discontinuities in the motion field at pixel resolution while maintaining the same amount of motion vectors as the conventional block matching technique.

A number of algorithms using this type of motion field segmentation technique have been developed and their performance has been tested and evaluated on some real video sequences (Orchard, 1993). Two of the 40-frame test sequences used were the "Table Tennis" and the "Football" sequences. The former contains fast ball motion and camera zooming, while the latter contains small objects with relatively moderate amounts of motion and camera panning. Several proposed algorithms were compared with conventional block matching in terms of average pixel prediction error energy and bits per frame required for coding prediction error. For the average pixel prediction error energy, the proposed algorithms achieve a significant reduction, ranging from –0.7 to –2.8 dB with respect to the "Table Tennis" sequence, and from –1.3 to –4.8 dB with the "Football" sequence. For bits per frame required for coding prediction error, a reduction of 20 to 30% was reported.

### 11.6.4 OVERLAPPED BLOCK MATCHING

All the techniques discussed so far in this section aim at more reliable motion estimation. As a result, they also alleviate annoying block artifacts to a certain extent. In this subsection we discuss a group of techniques, termed overlapped block matching, developed to alleviate or eliminate block artifacts (Watanabe, 1991; Nogaki and Ohta, 1992; Auyeung et al., 1992).

The idea is to relax the restriction of a nonoverlapped block partition imposed in the block-based model in block matching. After the nonoverlapped, fixed size, small rectangular block partition has been made, each block is enlarged along all four directions from the center of the block. Refer to Figure 11.21. Both motion estimation (block matching) and motion-compensated prediction are conducted in the same manner as that in block matching except for the inclusion of

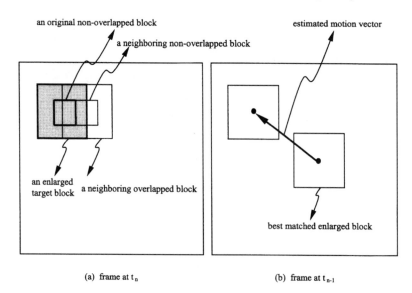

(a) frame at $t_n$                              (b) frame at $t_{n-1}$

**FIGURE 11.21**   Overlapped block matching.

a window function. That is, a 2-D window function is utilized in order to maintain an appropriate quantitative level along the overlapped portion. The window function decays towards the boundaries. In (Nogaki and Ohta, 1992) a sine-shaped window function was used.

Next, we use the algorithm proposed by Nogaki and Ohta as an example to specifically illustrate this type of technique. Consider one of the enlarged, overlapped original (also known as target) blocks, T(x,y), with a dimension of $l \times l$. Assume that a vector $v_i$ is one of the candidate displacement vectors under consideration. The predicted version of the target block with $v_i$ is denoted by $v_i$, $P_{v_i}$ $(x, y)$. Thus, the prediction error with $v_i$, $E_{v_i}(x, y)$ can be calculated according to the following equation

$$E_{v_i}(x,y) = P_{v_i}(x,y) - T(x,y) \qquad (11.8)$$

The window function W(x, y) is applied at this stage as follows, resulting in a window-operated prediction error with $v_i$, $WE_{v_i}$.

$$WE_{v_i}(x,y) = E_{v_i}(x,y) \times W(x,y) \qquad (11.9)$$

Assume that the MAD is used as the matching criterion. It can then be determined as usual by using the window-operated prediction error $WE_{v_i}(x, y)$. That is,

$$MAD = \frac{1}{l^2} \sum_{x=1}^{l} \sum_{y=1}^{l} \left| WE_{v_i}(x,y) \right|. \qquad (11.10)$$

The best match, which corresponds to the minimum MAD, produces the displacement vector $v$.

In motion-compensated prediction, the predicted version of the enlarged target block, $P_v$ (x, y) is derived from the frame at $t_{i-1}$ by using estimated vector $v$. The same window function W (x, y) is used to generate the final window-operated predicted version of the target block. That is,

$$WP_v(x,y) = P_v(x,y) \times W(x,y) \qquad (11.11)$$

It was reported by Nogaki (1992) that the luminance signal of an HDTV sequence was used in computer simulation. A block size of $16 \times 16$ was used for conventional block matching, while a block size of $32 \times 32$ was employed for the proposed overlapped block matching. The maximum displacement range d was taken as d = 15, i.e., from −15 to +15 in both the horizontal and vertical directions. The simulation indicated a reduction in the power of the prediction error by about 19%. Subjectively, it was observed that the blocking edges originally existing in the prediction error signal with conventional block matching was largely removed with the proposed overlapped block matching technique.

## 11.7  SUMMARY

By far, block matching is used more frequently than any other motion estimation technique in motion-compensated coding. By partitioning a frame into nonoverlapped, equally spaced, fixed size, small rectangular blocks and assuming that all the pixels in a block experience the same translational motion, block matching avoids the difficulty encountered in motion estimation of arbitrarily shaped blocks. Consequently, block matching is much simpler and involves less side information compared with motion estimation with arbitrarily shaped blocks.

Although this simple model considers translation motion only, other types of motions, such as rotation and zooming of large objects, may be closely approximated by the piecewise translation of these small blocks, provided that these blocks are small enough. This important observation, originally made by Jain and Jain, has been confirmed again and again since then.

Various issues related to block matching such as selection of block sizes, matching criteria, search strategies, matching accuracy, and its limitations and improvements are discussed in this chapter. Specifically, a block size of $16 \times 16$ is used most often. For more accurate motion estimation, the size of $8 \times 8$ is used sometimes. In the latter case, more accurate motion estimation is obtained at the cost of more side information and higher computational complexity.

There are several different types of matching criteria that can be used in block matching. Since it was shown that the different criteria do not cause significant differences in block matching, the mean absolute difference is hence preferred due to its simplicity in implementation.

On the one hand, a full-search procedure delivers good accuracy in searching for the best match. On the other hand, it requires a large amount of computation. In order to lower computational complexity, several fast searching procedures were developed: 2-D logarithmic search, coarse-fine three-step search, and conjugate direction search, to name a few.

Besides these suboptimum search procedures, there are some other measures developed to lower computation. One of them is subsampling in the original blocks and the correlation windows. By subsampling, the computational burden in block matching can be reduced drastically, while the accuracy of the estimated motion vectors may be affected. Therefore, the subsampling procedure is only recommended for the case with a large block size.

Naturally, the multiresolution structure, a powerful computational configuration in image processing, lends itself well to a fast search in block matching. It significantly reduces the computations involved. Thresholding multiresolution block matching further saves computation.

In terms of matching accuracy, several common choices are one-pixel, half-pixel, and quarter-pixel accuracies. Spatial interpolation is usually required for half-pixel and quarter-pixel accuracies. That is, a higher accuracy is achieved with more computation.

Limitations with block matching techniques are mainly an unreliable motion vector field and block artifacts. Both are caused by the simple model: each block is assumed to experience a uniform translation. Much efforts have been made to improve these drawbacks. Several techniques that are an improvement over the conventional block matching technique are discussed in this chapter.

In the hierarchical block matching technique, a set of different sizes for both the original block and the correlation window are used. The first level in the hierarchy with a large window size and a large displacement range determines a major portion of the displacement vector reliability. The successive levels with smaller window sizes and smaller displacement ranges are capable of adaptively estimating motion vectors more locally.

The multigrid block matching technique uses multigrid structure, another powerful computational structure in image processing, to provide a variable size block matching. With a split-and-merge strategy, the thresholding multigrid block matching technique segments an image into a set of variable size blocks, each of which experiences an approximately uniform motion. A tree structure (bin-tree or quad-tree) is used to record the relationship between these variable size blocks. With the flexibility provided through the variable-size methodology, the thresholding block matching technique is capable of making the motion model of the uniform motion within each block more accurate than fixed-size block matching can do.

As pointed out in Chapter 10, the ultimate goal of motion compensation in video coding is to achieve a high coding efficiency. In other words, accurate true motion estimation is not the final goal. From this point of view, in the above-mentioned multigrid block matching, the decision of splitting a block is made only when the bits used to encode extra motion vectors involved in the splitting are less than the bits saved from encoding reduced prediction error due to more accurate estimation. To this end, an adaptive entropy criterion is proposed and used in the optimal multigrid block matching technique. Not only is it optimal in the sense of bit saving, but it also eliminates the need for setting a threshold.

Apparently the block-based model encounters a more severe problem along moving boundaries. To solve the problem, the predictive motion field segmentation technique make the blocks involving moving boundaries have the motion field measured with pixel resolution instead of block resolution. In order to save shape overhead, segmentation is carried out backwards, i.e., based on previously decoded frames. In order to avoid a large increase of side information associated with extra motion vectors, the motion vectors applied to these segmented regions along moving boundaries are selected from a set of neighboring motion vectors. As a result, the technique is capable of reconstructing discontinuities in the motion field at pixel resolution while maintaining the same amount of motion vectors as the conventional block matching technique.

The last improvement over conventional block matching discussed in this chapter is overlapped block matching. In contrast to dealing with blocks independently of each other, the overlapped block matching technique enlarges blocks so as to make them overlap. A window function is then constructed and used in both motion estimation and motion compensation. Because it relaxes the restriction of a nonoverlapped block partition imposed by conventional block matching, it achieves better performance than the conventional block matching.

## 11.8 EXERCISES

**11-1.** Refer to Figure 11.2. It is said that there are a total of $(2d + 1) \times (2d + 1)$ positions that need to be examined in block matching with full search if one-pixel accuracy is required. How many positions are there that need to be exmined in block matching with full search if half-pixel and quarter-pixel accuracies are required?

**11-2.** What are the two effects that subsampling in the original block and the correlation block may bring out?

**11-3.** Read Burt and Adelson (1983) or Burt (1984), and explain why the pyramid is named after Gauss.

**11-4.** Read Burt and Adelson (1983) or Burt (1984), and explain why a pyramid structure is considered as a powerful computational configuration. Specifically, in multiresolutional block matching, how and to what extent does it save computation dramatically, compared with the conventional block matching technique? You may want to refer to Section 11.3.7.

**11-5.** How is the threshold determined in the thresholding multidimensional block matching technique (refer to Section 11.3.7). It is said that the square root of the MSE value, derived from the given PSNR according to Equation 11.6, is used as an initial threshold value. Justify the necessity of the square root operation.

**11-6.** Refer to Section 11.6.1 or the paper by Bierling (1988). State the different requirements in the applications of motion-compensated interpolation and motion-compensated coding. Discuss where a full resolution of the translational motion vector field may be used?

**11-7.** Read the paper Dufaux and Moscheni (1995), and explain the main feature of optimal multigrid block matching. State how the adaptive entropy criterion is established. Implement the algorithm and compare its performance with that presented by Chan et al. (1990).

**11-8.** Learn the predictive motion field segmentation technique (Orchard, 1993). Explain how the algorithms avoid a large increase in overhead due to motion field segmentation.

**11-9.** Implement the overlapped block matching algorithm introduced by Nogaki (1992). Compare its performance with that of the conventional block matching technique.

# REFERENCES

Anandan, P. Measurement Visual Motion From Image Sequences, Ph.D. thesis, COINS Department, University of Massachusetts, Amherst, 1987.

Anuta, P. F. Digital registration of multispectral video imagery, *Soc. Photo-Opt. Instrum. Eng. J.*, 7, 168-175, 1969.

Auyeung, C., J. Kosmach, M. Orchard, and T. Kalafatis, Overlapped block motion compensation, *SPIE Proc. Visual Commun. Image Process. '92*, Boston, MA, Nov. 1992, vol. 1818, 561-571.

Bierling, M. Displacement estimation by hierarchical blockmatching, *SPIE Proc. Visual Commun. Image Process.*, 1001, 942-951, 1988.

Brailean, J. Universal Accessibility and Object-Based Functionality, *ISCAS Tutorial on MPEG 4*, June 1997, Chap. 3.3.

Brofferio, S. and F. Rocca, Interframe redundancy reduction of video signals generated by translating objects, *IEEE Trans. Commun.*, COM-25, 448-455, 1977.

Burt, P. J. and E. H. Adelson, The Laplacian pyramid as a compact image code, *IEEE Trans. Commun.*, COM-31(4), 532-540, 1983.

Burt, P. J. The pyramid as a structure for efficient computation, in *Multiresolution Image Processing and Analysis*, A. Rosenfeld, Ed., Springer-Verlag, New York, 1984, 6.

Cafforio, C. and F. Rocca, Method for measuring small displacement of television images, *IEEE Trans. Inf. Theory*, IT-22, 573-579, 1976.

Chan, M. H., Y. B. Yu, and A. G. Constantinides, Variable size block matching motion compensation with applications to video coding, *IEEE Proc.*, 137(4), 205-212, 1990.

Dufaux, F. and M. Kunt, Multigrid block matching motion estimation with an adaptive local mesh refinement, *SPIE Proc. Visual Commun. Image Process. '92*, 1818, 97-109, 1992.

Dufaux, F. Multigrid Block Matching Motion Estimation for Generic Video Coding, Ph.D. dissertation, Swiss Federal Institute of Technology, Lausanne, Switzerland, 1994.

Dufaux, F. and F. Moscheni, Motion estimation techniques for digital TV: A review and a new contribution, *Proc. IEEE*, 83(6), 858-876, 1995.

Hackbusch, W. and U. Trottenberg, Eds., *Multigrid Methods*, Springer-Verlag, New York, 1982.

Haskell, B. G. and J. O. Limb, Predictive video encoding using measured subject velocity, U.S. Patent 3,632,865, January 1972.

Jain, J. R. and A. K. Jain, Displacement measurement and its application in interframe image coding, *IEEE Trans. Commun.*, COM-29(12), 1799-1808, 1981.

Jain, A. K. *Fundamentals of Digital Image Processing*, Prentice-Hall, Englewood Cliffs, NJ, 1989.

Koga, T., K. Linuma, A. Hirano, Y. Iijima, and T. Ishiguro, Motion-compensated interframe coding for video conferencing, *Proc. NTC'81*, G5.3.1-G5.3.5, New Orleans, LA, Dec. 1981.

Knuth, D. E. Searching and Sorting, *The Art of Computer Programming*, Vol. 3, Addison-Wesley, Reading, MA, 1973.

Limb, J. O. and J. A. Murphy, Measuring the speed of moving objects from television signals, *IEEE Trans. Commun.*, COM-23, 474-478, 1975.

Lin, S., Y. Q. Shi, and Y.-Q. Zhang, An optical flow-based motion compensation algorithm for very low bit-rate video coding, *Proc. 1997 IEEE Int. Conf. Acoustics, Speech Signal Process.*, 2869-2872, Munich, Germany, April 1997; *Int. J. Imaging Syst. Technol.*, 9(4), 230-237, 1998.

Moscheni, F., F. Dufaux, and H. Nicolas, Entropy criterion for optimal bit allocation between motion and prediction error information, in *SPIE 1993 Proc. Visual Commun. Image Process.*, 235-242, Cambridge, MA, Nov. 1993.

Musmann, H. G., P. Pirsch, and H. J. Grallert, Advances in picture coding, *Proc. IEEE*, 73(4), 523-548, 1985.

Netravali, A. N. and J. D. Robbins, Motion-compensated television coding: Part I, *Bell Syst. Tech. J.*, 58(3), 631-670, 1979.

Nogaki, S. and Ohta, M., An overlapped block motion compensation for high-quality motion picture coding, *Proc. IEEE Int. Symp. Circuits and Systems*, vol. 1, 184-187, San Diego, 1992.

Orchard, M. T. Predictive motion-field segmentation for image sequence coding, *IEEE Trans. Circuits and Syst. Video Technol.*, 3(1), 54-69, 1993.

Pratt, W. K. Correlation techniques of image registration, *IEEE Trans. Aerosp. Electron. Syst.*, AES-10(3), 353-358, 1974.

Rocca, F. and Zanoletti, S., Bandwidth reduction via movement compensation on a model of the random video process, *IEEE Trans. Comm.,* vol. COM-20, 960-965, Oct. 1972.

Shi, Y. Q. and X. Xia, A thresholding multidimensional block matching algorithm, *IEEE Trans. Circuits and Syst. Video Technol.,* 7(2), 437-440, April 1997.

Srinivasan, R. and K. R. Rao, Predictive coding based on efficient motion estimation, *Proc. of ICC,* 521-526, May 1984.

Tzovaras, D., M. G. Strintzis, and H. Sahinolou, Evaluation of multiresolution block matching techniques for motion and disparity estimation, *Signal Process. Image Commun.,* 6, 56-67, 1994.

Watanabe, H. and Singhal, S., Windowed motion compensation, SPIE, vol. 1605, in *Visual Communications and Image Processing, 1991: Visual Communication,* 582-589, November 1991.

Xia, X. and Y. Q. Shi, A thresholding hierarchical block matching algorithm, *Proc. IEEE 1996 Int. Symp. Circuits Syst.,* II, pp. 624-627, Atlanta, GA, May 1996; *J. Comput. Sci. Inf. Manage.,* 1(2), 83-90, 1998.

# 12 Pel Recursive Technique

As discussed in Chapter 10, the pel recursive technique is one of the three major approaches to two-dimensional displacement estimation in image planes for the signal processing community. Conceptually speaking, it is one type of region-matching technique. In contrast to block matching (which was discussed in the previous chapter), it *recursively* estimates displacement vectors for *each pixel* in an image frame. The displacement vector of a pixel is estimated by recursively minimizing a nonlinear function of the dissimilarity between two certain regions located in two consecutive frames. Note that *region* means a group of pixels, but it could be as small as a single pixel. Also note that the terms *pel* and *pixel* have the same meaning. Both terms are used frequently in the field of signal and image processing.

This chapter is organized as follows. A general description of the recursive technique is provided in Section 12.1. Some fundamental techniques in optimization are covered in Section 12.2. Section 12.3 describes the Netravali and Robbins algorithm, the pioneering work in this category. Several other typical pel recursive algorithms are introduced in Section 12.4. In Section 12.5, a performance comparison between these algorithms is made.

## 12.1 PROBLEM FORMULATION

In 1979 Netravali and Robbins published the first pel recursive algorithm to estimate displacement vectors for motion-compensated interframe image coding. Netravali and Robbins (1979) defined a quantity, called the displaced frame difference (DFD), as follows.

$$DFD\left(x, y; d_x, d_y\right) = f_n(x, y) - f_{n-1}\left(x - d_x, y - d_y\right), \qquad (12.1)$$

where the subscript $n$ and $n - 1$ indicate two moments associated with two successive frames based on which motion vectors are to be estimated; $x$, $y$ are coordinates in image planes, $d_x$, $d_y$ are the two components of the displacement vector, $\bar{d}$, along the horizontal and vertical directions in the image planes, respectively. $DFD(x, y; d_x, d_y)$ can also be expressed as $DFD(x, y; \bar{d})$. Whenever it does not cause confusion, it can be written as $DFD$ for the sake of brevity. Obviously, if there is no error in the estimation, i.e., the estimated displacement vector is exactly equal to the true motion vector, then $DFD$ will be zero.

A nonlinear function of the $DFD$ was then proposed as a dissimilarity measure by Netravali and Robbins (1979), which is a square function of $DFD$, i.e., $DFD^2$.

Netravali and Robbins thus converted displacement estimation into a minimization problem. That is, each pixel corresponds to a pair of integers $(x, y)$, denoting its spatial position in the image plane. Therefore, the $DFD$ is a function of $\bar{d}$. The estimated displacement vector $\bar{d} = (d_x, d_y)^T$, where $(\ )^T$ denotes the transposition of the argument vector or matrix, can be determined by minimizing the $DFD^2$. This is a typical nonlinear programming problem, on which a large body of research has been reported in the literature. In the next section, several techniques that rely on a method, called descent method, in optimization are introduced. The Netravali and Robbins algorithm can be applied to a pixel once or iteratively applied several times for displacement estimation. Then the algorithm moves to the next pixel. The estimated displacement vector of a pixel can be used as an initial estimate for the next pixel. This recursion can be carried out

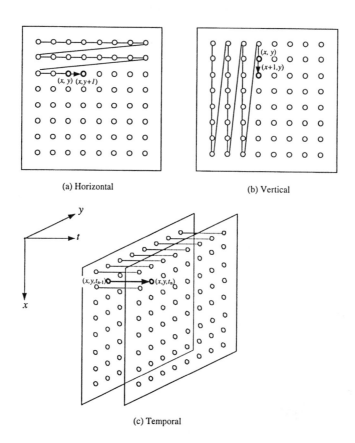

**FIGURE 12.1**   Three types of recursions: (a) horizontal; (b) vertical; (c) temporal.

horizontally, vertically, or temporally. By *temporally*, we mean that the estimated displacement vector can be passed to the pixel of the same spatial position within image planes in a temporally neighboring frame. Figure 12.1 illustrates these three different types of recursion.

## 12.2   DESCENT METHODS

Consider a nonlinear real-valued function $z$ of a vector variable $\bar{x}$,

$$z = f(\bar{x}), \tag{12.2}$$

with $\bar{x} \in R^n$, where $R^n$ represents the set of all $n$-tuples of real numbers. The question we face now is how to find such a vector denoted by $\bar{x}^*$ that the function $z$ is minimized. This is classified as an unconstrained nonlinear programming problem.

### 12.2.1   FIRST-ORDER NECESSARY CONDITIONS

According to the optimization theory, if $f(\bar{x})$ has continuous first-order partial derivatives, then the first-order necessary conditions that $\bar{x}^*$ has to satisfy are

$$\nabla f(\bar{x}^*) = 0, \tag{12.3}$$

where $\nabla$ denotes the gradient operation with respect to $\bar{x}$ evaluated at $\bar{x}^*$. Note that whenever there is only one vector variable in the function $z$ to which the gradient operator is applied, the sign $\nabla$ would remain without a subscript, as in Equation 12.3. Otherwise, i.e., if there is more than one vector variable in the function, we will explicitly write out the variable, to which the gradient operator is applied, as a subscript of the sign $\nabla$. In the component form, Equation 12.3 can be expressed as

$$\begin{cases} \dfrac{\partial f(\bar{x})}{\partial x_1} = 0 \\ \dfrac{\partial f(\bar{x})}{\partial x_2} = 0 \\ \quad \vdots \\ \dfrac{\partial f(\bar{x})}{\partial x_n} = 0. \end{cases} \tag{12.4}$$

## 12.2.2 SECOND-ORDER SUFFICIENT CONDITIONS

If $F(\bar{x})$ has second-order continuous derivatives, then the second-order sufficient conditions for $F(\bar{x}^*)$ to reach the minimum are known as

$$\nabla f(\bar{x}^*) = 0 \tag{12.5}$$

and

$$\mathbf{H}(\bar{x}^*) > 0, \tag{12.6}$$

where $\mathbf{H}$ denotes the Hessian matrix and is defined as follows.

$$\mathbf{H}(\bar{x}) = \begin{bmatrix} \dfrac{\partial^2 f(\bar{x})}{\partial^2 x_1} & \dfrac{\partial^2 f(\bar{x})}{\partial x_1 \partial x_2} & \cdots & \dfrac{\partial^2 f(\bar{x})}{\partial x_1 \partial x_n} \\ \dfrac{\partial^2 f(\bar{x})}{\partial x_2 \partial x_1} & \dfrac{\partial^2 f(\bar{x})}{\partial^2 x_2} & \cdots & \dfrac{\partial^2 f(\bar{x})}{\partial x_2 \partial x_n} \\ \vdots & & \vdots & \\ \dfrac{\partial^2 f(\bar{x})}{\partial x_n \partial x_1} & \dfrac{\partial^2 f(\bar{x})}{\partial x_n \partial x_2} & \cdots & \dfrac{\partial^2 f(\bar{x})}{\partial^2 x_n} \end{bmatrix}. \tag{12.7}$$

We can thus see that the Hessian matrix consists of all the second-order partial derivatives of $f$ with respect to the components of $\bar{x}$. Equation 12.6 means that the Hessian matrix $\mathbf{H}$ is positive definite.

## 12.2.3 UNDERLYING STRATEGY

Our aim is to derive an iterative procedure for the minimization. That is, we want to find a sequence

$$\bar{x}_0, \bar{x}_1, \bar{x}_2, \cdots, \bar{x}_n, \cdots, \tag{12.8}$$

such that

$$f(\bar{x}_0) > f(\bar{x}_1) > f(\bar{x}_2) > \cdots > f(\bar{x}_n) > \cdots \tag{12.9}$$

and the sequence converges to the minimum of $f(\bar{x}), f(\bar{x}^*)$.

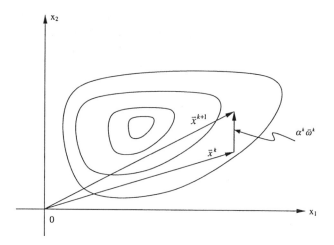

**FIGURE 12.2**   Descent method.

A fundamental underlying strategy for almost all the descent algorithms (Luenberger, 1984) is described next. We start with an initial point in the space; we determine a direction to move according to a certain rule; then we move along the direction to a relative minimum of the function $z$. This minimum point becomes the initial point for the next iteration.

This strategy can be better visualized using a 2-D example, shown in Figure 12.2. There, $\bar{x} = (x_1, x_2)^t$. Several closed curves are referred to as *contour curves* or *level curves*. That is, each of the curves represents

$$f(x_1, x_2) = c, \tag{12.10}$$

with $c$ being a constant.

Assume that at the $k$th iteration, we have a guess: $\bar{x}^k$. For the $(k + 1)$th iteration, we need to

- Find a search direction, pointed by a vector $\bar{\omega}^k$;
- Determine an optimal step size $\alpha^k$ with $\alpha^k > 0$,

such that the next guess $\bar{x}^{k+1}$ is

$$\bar{x}^{k+1} = \bar{x}^k + \alpha^k \vec{\omega}^k \tag{12.11}$$

and $\bar{x}^{k+1}$ satisfies $f(\bar{x}^k) > f(\bar{x}^{k+1})$.

In Equation 12.11, $\bar{x}^k$ can be viewed as a prediction vector for $\bar{x}^{k+1}$, while $\alpha^k \bar{\omega}^k$ an update vector, $\vec{v}^k$. Hence, using the Taylor series expansion, we can have

$$f(\bar{x}^{k+1}) = f(\bar{x}^k) + \langle \nabla f(\bar{x}^k), \alpha^k \bar{\omega}^k \rangle + \varepsilon, \tag{12.12}$$

where $\langle s, t \rangle$ denotes the inner product between vectors $\vec{s}$ and $\vec{t}$; and $\varepsilon$ represents the higher-order terms in the expansion. Consider that the increment of $\alpha^k \bar{\omega}^k$ is small enough and, thus, $\varepsilon$ can be ignored. From Equation 12.10, it is obvious that in order to have $f(\bar{x}^{k+1}) < F(\bar{x}^k)$ we must have $\langle \nabla f(\bar{x}^k), \alpha^k \bar{\omega}^k \rangle < 0$. That is,

$$f(\bar{x}^{k+1}) < f(\bar{x}^k) \Rightarrow \langle \nabla f(\bar{x}^k), \alpha^k \bar{\omega}^k \rangle < 0. \tag{12.13}$$

Choosing a different update vector, i.e., the product of the $\bar{\omega}^k$ vector and the step size $\alpha^k$, results in a different algorithm in implementing the descent method.

In the same category of the descent method, a variety of techniques have been developed. The reader may refer to Luenberger (1984) or the many other existing books on optimization. Two commonly used techniques of the descent method are discussed below. One is called the steepest descent method, in which the search direction represented by the $\bar{\omega}$ vector is chosen to be opposite to that of the gradient vector, and a real parameter of the step size $\alpha^k$ is used; the other is the Newton–Raphson method, in which the update vector in estimation, determined jointly by the search direction and the step size, is related to the Hessian matrix, defined in Equation 12.7. These two techniques are further discussed in Sections 12.2.5 and 12.2.6, respectively.

### 12.2.4 CONVERGENCE SPEED

Speed of convergence is an important issue in discussing the descent method. It is utilized to evaluate the performance of different algorithms.

**Order of Convergence** — Assume a sequence of vectors $\{\bar{x}^k\}$, with $k = 0, 1, \cdots, \infty$, converges to a minimum denoted by $\bar{x}^*$. We say that the convergence is of order $p$ if the following formula holds (Luenberger, 1984):

$$0 \leq \overline{\lim}_{k \to \infty} \frac{\left|\bar{x}^{k+1} - \bar{x}^*\right|}{\left|\bar{x}^k - \bar{x}^*\right|^p} < \infty, \tag{12.14}$$

where $p$ is positive, $\overline{\lim}$ denotes the limit superior, and $|\ |$ indicates the magnitude or norm of a vector argument. For the two latter notions, more descriptions follow.

The concept of the limit superior is based on the concept of supremum. Hence, let us first discuss the supremum. Consider a set of real numbers, denoted by $Q$, that is bounded above. Then there must exist a smallest real number $o$ such that for all the real numbers in the set $Q$, i.e., $q \in Q$, we have $q \leq o$. This real number $o$ is referred to as the least upper bound or the supremum of the set $Q$, and is denoted by

$$\sup\{q : q \in Q\} \quad \text{or} \quad \sup_{q \in Q}(q). \tag{12.15}$$

Now turn to a real bounded above sequence $r^k, k = 0, 1, \cdots, \infty$. If $s^k = \sup\{r^j : j \geq k\}$, then the sequence $\{s^k\}$ converges to a real number $s^*$. This real number $s^*$ is referred to as the limit superior of the sequence $\{r^k\}$ and is denoted by

$$\overline{\lim}_{k \to \infty}(r^k). \tag{12.16}$$

The magnitude or norm of a vector $\bar{x}$, denoted by $|\bar{x}|$, is defined as

$$|\bar{x}| = \langle \bar{x}, \bar{x} \rangle, \tag{12.17}$$

where $\langle s, t \rangle$ is the inner product between the vector $\bar{s}$ and $\bar{t}$. Throughout this discussion, when we say *vector* we mean column vector. (Row vectors can be handled accordingly.) The inner product is therefore defined as

$$\langle \bar{s}, \bar{t} \rangle = \bar{s}, \bar{t}^T, \tag{12.18}$$

with the superscript $T$ indicating the transposition operator.

With the definitions of the limit superior and the magnitude of a vector introduced, we are now in a position to understand easily the concept of the order of convergence defined in Equation 12.14. Since the sequences generated by the descent algorithms behave quite well in general (Luenberger, 1984), the limit superior is rarely necessary. Hence, roughly speaking, instead of the limit superior, the limit may be used in considering the speed of convergence.

**Linear Convergence** — Among the various orders of convergence, the order of unity is of importance, and is referred to as linear convergence. Its definition is as follows. If a sequence $\{\bar{x}^k\}$, $k = 0, 1, \cdots, \infty$, converges to $\bar{x}^*$ with

$$\lim_{k \to \infty} \frac{\left|\bar{x}^{k+1} - \bar{x}^*\right|}{\left|\bar{x}^k - \bar{x}^*\right|} = \gamma < 1, \tag{12.19}$$

then we say that this sequence converges linearly with a convergence ratio $\gamma$. The linear convergence is also referred to as geometric convergence because a linear convergent sequence with convergence ratio $\gamma$ converges to its limit at least as fast as the geometric sequences $c\gamma^k$, with $c$ being a constant.

## 12.2.5  STEEPEST DESCENT METHOD

The steepest descent method, often referred to as the gradient method, is the oldest and simplest one among various techniques in the descent method. As Luenberger pointed out in his book, it remains the fundamental method in the category for the following two reasons. First, because of its simplicity, it is usually the first method attempted for solving a new problem. This observation is very true. As we shall see soon, when handling the displacement estimation as a nonlinear programming problem in the pel recursive technique, the first algorithm developed by Netravali and Robbins is essentially the steepest descent method. Second, because of the existence of a satisfactory analysis for the steepest descent method, it continues to serve as a reference for comparing and evaluating various newly developed and more advanced methods.

**Formula** — In the steepest descent method, $\bar{\omega}^k$ is chosen as

$$\bar{\omega} = -\nabla f(\bar{x}^k), \tag{12.20}$$

resulting in

$$f(\bar{x}^{k+1}) = f(\bar{x}^k) - \alpha^k \nabla f(\bar{x}^k), \tag{12.21}$$

where the step size $\alpha^k$ is a real parameter, and, with our rule mentioned before, the sign $\nabla$ here denotes a gradient operator with respect to $\bar{x}^k$. Since the gradient vector points to the direction along which the function $f(\bar{x})$ has greatest increases, it is naturally expected that the selection of the negative direction of the gradient as the search direction will lead to the steepest descent of $f(\bar{x})$. This is where the term *steepest descent* originated.

**Convergence Speed** — It can be shown that if the sequence $\{\bar{x}\}$ is bounded above, then the steepest descent method will converge to the minimum. Furthermore, it can be shown that the steepest descent method is linear convergent.

**Selection of Step Size** — It is worth noting that the selection of the step size $\alpha^k$ has significant influence on the performance of the algorithm. In general, if it is small, it produces an accurate

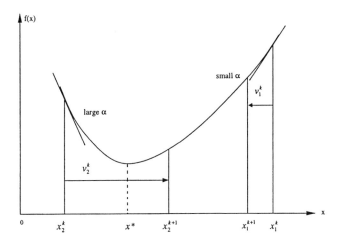

**FIGURE 12.3**  An illustration of effect of selection of step size on minimization performance. Too small $\alpha$ requires more steps to reach $x^*$. Too large $\alpha$ may cause overshooting.

estimate of $\bar{x}^*$. But a smaller step size means it will take longer for the algorithm to reach the minimum. Although a larger step size will make the algorithm converge faster, it may lead to an estimate with large error. This situation can be demonstrated in Figure 12.3. There, for the sake of an easy graphical illustration, $\bar{x}$ is assumed to be one dimensional. Two cases of too small (with subscript 1) and too large (with subscript 2) step sizes are shown for comparison.

### 12.2.6  NEWTON-RAPHSON'S METHOD

The Newton–Raphson method is the next most popular method among various descent methods.

**Formula** — Consider $\bar{x}^k$ at the $k$th iteration. The $k + 1$th guess, $\bar{x}^{k+1}$, is the sum of $\bar{x}^k$ and $\bar{v}^k$,

$$\bar{x}^{k+1} = \bar{x}^k + \bar{v}^k, \tag{12.22}$$

where $\bar{v}^k$ is an update vector as shown in Figure 12.4. Now expand the $\bar{x}^{k+1}$ into the Taylor series explicitly containing the second-order term.

$$f\left(\bar{x}^{k+1}\right) - f\left(\bar{x}^k\right) + \langle \nabla f, \bar{v} \rangle + \frac{1}{2}\left\langle H\left(\bar{x}^k\right)\bar{v}, \bar{v} \right\rangle + \varphi, \tag{12.23}$$

where $\varphi$ denotes the higher-order terms, $\nabla$ the gradient, and $\mathbf{H}$ the Hessian matrix. If $\bar{v}$ is small enough, we can ignore the $\varphi$. According to the first-order necessary conditions for $\bar{x}^{k+1}$ to be the minimum, discussed in Section 12.2.1, we have

$$\nabla_{\bar{v}} f\left(\bar{x}^k + \bar{v}\right) = \nabla f\left(\bar{x}^k\right) + \mathbf{H}\left(\bar{x}^k\right)\bar{v} = 0, \tag{12.24}$$

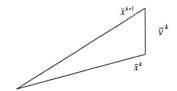

**FIGURE 12.4**  Derivation of the Newton–Raphson method.

where $\nabla_v$ denotes the gradient operator with respect to $\bar{v}$. This leads to

$$\bar{v} = -\mathbf{H}^{-1}(\bar{x}^k)\nabla f(\bar{x}^k). \tag{12.25}$$

The Newton–Raphson method is thus derived below.

$$f(\bar{x}^{k+1}) = f(\bar{x}^k) - \mathbf{H}^{-1}(\bar{x}^k)\nabla f(\bar{x}^k). \tag{12.26}$$

Another loose and intuitive way to view the Newton–Raphson method is that its format is similar to the steepest descent method, except that the step size $\alpha^k$ is now chosen as $\mathbf{H}^{-1}(\bar{x}^k)$, the inverse of the Hessian matrix evaluated at $\bar{x}^k$.

The idea behind the Newton–Raphson method is that the function being minimized is approximated locally by a quadratic function and this quadratic function is then minimized. It is noted that any function will behave like a quadratic function when it is close to the minimum. Hence, the closer to the minimum, the more efficient the Newton–Raphson method. This is the exact opposite of the steepest descent method, which works more efficiently at the beginning, and less efficiently when close to the minimum. The price paid with the Newton–Raphson method is the extra calculation involved in evaluating the inverse of the Hessian matrix at $\bar{x}^k$.

**Convergence Speed** — Assume that the second-order sufficient conditions discussed in Section 12.2.2 are satisfied. Furthermore, assume that the initial point $\bar{x}^0$ is sufficiently close to the minimum $\bar{x}^*$. Then it can be shown that the Newton–Raphson method converges with an order of at least two. This indicates that the Newton–Raphson method converges faster than the steepest descent method.

**Generalization and Improvements** — In Luenberger (1984), a general class of algorithms is defined as

$$\bar{x}^{k+1} = \bar{x}^k - \alpha^k G\nabla f(\bar{x}^k), \tag{12.27}$$

where $G$ denotes an $n \times n$ matrix, and $\alpha^k$ a positive parameter. Both the steepest descent method and the Newton–Raphson method fall into this framework. It is clear that if $G$ is an $n \times n$ identical matrix $\mathbf{I}$, this general form reduces to the steepest descent method. If $G = \mathbf{H}$ and $\alpha = 1$ then this is the Newton–Raphson method.

Although it descends rapidly near the solution, the Newton–Raphson method may not descend for points far away from the minimum because the quadratic approximation may not be valid there. The introduction of the $\alpha^k$, which minimizes $f$, can guarantee the descent of $f$ at the general points. Another improvement is to set $G = [\zeta^k I + H(\bar{x}^k)]^{-1}$ with $\zeta \geq 0$. Obviously, this is a combination of the steepest descent method and the Newton–Raphson method. Two extreme ends are that the steepest method (very large $\zeta^k$) and the Newton–Raphson method ($\zeta^k = 0$). For most cases, the selection of the parameter $\zeta^k$ aims at making the $G$ matrix positive definite.

### 12.2.7 OTHER METHODS

There are other gradient methods such as the Fletcher–Reeves method (also known as the conjugate gradient method) and the Fletcher–Powell–Davidon method (also known as the variable metric method). Readers may refer to Luenberger (1984) or other optimization text.

## 12.3 THE NETRAVALI–ROBBINS PEL RECURSIVE ALGORITHM

Having had an introduction to some basic nonlinear programming theory, we now turn to the pel recursive technique in displacement estimation from the perspective of the descent methods. Let us take a look at the first pel recursive algorithm, the Netravali–Robbins pel recursive algorithm. It actually estimates displacement vectors using the steepest descent method to minimize the squared DFD. That is,

$$\vec{d}^{\,k+1} = \vec{d}^{\,k} - \frac{1}{2}\,\alpha \nabla_{\vec{d}}\, DFD^2\!\left(x, y, \vec{d}^{\,k}\right),$$
(12.28)

where $\nabla_{\vec{d}} DFD^2(x,\, y,\, \vec{d}^k)$ denotes the gradient of $DFD^2$ with respect to $\vec{d}$ evaluated at $\vec{d}^k$, the displacement vector at the $k$th iteration, and $\alpha$ is positive. This equation can be further written as

$$\vec{d}^{\,k+1} = \vec{d}^{\,k} - \alpha DFD\!\left(x, y, \vec{d}^{\,k}\right)\nabla_{\vec{d}}\, DFD\!\left(x, y, \vec{d}^{\,k}\right).$$
(12.29)

A a result of Equation 12.1, the above equation leads to

$$\vec{d}^{\,k+1} = \vec{d}^{\,k} - \alpha DFD\!\left(x, y, \vec{d}^{\,k}\right)\nabla_{x,y}\, f_{n-1}\!\left(x - d_x, y - d_y\right),$$
(12.30)

where $\nabla_{x,y}$ means a gradient operator with respect to $x$ and $y$. Netravali and Robbins (1979) assigned a constant of $1/1024$ to $\alpha$, i.e., $1/1024$.

### 12.3.1 INCLUSION OF A NEIGHBORHOOD AREA

To make displacement estimation more robust, Netravali and Robbins considered an area for evaluating the $DFD^2$ in calculating the update term. More precisely, they assume the displacement vector is constant within a small neighborhood $\Omega$ of the pixel for which the displacement is being estimated. That is,

$$\vec{d}^{\,k+1} = \vec{d}^{\,k} - \frac{1}{2}\alpha \nabla_{\vec{d}} \sum_{i,x,y,\in\Omega} w_i DFD^2\!\left(x, y,; \vec{d}^{\,k}\right),$$
(12.31)

where $i$ represents an index for the $i$th pixel $(x,\, y)$ within $\Omega$, and $w_i$ is the weight for the $i$th pixel in $\Omega$. All the weights satisfy the following two constraints.

$$\begin{cases} w_i \geq 0 \\ \displaystyle\sum_{i\in\Omega} w_i = 1. \end{cases}$$
(12.32)
(12.33)

This inclusion of a neighborhood area also explains why pel recursive technique is classified into the category of region-matching techniques as we discussed at the beginning of this chapter.

### 12.3.2 INTERPOLATION

It is noted that interpolation will be necessary when the displacement vector components $d_x$ and $d_y$ are not integer numbers of pixels. A bilinear interpolation technique is used by Netravali and Robbins (1979). For the bilinear interpolation, readers may refer to Chapter 10.

### 12.3.3 SIMPLIFICATION

To make the proposed algorithm more efficient in computation, Netravali and Robbins also proposed simplified versions of the displacement estimation and interpolation algorithms in their paper.

One simplified version of the Netravali and Robbins algorithm is as follows:

$$\bar{d}^{k+1} = \bar{d}^k - \alpha \, sign\left\{DFD\left(x,,\bar{d}^k\right)\right\} sign\left\{\nabla_{x,y} f_{n-1}\left(x-d_x, y-d_y\right)\right\}, \tag{12.34}$$

where $sign\{s\} = 0, 1, -1$, depending on $s = 0$, $s > 0$, $s < 0$, respectively, while the sign of a vector quantity is the vector of signs of its components. In this version the update vectors can only assume an angle which is an integer multiple of $45°$. As shown in Netravali and Robbins (1979), this version is effective.

### 12.3.4 PERFORMANCE

The performance of the Netravali and Robbins algorithm has been evaluated using computer simulation (Netravali and Robbins, 1979). Two video sequences with different amounts and different types of motion are tested. In either case, the proposed pel recursive algorithm displays superior performance over the replenishment algorithm (Mounts, 1969; Haskell, 1979), which was discussed briefly in Chapter 10. The Netravali and Robbins algorithm achieves a bit rate which is 22 to 50% lower than that required by the replenishment technique with the simple frame difference prediction.

## 12.4 OTHER PEL RECURSIVE ALGORITHMS

The progress and success of the Netravali and Robbins algorithm stimulated great research interests in pel recursive techniques. Many new algorithms have been developed. Some of them are discussed in this section.

### 12.4.1 THE BERGMANN ALGORITHM (1982)

Bergmann modified the Netravali and Robbins algorithm by using the Newton–Raphson method (Bergmann, 1982). In doing so, the following difference between the fundamental framework of the descent methods discussed in Section 12.2 and the minimization problem in displacement estimation discussed in Section 12.3 need to be noticed. That is, the object function $f(\bar{x})$ discussed in Section 12.2 now becomes $DFD^2(x, y, \bar{d})$. The Hessian matrix $\mathbf{H}$, consisting of the second-order partial derivatives of the $f(\bar{x})$ with respect to the components of $\bar{x}$ now become the second-order derivatives of $DFD^2$ with respect to $d_x$ and $d_y$. Since the vector $\bar{d}$ is a 2-D column vector now, the $\mathbf{H}$ matrix is hence a $2 \times 2$ matrix. That is,

$$\mathbf{H} = \begin{bmatrix} \dfrac{\partial^2 DFD^2\left(x,y,\bar{d}\right)}{\partial^2 d_x} & \dfrac{\partial^2 DFD^2\left(x,y,\bar{d}\right)}{\partial d_x \partial d_y} \\[2ex] \dfrac{\partial^2 DFD^2\left(x,y,\bar{d}\right)}{\partial d_y \partial d_x} & \dfrac{\partial^2 DFD^2\left(x,y,\bar{d}\right)}{\partial^2 d_y} \end{bmatrix}. \tag{12.35}$$

As expected, the Bergmann algorithm (1982) converges to the minimum faster than the steepest descent method since the Newton–Raphson method converges with an order of at least two.

### 12.4.2 THE BERGMANN ALGORITHM (1984)

Based on the Burkhard and Moll algorithm (Burkhard and Moll, 1979), Bergmann developed an algorithm that is similar to the Newton–Raphson algorithm. The primary difference is that an average of two second-order derivatives is used to replace those in the Hessian matrix. In this sense, it can be considered as a variation of the Newton–Raphson algorithm.

### 12.4.3   THE CAFFORIO AND ROCCA ALGORITHM

Based on their early work (Cafforio and Rocca, 1975), Cafforio and Rocca proposed an algorithm in 1982, which is essentially the steepest descent method. That is, the step size $\alpha$ is defined as follows (Cafforio and Rocca, 1982):

$$\alpha = \frac{1}{\left|\nabla f_{n-1}\left(x - d_x, y - d_y\right)\right|^2 + \eta^2}, \tag{12.36}$$

with $\eta^2 = 100$. The addition of $\eta^2$ is intended to avoid the problem that would have occurred in a uniform region where the gradients are very small.

### 12.4.4   THE WALKER AND RAO ALGORITHM

Walker and Rao developed an algorithm based on the steepest descent method (Walker and Rao, 1984; Tekalp, 1995), and also with a variable step size. That is,

$$\alpha = \frac{1}{2\left|\nabla f_{n-1}\left(x - d_x, y - d_y\right)\right|^2}, \tag{12.37}$$

where

$$\left|\nabla f_{n-1}\left(x - d_x, y - d_y\right)\right|^2 = \left(\frac{\partial f_{n-1}\left(x - d_x, y - d_y\right)}{\partial d_x}\right)^2 + \left(\frac{\partial f_{n-1}\left(x - d_x, y - d_y\right)}{\partial d_y}\right)^2. \tag{12.38}$$

It is observed that this step size is variable instead of being a constant. Furthermore, this variable step size is reverse proportional to the norm square of the gradient of $f_{n-1}$ $(x - d_x, y - d_y)$ with respect to $x, y$. That means this type of step size will be small in the edge or rough area, and will be large in the relatively smooth area. These features are desirable.

Although it is quite similar to the Cafforio and Rocca algorithm, the Walker and Rao algorithm differs in the following two aspects. First, the $\alpha$ is selected differently. Second, implementation of the algorithm is different. For instance, instead of putting an $\eta^2$ in the denominator of $\alpha$, the Walker and Rao algorithm uses a logic.

As a result of using the variable step size $\alpha$, the convergence rate is improved substantially. This implies fast implementation and accurate displacement estimation. It was reported that usually one to three iterations are able to achieve quite satisfactory results in most cases.

Another contribution is that the Walker and Rao algorithm eliminates the need to transmit explicit address information to bring out higher coding efficiency.

## 12.5   PERFORMANCE COMPARISON

A comprehensive survey of various algorithms using the pel recursive technique can be found in a paper by Musmann, Pirsch, and Grallert (1985). There, two performance features are compared among the algorithms. One is the convergence rate and hence the accuracy of displacement estimation. The other is the stability range. By *stability range*, we mean a range starting from which an algorithm can converge to the minimum of $DFD^2$, or the true displacement vector.

Compared with the Netravali and Robbins algorithm, those improved algorithms discussed in the previous section do not use a constant step size, thus providing better adaptation to local image

**TABLE 12.1**
**Classification of Several Pel Recursive Algorithms**

| Algorithms | Category I Steepest Descent Based | Category II Newton–Raphson Based |
|---|---|---|
| Netravali and Robbins | Steepest descent | |
| Bergmann (1982) | | Newton–Raphson |
| Walker and Rao | Variation of steepest descent | |
| Cafforio and Rocca | Variation of steepest descent | |
| Bergmann (1984) | | Variation of Newton–Raphson |

statistics. Consequently, they achieve a better convergence rate and more accurate displacement estimation. According to Bergmann (1984) and Musmann et al. (1985), the Bergmann algorithm (1984) performs best among these various algorithms in terms of convergence rate and accuracy.

According to Musmann et al. (1985), the Newton–Raphson algorithm has a relatively smaller stability range than the other algorithms. This agrees with our discussion in Section 12.2.2. That is, the performance of the Newton–Raphson method improves when it works in the area close to the minimum. The choice of the initial guess, however, is relatively more restricted.

## 12.6 SUMMARY

The pel recursive technique is one of three major approaches to displacement estimation for motion compensation. It recursively estimates displacement vectors in a pixel-by-pixel fashion. There are three types of recursion: horizontal, vertical, and temporal. Displacement estimation is carried out by minimizing the square of the displaced frame difference (DFD). Therefore, the steepest descent method and the Newton–Raphson method, the two most fundamental methods in optimization, naturally find their application in pel recursive techniques. The pioneering Netravali and Robbins algorithm and several other algorithms such as the Bergmann (1982), the Cafforio and Rocca, the Walker and Rao, and the Bergmann (1984) are discussed in this chapter. They can be classified into one of two categories: the steepest-descent-based algorithms or the Newton–Raphson-based algorithms. Table 12.1 contains a classification of these algorithms.

Note that the DFD can be evaluated within a neighborhood of the pixel for which a displacement vector is being estimated. The displacement vector is assumed constant within this neighborhood. This makes the displacement estimation more robust against various noises.

Compared with the replenishment technique with simple frame difference prediction (the first real interframe coding algorithm), the Netravali and Robbins algorithm (the first pel recursive technique) achieves much higher coding efficiency. Specifically, a 22 to 50% savings in bit rate has been reported for some computer simulations. Several new pel recursive algorithms have made further improvements in terms of the convergence rate and the estimation accuracy through replacement of the fixed step size utilized in the Netravali and Robbins algorithm, which make these algorithms more adaptive to the local statistics in image frames.

## 12.7 EXERCISES

**12-1.** What is the definition of the displaced frame difference? Justify Equation 12.1.

**12-2.** Why does the inclusion of a neighborhood area make the pel recursive algorithm more robust against noise?

**12-3.** Compare the performance of the steepest descent method with that of the Newton–Raphson method.

**12-4.** Explain the function of $\eta^2$ in the Cafforio and Rocca algorithm.

**12-5.** What is the advantage you expect to have from the Walker and Rao algorithm?

**12-6.** What is the difference between the Bergmann algorithm (1982) and the Bergmann algorithm (1984)?

**12-7.** Why does the Newton–Raphson method have a smaller stability range?

## REFERENCES

Bergmann, H. C. Displacement estimation based on the correlation of image segments, *IEEE Proceedings of International Conference on Electronic Image Processing*, 215-219, York, U.K., July 1982.

Bergmann, H. C. Ein Schnell Konvergierendes Displacement-Schätzverfahrenfür die Interpolation von Fernsehbildsequenzen, Ph.D. dissertation, Technical University of Hannover, Hannover, Germany, February 1984.

Biemond, J., L. Looijenga, D. E. Boekee, and R. H. J. M. Plompen, A pel recursive Wiener-based displacement estimation algorithm, *Signal Processing*, 13, 399-412, December 1987.

Burkhard, H. and H. Moll, A modified Newton–Raphson search for the model-adaptive identification of delays, in *Identification and System Parameter Identification*, R. Isermann, Ed., Pergamon Press, New York, 1979, 1279-1286.

Cafforio, C. and F. Rocca, The differential method for image motion estimation, in *Image Sequence Processing and Dynamic Scene Analysis*, T. S. Huang, Ed., Berlin, Germany: Springer-Verlag, New York, 1983, 104-124.

Haskell, B. G. Frame replenishment coding of television, a chapter in *Image Transmission Techniques*, W. K. Pratt, Ed., Academic Press, New York, 1979.

Luenberger, D. G. *Linear and Nonlinear Programming*, Addison Wesley, Reading, MA, 1984.

Mounts, F. W. A video encoding system with conditional picture-element replenishment, *Bell Syst. Tech. J.*, 48(7), 2545-1554, 1969.

Musmann, H. G., P. Pirsch, and H. J. Grallert, Advances in picture coding, *Proc. IEEE*, 73(4), 523-548, 1985.

Netravali, A. N. and J. D. Robbins, Motion-compensated television coding: Part I, *Bell Syst. Tech. J.*, 58(3), 631-670, 1979.

Tekalp, A. M. *Digital Video Processing*, Prentice-Hall, Englewood Cliffs, NJ, 1995.

Walker, D. R. and K. R. Rao, Improved pel-recursive motion compensation, *IEEE Trans. Commun.*, COM-32, 1128-1134, 1984.

# 13  Optical Flow

As mentioned in Chapter 10, optical flow is one of three major techniques that can be used to estimate displacement vectors from successive image frames. As opposed to the other two displacement estimation techniques discussed in Chapters 11 and 12, block matching and pel recursive method, however, the optical flow technique was developed primarily for 3-D motion estimation in the computer vision community. Although it provides a relatively more accurate displacement estimation than the other two techniques, as we shall see in this and the next chapter, optical flow has not yet found wide applications for motion-compensated video coding. This is mainly due to the fact that there are a large number of motion vectors (one vector per pixel) involved, hence, the more side information that needs to be encoded and transmitted. As emphasized in Chapter 11, we should not forget the ultimate goal in motion-compensated video coding: to encode video data with a *total* bit rate as low as possible, while maintaining a satisfactory quality of reconstructed video frames at the receiving end. If the extra bits required for encoding a large amount of optical flow vectors counterbalance the bits saved in encoding the prediction error (as a result of more accurate motion estimation), then the usage of optical flow in motion-compensated coding is not worthwhile. Besides, more computation is required in optical flow determination. These factors have prevented optical flow from being practically utilized in motion-compensated video coding. With the continued advance in technologies, however, we believe this problem may be resolved in the near future. In fact, an initial, successful attempt has been made (Shi et al., 1998).

On the other hand, in theory, the optical flow technique is of great importance in understanding the fundamental issues in 2-D motion determination, such as the aperture problem, the conservation and neighborhood constraints, and the distinction and relationship between 2-D motion and 2-D apparent motion.

In this chapter we focus on the optical flow technique. In Section 13.1, as stated above, some fundamental issues associated with optical flow are addressed. Section 13.2 discusses the differential method. The correlation method is covered in Section 13.3. In Section 13.4, a multiple attributes approach is presented. Some performance comparisons between various techniques are included in Sections 13.3 and 13.4. A summary is given in Section 13.5.

## 13.1  FUNDAMENTALS

Optical flow is referred to as the 2-D distribution of apparent velocities of movement of intensity patterns in an image plane (Horn and Schunck, 1981). In other words, an optical flow field consists of a dense velocity field with one velocity vector for each pixel in the image plane. If we know the time interval between two consecutive images, which is usually the case, then velocity vectors and displacement vectors can be converted from one to another. In this sense, optical flow is one of the techniques used for displacement estimation.

### 13.1.1  2-D Motion and Optical Flow

In the above definition, it is noted that the word *apparent* is used and nothing about 3-D motion in the scene is stated. The implication behind this observation is discussed in this subsection. We start with the definition of 2-D motion. 2-D motion is referred to as motion in a 2-D image plane caused by 3-D motion in the scene. That is, 2-D motion is the projection (commonly perspective projection) of 3-D motion in the scene onto the 2-D image plane. This can be illustrated by using

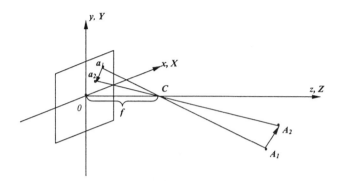

**FIGURE 13.1**   2-D motion vs. 3-D motion.

a very simple example, shown in Figure 13.1. There the world coordinate system $O$-$XYZ$ and the camera coordinate systems $o$-$xyz$ are aligned. The point $C$ is the optical center of the camera. A point $A_1$ moves to $A_2$, while its perspective projection moves correspondingly from $a_1$ to $a_2$. We then see that a 2-D motion (from $a_1$ to $a_2$) in image plane is invoked by a 3-D motion (from $A_1$ to $A_2$) in 3-D space. By a 2-D motion field, or sometimes image flow, we mean a dense 2-D motion field: One velocity vector for each pixel in the image plane.

Optical flow, according to its definition, is caused by movement of intensity patterns in an image plane. Therefore 2-D motion (field) and optical flow (field) are generally different. To support this conclusion, let us consider the following two examples. One is given by Horn and Schunck (1981). Imagine a uniform sphere rotating with a constant speed in the scene. Assume the luminance and all other conditions do not change at all when pictures are taken. Then, there is no change in brightness patterns in the images. According to the definition of optical flow, the optical flow is zero, whereas the 2-D motion field is obviously not zero. At the other extreme, consider a stationary scene; all objects in 3-D world space are still. If illuminance changes when pictures are taken in such a way that there is movement of intensity patterns in image planes, as a consequence, optical flow may be nonzero. This confirms a statement made by Singh (1991): the scene does not have to be in motion relative to the image for the optical flow field to be nonzero. It can be shown that the 2-D motion field and the optical flow field are equal under certain conditions. Understanding the difference between the two quantities and the conditions under which they are equal is important.

This understanding can provide us with some sort of guide to evaluate the reliability of estimating 3-D motion from optical flow. This is because, in practice, time-varying image sequences are only what we have at hand. The task in computer vision is to interpret 3-D motion from time-varying sequences. Therefore, we can only work with optical flow in estimating 3-D motion. Since the main focus of this book is on image and video coding, we do not cover these equality conditions here. Interested readers may refer to Singh (1991). In motion-compensated video coding, it is likewise true that the image frames and video data are only what we have at hand. We also, therefore, have to work with optical flow. Our attention is thus turned to optical flow determination and its usage in video data compression.

### 13.1.2 APERTURE PROBLEM

The aperture problem is an important issue, originating in optics. Since it is inherent in the local estimation of optical flow, we address this issue in this subsection. In optics, apertures are openings in flat screens (Bracewell, 1995). Therefore, apertures can have various shapes, such as circular, semicircular, and rectangular. Examples of apertures include a thin slit or array of slits in a screen. A circular aperture, a round hole made on the shutter of a window, was used by Newton to study the composition of sunlight. It is also well known that the circular aperture is of special interest in studying the diffraction pattern (Sears et al., 1986).

Roughly speaking, the aperture problem in motion analysis refers to the problem that occurs when viewing motion via an aperture, i.e., a small opening in a flat screen. Marr (1982) states that when a straight moving edge is observed through an aperture, only the component of motion orthogonal to the edge can be measured. Let us examine some simple examples depicted in Figure 13.2. In Figure 13.2(a), a large rectangular $ABCD$ is located in the $XOZ$ plane. A rectangular screen $EFGH$ with a circular aperture is perpendicular to the $OY$ axis. Figure 13.2(b) and (c) show, respectively, what is observed through the aperture when the rectangular $ABCD$ is moving along the positive $X$ and $Z$ directions with a uniform speed. Since the circular opening is small and the line $AB$ is very long, no motion will be observed in Figure 13.2(b). Obviously, in Figure 13.2(c) the upward movement can be observed clearly. In Figure 13.2(d), the upright corner of the rectangle $ABCD$, angle $B$, appears. At this time the translation along any direction in the $XOZ$ plane can be observed clearly. The phenomena observed in this example demonstrate that it is sometimes impossible to estimate motion of a pixel by only observing a small neighborhood surrounding it. The only motion that can be estimated from observing a small neighborhood is the motion orthogonal to the underlying moving contour. In Figure 13.2(b), there is no motion orthogonal to the moving contour $AB$; the motion is aligned with the moving contour $AB$, which cannot be observed through the aperture. Therefore, no motion can be observed through the aperture. In Figure 13.2(c), the observed motion is upward, which is perpendicular to the horizontal moving contour $AB$. In Figure 13.2(d), any translation in the $XOZ$ plane can be decomposed into horizontal and vertical components. Either of these two components is orthogonal to one of the two moving contours: $AB$ or $BC$.

A more accurate statement on the aperture problem needs a definition of the so-called normal optical flow. The normal optical flow refers to the component of optical flow along the direction pointed by the local intensity gradient. Now we can make a more accurate statement: the only motion in an image plane that can be determined is the normal optical flow.

In general, the aperture problem becomes severe in image regions where strong intensity gradients exist, such as at the edges. In image regions with strong higher-order intensity variations, such as corners or textured areas, the true motion can be estimated. Singh (1991) provides a more elegant discussion on the aperture problem, in which he argues that the aperture problem should be considered as a continuous problem (it always exists, but in varying degrees of acuteness) instead of a binary problem (either it exists or it does not).

### 13.1.3  ILL-POSED INVERSE PROBLEM

Motion estimation from image sequences, including optical flow estimation, belongs in the category of inverse problems. This is because we want to infer motion from given 2-D images, which is the perspective projection of 3-D motion. According to Hadamard (Bertero et al., 1988), a mathematical problem is well posed if it possesses the following three characteristics:

1. Existence. That is, the solution exists.
2. Uniqueness. That is, the solution is unique.
3. Continuity. That is, when the error in the data tends toward zero, then the induced error in the solution tends toward zero as well.

Inverse problems usually are not well posed in that the solution may not exist. In the example discussed in Section 13.1.1, i.e., a uniform sphere rotated with illuminance fixed, the solution to motion estimation does not exist since no motion can be inferred from given images. The aperture problem discussed in Section 13.1.2 is the case in which the solution to the motion may not be unique. Let us take a look at Figure 13.2(b). From the given picture, one cannot tell whether the straight line $AB$ is static, or is moving horizontally. If it is moving horizontally, one cannot tell the moving speed. In other words, infinitely many solutions exist for the case. In optical flow determination, we will

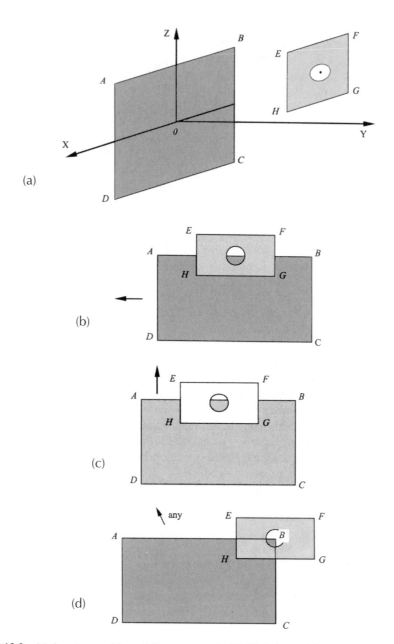

**FIGURE 13.2**    (a) Aperture problem: A large rectangle ABCD is located in the *XOZ* plane. A rectangular screen *EFGH* with a circular aperture is perpendicular to the *OY* axis. (b) Aperture problem: No motion can be observed through the circular aperture when the rectangular *ABCD* is moving along the positive X direction. (c) Aperture problem: The motion can be observed through the circular aperture when the *ABCD* is moving along the positive Z direction. (d) Aperture problem: The translation of *ABCD* along any direction in the *XOZ* plane can be observed through the circular aperture when the upright corner of the rectangle *ABCD*, angle *B*, appears in the aperture.

see that computations are noise sensitive. That is, even a small error in the data can produce an extremely large error in the solution. Hence, we see that the motion estimation from image sequences suffers from all three aspects just mentioned: nonexistence, nonuniqueness, and discontinuity. The last term is also referred to as the instability of the solution.

It is pointed out by Bertero et al. (1988) that all the low-level processing (also known as early vision) in computational vision are inverse problems and are often ill posed. Examples in low-level processing include motion recovery, computation of optical flow, edge detection, structure from stereo, structure from motion, structure from texture, shape from shading, and so on. Fortunately, the problem with early vision is mildly ill posed in general. By *mildly*, we mean that a reduction of errors in the data can significantly improve the solution.

Since the early 1960s, the demand for accurate approximates and stable solutions in areas such as optics, radioastronomy, microscopy, and medical imaging has stimulated great research efforts in inverse problems, resulting in a unified theory: the regularization theory of ill-posed problems (Tikhonov and Arsenin, 1977). In the discussion of optical flow methods, we shall see that some regularization techniques have been posed and have improved accuracy in flow determination. More-advanced algorithms continue to come.

### 13.1.4 Classification of Optical Flow Techniques

Optical flow in image sequences provides important information regarding both motion and structure, and it is useful in such diverse fields as robot vision, autonomous navigation, and video coding. Although this subject has been studied for more than a decade, reducing the error in the flow estimation remains a difficult problem. A comprehensive review and a comparison of the accuracy of various optical flow techniques have recently been made (Barron et al., 1994). So far, most of the techniques in the optical flow computations use one of the following basic approaches:

- Gradient-based (Horn and Schunck, 1981; Lucas and Kanade, 1981; Nagel and Enkelman, 1986; Uras et al., 1988; Szeliski et al., 1995; Black and Anandan, 1996),
- Correlation-based (Anandan, 1989; Singh, 1992; Pan et al., 1998),
- Spatiotemporal energy-based (Adelson and Bergen, 1985; Heeger, 1988; Bigun et al., 1991),
- Phase-based (Waxman et al., 1988; Fleet and Jepson, 1990).

Besides these deterministic approaches, there is the stochastic approach to optical flow computation (Konrad and Dubois, 1992). In this chapter we focus our discussion of optical flow on the gradient-based and correlation-based techniques because of their frequent applications in practice and fundamental importance in theory. We also discuss multiple attribute techniques in optical flow determination. The other two approaches will be briefly touched upon when we discuss new techniques in motion estimation in the next chapter.

## 13.2 GRADIENT-BASED APPROACH

It is noted that before the methods of optical flow determination were actually developed, optical flow had been discussed and exploited for motion and structure recovery from image sequences in computer vision for years. That is, the optical flow field was assumed to be available in the study of motion recovery. The first type of methods in optical flow determination is referred to as gradient-based techniques. This is because the spatial and temporal partial derivatives of intensity function are utilized in these techniques. In this section, we present the Horn and Schunck algorithm. It is regarded as the most prominent representative of this category. After the basic concepts are presented, some other methods in this category are briefly discussed.

### 13.2.1 The Horn and Schunck Method

We shall begin with a very general framework (Shi et al., 1994) to derive a brightness time-invariance equation. We then introduce the Horn and Schunck method.

### 13.2.1.1    Brightness Invariance Equation

As stated in Chapter 10, the imaging space can be represented by

$$f(x, y, t, \bar{s}),\tag{13.1}$$

where $\bar{s}$ indicates the sensor's position in 3-D world space, i.e., the coordinates of the sensor center and the orientation of the optical axis of the sensor. The $\bar{s}$ is a 5-D vector. That is, $\bar{s}$ where $(\tilde{x}, \tilde{y}, \tilde{z}, \beta, \gamma)$, where $\tilde{x}, \tilde{y},$ and $\tilde{z}$ represent the coordinate of the optical center of the sensor in 3-D world space; and $\beta$ and $\gamma$ represent the orientation of the optical axis of the sensor in 3-D world space, the Euler angles, pan and tilt, respectively.

With this very general notion, each picture, which is taken by a sensor located on a particular position at a specific moment, is merely a special cross section of this imaging space. Both temporal and spatial image sequences become a proper subset of the imaging space.

Assume now a world point $P$ in 3-D space that is perspectively projected onto the image plane as a pixel with the coordinates $x_P$ and $y_P$. Then, $x_P$ and $y_P$ are also dependent on $t$ and $\bar{s}$. That is,

$$f = f\left(x_P(t, \bar{s}), y_P(t, \bar{s}), t, \bar{s}\right).\tag{13.2}$$

If the optical radiation of the world point $P$ is invariant with respect to the time interval from $t_1$ to $t_2$, we then have

$$f\left(x_P(t_1, \bar{s}_1), y_P(t_1, \bar{s}_1), t_1, \bar{s}_1\right) = f\left(x_P(t_2, \bar{s}_1), y_P(t_2, \bar{s}_1), t_2, \bar{s}_1\right).\tag{13.3}$$

This is the brightness time-invariance equation.

At a specific moment $t_1$, if the optical radiation of $P$ is isotropical we then get

$$f\left(x_P(t_1, \bar{s}_1), y_P(t_1, \bar{s}_1), t_1, \bar{s}_1\right) = f\left(x_P(t_1, \bar{s}_2), y_P(t_1, \bar{s}_2), t_1, \bar{s}_2\right).\tag{13.4}$$

This is the brightness space-invariance equation.

If both conditions are satisfied, we get the brightness time-and-space-invariance equation, i.e.,

$$f\left(x_P(t_1, \bar{s}_1), y_P(t_1, \bar{s}_1), t_1, \bar{s}_1\right) = f\left(x_P(t_2, \bar{s}_2), y_P(t_2, \bar{s}_2), t_2, \bar{s}_2\right).\tag{13.5}$$

Consider two brightness functions $f(x(t, \bar{s}), y(t, \bar{s}), t, \bar{s})$ and $f(x(t + \Delta t, \bar{s} + \Delta \bar{s}), y(t + \Delta t, \bar{s} + \Delta \bar{s}), t + \Delta t, \bar{s} + \Delta \bar{s})$ in which the variation in time, $\Delta t$, and the variation in the spatial position of the sensor, $\Delta \bar{s}$, are very small. Due to the time-and-space-invariance of brightness, we can get

$$f\left(x(t, \bar{s}), y(t, \bar{s}), t, \bar{s}\right) = f\left(x(t + \Delta t, \bar{s} + \Delta \bar{s}), y(t + \Delta t, \bar{s} + \Delta s), t + \Delta t, \bar{s} + \Delta \bar{s}\right).\tag{13.6}$$

The expansion of the right-hand side of the above equation in the Taylor series at $(t, \bar{s})$ and the use of Equation 13.5 lead to

$$\left(\frac{\partial f}{\partial x} u + \frac{\partial f}{\partial y} v + \frac{\partial f}{\partial t}\right)\Delta t + \left(\frac{\partial f}{\partial x} u^{\bar{s}} + \frac{\partial f}{\partial y} v^{\bar{s}} + \frac{\partial f}{\partial \bar{s}}\right)\Delta \bar{s} + \varepsilon = 0,\tag{13.7}$$

where

$$u = \frac{\partial x}{\partial t}, \quad v \overset{\triangle}{=} \frac{\partial y}{\partial t}, \quad u^{\bar{s}} \overset{\triangle}{=} \frac{\partial x}{\partial \bar{s}}, \quad u^{\bar{s}} \overset{\triangle}{=} \frac{\partial y}{\partial \bar{s}}.$$

If $\Delta \bar{s} = 0$, i.e., the sensor is static in a fixed spatial position (in other words, both the coordinate of the optical center of the sensor and its optical axis direction remain unchanged), dividing both sides of the equation by $\Delta t$ and evaluating the limit as $\Delta t \to 0$ degenerate Equation 13.7 into

$$\frac{\partial f}{\partial x} u + \frac{\partial f}{\partial y} v + \frac{\partial f}{\partial t} = 0. \tag{13.8}$$

If $\Delta t = 0$, both its sides are divided by $\Delta \bar{s}$, and $\Delta \bar{s} \to 0$ is examined. Equation 13.7 then reduces to

$$\frac{\partial f}{\partial x} u^{\bar{s}} + \frac{\partial f}{\partial y} v^{\bar{s}} + \frac{\partial f}{\partial \bar{s}} = 0. \tag{13.9}$$

When $\Delta t = 0$, i.e., at a specific time moment, the images generated with sensors at different spatial positions can be viewed as a spatial sequence of images. Equation 13.9 is, then, the equation for the spatial sequence of images.

For the sake of brevity, we will focus on the gradient-based approach to optical flow determination with respect to temporal image sequences. That is, in the rest of this section we will address only Equation 13.8. It is noted that the derivation can be extended to spatial image sequences. The optical flow technique for spatial image sequences is useful in stereo image data compression. It plays an important role in motion and structure recovery. Interested readers are referred to Shi et al. (1994) and Shu and Shi (1993).

### 13.2.1.2 Smoothness Constraint

A careful examination of Equation 13.8 reveals that we have two unknowns: $u$ and $v$, i.e., the horizontal and vertical components of an optical flow vector at a three-tuple $(x, y, t)$, but only one equation to relate them. This once again demonstrates the ill-posed nature of optical flow determination. This also indicates that there is no way to compute optical flow by considering a single point of the brightness pattern moving independently. As stated in Section 13.1.3, some regularization measure — here an extra constraint — must be taken to overcome the difficulty.

A most popularly used constraint was proposed by Horn and Schunck and is referred to as the smoothness constraint. As the name implies, it constrains flow vectors to vary from one to another smoothly. Clearly, this is true for points in the brightness pattern most of the time, particularly for points belonging to the same object. It may be violated, however, along moving boundaries. Mathematically, the smoothness constraint is imposed in optical flow determination by minimizing the square of the magnitude of the gradient of the optical flow vectors:

$$\left(\frac{\partial u}{\partial x}\right)^2 + \left(\frac{\partial u}{\partial y}\right)^2 + \left(\frac{\partial v}{\partial x}\right)^2 + \left(\frac{\partial v}{\partial y}\right)^2. \tag{13.10}$$

It can be easily verified that the smoother the flow vector field, the smaller these quantities. Actually, the square of the magnitude of the gradient of intensity function with respect to the spatial coordinates, summed over a whole image or an image region, has been used as a smoothness measure of the image or the image region in the digital image processing literature (Gonzalez and Woods, 1992).

### 13.2.1.3 Minimization

Optical flow determination can then be converted into a minimization problem.

The square of the left-hand side of Equation 13.8, which can be derived from the brightness time-invariance equation, represents one type of error. It may be caused by quantization noise or other noises and can be written as

$$\varepsilon_b^2 = \left( \frac{\partial f}{\partial x} u + \frac{\partial f}{\partial y} v + \frac{\partial f}{\partial t} \right)^2.$$ (13.11)

The smoothness measure expressed in Equation 13.10 denotes another type of error, which is

$$\varepsilon_s^2 = \left( \frac{\partial u}{\partial x} \right)^2 + \left( \frac{\partial u}{\partial y} \right)^2 + \left( \frac{\partial v}{\partial x} \right)^2 + \left( \frac{\partial v}{\partial y} \right)^2.$$ (13.12)

The total error to be minimized is

$$\varepsilon^2 = \sum_x \sum_y \varepsilon_b^2 + \alpha^2 \varepsilon_s^2$$

$$= \sum_x \sum_y \left( \frac{\partial f}{\partial x} u + \frac{\partial f}{\partial y} v + \frac{\partial f}{\partial t} \right)^2 + \alpha^2 \left[ \left( \frac{\partial u}{\partial x} \right)^2 + \left( \frac{\partial u}{\partial y} \right)^2 + \left( \frac{\partial v}{\partial x} \right)^2 + \left( \frac{\partial v}{\partial y} \right)^2 \right],$$ (13.13)

where $\alpha$ is a weight between these two types of errors. The optical flow quantities $u$ and $v$ can be found by minimizing the total error. Using the calculus of variation, Horn and Schunck derived the following pair of equations for two unknown $u$ and $v$ at each pixel in the image.

$$\begin{cases} f_x^2 u + f_x f_y v = \alpha^2 \nabla^2 u - f_x f_t \\ f_x f_y u + f_y^2 v = \alpha^2 \nabla^2 v - f_y f_t \end{cases},$$ (13.14)

where

$$f_x = \frac{\partial f}{\partial x}, \quad f_y = \frac{\partial f}{\partial y}, \quad f_t = \frac{\partial f}{\partial t};$$

$\nabla^2$ denotes the Laplacian operator. The Laplacian operator of $u$ and $v$ are defined below.

$$\nabla^2 u = \frac{\partial^2 u}{\partial x^2} + \frac{\partial^2 u}{\partial y^2}$$

$$\nabla^2 v = \frac{\partial^2 v}{\partial x^2} + \frac{\partial^2 v}{\partial y^2}.$$ (13.15)

### 13.2.1.4 Iterative Algorithm

Instead of using the classical algebraic method to solve the pair of equations for $u$ and $v$, Horn and Schunck adopted the Gaussian Seidel (Ralston and Rabinowitz, 1978) method to have the following iterative procedure:

$$u^{k+1} = \bar{u}^k - \frac{f_x \left[ f_x \bar{u}^k + f_y \bar{v}^k + f_t \right]}{\alpha^2 + f_x^2 + f_y^2}$$

$$v^{k+1} = \bar{v}^k - \frac{f_y \left[ f_x \bar{u}^k + f_y \bar{v}^k + f_t \right]}{\alpha^2 + f_x^2 + f_y^2},$$

(13.16)

where the superscripts $k$ and $k + 1$ are indexes of iteration and $\bar{u}$, $\bar{v}$ are the local averages of $u$ and $v$, respectively.

Horn and Schunck define $\bar{u}$, $\bar{v}$ as follows:

$$\bar{u} = \frac{1}{6} \{ u(x, y+1) + u(x, y-1) + u(x+1, y) + u(x-1, y) \}$$

$$+ \frac{1}{12} \{ u(x-1, y-1) + u(x-1, y+1) + u(x+1, y-1) + u(x+1, y+1) \}$$

$$\bar{v} = \frac{1}{6} \{ v(x, y+1) + v(x, y-1) + v(x+1, y) + v(x-1, y) \}$$

$$+ \frac{1}{12} \{ v(x-1, y-1) + v(x-1, y+1) + v(x+1, y-1) + v(x+1, y+1) \}.$$

(13.17)

The estimation of the partial derivatives of intensity function and the Laplacian of flow vectors need to be addressed. Horn and Schunck considered a $2 \times 2 \times 2$ spatiotemporal neighborhood, shown in Figure 13.3, for estimation of partial derivatives $f_x, f_y$, and $f_t$. Note that replacing the first-order differentiation by the first-order difference is a common practice in managing digital images. The arithmetic average can remove the noise effect, thus making the obtained first-order differences less sensitive to various noises.

The Laplacian of $u$ and $v$ are approximated by

$$\nabla^2 u = \bar{u}(x, y) - u(x, y)$$

$$\nabla^2 v = \bar{v}(x, y) - v(x, y).$$

(13.18)

Equivalently, the Laplacian of $u$ and $v$, $\nabla^2(u)$ and $\nabla^2(v)$, can be obtained by applying a $3 \times 3$ window operator, shown in Figure 13.4, to each point in the $u$ and $v$ planes, respectively.

Similar to the pel recursive technique discussed in the previous chapter, there are two different ways to iterate. One way is to iterate at a pixel until a solution is steady. Another way is to iterate only once for each pixel. In the latter case, a good initial flow vector is required and is usually derived from the previous pixel.

### 13.2.2 Modified Horn and Schunck Method

Observing that the first-order difference is used to approximate the first-order differentiation in Horn and Schunck's original algorithm, and regarding this as a relatively crude form and a source

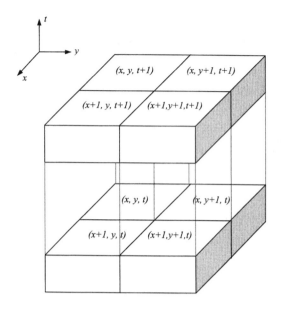

$$f_x = \frac{1}{4}\left\{\left[f(x+1,y,t)-f(x,y,t)\right]+\left[f(x+1,y,t+1)-f(x,y,t+1)\right]\right.$$

$$\left.+\left[f(x+1,y+1,t)-f(x,y,t)\right]+\left[f(x+1,y+1,t+1)-f(x,y+1,t+1)\right]\right\}$$

$$f_y = \frac{1}{4}\left\{\left[f(x,y+1,t)-f(x,y,t)\right]+\left[f(x+1,y+1,t)-f(x+1,y,t)\right]\right.$$

$$\left.+\left[f(x,y+1,t+1)-f(x,y,t+1)\right]+\left[f(x+1,y+1,t+1)-f(x+1,y,t+1)\right]\right\}$$

$$f_x = \frac{1}{4}\left\{\left[f(x,y,t+1)-f(x,y,t)\right]+\left[f(x+1,y,t+1)-f(x+1,y,t)\right]\right.$$

$$\left.+\left[f(x,y+1,t+1)-f(x,y+1,t)\right]+\left[f(x+1,y+1,t+1)-f(x+1,y+1,t)\right]\right\}$$

**FIGURE 13.3**   Estimation of $f_x$, $f_y$, and $f_t$.

of error, Barron, Fleet, and Beauchemin developed a modified version of the Horn and Schunck method (Barron et al., 1994).

It features a spatiotemporal presmoothing and a more-advanced approximation of differentiation. Specifically, it uses a Gaussian filter as a spatiotemporal prefilter. By the term *Gaussian filter*, we mean a low-pass filter with a mask shaped similar to that of the Gaussian probability density function. This is similar to what was utilized in the formulation of the Gaussian pyramid, which was discussed in Chapter 11. The term *spatiotemporal* means that the Gaussian filter is used for low-pass filtering in both spatial and temporal domains.

With respect to the more-advanced approximation of differentiation, a four-point central difference operator is used, which has a mask, shown in Figure 13.5.

As we will see later in this chapter, this modified Horn and Schunck algorithm has achieved better performance than the original one as a result of the two above-mentioned measures. This success indicates that a reduction of noise in image (data) leads to a significant reduction of noise in optical flow (solution). This example supports the statement we mentioned earlier that the ill-posed problem in low-level computational vision is mildly ill posed.

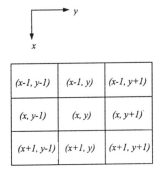

$$\nabla^2 u \approx \frac{1}{6}\left[u(x-1,y)+u(x,y-1)+u(x,y+1)+u(x+1,y)\right]$$

$$+\frac{1}{12}\left[u(x-1,y-1)+u(x-1,y+1)+u(x+1,y-1)+u(x+1,y+1)\right]$$

$$-u(x,y)$$

$$\nabla^2 v \approx \frac{1}{6}\left[v(x-1,y)+v(x,y-1)+v(x,y+1)+v(x+1,y)\right]$$

$$+\frac{1}{12}\left[v(x-1,y-1)+v(x-1,y+1)+v(x+1,y-1)+v(x+1,y+1)\right]$$

$$-v(x,y)$$

**FIGURE 13.4**  A $3\times3$ window operation for estimation of the Laplacian of flow vector.

| $-\dfrac{1}{12}$ | $\dfrac{8}{12}$ | 0 | $-\dfrac{8}{12}$ | $\dfrac{1}{12}$ |
|---|---|---|---|---|

**FIGURE 13.5**  Four-point central difference operator mask.

### 13.2.3  THE LUCAS AND KANADE METHOD

Lucas and Kanade assume a flow vector is constant within a small neighborhood of a pixel, denoted by $\Omega$. Then they form a weighted object function as follows.

$$\sum_{(x,y)\in\Omega} w^2(x,y)\left[\frac{\partial f(x,y,t)}{\partial x}u+\frac{\partial f(x,y,t)}{\partial v}v+\frac{\partial f(x,y,t)}{\partial t}\right]^2, \tag{13.19}$$

where $w(x,y)$ is a window function, which gives more weight to the central portion than the surrounding portion of the neighborhood $\Omega$.

The flow determination thus becomes a problem of a least-square fit of the brightness invariance constraint. We observe that the smoothness constraint has been implied in Equation 13.19, where the flow vector is assumed to be constant within $\Omega$.

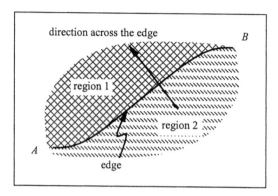

**FIGURE 13.6**    Oriented-smoothness constraint.

### 13.2.4 The Nagel Method

Nagel first used the second-order derivatives in optical flow determination in the very early days (Nagel, 1983). Since the brightness function $f(x, y, t, \bar{s})$ is a real-valued function of multiple variables (or a vector of variables), the Hessian matrix, discussed in Chapter 12, is used for the second-order derivatives.

An oriented-smoothness constraint was developed by Nagel that prohibits imposition of the smoothness constraint across edges, as illustrated in Figure 13.6. In the figure, an edge $AB$ separates two different moving regions: region 1 and region 2. The smoothness constraint is imposed in these regions separately. That is, no smoothness constraint is imposed across the edge. Obviously, it would be a disaster if we smoothed the flow vectors across the edge. As a result, this reasonable treatment effectively improves the accuracy of optical flow estimation (Nagel, 1989).

### 13.2.5 The Uras, Girosi, Verri, and Torre Method

The Uras, Girosi, Verri, and Torre method is another method that uses second-order derivatives. Based on a local procedure, it performs quite well (Uras et al., 1988).

## 13.3 CORRELATION-BASED APPROACH

The correlation-based approach to optical flow determination is similar to block matching, covered in Chapter 11. As may be recalled, the conventional block-matching technique partitions an image into nonoverlapped, fixed-size, rectangular blocks. Then, for each block, the best matching in the previous image frame is found. In doing so, a search window is opened in the previous frame according to some *a priori* knowledge: the time interval between the two frames and the maximum possible moving velocity of objects in frames. Centered on each of the candidate pixels in the search window, a rectangle correlation window of the same size as the original block is opened. The best-matched block in the search window is chosen such that either the similarity measure is maximized or the dissimilarity measure is minimized. The relative spatial position between these two blocks (the original block in the current frame and the best-matched one in the previous frame) gives a translational motion vector to the original block. In the correlation-based approach to optical flow computation, the mechanism is very similar to that in conventional block matching. The only difference is that for each pixel in an image, we open a rectangle correlation window centered on this pixel for which an optical flow vector needs to be determined. It is for this correlation window that we find the best match in the search window in its temporal neighboring image frame. This is shown in Figure 13.7. A comparison between Figures 13.7 and 11.1 can convince us about the

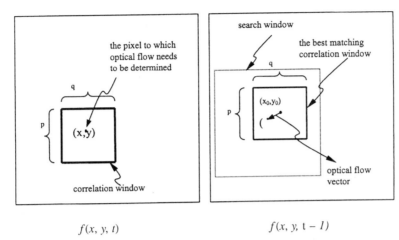

$$f(x, y, t) \qquad\qquad\qquad f(x, y, t-1)$$

**FIGURE 13.7**   Correlation-based approach to optical flow determination.

above observation. In this section, we first briefly discuss Anandan's method, which is pioneer work in this category. Then Singh's method is described. His unified view of optical flow computation is introduced. We then present a correlation-feedback method by Pan, Shi, and Shu, which uses the feedback technique in flow calculation.

### 13.3.1   THE ANANDAN METHOD

As mentioned in Chapter 11, the sum of squared difference (SSD) is used as a dissimilarity measure in (Anandan, 1987). It is essentially a simplified version of the well-known mean square error (MSE). Due to its simplicity, it is used in the methods developed by Singh (1992), and Pan, Shi, and Shu (1998).

In the Anandan method (Anandan, 1989), a pyramid structure is formed, and it can be used for an efficient coarse-fine search. This is very similar to the multiresolution block-matching techniques discussed in Chapter 11. In the higher levels (with lower resolution) of the pyramid, a full search can be performed without a substantial increase in computation. The estimated velocity (or displacement) vector can be propagated to the lower levels (with higher resolution) for further refinement. As a result, a relatively large motion vector can be estimated with a certain degree of accuracy.

Instead of the Gaussian pyramid discussed in Chapter 11, however, a Laplacian pyramid is used here. To understand the Laplacian pyramid, let us take a look at Figure 13.8(a). There two consecutive levels are shown in a Gaussian pyramid structure: level $k$, denoted by $f^k(x, y)$, and level $k + 1$, $f^{k+1}(x, y)$. Figure 13.8(b) shows how level $k + 1$ can be derived from level $k$ in the Gaussian pyramid. That is, as stated in Chapter 11, level $k + 1$ in the Gaussian pyramid can be obtained through low-pass filtering applied to level $k$, followed by subsampling. Figure 13.8(c), level $k + 1$ is first interpolated, thus producing an estimate of level $k$, $\hat{f}^k(x, y)$. The difference between the original level $k$ and the interpolated estimate of level $k$ generates an error at level $k$, denoted by $e^k(x, y)$. If there are no quantization errors involved, then level $k$, $f^k(x, y)$ can be recovered completely from the interpolated estimate of level $k$, $\hat{f}^k(x, y)$, and the error at level $k$, $e^k(x, y)$. That is,

$$f^k(x, y) = \hat{f}^k(x, y) + e^k(x, y). \tag{13.20}$$

With quantization errors, however, the recovery of level $k$, $f^k(x, y)$ is not error free. It can be shown that coding $\hat{f}^k(x, y)$ and $e^k(x, y)$ is more efficient than directly coding $f^k(x, y)$.

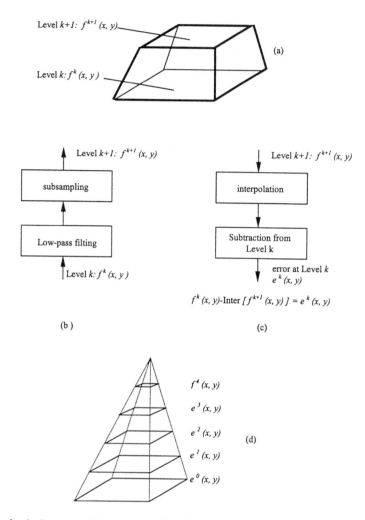

**FIGURE 13.8** Laplacian pyramid (level $k$ in a Gaussian pyramid). (a) Two consecutive levels in a pyramid structure. (b) Derivation of level $k + 1$ from level K. (c) Derivation of error at level $k$ in a Laplacian pyramid. (d) Structure of Laplacian pyramid.

A set of images $e^k(x, y)$, $k = 0, 1, ..., K - 1$ and $f^K(x, y)$ forms a Laplacian pyramid. Figure 13.8(d) displays a Laplacian pyramid with $K = 5$. It can be shown that Laplacian pyramids provide an efficient way for image coding (Burt and Adelson, 1983). A more-detailed description of Gaussian and Laplacian pyramids can be found in Burt (1984) and Lim (1990).

## 13.3.2 THE SINGH METHOD

Singh (1991, 1992) presented a unified point of view on optical flow computation. He classified the information available in image sequences for optical flow determination into two categories: conservation information and neighborhood information. Conservation information is the information assumed to be conserved from one image frame to the next in flow estimation. Intensity is an example of conservation information, which is used most frequently in flow computation. Clearly, the brightness invariance constraint in the Horn and Schunck method is another way to state this type of conservation. Some functions of intensity may be used as conservation information as well.

In fact, Singh uses the Laplacian of intensity as conservation information for computational simplicity. More examples can be found later in Section 13.4. Other information, different from intensity, such as color, can be used as conservation information. Neighborhood information is the information available in the neighborhood of the pixel from which optical flow is estimated.

These two different types of information correspond to two steps in flow estimation. In the first step, conservation information is extracted, resulting in an initial estimate of flow vector. In the second step, this initial estimate is propagated into a neighborhood area and is iteratively updated. Obviously, in the Horn and Schunck method, the smoothness constraint is essentially one type of neighborhood information. Iteratively, estimates of flow vectors are refined with neighborhood information so that flow estimators from areas having sufficient intensity variation, such as the intensity corners as shown in Figure 13.2(d) and areas with strong texture, can be propagated into areas with relatively small intensity variation or uniform intensity distribution.

With this unified point of view on optical flow estimation, Singh treated flow computation as parameter estimation. By applying estimation theory to flow computation, he developed an estimation-theoretical method to determine optical flow. It is a correlation-based method and consists of the above-mentioned two steps.

### 13.3.2.1 Conservation Information

In the first step, for each pixel $(x, y)$ in the current frame $f_n(x, y)$, a correlation window of $(2l + 1) \times (2l + 1)$ is opened, centered on the pixel. A search window of $(2N+1) \times (2N+1)$ is opened in the previous frame $f_{n-1}(x, y)$ centered on $(x, y)$. An error distribution of those $(2N + 1) \times (2N + 1)$ samples are calculated by using SSD as follows:

$$E_c(u,v) = \sum_{s=-l}^{l}\sum_{t=-l}^{l}\left[f_n(x+s,y+t)-f_{n-1}(x-u+s,y-v+t)\right]^2 \quad -N \le u,v \le N. \quad (13.21)$$

A response–distribution for these $(2N + 1) \times (2N + 1)$ samples is then calculated.

$$R_c(u,v) = e^{-\beta E_c(u,v)}, \quad (13.22)$$

where $\beta$ is a parameter, whose function and selection will be described in Section 13.3.3.1.

According to the weighted-least-square estimation, the optical flow can be estimated in this step as follows:

$$u_c = \frac{\sum_u \sum_v R_c(u,v)u}{\sum_u \sum_v R_c(u,v)}$$

$$\quad (13.23)$$

$$v_c = \frac{\sum_u \sum_v R_c(u,v)v}{\sum_u \sum_v R_c(u,v)}.$$

Assuming errors are additive and zero-mean random noise, we can also find the covariance matrix associated with the above estimate:

$$S_c = \begin{pmatrix} \dfrac{\sum_u \sum_v R_c(u,v)(u-u_c)^2}{\sum_u \sum_v R_c(u,v)} & \dfrac{\sum_u \sum_v R_c(u,v)(u-u_c)(v-v_c)}{\sum_u \sum_v R_c(u,v)} \\[4mm] \dfrac{\sum_u \sum_v R_c(u,v)(u-u_c)(v-v_c)}{\sum_u \sum_v R_c(u,v)} & \dfrac{\sum_u \sum_v R_c(u,v)(v-v_c)^2}{\sum_u \sum_v R_c(u,v)} \end{pmatrix}. \tag{13.24}$$

### 13.3.2.2  Neighborhood Information

After step 1, all initial estimates are available. In step 2, they need to be refined according to neighborhood information. For each pixel, the method considers a $(2w+1) \times (2w+1)$ neighborhood centered on it. The optical flow of the center pixel is updated from the estimates in the neighborhood. A set of Gaussian coefficients is used in the method such that the closer the neighbor pixel to the center pixel, the more influence the neighbor pixel has on the flow vector of the center pixel. The weighted-least-square based estimate in this step is

$$\bar{u} = \frac{\sum_u \sum_v R_n(u,v)u}{\sum_u \sum_v R_n(u,v)}$$

$$\bar{v} = \frac{\sum_u \sum_v R_n(u,v)v}{\sum_u \sum_v R_n(u,v)}, \tag{13.25}$$

and the associated covariance matrix is

$$S_c = \begin{pmatrix} \dfrac{\sum_i R_n(u_i,v_i)(u_i-\bar{u})^2}{\sum_i R_n(u_i,v_i)} & \dfrac{\sum_i R_n(u_i,v_i)(u_i-\bar{u})(v_i-\bar{v})}{\sum_i R_n(u_i,v_i)} \\[4mm] \dfrac{\sum_i R_n(u_i,v_i)(u_i-\bar{u})(v_i-\bar{v})}{\sum_i R_n(u_i,v_i)} & \dfrac{\sum_i R_n(u_i,v_i)(v_i-\bar{v})^2}{\sum_i R_n(u_i,v_i)} \end{pmatrix}, \tag{13.26}$$

where $1 \leq i \leq (2w+1)^2$.

In implementation, Singh uses a $3 \times 3$ neighborhood (i.e., $w = 1$) centered on the pixel under consideration. The weights are depicted in Figure 13.9.

### 13.3.2.3  Minimization and Iterative Algorithm

According to estimation theory (Beck and Arnold, 1977), two covariance matrices, expressed in Equations 13.24 and 13.26, respectively, are related to the confidence measure. That is, the reciprocals of the eigenvalues of the covariance matrix reveal confidence of the estimate along the

| ( 0.25×0.25 ) $\dfrac{1}{16}$ | ( 0.5×0.25 ) $\dfrac{1}{8}$ | ( 0.25×0.25 ) $\dfrac{1}{16}$ |
|---|---|---|
| ( 0.5×0.25 ) $\dfrac{1}{8}$ | ( 0.5×0.5 ) $\dfrac{1}{4}$ | ( 0.5×0.25 ) $\dfrac{1}{8}$ |
| ( 0.25×0.25 ) $\dfrac{1}{16}$ | ( 0.5×0.25 ) $\dfrac{1}{8}$ | ( 0.25×0.25 ) $\dfrac{1}{16}$ |

**FIGURE 13.9**   $3 \times 3$ Gaussian mask.

direction represented by the corresponding eigenvectors. Moreover, conservation error and neighborhood error can be represented as the following two quadratic terms, respectively.

$$\left(U - U_c\right)^T S_c^{-1}\left(U - U_c\right) \tag{13.27}$$

$$\left(U - \overline{U}\right)^T S_n^{-1}\left(U - \overline{U}\right), \tag{13.28}$$

where $\overline{U} = (\bar{u}, \bar{v})$, $U_c = (u_c, v_c)$, $U = (u, v)$.

The minimization of the sum of these two errors over the image area leads to an optimal estimate of optical flow. That is, find $(u, v)$ such that the following error is minimized.

$$\sum_x \sum_y \left[\left(U - U_c\right)^T S_c^{-1}\left(U - U_c\right) + \left(U - \overline{U}\right)^T S_n^{-1}\left(U - \overline{U}\right)\right]. \tag{13.29}$$

An iterative procedure according to the Gauss–Siedel algorithm (Ralston and Rabinowitz, 1978) is used by Singh:

$$U^{k+1} = \left[S_c^{-1} + S_n^{-1}\right]^{-1}\left[S_c^{-1}U_c + S_n^{-1}\overline{U}^k\right] \tag{13.30}$$

$$U^0 = U_c.$$

Note that $U_c$, $S_c$ are calculated once and remain unchanged in all the iterations. On the contrary, $\overline{U}$ and $S_n$ vary with each iteration. This agrees with the description of the method in Section 13.3.2.2.

### 13.3.3   THE PAN, SHI, AND SHU METHOD

Applying feedback (a powerful technique widely used in automatic control and many other fields) to a correlation-based algorithm, Pan, Shi, and Shu developed a correlation-feedback method to compute optical flow. The method is iterative in nature. In each iteration, the estimated optical flow and its several variations are fed back. For each of the varied optical flow vectors, the corresponding sum of squared displaced frame difference (DFD), which was discussed in Chapter 12 and which often involves bilinear interpolation, is calculated. This useful information is then utilized in a revised version of a correlation-based algorithm (Singh, 1992). They choose to work with this

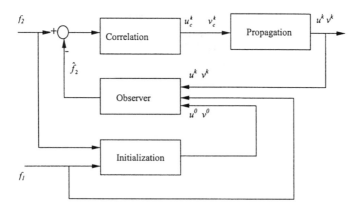

**FIGURE 13.10**   Block diagram of correlation feedback technique.

algorithm because it has several merits, and its estimation-theoretical computation framework lends itself to the application of the feedback technique.

As expected, the repeated usage of two given images via the feedback iterative procedure improves the accuracy of optical flow considerably. Several experiments on real image sequences in the laboratory and some synthetic image sequences demonstrate that the correlation-feedback algorithm performs better than some standard gradient- and correlation-based algorithms in terms of accuracy.

### 13.3.3.1   Proposed Framework

The block diagram of the proposed framework is shown in Figure 13.10 and described next.

**Initialization** — Although any flow algorithms can be used to generate an initial optical flow field $\bar{u}^o = (u^o, v^o)$ (even a nonzero initial flow field without applying any flow algorithm may work, but slowly), the Horn and Schunck algorithm (Horn and Schunck, 1981), discussed in Section 13.2.1 (usually 5 to 10 iterations) is used to provide an appropriate starting point after preprocessing (involving low-pass filtering), since the algorithm is fast and the problem caused by the smoothness constraint is not serious in the first 10 to 20 iterations. The modified Horn and Schunck method, discussed in Section 13.2.2, may also be used for the initialization.

**Observer** — The DFD at the $k$th iteration is observed as $f_n(\bar{x}) - f_{n-1}(\bar{x} - \bar{u}^k)$, where $f_n$ and $f_{n-1}$ denote two consecutive digital images, $\bar{x} = (x, y)$ denotes the spatial coordinates of the pixel under consideration, and $\bar{u}^k = (u^k, v^k)$ denotes the optical flow of this pixel estimated at the $k$th iteration. (Note that the vector representation of the spatial coordinates in image planes is used quite often in the literature, because of its brevity in notation.) Demanding fractional pixel accuracy usually requires interpolation. In the Pan et al. work, the bilinear interpolation is adopted. The bilinearly interpolated image is denoted by $\hat{f}_{n-1}$.

**Correlation** — Once the bilinearly interpolated image is available, a correlation measure needs to be selected to search for the best match of a given pixel in $f_n(\bar{x})$ in a search area in the interpolated image. In their work, the sum-of-square-differences (SSD) is used. For each pixel in $f_n$, a correlation window $W_c$ of size $(2l + 1) \times (2l + 1)$ is formed, centered on the pixel.

The search window in the proposed approach is quite different from that used in the correlation-based approach, say, that of Singh (1992). Let $u$ be a quantity chosen from the following five quantities:

$$u \in \left\{ u^k - \frac{1}{2}u^k, u^k - \frac{1}{4}u^n, u^k, u^k + \frac{1}{4}u^k, u^k + \frac{1}{2}u^k \right\}. \tag{13.31}$$

Let $v$ be a quantity chosen from the following five quantities:

$$v \in \left\{ v^k - \frac{1}{2}v^k, v^k - \frac{1}{4}v^n, v^k, v^k + \frac{1}{4}v^k, v^k + \frac{1}{2}v^k \right\}. \tag{13.32}$$

Hence, there are 25 (i.e., $5 \times 5$) possible combinations for $(u, v)$. (It is noted that the restriction of the nonzero initial flow field mentioned above in part A comes from here). Note that other choices of variations around $(u^k, v^k)$ are possible. Each of them corresponds to a pixel, $(x - u, y - v)$, in the bilinearly interpolated image plane. A correlation window is formed and centered in this pixel. The 25 samples of error distribution around $(u^k, v^k)$ can be computed by using the SSD. That is,

$$E(u,v) = \sum_{s=-l}^{l} \sum_{t=-l}^{l} \left( f_n(x+s, y+t) - \hat{f}_{n-1}(x-u+s, y-v+t) \right)^2. \tag{13.33}$$

The 25 samples of response distribution can be computed as follows:

$$R_c(u,v) = e^{-\beta E(u,v)}, \tag{13.34}$$

where $\beta$ is chosen so as to make the maximum $R_c$ among the 25 samples of response distribution be a number close to unity. The choice of an exponential function for converting the error distribution into the response distribution is based primarily on the following consideration: the exponential function is well behaved when the error approaches zero and all the response distribution values are positive. The choice of $\beta$ mentioned above is motivated by the following observation: in this way, the $R_c$ values, which are the weights used in Equation 13.35, will be more effective. That is, the computation in Equation 13.35 will be more sensitive to the variation of the error distribution defined in Equation 13.33.

The optical flow vector derived at this correlation stage is then calculated as follows, according to the weighted-least-squares estimation (Singh, 1992).

$$u^k(x,y) = \frac{\sum_u \sum_v R_c(u,v)u}{\sum_u \sum_v R_c(u,v)}, \quad v_c^k(x,y) = \frac{\sum_u \sum_v R_c(u,v)v}{\sum_u \sum_v R_c(u,v)}. \tag{13.35}$$

**Propagation** — Except in the vicinity of motion boundaries, the motion vectors associated with neighboring pixels are expected to be similar. Therefore, this constraint can be used to regularize the motion field. That is,

$$u^{k+1}(x,y) = \sum_{i=-w}^{w} \sum_{j=-w}^{w} w_1(i,j)u_c^k(x+i,y+j), v^{k+1}(x,y) = \sum_{i=-w}^{w} \sum_{j=-w}^{w} w_1(i,j)u_c^k(x+i,y+j), \tag{13.36}$$

where $w_1(i,j)$ is a weighting function. The Gaussian mask shown in Figure 13.9 is chosen as the weighting function $w_1(i,j)$ used in our experiments. By using this mask, the velocity of various pixels in the neighborhood of a pixel will be weighted according to their distance from the pixel: the larger the distance, the smaller the weight. The mask smooths the optical flow field as well.

**Convergence** — Under the assumption of the symmetric response distribution with a single maximum value assumed by the ground-truth optical flow, the convergence of the correlation-feedback technique is justified by Pan et al. (1995).

### 13.3.3.2   Implementation and Experiments

**Implementation** — To make the algorithm more robust against noise, three consecutive images in an image sequence, denoted by $f_1, f_2$, and $f_3$, respectively, are used to implement their algorithm instead of the two images in the above principle discussion. This implementation was proposed by Singh (1992). Assume the time interval between $f_1$ and $f_2$ is the same as that between $f_2$ and $f_3$. Also assume the apparent 2-D motion is uniform during these two intervals along the motion trajectories. From images $f_1$ and $f_2$, $(u^o, v^o)$ can be computed. From $(u^k, v^k)$, the optical flow estimated during the $k$th iteration, and $f_1$ and $f_2$, the response distribution, $R_c^+(u^k, v^k)$, can be calculated as

$$R_c^+\left(u^k,v^k\right)=\exp\left\{-\beta\sum_{s=-l}^{l}\sum_{t=-l}^{l}\left[f_2(x+s,y+t)-\hat{f}_1\left(x-u^k+s,y-v^k+t\right)\right]^2\right\}. \qquad (13.37)$$

Similarly, from images $f_3$ and $f_2$, $(-u^k, -v^k)$ can be calculated. Then $R_c^-(-u^k, -v^k)$ can be calculated as

$$R_c^-\left(-u^k,-v^k\right)=\exp\left\{-\beta\sum_{s=-l}^{l}\sum_{t=-l}^{l}\left[f_2(x+s,y+t)-\hat{f}_3\left(x-u^k+s,y+v^k+t\right)\right]^2\right\}. \qquad (13.38)$$

The response distribution $R_c(u^k, v^k)$ can then be determined as the sum of $R_c^+(u^k, v^k)$ and $R_c^-(-u^k, -v^k)$. The size of the correlation window and the weighting function is chosen to be $3 \times 3$, i.e., $l = 1$, $w = 1$. In each search window, $\beta$ is chosen so as to make the larger one among $R_c^+$ and $R_c^-$ a number close to unity. In the observer stage, the bilinear interpolation is used, which is shown to be faster and better than the B-spline in the many experiments of Pan et al.

**Experiment I** — Figure 13.11 shows the three successive image frames $f_1, f_2$, and $f_3$ about a square post. They were taken by a CCD video camera and a DATACUBE real-time image processing system supported by a Sun workstation. The square post is moving horizontally, perpendicular to the optical axis of the camera, in a uniform speed of 2.747 pixels per frame. To remove various noises to a certain extent and to speed up processing, these three $256 \times 256$ images are low-pass filtered and then subsampled prior to optical flow estimation. That is, the intensities of every 16 pixels in a block of $4 \times 4$ are averaged and the average value is assigned to represent this block. Note that the choice of other low-pass filters is also possible. In this way, these three images are compressed into three $64 \times 64$ images. The "ground-truth" 2-D motion velocity vector is hence known as $u^a = -0.6868$; $v^a = 0$.

To compare the performance of the correlation-feedback approach with that of the gradient-based and correlation-based approaches, the Horn and Schunck algorithm is chosen to represent the gradient-based approach and Singh's framework to represent the correlation-based approach. Table 13.1 shows the results of the comparison. There, $l$, $w$, and $N$ indicate the sizes of the correlation window, weighting function, and search window, respectively. The program that implements Singh's algorithm is provided by Barron et al. (1994). In the correlation-feedback algorithm, ten iterations of the Horn and Schunck algorithm with $\alpha = 5$ are used in the initialization. (Recall that the $\alpha$ is a regularization parameter used by Horn and Schunck, 1981). Only the central $40 \times 40$ flow vector array is used to compute $u_{error}$, which is the root mean square (RMS) error in the vector magnitudes between the ground-truth and estimated optical flow vectors. It is noted that the relative error in Experiment I is greater than 10%. This is because the denominator in the formula calculating the RMS error is too small due to the static background and, hence, there are many zero ground-truth 2-D motion velocity vectors in this experiment. Relatively speaking, the correlation-feedback algorithm performs best in determining optical flow for a texture post in translation. The correct optical flow field and those calculated by using three different algorithms are shown in Figure 13.12.

 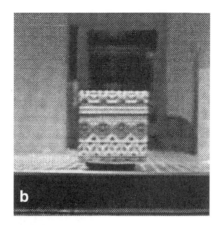

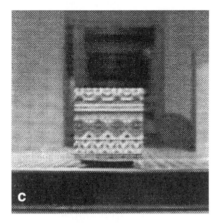

**FIGURE 13.11**   Texture square (a). Texture square (b). Texture square (c).

**TABLE 13.1**
**Comparison in Experiment I**

| Techniques | Gradient-Based Approach | Correlation-Based Approach | Correlation-Feedback Approach |
|---|---|---|---|
| 13.3.3.3 Conditions | *Iteration no.* = 128<br>$\alpha = 5$ | *Iteration no.* = 25<br>$l = 2, w = 2$<br>N = 4 | *Iteration no.* = 10<br>*Iteration no.* (*Horn*) = 10<br>$l = 1, w = 1, N = 5$ |
| $u_{error}$ | 56.37% | 80.97% | 44.56% |

**Experiment II** — The images in Figure 13.13 were obtained by rotating a CCD camera with respect to the center of a ball. The rotating velocity is 2.5° per frame. Similarly, three $256 \times 256$ images are compressed into three $64 \times 64$ images by using the averaging and subsampling discussed above. Only the central $40 \times 40$ optical vector arrays are used to compute $u_{error}$. Table 13.2 reports the results for this experiment. There, $u_{error}$, $l$, $w$, and $N$ have the same meaning as that discussed in Experiment I. It is obvious that our correlation-feedback algorithm performs best in determining optical flow for this rotating ball case.

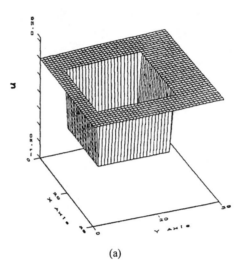

(a)

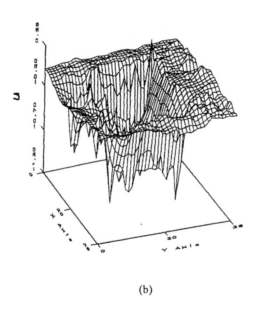

(b)

**FIGURE 13.12**  (a) Correct optical flow field. (b) Optical flow field calculated by the gradient-based approach. (c) Optical flow field calculated by the correlation-based approach. (d) Optical flow field calculated by the correlation-feedback approach.

**Experiment III** — To compare the correlation-feedback algorithm with other existing techniques in a more objective, quantitative manner, Pan et al. cite some results reported by Barron et al. (1994), which were obtained by applying some typical optical flow techniques to some image sequences chosen with deliberation. In the meantime they report the results obtained by applying their feedback technique to the identical image sequences with the same accuracy measurement as used by Barron et al. (1994).

Three image sequences used by Barron et al. (1994) were utilized here. They are named "Translating Tree," "Diverging Tree," and "Yosemite." The first two simulate translational camera motion with respect to a textured planar surface (Figure 13.14), and are sometimes referred to as

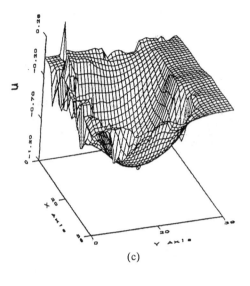

(c)

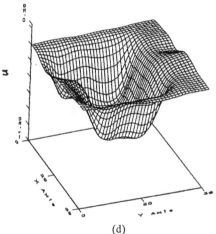

(d)

**FIGURE 13.12** (continued)

"Tree 2-D" sequence. Therefore, there are no occlusions and no motion discontinuities in these two sequences. In the "Translating Tree" sequence, the camera moves normally to its line of sight, with velocities between 1.73 and 2.26 pixels/frame parallel to the x-axis in the image plane. In the "Diverging Tree" sequence, the camera moves along its line of sight. The focus of expansion is at the center of the image. The speeds vary from 1.29 pixels/frame on left side to 1.86 pixels/frame on the right. The "Yosemite" sequence is a more complex test case (see Figure 13.15). The motion in the upper right is mainly divergent. The clouds translate to the right with a speed of 1 pixel/frame, while velocities in the lower left are about 4 pixels/frame. This sequence is challenging because of the range of velocities and the occluding edges between the mountains and at the horizon. There is severe aliasing in the lower portion of the images, causing most methods to produce poorer velocity measurements. Note that this synthetic sequence is for quantitative study purposes since its ground-truth flow field is known and is, otherwise, far less complex than many real-world outdoor sequences processed in the literature.

The angular measure of the error used by Barron et al. (1994) is utilized here, as well. Let image velocity $\bar{u} = (u, v)$ be represented as 3-D direction vectors,

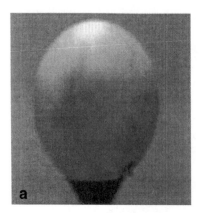
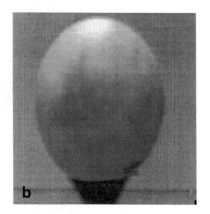

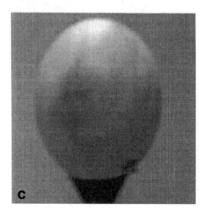

**FIGURE 13.13**   A rotating ball in three different frames — a, b, c. The rotating velocity is 2.5° per frame.

**TABLE 13.2**
**Comparison in Experiment II**

| Techniques | Gradient-Based Approach | Correlation-Based Approach | Correlation-Feedback Approach |
|---|---|---|---|
| Conditions | *Iteration no.* = 128 | *Iteration no.* = 25 | *Iteration no.* = 10 |
|  | $\alpha = 5$ | $l = 2, w = 2$ | *Iteration no.* (*Horn*) = 10 |
|  |  | N = 4 | $l = 1, w = 1, N = 5$ |
| $u_{error}$ | 65.67% | 55.29% | 49.80% |

$$\bar{V} \equiv \frac{1}{\sqrt{u^2 + v^2 + 1}}(u, v, 1). \tag{13.39}$$

The angular error between the correct image velocity $\bar{V}$ and an estimate $\bar{V}_e$ is $\psi_E$ = across $(\bar{V}_c \cdot \bar{V}_e)$. It is obvious that the smaller the angular error $\psi_E$, the more accurate the estimation of the optical flow field will be. Despite the fact that the confidence measurement can be used in the correlation-feedback algorithm, as well, Pan et al. did not consider the usage of the confidence measurement in their work. Therefore, only the results with 100% density in Tables 4.6, 4.7, and 4.10 in the Barron et al. (1994) paper were used in Tables 13.3, 13.4, and 13.5, respectively.

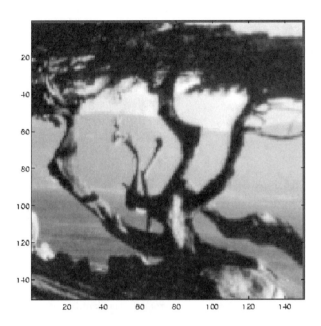

**FIGURE 13.14**   A frame of the "Tree 2-D" sequence.

**FIGURE 13.15**   A frame of the "Yosemite" sequence.

Prior to computation of the optical flow field, the "Yosemite" and "Tree 2-D" test sequences were compressed by a factor of 16 and 4, respectively, using the averaging and subsampling method discussed earlier.

As mentioned by Barron et al. (1994) the optical flow field for the "Yosemite" sequence is complex, and Table 13.5 indicates that the correlation-feedback algorithm evidently performs best. A robust method was developed and applied to a cloudless Yosemite sequence (Black and Anandan, 1996). It is noted that the performance of flow determination algorithms will be improved if the sky is removed from consideration (Barron et al., 1994; Black and Anandan, 1996). Still, it is clear

**TABLE 13.3**
**Summary of the "Translating Tree" 2-D Velocity Results**

| Techniques | Average Error, ° | Standard Deviation, ° | Density, % |
|---|---|---|---|
| Horn and Schunck (original) | 38.72 | 27.67 | 100 |
| Horn and Schunck (modified) | 2.02 | 2.27 | 100 |
| Uras et al. (unthresholded) | 0.62 | 0.52 | 100 |
| Nagel | 2.44 | 3.06 | 100 |
| Anandan | 4.54 | 3.10 | 100 |
| Singh (step 1, $l = 2$, $w = 2$) | 1.64 | 2.44 | 100 |
| Singh (step 2, $l = 2$, $w = 2$) | 1.25 | 3.29 | 100 |
| Correlation feedback ($l = 1$, $w = 1$) | 1.07 | 0.48 | 100 |

**TABLE 13.4**
**Summary of the "Diverging Tree" 2-D Velocity Results**

| Techniques | Average Error, ° | Standard Deviation, ° | Density, % |
|---|---|---|---|
| Horn and Schunck (original) | 12.02 | 11.72 | 100 |
| Horn and Schunck (modified) | 2.55 | 3.67 | 100 |
| Uras et al. (unthresholded) | 4.64 | 3.48 | 100 |
| Nagel | 2.94 | 3.23 | 100 |
| Anandan (frames 19 and 21) | 7.64 | 4.96 | 100 |
| Singh (step 1, $l = 2$, $w = 2$) | 17.66 | 14.25 | 100 |
| Singh (step 2, $l = 2$, $w = 2$) | 8.60 | 5.60 | 100 |
| Pan, Shi, and Shu ($l = 1$, $w = 1$) | 5.12 | 2.16 | 100 |

**TABLE 13.5**
**Summary of the "Yosemite" 2-D Velocity Results**

| Techniques | Average Error, ° | Standard Deviation, ° | Density, % |
|---|---|---|---|
| Horn and Schunck (original) | 32.43 | 30.28 | 100 |
| Horn and Schunck (modified) | 11.26 | 16.41 | 100 |
| Uras et al. (unthresholded) | 10.44 | 15.00 | 100 |
| Nagel | 11.71 | 10.59 | 100 |
| Anandan (frames 19 and 21) | 15.84 | 13.46 | 100 |
| Singh (step 1, $l = 2$, $w = 2$) | 18.24 | 17.02 | 100 |
| Singh (step 2, $l = 2$, $w = 2$) | 13.16 | 12.07 | 100 |
| Pan, Shi, and Shu ($l = 1$, $w = 1$) | 7.93 | 6.72 | 100 |

that the algorithm in the Black and Anandan (1996) paper achieved very good performance in terms of accuracy. In order to make a comparison with their algorithm, the correlation-feedback algorithm was applied to the same cloudless Yosemite sequence. The results were reported in Table 13.6, from which it can be observed that the results obtained by Pan et al. are slightly better. Tables 13.3 and 13.4 indicate that the feedback technique also performs very well in translating and diverging texture post cases.

**Experiment IV** — Here, the correlation-feedback algorithm is applied to a real sequence named *Hamburg Taxi*, which is used as a testing sequence by Barron et al. (1994). There are four moving

**TABLE 13.6**
**Summary of the cloudless "Yosemite" 2-D Velocity Results**

| Techniques | Average Error, ° | Standard Deviation, ° | Density, % |
|---|---|---|---|
| Robust formulation | 4.46 | 4.21 | 100 |
| Pan, Shi, and Shu ($l = 1$, $w = 1$) | 3.79 | 3.44 | 100 |

**FIGURE 13.16**   Hamburg Taxi.

objects in the scene: a moving pedestrian in the upper left portion, a turning car in the middle, a car moving toward right at the left side and a car moving toward left at the right side. A frame of the sequence and the needle diagram of flow vectors estimated by using ten iterations of the correlation-feedback algorithm (with ten iterations of the Horn and Schunck algorithm for initialization) are shown in Figures 13.16 and 13.17, respectively. The needle diagram is printed in the same fashion as those shown by Barron et al. (1994). It is noted that the moving pedestrian in the upper left portion cannot be shown because of the scale used in the needle diagram. The other three moving vehicles in the sequence are shown very clearly. The noise level is low. Compared with those diagrams reported by Barron et al. (1994), the correlation-feedback algorithm achieves very good results.

For a comparison on a local basis, the portion of the needle diagram associated with the area surrounding the turning car (a sample of the velocity fields), obtained by 50 iterations of the correlation-feedback algorithm with five iterations of the Horn and Schunck algorithm as initialization, is provided in Figure 13.18(c). Its counterparts obtained by applying the Horn and Schunck (50 iterations) and the Singh (50 iterations) algorithms are displayed in Figure 13.18(a) and (b), respectively. It is observed that the correlation-feedback algorithm achieves the best results among the three algorithms.

### 13.3.3.4   Discussion and Conclusion

Although it uses a revised version of a correlation-based algorithm (Singh, 1992), the correlation-feedback technique is quite different from the correlation-based algorithm (Singh, 1992) in the following four aspects. First, different optimization criteria: the algorithm does not use the iterative minimization procedure used in (Singh, 1992). Instead, some variations of the estimated optical

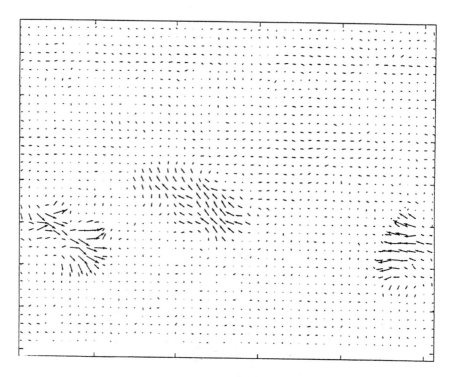

**FIGURE 13.17**  Needle diagram of flow field of Hamburg Taxi sequence obtained by using the correlation-feedback algorithm.

flow vectors are generated and fed back. The associated bilinearly interpolated displaced frame difference for each variation is calculated and utilized. In essence, the feedback approach utilizes two given images repeatedly, while the Singh method uses two given images only once ($u_c$ and $v_c$ derived from the two given images are only calculated once). The best local matching between the displaced image, generated via feedback of the estimated optical flow, and the given image is actually used as the ultimate criterion for improving optical flow accuracy in the iterative process. Second, the search window in the algorithm is an adaptive "rubber" window, having a variable size depending on ($u^k$, $v^k$). In the correlation-based approaches (Singh, 1992), the search window has a fixed size. Third, the algorithm uses a bilinear interpolation technique in the observation stage and provides the correlation stage with a virtually continuous image field for more accurate motion vector computation, while that of Singh (1992) does not. Fourth, different performances are achieved when image intensity is a linear function of image coordinates. In fact, in the vicinity of a pixel, the intensity can usually be considered as such a linear function. Except if the optical flow vectors happen to have only an integer multiple of pixels as their components, an analysis by Pan (1994) shows that the correlation-based approach (Singh, 1992) will not converge to the apparent 2-D motion vectors and will easily have error much greater than 10%. Pan (1994) also shows that the linear intensity function guarantees the assumption of the symmetric response distribution with a single maximum value assumed by the ground-truth optical flow. As discussed in Section 13.3.3.1, under this assumption the convergence of the correlation-feedback technique is justified.

Numerous experiments have demonstrated the convergence and accuracy of the correlation-feedback algorithm, and usually it is more accurate than some standard gradient- and correlation-based approaches. In the complicated optical flow cases, specifically in the case of the "Yosemite" image sequence (regarded as the most challenging quantitative test image sequence by Barron et al. (1994), it performs better than all other techniques.

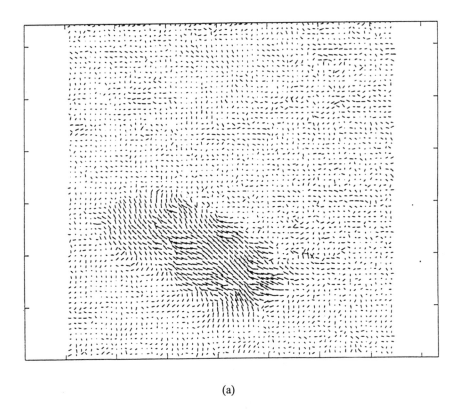

(a)

**FIGURE 13.18**   A portion of the needle diagram obtained by using (a) the Horn and Schunk algorithm, (b) the Singh algorithm, and (c) the correlation-feedback algorithm.

## 13.4   MULTIPLE ATTRIBUTES FOR CONSERVATION INFORMATION

As stated at the beginning of this chapter, there are many algorithms in optical flow computation reported in the literature. Many more new algorithms continue to be developed. In Sections 13.2 and 13.3, we introduced some typical algorithms using gradient- and correlation-based approaches. We will not explore various algorithms any further here. It is hoped that the fundamental concepts and algorithms introduced above have provided a solid base for readers to study more-advanced techniques.

We would like to discuss optical flow from another point of view, however: multiple image attributes vs. a single image attribute. All of the methods we have discussed so far use only one kind of image attributes as conservation information in flow determination. Most methods use intensity. Singh's method uses the Laplacian of intensity, which is calculated by using the difference of the Gaussian operation (Burt, 1984). It was reported by Weng, Ahuja, and Huang (1992) that using a single attribute as conservation information may result in ambiguity in matching two perspective views, while multiple attributes, which are motion insensitive, may reduce ambiguity remarkably, resulting in better matching. An example is shown in Figure 13.19 to illustrate this argument. In this section, the Weng et al. method is discussed first. Then we introduce the Xia and Shi method, which uses multiple attributes in a framework based on weighted-least-square estimation and feedback techniques.

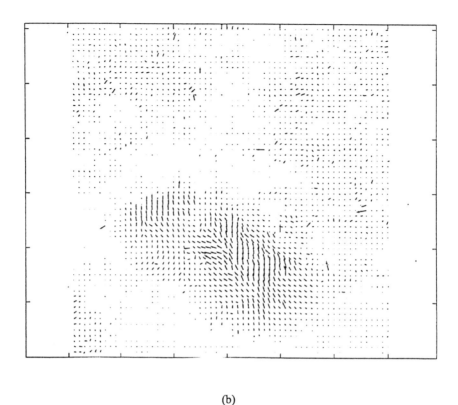

(b)

**FIGURE 13.18** (continued)

### 13.4.1 THE WENG, AHUJA, AND HUANG METHOD

Weng, Ahuja, and Huang proposed a quite different approach to image point matching (Weng et al., 1992). Note that the image matching amounts to flow field computation since it calculates a displacement field for each point in image planes, which is essentially a flow field if the time interval between two image frames is known.

Based on an analysis indicating that using image intensity as a single attribute is not enough in accurate image matching, Weng, Ahuja, and Huang utilize multiple attributes associated with images in estimation of the dense displacement field. These image attributes are motion insensitive; i.e., they generally sustain only small change under motion assumed to be locally rigid. The image attributes used are image intensity, edgeness, and cornerness. For each image attribute, the algorithm forms a residual function, reflecting the inaccuracy of the estimated matching. The matching is then determined via an iterative procedure to minimize the weighted sum of these residual functions. In handling neighborhood information, a more-advanced smoothness constraint is used to take care of moving discontinuities. The method considers uniform regions and the occlusion issue as well.

In addition to using multiple image attributes, the method is pointwise processing. There is no need for calculation of correlation within two correlation windows, which saves computation dramatically. However, the method also has some drawbacks. First, the edgeness and cornerness involve calculation of the spatial gradient, which is noise sensitive. Second, in solving for minimization, the method resorts to numerical differentiation again: the estimated displacement vectors are updated based on the partial derivatives of the noisy attribute images. In a word, the computational framework heavily relies on numerical differentiation, which is considered to be impractical for accurate computation (Barron et al., 1994).

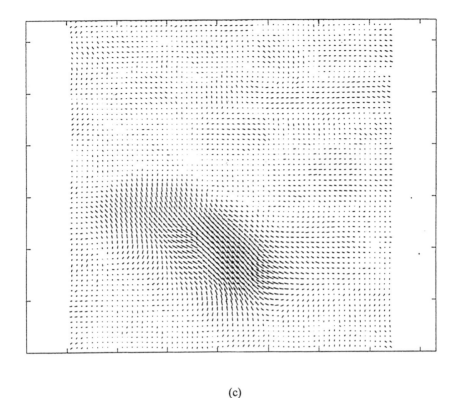

(c)

**FIGURE 13.18** (continued)

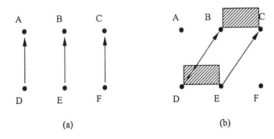

(a)                                                              (b)

**FIGURE 13.19** Multiple attributes vs. single attribute. (a) With intensity information only, points D, E, and F tend to match to points A, B, and C, respectively. (b) With intensity, edge and corner information points D and E tend to match points B and C, respectively.

On the other hand, the Pan, Shi, and Shu method, discussed in Section 13.3.3 in the category of correlation based approaches, seems to have some complementary features. It is correlation-based. It uses intensity as a single attribute. In these two aspects the Pan et al. method is inferior to the method by Weng, Ahuja, and Huang. The feedback technique and the weighted least-square computation framework used in the Pan et al. method are superior, however, compared with the method by Weng et al. Motivated by the above observations, an efficient, multiattribute feedback method was developed by Xia and Shi (Xia and Shi, 1995; Xia, 1996), and is discussed in the next subsection. It is expected that more insight into the Weng, Ahuja, and Huang method will become clear in the discussion as well.

### 13.4.2 THE XIA AND SHI METHOD

This method uses multiple attributes that are motion insensitive. The following five attributes are used: image intensity, horizontal edgeness, vertical edgeness, contrast, and entropy. The first three are used by Weng et al. (1992) as well, and can be considered as structural attributes, while the last two, which are not used by Weng et al. (1992), can be considered as textural attributes according to Haralick (1979).

Instead of the computational framework presented by Weng et al. (1992), which, as discussed above, may not be practical for accurate computation, the method uses the computational framework of Pan (1994; 1998). That is, the weighted-least-squared estimation technique used by Singh (1992) and the feedback technique used by Pan (1994; 1998) are utilized here. Unlike in the Weng et al. (1992) method, subpixel accuracy is considered and a confidence measure is generated in the method.

The Xia and Shi method is also different from those algorithms presented by Singh (1992) and Pan et al. (1995; 1998). First, there is no correlation in the method, while both Singh (1992) and Pan et al. (1995; 1998) are correlation based. Specifically, the method is a point-wise processing. Second, the method uses multiple attributes, while both Singh (1992) and Pan et al. (1995; 1998) use image intensity as a single attribute.

In summary, the Xia and Shi method to compute optical flow is motivated by several existing algorithms mentioned above. It does, however, differ from each of them significantly.

#### 13.4.2.1 Multiple Image Attributes

As mentioned before, there are five image attributes in the Xia and Shi method. They are defined below.

**Image Intensity** — The intensity at a pixel $(x, y)$ in an image $f_n(x, y)$, denoted by $A_i(x, y)$, i.e., $A_i(x, y) = f_n(x, y)$.

**Horizontal Edgeness** — The horizontal edgeness at a pixel $(x, y)$, denoted by $A_h(x, y)$, is defined as

$$A_h(x, y) = \frac{\partial f(x, y)}{\partial y}, \tag{13.40}$$

i.e., the partial derivative of $f(x, y)$ with respect to $y$, the second component of the gradient of intensity function at the pixel.

**Vertical Edgeness** — The vertical edgeness at a pixel $(x, y)$, denoted by $A_v(x, y)$, is defined as

$$A_v(x, y) = \frac{\partial f(x, y)}{\partial x}, \tag{13.41}$$

i.e., the first component of the gradient of intensity function at the pixel. Note that the partial derivatives in Equations 13.40 and 13.41 are computed by applying a Sobel operator (Gonzalez and Woods, 1992) in a $3 \times 3$ neighborhood of the pixel.

**Contrast** — The local contrast at a pixel $(x, y)$, denoted by $A_c(x, y)$, is defined as

$$A_c(x, y) = \sum_{i,j \in S} (i - j)^2 C_{i,j}, \tag{13.42}$$

where $S$ is a set of all the distinct gray levels within a $3 \times 3$ window centered at pixel $(x, y)$. $C_{i,j}$ specifies a relative frequency with which two neighboring pixels separated horizontally by a distance of 1 occur in the $3 \times 3$ window, one with gray level $i$ and the other with gray level $j$.

**Entropy** — The local entropy at a point $(x, y)$, denoted by $A_e(x, y)$, is given by

$$A_e(x, y) = -\sum_{i \in S} p_i \log p_i, \qquad (13.43)$$

where $S$ was defined above, and $p_i$ is the probability of occurrence of the gray level $i$ in the $3 \times 3$ window.

Since the intensity is assumed to be invariant to motion, so are the horizontal edgeness, vertical edgeness, contrast, and entropy.

As mentioned above, the intensity and edgeness are used as attributes in the Weng et al. algorithm as well. Compared with the negative and positive cornerness used in the Weng et al. algorithm, the local contrast and entropy need no differentiation and therefore are less sensitive to various noises in original images. In addition, these two attributes are inexpensive in terms of computation. They reflect the textural information about the local neighborhood of the pixel for which the flow vector is to be estimated.

### 13.4.2.2  Conservation Stage

In the Xia and Shi algorithm, this stage is similar to that in the Pan et al. algorithm. That is, for a flow vector estimated at the $k$th iteration, denoted by $(u^k, v^k)$, we find its 25 variations, $(u, v)$, according to

$$u \in \left\{ u^k - \frac{u^k}{2}, u^k - \frac{u^k}{4}, u^k, u^k + \frac{u^k}{4}, u^k + \frac{u^k}{2} \right\}$$

$$\qquad (13.44)$$

$$v \in \left\{ v^k - \frac{v^k}{2}, v^k - \frac{v^k}{4}, u^k, v^k + \frac{v^k}{4}, v^k + \frac{v^k}{2} \right\}.$$

For each of these 25 variations, the matching error is computed as

$$E(u, v) = r_{A_i}^2(x, y, u, v) + r_{A_h}^2(x, y, u, v) + r_{A_v}^2(x, y, u, v) + r_{A_c}^2(x, y, u, v) + r_{A_e}^2(x, y, u, v), \qquad (13.45)$$

where $r_{A_i}, r_{A_h}, r_{A_v}, r_{A_c}, r_{A_e}$ denote the residual function with respect to the five attributes, respectively. The residual function of intensity is defined as

$$r_{A_i}(x, y, u, v) = A_{i_n}(x, y) - A_{i_{n-1}}(x - u, y - v) = f_n(x, y) - f_{n-1}(x - u, y - v), \qquad (13.46)$$

where $f_n(x, y)$, $f_{n-1}(x, y)$ is defined as before, i.e., the intensity function at $t_n$ and $t_{n-1}$, respectively; $A_{i_n}$, $A_{i_{n-1}}$ denote the intensity attributes on $f_n$ and $f_{n-1}$, respectively.

It is observed that the residual error of intensity is essentially the DFD discussed in Chapter 12. The rest of the residual functions are defined similarly. When subpixel accuracy is required, spatial interpolation in the attribute images generally is necessary. Thus, the flow vector estimation is now converted to a minimization problem. That is, find $u$ and $v$ at pixel $(x, y)$ such that the matching error defined in Equation 13.45 is minimized. The weighted least-square method (Singh, 1992; Pan et al., 1998) is then used. That is,

$$R(u, v) = e^{-\beta E(u, v)} \qquad (13.47)$$

$$u_c^{k+1} = \frac{\sum_u \sum_v R(u,v)u}{\sum_u \sum_v R(u,v)}, \quad v_c^{k+1} = \frac{\sum_u \sum_v R(u,v)v}{\sum_u \sum_v R(u,v)}. \tag{13.48}$$

Since the weighted least-square method has been discussed in detail in Sections 13.3.2 and 13.3.3, we will not go into more detail here.

### 13.4.2.3 Propagation Stage

Similar to what was proposed in the Pan et al. algorithm, in this stage Xia and Shi form a window $W$ of size $(2w + 1) \times (2w + 1)$ centered at the pixel $(x, y)$ in the image $f_n(x, y)$. The flow estimate at the pixel $(x, y)$ in this stage, denoted by $(u^{k+1}, v^{k+1})$, is calculated as a weighted sum of the flow vectors of the pixel within the window $W$.

$$u^{k+1} = \sum_{s=-w}^{w} \sum_{t=-w}^{w} w_1[f_n(x,y), f_n(x+s,y+t)] \cdot u_c^{k+1}(x+s,y+t)$$

$$\tag{13.49}$$

$$v^{k+1} = \sum_{s=-w}^{w} \sum_{t=-w}^{w} w_1[f_n(x,y), f_n(x+s,y+t)] \cdot v_c^{k+1}(x+s,y+t),$$

where $w_1[.,.]$ is a weight function. For each point in the window $W$, a weight is assigned according to the weight function. Let $(x + s, y + t)$ denote a pixel within the window $W$; then the weight of the pixel $(x + s, y + t)$ is given by

$$w_1[f_n(x,y), f_n(x+s;y+t)] = \frac{c}{\varepsilon + |f_n(x,y) - f_n(x+s,y+t)|}, \tag{13.50}$$

where $\varepsilon$ is a small positive number to prevent the denominator from vanishing, $c$ is a normalization constant that makes the summation of all the weights in the $W$ equal 1.

From the above equation, we see that the weight is determined based on the intensity difference between the pixel under consideration and its neighboring pixel. The larger the difference in the intensity, the more likely the two points belong to different regions. Therefore, the weight will be small in this case. On the other hand, the flow vector in the same region will be similar since the corresponding weight is large. Thus, the weighting function implicitly takes flow discontinuity into account and is more advanced than that of Singh (1992) and Pan et al. (1994; 1998).

### 13.4.2.4 Outline of Algorithm

The following summarizes the procedures of the algorithm.

1. Perform a low-pass prefiltering on two input images to remove various noises.
2. Generate attribute images: intensity, horizontal edgeness, vertical edgeness, local contrast, and local entropy. Those attributes are computed at each grid point of both images.
3. Set the initial flow vectors to zero. Set the maximum iteration number and/or estimation accuracy.
4. For each pixel under consideration, generate flow variations according to Equation 13.44. Compute matching error for each flow variation according to Equation 13.45 and transform them to the corresponding response distribution $R$ using Equation 13.47. Compute the flow estimation $u^c$, $v^c$ using Equation 13.48.

5. Form a $(2w + 1) \times (2w + 1)$ neighborhood window $W$ centered at the pixel. Compute the weight for each pixel within the window $W$ using Equation 13.50. Update the flow vector using Equation 13.49.

6. Decrease the preset iteration number. If the iteration number is zero, the algorithm returns with the resultant optical flow field. Otherwise, go to the next step.

7. If the change in flow vector over two successive iterations is less than the predefined threshold, the algorithm returns with the estimated optical flow field. Otherwise, go to step 4.

### 13.4.2.5 Experimental Results

To compare the method with other methods existing in the literature, similar to what has been done by Pan et al. (1998) (discussed above in Section 13.3.3), the method was applied to three test sequences used by Barron et al. (1994): the "Translating Tree" sequence, the "Diverging Tree" sequence, and the "Yosemite" sequence. The same accuracy criterion is used as that by Barron et al. (1994). Only those results reported by Barron et al. (1994) with 100% density are listed in Tables 13.7, 13.8, and 13.9 for a fair and easy comparison. The Weng et al. algorithm was implemented by Xia and Shi and the results were reported by Xia and Shi (1995).

**TABLE 13.7**
**Summary of the "Translating Tree" 2D Velocity Results**

| Techniques | Average Error, ° | Standard Deviation, ° | Density, % |
|---|---|---|---|
| Horn and Schunck (original) | 38.72 | 27.67 | 100 |
| Horn and Schunck (modified) | 2.02 | 2.27 | 100 |
| Uras et al. (unthresholded) | 0.62 | 0.52 | 100 |
| Nagel | 2.44 | 3.06 | 100 |
| Anandan | 4.54 | 3.10 | 100 |
| Singh (step 1, $n = 2$, $w = 2$) | 1.64 | 2.44 | 100 |
| Singh (step 2, $n = 2$, $w = 2$) | 1.25 | 3.29 | 100 |
| Pan, Shi, and Shu ($n = 1$, $w = 1$) | 1.07 | 0.48 | 100 |
| Weng, Ahuja, and Huang | 1.81 | 2.03 | 100 |
| Xia and Shi | 0.55 | 0.52 | 100 |

**TABLE 13.8**
**Summary of the "Diverging Tree" 2D Velocity Results**

| Techniques | Average Error, ° | Standard Deviation, ° | Density, % |
|---|---|---|---|
| Horn and Schunck (original) | 32.43 | 30.28 | 100 |
| Horn and Schunck (modified) | 11.26 | 16.41 | 100 |
| Uras et al. (unthresholded) | 10.44 | 15.00 | 100 |
| Nagel | 11.71 | 10.59 | 100 |
| Anandan | 15.84 | 13.46 | 100 |
| Singh (step 1, $n = 2$, $w = 2$, $N = 4$) | 18.24 | 17.02 | 100 |
| Singh (step 2, $n = 2$, $w = 2$, $N = 4$) | 13.16 | 12.07 | 100 |
| Pan, Shi, and Shu ($n = 1$, $w = 1$) | 7.93 | 6.72 | 100 |
| Weng, Ahuja, and Huang | 8.41 | 8.22 | 100 |
| Xia and Shi | 7.54 | 6.61 | 100 |

**TABLE 13.9**
**Summary of the "Yosemite" 2D Velocity Results**

| Techniques | Average Error, ° | Standard Deviation, ° | Density, % |
|---|---|---|---|
| Horn and Schunck (original) | 12.02 | 11.72 | 100 |
| Horn and Schunck (modified) | 2.55 | 3.67 | 100 |
| Uras et al. (unthresholded) | 4.64 | 3.48 | 100 |
| Nagel | 2.94 | 3.23 | 100 |
| Anandan (frame 19 and 21) | 7.64 | 4.96 | 100 |
| Singh (step 1, $n = 2$, $w = 2$, $N = 4$) | 17.66 | 14.25 | 100 |
| Singh (step 2, $n = 2$, $w = 2$, $N = 4$) | 8.60 | 5.60 | 100 |
| Pan, Shi, and Shu ($n = 1$, $w = 1$) | 5.12 | 2.16 | 100 |
| Weng, Ahuja, and Huang | 8.01 | 9.71 | 100 |
| Xia and Shi | 4.04 | 3.82 | 100 |

### 13.4.2.6 Discussion and Conclusion

The above experimental results demonstrate that the Xia and Shi method outperforms both the Pan, Shi, and Shu method and the Weng, Ahuja, and Huang method in terms of accuracy of optical flow determined. Computationally speaking, the Xia and Shi method is less expensive than the Pan et al., since there is no correlation involved and the correlation is known to be computationally expensive.

## 13.5 SUMMARY

The optical flow field is a dense 2-D distribution of apparent velocities of movement of intensity patterns in image planes, while the 2-D motion field can be understood as the perspective projection of 3-D motion in the scene onto image planes. They are different. Only under certain circumstances are they equal to each other. In practice, however, they are closely related in that image sequences are usually the only data we have in motion analysis. Hence, we can only deal with the optical flow in motion analysis, instead of the 2-D motion field. The aperture problem in motion analysis refers to the problem that occurs when viewing motion via an aperture. Specifically, the only motion we can observe from local measurement is the motion component orthogonal to the underlying moving contour. That is another way to manifest the ill-posed nature of optical flow computation. In general, motion analysis from image sequences is an inverse problem, which is ill posed. Fortunately, low-level computational vision problems are only mildly ill posed. Hence, lowering the noise in image data leads to a possible significant reduction of errors in flow determination.

Numerous flow determination algorithms have appeared over the course of more than a decade. Most of the techniques take one of the following approaches: the gradient-based approach, the correlation-based approach, the energy-based approach, or the phase-based approach. In addition to these deterministic approaches, there is also a stochastic approach. A unification point of view of optical flow computation is presented in Section 13.3. That is, for any algorithm in optical flow computation, there are two types of information that need to be extracted — conservation information and neighborhood information.

Several techniques are introduced for the gradient-based approach, particularly the Horn and Schunck algorithm, which is a pioneer work in flow determination. There, the brightness invariant equation is used to extract conservation information and the smoothness constraint is used to extract neighborhood information. The modified Horn and Schunck algorithm shows significant error reduction in flow determination, because of a reduction of noise in image data, which confirms the mildly ill-posed nature of optical flow computation.

Several techniques are discussed for the correlation-based approach. The Singh algorithm is given emphasis due to its estimation-theoretical framework. The Pan, Shi, and Shu algorithm, which applies the feedback technique to the correlation method, demonstrates an accuracy enhancement in flow estimation.

Section 13.4 addresses the usage of multiple image attributes vs. that of a single image attribute in the flow determination technique. It is found that the use of multiple motion-insensitive attributes can help reduce the ambiguity in motion analysis. The application of multiple image attributes to conservation information turns out to be promising for flow computation.

Some experimental works are presented in Sections 13.3 and 13.4. With Barron et al.'s recent comprehensive survey of various existing optical flow algorithms, we can have a quantitative assessment on various optical flow techniques.

Optical flow finds application in areas such as computer vision, image interpolation, temporal filtering, and video coding. In computational vision, raising the accuracy of optical flow estimation is important. In video coding, however, lowering the bit rate for both prediction error and motion overhead, while keeping certain quality of reconstructed frames, is the ultimate goal. Properly handling the large amount of velocity vectors is a key issue in this regard. It is noted that the optical flow-based motion estimation for video compression has been applied for many years. However, the high bit overhead and computational complexity prevent it from practical usage in video coding. With the continued advance in technologies, however, we believe this problem may be resolved in the near future. In fact, an initial, successful attempt has been made and reported by Shi et al. (1998). There, based on a study that demonstrates that flow vectors are highly correlated and can be modeled by a first-order autoregressive (AR) model, the discrete cosine transform (DCT) is applied to flow vectors. An adaptive threshold technique is developed to match optical flow motion prediction and to minimize the residual errors. Consequently, this optical flow-based motion-compensated video coding algorithm achieves good performance for very low bit rate video coding. It obtains a bit rate compatible with that obtained by an H.263 standard algorithm, which uses block matching for motion estimation. (Note that the video coding standard H.263 is covered in Chapter 19.) Furthermore, the reconstructed video frames by using this flow-based algorithm are free of annoying blocking artifacts. This effect is demonstrated in Figure 13.20. Note that Figure 13.20 (b) has appeared in Figure 11.12, where the same picture is displayed in a larger size and the blocking artifacts are hence clearer.

## 13.6   EXERCISES

**13-1.**  What is an optical flow field? What is a 2-D motion field? What is the difference between the two? How are they related to each other?

**13-2.**  What is an aperture problem? Give two of your own examples.

**13-3.**  What is the ill-posed problem? Why do we consider motion analysis from image sequences an ill-posed problem?

**13-4.**  Is the relationship between the optical flow in an image plane and the velocities of objects in 3-D world space necessarily obvious? Justify your answer.

**13-5.**  What does the smoothness constraint imply? Why is it required?

**13-6.**  How are the derivatives of intensity function and the Laplacian of flow components estimated in the Horn and Schunck method?

**13-7.**  What are the differences between the Horn and Schunck original method and the modified Horn and Schunck method? What do you observe from these differences?

**13-8.**  What is the difference between the smoothness constraint proposed by Horn and Schunck and the oriented smoothness constraint proposed by Nagel? Provide comments.

**13-9.**  In your own words, describe the Singh method. What is the weighted-least-square estimation technique?

**FIGURE 13.20** (a) The 21st original frame of the Miss America sequence; (b) the reconstructed 21st frame with H.263; (c) the reconstructed 21st frame with the proposed algorithm.

**13-10.** In your own words, describe conservation information and neighborhood information. Using this perspective, take a new look at the Horn and Schunck algorithm.

**13-11.** How is the feedback technique applied in the Pan et al. algorithm?

**13-12.** In your own words, tell the difference between the Singh method and the Pan et al. method.

**13-13.** Give two of your own examples to show that multiple image attributes are able to reduce ambiguity in image matching.

**13-14.** How does the Xia and Shi method differ from the Weng et al. method?

**13-15.** How does the Xia and Shi method differ from the Pan et al. method?

## REFERENCES

Adelson, E. H. and J. R. Bergen, Spatiotemporal energy model for the perception of motion, *J. Opt. Soc. Am. A*, 2(2), 284-299, 1985.

Aggarwal, J. K. and N. Nandhakumar, On the computation of motion from sequences of images — a review, *Proc. IEEE*, 76(8), 917-935, 1988.

Anandan, P. Measurement Visual Motion from Image Sequences, Ph.D. thesis, COINS Department, University of Massachusetts, Amherst, 1987.

Anandan, P. A computational framework and an algorithm for the measurement of visual motion, *Int. J. Comput. Vision*, 2, 283-310, 1989.

Barron, J. L., D. J. Fleet, and S. S. Beauchemin, Systems and experiment performance of optical flow techniques, *Int. J. Comput. Vision*, 12(1), 43-77, 1994.

Beck, J. V. and K. J. Arnold, *Parameter Estimation Engineering and Science*, John Wiley & Sons, New York, 1977.

Bertero, M., T. A. Poggio, and V. Torre, Ill-posed problems in early vision, *Proc. IEEE*, 76(8), 869-889, 1988.

Bigun, J., G. Granlund, and J. Wiklund, Multidimensional orientation estimation with applications to texture analysis and optical flow, *IEEE Trans. Pattern Anal. Machine Intell.*, 13, 775-790, 1991.

Black, M. J. and P. Anandan, The robust estimation of multiple motions: parametric and piecewise-smooth flow fields, *Comp. Vision and Image Understanding*, 63(1), 75-104, 1996.

Bracewell, R. N. *Two-Dimensional Imaging*, Prentice-Hall, Englewood Cliffs, NJ, 1995.

Burt, P. J. and E. H. Adelson, The Laplacian pyramid as a compact image code, *IEEE Trans. Commun.*, 31(4), 532-540, 1983.

Burt, P. J. The pyramid as a structure for efficient computation, in A. Rosenfeld, Ed., *Multires. Image Proc. Anal.*, 6-37, Springer Verlag, New York, 1984.

Gonzalez, R. C. and R. E. Woods, *Digital Image Processing*, Addison-Wesley, Reading, MA, 1992.

Fleet, D. J. and A. D. Jepson, Computation of component image velocity from local phase information, *Int. J. Comput. Vision*, 5, 77-104, 1990.

Haralick, R. M. Statistical and structural approaches to texture, *Proc. IEEE*, 67(5), 786-804, 1979.

Heeger, D. J. Optical flow using spatiotemporal filters, *Int. J. Comput. Vision*, 1, 279-302, 1988.

Horn, B. K. P. and B. G. Schunck, Determining optical flow, *Artif. Intell.*, 17, 185-203, 1981.

Konrad, J. and E. Dubois, Bayesian estimation of motion vector fields, *IEEE Trans. Pattern Anal. Mach. Intell.*, 14(9), 910-927, 1992.

Lim, J. S. *Two-Dimensional Signal and Image Processing*, Prentice-Hall, Englewood Cliffs, NJ, 1990.

Lucas, B. and T. Kanade, An iterative image registration technique with an application to stereo vision, *Proc. DARPA Image Understanding Workshop*, 121-130, 1981.

Marr, D. *Vision*, Freeman, Boston, MA, 1982.

Nagel, H. H. Displacement vectors derived from second-order intensity variations in image sequences, *Comp. Graphics Image Proc.*, 21, 85-117, 1983.

Nagel, H. H. and W. Enkelmann, An investigation of smoothness constraints for the estimation of displacement vector fields from image sequences, *IEEE Trans. Pattern Anal. Machine Intell.*, 8, 565-593, 1986.

Pan, J. N. Motion Estimation Using Optical Flow Field, Ph.D. dissertation, Electrical and Computer Engineering, New Jersey Institute of Technology, Newark, NJ, 1994.

Pan, J. N., Y. Q. Shi, and C. Q. Shu, A convergence justification of the correlation-feedback algorithm in optical flow determination, Technical Report, Electronic Imaging Laboratory, Electrical and Computer Engineering Department, New Jersey Institute of Technology, Newark, NJ, 1995.

Pan, J. N., Y. Q. Shi, and C. Q. Shu, Correlation-feedback technique in optical flow determination, *IEEE Trans. Image Process.*, 7(7), 1061-1067, 1998.

Ralston, A. and P. Rabinowitz, *A First Course in Numerical Analysis*, McGraw-Hill, New York, 1978.

Sears, F. W., M. W. Zemansky, and H. D. Young, *University Physics*, Addison-Wesley, Reading, MA, 1986.

Shi, Y. Q., C. Q. Shu, and J. N. Pan, Unified optical flow field approach to motion analysis from a sequence of stereo images, *Patt. Recog.*, 27(12), 1577-1590, 1994.

Shi, Y. Q., S. Lin, and Y. Q. Zhang, Optical flow-based motion compensation algorithm for very low-bit-rate video coding, *Int. J. Imaging Syst. Technol.*, 9(4), 230-237, 1998.

Shu, C. Q. and Y. Q. Shi, Direct recovering of Nth order surface structure using UOFF approach, *Patt. Recog.*, 26(8), 1137-1148, 1993.

Singh, A. *Optical Flow Computation: A Unified Perspective*, IEEE Computer Society Press, Los Alamitos, CA, 1991.

Singh, A. An estimation-theoretic framework for image-flow computation, *CVGIP: Image Understanding*, 56(2), 152-177, 1992.

Szeliski, R., S. B. Kang, and H.-Y. Shum, A parallel feature tracker for extended image sequences, *Proc. Int. Symp. Computer Vision*, 241-246, Florida, November 1995.

Tekalp, A. M. *Digital Video Processing*, Prentice-Hall PTR, Upper Saddle River, NJ, 1995.

Tikhonov, A. N. and V. Y. Arsenin, *Solutions of Ill-posed Problems*, Winston & Sons, Washington, D.C., 1977.

Uras, S., F. Girosi, A. Verri, and V. Torre, A computational approach to motion perception, *Biol. Cybern.*, 60, 79-97, 1988.

Waxman, A. M., J. Wu, and F. Bergholm, Convected activation profiles and receptive fields for real time measurement of short range visual motion, *Proc. IEEE Computer Vision and Pattern Recognition*, 717-723, Ann Arbor, 1988.

Weng, J., N. Ahuja, and T. S. Huang, Matching two perspective views, *IEEE Trans. PAMI*, 14(8), 806-825, 1992.

Xia, X. and Y. Q. Shi, A multiple attributes algorithm to compute optical flow, *Proc. Twenty-ninth Annual Conf. Information Sciences and Systems*, p. 480, The Johns Hopkins University, Baltimore, MD, March 1995.

Xia, X. Motion Estimation and Video Coding, Ph.D. dissertation, Electrical and Computer Engineering, New Jersey Institute of Technology, Newark, NJ, October, 1996.

# 14 Further Discussion and Summary on 2-D Motion Estimation

Since Chapter 10, we have been devoting our discussion to motion analysis and motion-compensated coding. Following a general description in Chapter 10, three major techniques — block matching, pel recursion, and optical flow — are covered in Chapters 11, 12, and 13, respectively.

In this chapter, before concluding this subject, we provide further discussion and a summary. A general characterization for 2-D motion estimation, thus for all three techniques, is given in Section 14.1. In Section 14.2, different classifications of various methods for 2-D motion analysis are given in a wider scope. Section 14.3 is concerned with a performance comparison among the three major techniques. More-advanced techniques and new trends in motion analysis and motion compensation are introduced in Section 14.4.

## 14.1 GENERAL CHARACTERIZATION

A few common features characterizing all three major techniques are discussed in this section.

### 14.1.1 APERTURE PROBLEM

The aperture problem, discussed in Chapter 13, describes phenomena that occur when observing motion through a small opening in a flat screen. That is, one can only observe normal velocity. It is essentially a form of ill-posed problem since it is concerned with existence and uniqueness issues, as illustrated in Figure 13.2(a) and (b). This problem is inherent with the optical flow technique.

We note, however, that the aperture problem also exists in block matching and pel recursive techniques. Consider an area in an image plane having strong intensity gradients. According to our discussion in Chapter 13, the aperture problem does exist in this area no matter what type of technique is applied to determine local motion. That is, motion perpendicular to the gradient cannot be determined as long as only a local measure is utilized. It is noted that, in fact, the steepest descent method of the pel recursive technique only updates the estimate along the gradient direction (Tekalp, 1995).

### 14.1.2 ILL-POSED INVERSE PROBLEM

In Chapter 13, when we discuss the optical flow technique, a few fundamental issues are raised. It is stated that optical flow computation from image sequences is an inverse problem, which is usually ill-posed. Specifically, there are three problems: nonexistence, nonuniqueness, and instability. That is, the solution may not exist; if it exists, it may not be unique. The solution may not be stable in the sense that a small perturbation in the image data may cause a huge error in the solution.

Now we can extend our discussion to both block matching and pel recursion. This is because both block matching and pel recursive techniques are intended for determining 2-D motion from image sequences, and are therefore inverse problems.

### 14.1.3 CONSERVATION INFORMATION AND NEIGHBORHOOD INFORMATION

Because of the ill-posed nature of 2-D motion estimation, a unified point of view regarding various optical flow algorithms is also applicable for block matching and pel recursive techniques. That is, all three major techniques involve extracting conservation information and extracting neighborhood information.

Take a look at the block-matching technique. There, conservation information is a distribution of some sort of features (usually intensity or functions of intensity) within blocks. Neighborhood information manifests itself in that all pixels within a block share the same displacement. If the latter constraint is not imposed, block matching cannot work. One example is the following extreme case. Consider a block size of $1 \times 1$, i.e., a block containing only a single pixel. It is well known that there is no way to estimate the motion of a pixel whose movement is independent of all its neighbors (Horn and Schunck, 1981).

With the pel recursive technique, say, the steepest descent method, conservation information is the intensity of the pixel for which the displacement vector is to be estimated. Neighborhood information manifests itself as recursively propagating displacement estimates to neighboring pixels (spatially or temporally) as initial estimates.

In Section 12.3, it is pointed out that Netravali and Robbins suggested an alternative, called "inclusion of a neighborhood area." That is, in order to make displacement estimation more robust, they consider a small neighborhood $\Omega$ of the pixel for evaluating the square of displaced frame difference (DFD) in calculating the update term. They assume a constant displacement vector within the area. The algorithm thus becomes

$$\vec{d}^{\,k+1} = \vec{d}^{\,k} - \frac{1}{2}\alpha\nabla_{\vec{d}} \sum_{i,x,y\in Q} w_i DFD^2\left(x,y,;\vec{d}^{\,k}\right), \tag{14.1}$$

where $i$ represents an index for the $i$th pixel $(x, y)$ within $\Omega$, and $w_i$ is the weight for the $i$th pixel in $\Omega$. All the weights satisfy certain conditions; i.e., they are nonnegative, and their sum equals 1. Obviously, in this more-advanced algorithm, the conservation information is the intensity distribution within the neighborhood of the pixel, the neighborhood information is imposed more explicitly, and it is stronger than that in the steepest descent method.

### 14.1.4 OCCLUSION AND DISOCCLUSION

The problems of occlusion and disocclusion make motion estimation more difficult and hence more challenging. Here we give a brief description about these and other related concepts.

Let us consider Figure 14.1. There, the rectangle *ABCD* represents an object in an image taken at the moment of $t_{n-1}$, $f(x, y, t_{n-1})$. The rectangle EFGH denotes the same object, which has been translated, in the image taken at $t_n$ moment, $f(x, y, t_n)$. In the image $f(x, y, t_n)$, the area *BFDH* is occluded by the object that newly moves in. On the other hand, in $f(x, y, t_n)$, the area of *AECG* resurfaces and is referred to as a newly visible area, or a newly exposed area.

Clearly, when occlusion and disocclusion occur, all three major techniques discussed in this part will encounter a fatal problem, since conservation information may be lost, making motion estimation fail in the newly exposed areas. If image frames are taken densely enough along the temporal dimension, however, occlusion and disocclusion may not cause serious problems, since the failure in motion estimation may be restricted to some limited areas. An extra bit rate paid for the corresponding increase in encoding prediction error is another way to resolve the problem. If high quality and low bit rate are both desired, then some special measures have to be taken.

One of the techniques suitable for handling the situation is Kalman filtering, which is known as the best, by almost any reasonable criterion, technique working in the Gaussian white noise case

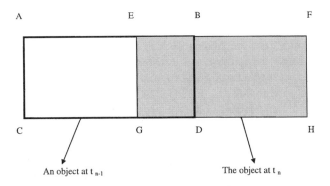

A    E    B    F

C    G    D    H

An object at $t_{n-1}$          The object at $t_n$

**FIGURE 14.1**   Occlusion and disocclusion.

(Brown and Hwang, 1992). If we consider the system that estimates the 2-D motion to be contaminated by Gaussian white noise, we can use Kalman filtering to increase the accuracy of motion estimation, particularly along motion discontinuities. It is powerful in doing incremental, dynamic, and real-time estimation.

In estimating 3-D motion, Kalman filtering was applied by Matthies et al. (1989) and Pan et al. (1994). Kalman filters were also utilized in optical flow computation (Singh, 1992; Pan and Shi, 1994). In using the Kalman filter technique, the question of how to handle the newly exposed areas was raised by Matthies et al. (1989). Pan et al. (1994) proposed one way to handle this issue, and some experimental work demonstrated its effectiveness.

## 14.1.5   RIGID AND NONRIGID MOTION

There are two types of motion: rigid motion and nonrigid motion. Rigid motion refers to motion of rigid objects. It is known that our human vision system is capable of perceiving 2-D projections of 3-D moving rigid bodies as 2-D moving rigid bodies. Most cases in computer vision are concerned with rigid motion. Perhaps this is due to the fact that most applications in computer vision fall into this category. On the other hand, rigid motion is easier to handle than nonrigid motion. This can be seen in the following discussion.

Consider a point $P$ in 3-D world space with the coordinates $(X, Y, Z)$, which can be represented by a column vector $\bar{v}$:

$$\bar{v} = (X, Y, Z)^T. \tag{14.2}$$

Rigid motion involves rotation and translation, and has six free motion parameters. Let $R$ denote the rotation matrix and $T$ the translational vector. The coordinates of point $P$ in the 3-D world after the rigid motion are denoted by $\bar{v}'$. Then we have

$$\bar{v}' = R\bar{v} + T. \tag{14.3}$$

Nonrigid motion is more complicated. It involves deformation in addition to rotation and translation, and thus cannot be characterized by the above equation. According to the Helmholtz theory (Sommerfeld, 1950), the counterpart of the above equation becomes

$$\bar{v}' = R\bar{v} + T + D\bar{v}, \tag{14.4}$$

where $D$ is a deformation matrix. Note that $R$, $T$, and $D$ are pixel dependent. Handling nonrigid motion, hence, is very complicated.

In videophony and videoconferencing applications, a typical scene might be a head-and-shoulder view of a person imposed on a background. The facial expression is nonrigid in nature. Model-based facial coding has been studied extensively (Aizawa and Harashima, 1994; Li et al., 1993; Arizawa and Huang, 1995). There, a 3-D wireframe model is used for handling rigid head motion. Li (1993) analyzes the facial nonrigid motion as a weighted linear combination of a set of *action units*, instead of determining $D\bar{v}$ directly. Since the number of action units is limited, the compuatation becomes less expensive. In the Aizawa and Harashima (1989) paper, the portions in the human face with rich expression, such as lips, are *cut* and then transmitted out. At the receiving end, the portions are *pasted* back in the face.

Among the three types of techniques, block matching may be used to manage rigid motion, while pel recursive and optical flow may be used to handle either rigid or nonrigid motion.

## 14.2   DIFFERENT CLASSIFICATIONS

There are various methods in motion estimation, which can be classified in many different ways. We discuss some of the classifications in this section.

### 14.2.1   DETERMINISTIC METHODS VS. STOCHASTIC METHODS

Most algorithms are deterministic in nature. To see this, let us take a look at the most prominent algorithm for each of the three major 2-D motion estimation techniques. That is, the Jain and Jain algorithm for the block matching technique (Jain and Jain, 1981); the Netravali and Robbins algorithm for the pel recursive technique (Netravali and Robbins, 1979); and the Horn and Schunck algorithm for the optical flow technique (Horn and Schunck, 1981). All are deterministic methods. There are also stochastic methods in 2-D motion estimation, such as the Konrad and Dubois algorithm (Konrad and Dubois, 1992), which estimates 2-D motion using the maximum *a posteriori* probability (MAP).

### 14.2.2   SPATIAL DOMAIN METHODS VS. FREQUENCY DOMAIN METHODS

While most techniques in 2-D motion analysis are spatial domain methods, there are also frequency domain methods (Kughlin and Hines, 1975; Heeger, 1988; Porat and Friedlander, 1990; Girod, 1993; Kojima et al., 1993; Koc and Liu, 1998). Heeger (1988) developed a method to determine optical flow in the frequency domain, which is based on spatiotemporal filters. The basic idea and principle of the method is introduced in this subsection. A very new and effective frequency method for 2-D motion analysis (Koc and Liu, 1998) is presented in Section 14.4, where we discuss new trends in 2-D motion estimation.

#### 14.2.2.1   Optical Flow Determination Using Gabor Energy Filters

The frequency domain method of optical flow computation developed by Heeger is suitable for highly textured image sequences. First, let us take a look at how motion can be detected in the frequency domain.

**Motion in the spatiotemporal frequency domain** — We initiate our discussion with a 1-D case. The spatial frequency of a (translationally) moving sinusoidal signal, $\omega_x$, is defined as cycles per distance (usually cycles per pixel), while temporal frequency, $\omega_t$, is defined as cycles per time unit (usually cycles per frame). Hence, the velocity of (translational) motion, defined as distance per time unit (usually pixels per frame), can be related to the spatial and temporal frequencies as follows.

$$v = \omega_t / \omega_x .$$

(14.5)

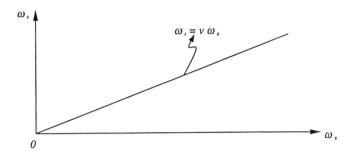

**FIGURE 14.2**  Velocity in 1-D spatiotemporal frequency domain.

A 1-D moving signal with a velocity $v$ may have multiple spatial frequency components. Each spatial frequency component $\omega_{xi}$, $i = 1,2,\ldots$ has a corresponding temporal frequency component $\omega_{ti}$ such that

$$\omega_{ti} = v\omega_{xi}. \tag{14.6}$$

This relation is shown in Figure 14.2. Thus, we see that in the spatiotemporal frequency domain, velocity is the slope of a straight line relating temporal and spatial frequencies.

For 2-D moving signals, we denote spatial frequencies by $\omega_x$ and $\omega_y$, and velocity vector by $\bar{v} = (v_x, v_y)$. The above 1-D result can be extended in a straightforward manner as follows:

$$\omega_t = v_x\omega_x + v_y\omega_y. \tag{14.7}$$

The interpretation of Equation 14.7 is that a 2-D translating texture pattern occupies a plane in the spatiotemporal frequency domain.

**Gabor Energy Filters** — As Adelson and Bergen (1985) pointed out, the translational motion of image patterns is characterized by orientation in the spatiotemporal domain. This can be seen from Figure 14.3. Therefore, motion can be detected by using spatiotemporally oriented filters. One filter of this type, suggested by Heeger, is the Gabor filter.

A 1-D sine-phase Gabor filter is defined as follows:

$$g(t) = \frac{1}{\sqrt{2\pi}\sigma} \sin(2\pi\omega t)\exp\left\{-\frac{t^2}{2\sigma^2}\right\}. \tag{14.8}$$

Obviously, this is a product of a sine function and a Gaussian probability density function. In the frequency domain, this is the convolution between a pair of impulses located in $\omega$ and $-\omega$, and the Fourier transform of the Gaussian, which is itself again a Gaussian function. Hence, the Gabor function is localized in a pair of Gaussian windows in the frequency domain. This means that the Gabor filter is able to pick up some frequency components selectively.

A 3-D sine Gabor function is

$$g(x,y,t) = \frac{1}{\sqrt{2}\pi^{3/2}\sigma_x\sigma_y\sigma_t} \cdot \exp\left\{-\frac{1}{2}\left(\frac{x^2}{\sigma_x^2} + \frac{y^2}{\sigma_y^2} + \frac{t^2}{\sigma_t^2}\right)\right\}$$

$$\cdot \sin\left[2\pi\left(\omega_{x_0}x + \omega_{y_0}y + \omega_{t_0}t\right)\right], \tag{14.9}$$

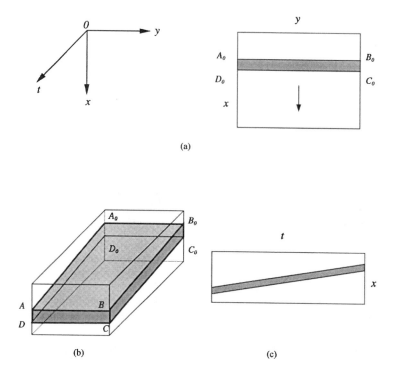

**FIGURE 14.3**   Orientation in spatiotemporal domain. (a) A horizontal bar translating downward. (b) A spatiotemporal cube. (c) A slice of the cube perpendicular to $y$ axis. The orientation of the slant edges represents the motion.

where $\sigma_x$, $\sigma_y$, and $\sigma_t$ are, respectively, the spreads of the Gaussian window along the spatiotemporal dimensions; and $\omega_{x_0}$, $\omega_{y_0}$, and $\omega_{t_0}$ are, respectively, the central spatiotemporal frequencies. The actual Gabor energy filter used by Heeger is the sum of a sine-phase filter (which is defined above), and a cosine-phase filter (which shares the same spreads and central frequencies as that in the sine-phase filter, and replaces sine by cosine in Equation 14.9). Its frequency response, therefore, is as follows.

$$G\left(\omega_x, \omega_y, \omega_t\right) = \frac{1}{4}\exp\left\{-4\pi^2\left[\sigma_x^2\left(\omega_x - \omega_{x_0}\right)^2 + \sigma_y^2\left(\omega_y - \omega_{y_0}\right)^2 + \sigma_t^2\left(\omega_t - \omega_{t_0}\right)^2\right]\right\}$$

$$+ \frac{1}{4}\exp\left\{-4\pi^2\left[\sigma_x^2\left(\omega_x + \omega_{x_0}\right)^2 + \sigma_y^2\left(\omega_y + \omega_{y_0}\right)^2 + \sigma_t^2\left(\omega_t + \omega_{t_0}\right)^2\right]\right\}. \tag{14.10}$$

This indicates that the Gabor filter is motion sensitive in that it responds largely to motion that has more power distributed near the central frequencies in the spatiotemporal frequency domain, while it responds poorly to motion that has little power near the central frequencies.

**Flow extraction with motion energy** — Using a vivid example, Heeger explains in his paper why one such filter is not sufficient in detection of motion. Multiple Gabor filters must be used. In fact, a set of 12 Gabor filters are utilized in Heeger's algorithm. The 12 Gabor filters in the set have one thing in common:

$$\omega_0 = \sqrt{\omega_{x0}^2 + \omega_{y0}^2}\,. \tag{14.11}$$

In other words, the 12 filters are tuned to the same spatial frequency band but to different spatial orientation and temporal frequencies.

Briefly speaking, optical flow is determined as follows. Denote the measured motion energy by $n_i, i = 1, 2 \ldots, 12$. Here $i$ indicates one of the 12 Gabor filters. The summation of all $n_i$ is denoted by

$$\bar{n} = \sum_{i=1}^{12} n_i. \tag{14.12}$$

Denote the predicted motion energy by $P_i(v_x, v_y)$, and the sum of predicted motion energy by

$$\bar{P} = \sum_{i=1}^{12} P_i(v_x, v_y). \tag{14.13}$$

Similar to what many algorithms do, optical flow determination is then converted to a minimization problem. That is, optical flow should be able to minimize error between the measured and predicted motion energies:

$$J(v_x, v_y) = \sum_{i=1}^{12} \left[ n_i - \bar{n}_i \frac{P_i(v_x, v_y)}{\bar{P}(v_x, v_y)} \right]^2. \tag{14.14}$$

Similarly, many readily available numerical methods can be used for solving this minimization problem.

### 14.2.3 REGION-BASED APPROACHES VS. GRADIENT-BASED APPROACHES

As stated in Chapter 10, methodologically speaking, there are generally two approaches to 2-D motion analysis for video coding: region based and gradient based. Now that we have gone through three major techniques, we can see this classification more clearly.

The region-based approach can be characterized as follows. For a region in an image frame, we find its best match in another image frame. The relative spatial position between these two regions produces a displacement vector. The best matching is found by minimizing a dissimilarity measure between the two regions, which is defined as

$$\sum_{(x,y) \in R} \sum M[f(x, y, t), f(x - dx, y - dy, t - \Delta t)], \tag{14.15}$$

where $R$ denotes a spatial region, on which the displacement vector $(d_x, d_y)^T$ estimate is based; $M[\alpha, \beta]$ denotes a dissimilarity measure between two arguments $\alpha$ and $\beta$; $\Delta t$ is the time interval between two consecutive frames.

Block matching certainly belongs to the region-based approach. By region we mean a rectangle block. For an original block in a (current) frame, block matching searches for its best match in another (previous) frame among candidates. Several dissimilarity measures are utilized, among which the mean absolute difference (MAD) is used most often.

Although it uses the spatial gradient of intensity function, the pel recursive method with inclusion of a neighborhood area assumes the same displacement vector within a neighborhood region. A weighted sum of the squared DFD within the region is used as a dissimilarity measure.

By using numerical methods such as various descent methods, the pel recursive method iteratively minimizes the dissimilarity measure, thus delivering displacement vectors. The pel recursive technique is therefore in the category of region-based approaches.

In optical flow computation, the two most frequently used techniques discussed in Chapter 13 are the gradient method and the correlation method. Clearly, the correlation method is region based. In fact, as we pointed out in Chapter 13, it is very similar to block matching.

As far as the gradient-based approach is concerned, we start its characterization with the brightness invariant equation, covered in Chapter 13. That is, we assume that brightness is conserved during the time interval between two consecutive image frames.

$$f(x,y,t) = f\left(x - d_x, y - d_y, t - \Delta t\right). \tag{14.16}$$

By expanding the right-hand side of the above equation into the Taylor series, applying the above equation, and some mathematical manipulation, we can derive the following equation.

$$f_x u + f_y v + f_t = 0, \tag{14.17}$$

where $f_x$, $f_y$, $f_t$ are partial derivatives of intensity function with respect to $x$, $y$, and $t$, respectively; and $u$ and $v$ are two components of pixel velocity. This equation contains gradients of intensity function with respect to spatial and temporal variables and links two components of the displacement vector. The square of the left-hand side in the above equation is an error that needs to be minimized. Through the minimization, we can estimate displacement vectors.

Clearly, the gradient method in optical flow determination, discussed in Chapter 13, falls into the above framework. There, an extra constraint is imposed and included into the error represented in Equation 14.17.

Table 14.1 summarizes what we discussed in this subsection.

**TABLE 14.1**
**Region-Based vs. Gradient-Based Approaches**

|  |  |  | Optical Flow | |
| --- | --- | --- | --- | --- |
|  | Block Matching | Pel Recursion | Gradient-Based Method | Correlation-Based Method |
| Regional-based approaches | √ | √ |  | √ |
| Gradient-based approaches |  |  | √ |  |

### 14.2.4 FORWARD VS. BACKWARD MOTION ESTIMATION

Motion-compensated predictive video coding may be done in two different ways: forward and backward (Boroczky, 1991). These ways are depicted in Figures 14.4 and 14.5, respectively. With the forward manner, motion estimation is carried out by using the original input video frame and the reconstructed previous input video frame. With the backward manner, motion estimation is implemented with two successive reconstructed input video frames.

The former provides relatively higher accuracy in motion estimation and hence more efficient motion compensation than the latter, owing to the fact that the original input video frames are utilized. However, the latter does not need to transmit motion vectors to the receiving end as an overhead, while the former does.

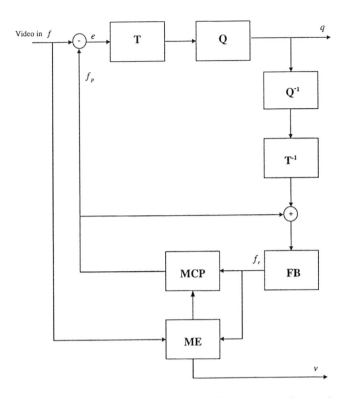

**FIGURE 14.4** Forward motion estimation and compensation, T: transformer, Q: quantizer, FB: frame buffer, MCP: motion-compensated predictor, ME: motion estimator, $e$: prediction error, $f$: input video frame, $f_p$: predicted video frame, $f_r$: reconstructed video frame, $q$: quantized transform coefficients, $v$: motion vector.

Block matching is used in almost all the international video coding standards, such as H.261, H.263, MPEG 1, and MPEG 2 (which are covered in the next part of this book), as forward-motion estimation. The pel recursive technique is used as backward-motion estimation. In this way, the pel recursive technique avoids encoding a large amount of motion vectors. On the other hand, however, it provides relatively less accurate motion estimation than block matching. Optical flow is usually used as forward-motion estimation in motion-compensated video coding. Therefore, as expected, it achieves higher motion estimation accuracy on the one hand and it needs to handle a large amount of motion vectors as overhead on the other hand. These will be discussed in the next section.

It is noted that one of the new improvements in the block-matching technique is described in Section 11.6.3. It is called the predictive motion field segmentation technique (Orchard, 1993), and it is motivated by backward-motion estimation. There, segmentation is conducted *backward*, i.e., based on previously decoded frames. The purpose of this is to save overhead for shape information of motion discontinuities.

## 14.3 PERFORMANCE COMPARISON AMONG THREE MAJOR APPROACHES

### 14.3.1 THREE REPRESENTATIVES

A performance comparison among the three major approaches; block matching, pel recursion, and optical flow, was provided in a review paper by Dufaux and Moscheni (1995). Experimental work was carried out as follows. The conventional full-search block matching is chosen as a representative

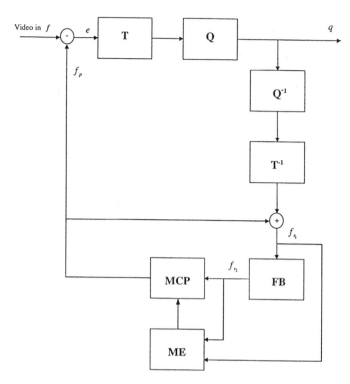

**FIGURE 14.5**   Backward-motion estimation and compensation, T: transformer, Q: quantizer, FB: frame buffer, MCP: motion-compensated predictor, ME: motion estimator, $e$: prediction error, $f$: input video frame, $f_p$: predicted video frame, $f_{r1}$: reconstructed video frame, $f_{r2}$: reconstructed previous video frame, $q$: quantized transform coefficients.

for the block-matching approach, while the Netravali and Robbins algorithm and the modified Horn and Schunck algorithm are chosen to represent the pel recursion and optical flow approaches, respectively.

## 14.3.2   Algorithm Parameters

In full-search block matching, the block size is chosen as $16 \times 16$ pixels, the maximum displacement is $\pm 15$ pixels, and the accuracy is half-pixel. In the Netravali and Robbins pel recursion, $\varepsilon = 1/1024$, the update term is averaged in an area of $5 \times 5$ pixels and clipped to a maximum of $1/16$ pixels per frame, and the algorithm iterates one iteration per pixel. In the modified Horn and Schunck algorithm, the weight $\alpha^2$ is set to 100, and 100 iterations of the Gauss and Seidel procedure are carried out.

## 14.3.3   Experimental Results and Observations

The three test video sequences are the "Mobile and Calendar," "Flower Garden," and "Table Tennis." Both subjective criteria (in terms of needle diagrams showing displacement vectors) and objective criteria (in terms of DFD error energy) are applied to access the quality of motion estimation.

It turns out that the pel recursive algorithm gives the worst accuracy in motion estimation. In particular, it cannot follow fast and large motions. Both block-matching and optical flow algorithms give better motion estimation.

It is noted that we must be cautious in drawing conclusions from these tests. This is because different algorithms in the same category and the same algorithm under different implementation conditions will provide quite different performances. In the above experiments, the full-search block matching with half-pixel accuracy is one of the better block-matching techniques. On the contrary, there are many improved pel recursive and optical flow algorithms, which outperform the chosen representatives in the reported experiments.

The experiments do, however, provide an insight about the three major approaches. Pel recursive algorithms are seldom used in video coding now, mainly because of their inaccurate motion estimation, although they do not require transmitting motion vectors to the receiving end. Although they can provide relatively accurate motion estimation, optical flow algorithms require a large amount of overhead for handling dense motion vectors. This prevents the optical flow techniques from wide and practical usage in video coding. Block matching is simple, yet very efficient for motion estimation. It provides quite accurate and reliable motion estimation for most practical video sequences in spite of its simple piecewise translational model. At the same time it does not require much overhead. Therefore, for first-generation video coding, block matching is considered to be the most suitable among the three approaches.

## 14.4   NEW TRENDS

In Chapters 11, 12, and 13, many new, effective improvements within the three major approaches were discussed. These techniques include multiresolution block matching, (locally adaptive) multigrid block matching, overlapped block matching, thresholding techniques, (predictive) motion field segmentation, feedback and multiple attributes in optical flow computation, subpixel accuracy, and so on. Some improvements will be discussed in Section IV, where various international video coding standards such as H.263 and MPEG 2, and 4 are introduced.

As pointed out by Orchard (1998), today our understanding of motion analysis and video compression is still based on an ad hoc framework, in general. What today's standards have achieved is not near the ideally possible performance. Therefore, more efforts are continuously made in this field, seeking much simpler and more practical, and efficient algorithms.

As an example of such developments, we conclude this chapter by presenting a novel method for 2-D motion estimation: the DCT-based motion estimation (Koc and Liu, 1998).

### 14.4.1   DCT-BASED MOTION ESTIMATION

As pointed out in Section 14.2.2, as opposed to the conventional 2-D motion estimation techniques, this method is carried out in the frequency domain. It is also different from the Gabor energy filter method by Heeger, discussed in Section 14.2.2.1. Without introducing Gabor filters, this mehtod is directly DCT based. The fundamental concepts and techniques of this method are discussed below.

#### 14.4.1.1   DCT and DST Pseudophases

The underlying idea behind this method is to estimate 2-D translational motion by determining the DCT and DST (discrete sine transform) *pseudophases*. Let us use the simpler 1-D case to illustrate this concept. Once it is established, it can be easily extended to the 2-D case.

Consider a 1-D signal sequence $\{f(n), n \in (0, 1, \cdots, N-1\}$ of length $N$. Its translated version is denoted by $\{g(n), n \in (0, 1, \cdots, N-1\}$. The translation is defined as follows.

$$g(n) = \begin{cases} f(n-d), & \text{if } (n-d) \in (0, 1, \cdots, N-1) \\ 0, & \text{otherwise} \end{cases}. \tag{14.18}$$

In the above equation, $d$ is the amount of the translation and it needs to be estimated. Let us define the following several functions before introducing the pseudophases. The DCT and the DST of the second kind of $g(n)$, $G^C(k)$, and $G^S(k)$ are defined as follows.

$$G^C(k) = \frac{2}{N} C(k) \sum_{n=0}^{N-1} g(n) \cos\left[\frac{k\pi}{N}(n+0.5)\right] \quad k \in \{0, 1, \cdots N-1\}$$ (14.19)

$$G^S(k) = \frac{2}{N} C(k) \sum_{n=0}^{N-1} g(n) \sin\left[\frac{k\pi}{N}(n+0.5)\right] \quad k \in \{1, \cdots N\}.$$ (14.20)

The DCT and DST of the first kind of $f(n)$, $F^C(k)$, and $F^S(k)$ are defined as

$$F^C(k) = \frac{2}{N} C(k) \sum_{n=0}^{N-1} f(n) \cos\left[\frac{k\pi}{N} n\right] \quad k \in \{0, 1, \cdots N-1\}$$ (14.21)

$$F^S(k) = \frac{2}{N} C(k) \sum_{n=0}^{N-1} f(n) \sin\left[\frac{k\pi}{N} n\right] \quad k \in \{1, \cdots N\}.$$ (14.22)

In the above equations, $C(k)$ is defined as

$$C(k) = \begin{cases} \dfrac{1}{\sqrt{2}} & \text{for } n = 0 \text{ or } N \\ 1 & \text{otherwise} \end{cases}.$$ (14.23)

Now we are in a position to introduce the following equation, which relates the translational amount $d$ to the DCT and DST of the original sequence and its translated version, defined above. That is,

$$\begin{bmatrix} G^C(k) \\ G^S(k) \end{bmatrix} = \begin{bmatrix} F^C(k) & -F^S(k) \\ F^C(k) & F^C(k) \end{bmatrix} \begin{bmatrix} D^C(k) \\ D^S(k) \end{bmatrix},$$ (14.24)

where $D^C(k)$ and $D^C(k)$ are referred to as the pseudophases and defined as follows:

$$D^C(k) \triangleq \cos\left[\frac{k\pi}{N}\left(d + \frac{1}{2}\right)\right]$$

$$D^S(k) \triangleq \sin\left[\frac{k\pi}{N}\left(d + \frac{1}{2}\right)\right].$$ (14.25)

Equation 14.24 can be solved for the amount of translation $d$, thus motion estimation. This becomes clearer when we rewrite the equation in a matrix-vector format. Denote the $2 \times 2$ matrix in Equation 14.24 by $\mathbf{F}(k)$, the $2 \times 1$ column vector at the left-hand side of the equation by $G(k)$, and the $2 \times 1$ column vector at the right-hand side by $D(k)$. It is easy to verify that the matrix $\mathbf{F}(k)$ is orthogonal by observing the following.

$$\lambda \mathbf{F}^T(k)\mathbf{F}(k) = \mathbf{I}, \qquad (14.26)$$

where $\mathbf{I}$ is a $2 \times 2$ identity matrix and the constant $\lambda$ is

$$\lambda = \frac{1}{\left[\mathbf{F}^C(k)\right]^2 + \left[\mathbf{F}^S(k)\right]^2}. \qquad (14.27)$$

We then derive the matrix-vector format of Equation 14.24 as follows:

$$\bar{D}(k) = \lambda \mathbf{F}^T(k)\bar{G}(k) \quad k \in \{1, \cdots, N-1\}. \qquad (14.28)$$

#### 14.4.1.2 Sinusoidal Orthogonal Principle

It was shown above that the pseudophases, which contain the translation information, can be determined in the DCT and DST frequency domain. But how the amount of the translation can be found has not been mentioned. Here, the algorithm uses the sinusoidal principle to pick up this information. That is, the inverse DST of the second kind of scaled pseudophase, $C(k)D^s(k)$, is found to equal an algebraic sum of the following two discrete impulses according to the sinusoidal orthogonal principle:

$$ISDT\{C(k)D^S(k)\} \frac{2}{N} \sum_{k=1}^{N} C^2(k)D^S(k)\sin\left[\frac{k\pi}{N}\left(n+\frac{1}{2}\right)\right] = \delta(d-n) - \delta(d+n+1). \quad (14.29)$$

Since the inverse DST is limited to $n \in \{0, 1, \cdots, N-1\}$, the only peak value among this set of $N$ values indicates the amount of the translation $d$. Furthermore, the direction of the translation (positive or negative) can be determined from the polarity (positive or negative) of the peak value.

The block diagram of the algorithm is shown in Figure 14.6. This technique can be extended to the 2-D case in a straightforward manner. Interested readers should refer to Koc and Liu (1998).

#### 14.4.1.3 Performance Comparison

The algorithm was applied to several typical testing video sequences, such as the "Miss America" and "Flower Garden" sequences, and an "Infrared Car" sequence. The results were compared with the conventional full-search block-matching technique and several fast-search block-matching techniques such as the 2-D logarithm search, three step search, search with subsampling in the original block, and the correlation windows.

Prior to applying the algorithm, one of the following preprocessing procedures is implemented: frame differentiation or edge extraction. It was reported that for the "Flower Garden" and "Infrared Car" sequences, the DCT-based algorithm achieves a higher coding efficiency than all three fast-search block-matching methods, while for the Miss America sequence it obtains a lower efficiency. It was also reported that it performs well even in a noisy situation.

A lower computational complexity, $O(M^2)$ for an $M \times M$ search range, is one of the major advantages possessed by the DCT-based motion estimation algorithm compared with conventional full-search block matching, $O(M^2 \cdot N^2)$ for an $M \times M$ search range and an $N \times N$ block size.

With DCT-based motion estimation, a fully DCT-based motion-compensated coder structure becomes possible, which is expected to achieve a higher throughput and a lower system complexity.

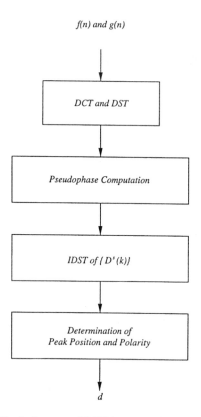

$f(n)$ and $g(n)$

DCT and DST

Pseudophase Computation

IDST of $\{D'(k)\}$

Determination of
Peak Position and Polarity

$d$

**FIGURE 14.6** Block diagram of DCT-based motion estimation (1-D case).

## 14.5 SUMMARY

In this chapter, which concludes the motion analysis and compensation portion of the book, we first generalize the discussion of the aperture problem, the ill-posed nature, and the conservation-and-neighborhood-information unified point of view, previously made with respect to the optical flow technique in Chapter 13, to cover block-matching and pel recursive techniques. Then, occlusion and disocclusion, and rigidity and nonrigidity are discussed with respect to the three techniques. The difficulty of nonrigid motion estimation is analyzed. Its relevance in visual communications is addressed.

Different classifications of various methods in the three major 2-D motion estimation techniques; block matching, pel recursion, and optical flow, are presented. Besides the frequently utilized deterministic methods, spatial domain methods, region-based methods, and forward-motion estimation, their counterparts — stochastic methods, frequency domain methods, gradient methods, and backward motion estimation — are introduced. In particular, two frequency domain methods are presented with some detail. They are the method using the Gabor energy filter and the DCT-based method.

A performance comparison among the three techniques is also introduced in this chapter, based on which observations are drawn. A main point is that block matching is at present the most suitable technique for 2-D motion estimation among the three techniques.

## 14.6   EXERCISES

**14-1.** What is the difference between rigid motion and nonrigid motion? In facial encoding, what is the nonrigid motion? How is the nonrigid motion handled?

**14-2.** How is 2-D motion estimation carried out in the frequency domain? What are the underlying ideas behind the Heeger method and the Koc and Liu method?

**14-3.** Why is one Gabor energy filter not sufficient in motion estimation? Draw the power spectrum of a 2-D sine-phase Gabor function.

**14-4.** Show the correspondence of a positive (negative) peak value in the inverse DST of the second kind of DST pseudophase to a positive (negative) translation in the 1-D spatial domain.

**14-5.** How does neighborhood information manifest itself in the pel recursive technique?

**14-6.** Using your own words and some diagrams, state that the translational motion of an image pattern is characterized by orientation in the spatiotemporal domain.

## REFERENCES

Adelson, E. H. and J. R. Bergen, Spatiotemporal energy models for the perception of motion, *J. Opt. Soc. Am.*, A2(2), 284-299, 1985.

Aizawa, K. and H. Harashima, Model-based analysis synthesis image coding (MBASIC) system for a person's face, *Signal Process. Image Commun.*, 139-152, 1989.

Aizawa, K. and T. S. Huang, Model-based image coding: advanced video coding techniques for very low bit rate applications, *Proc. IEEE*, 83(2), 259-271, 1995.

Boroczky, L. Pel-Recursive Motion Estimation for Image Coding, *Ph.D. dissertation*, Delft University of Technology, Netherlands, 1991.

Brown, R. G. and P. Y. C. Hwang, *Introduction to Random Signals*, 2nd ed., John Wiley & Sons, New York, 1992.

Dufaux, F. and F. Moscheni, Motion estimation techniques for digital TV: A review and a new contribution, *Proc. IEEE*, 83(6), 858-876, 1995.

Girod, B., Motion-compensating prediction with fractional-pel accuracy, *IEEE Trans. Commun.*, 41, 604, 1993.

Heeger, D. J. Optical flow using spatiotemporal filters, *Int. J. Comput. Vision*, 1, 279-302, 1988.

Horn, B. K. P. and B. G. Schunck, Determining optical flow, *Artif. Intell.*, 17, 185-203, 1981.

Jain, J. R. and A. K. Jain, Displacement measurement and its application in interframe image coding, *IEEE Trans. Commun.*, COM-29(12), 1799-1808, 1981.

Koc, U.-V. and K. J. R. Liu, DCT-based motion estimation, *IEEE Trans. Image Process.*, 7(7), 948-865, 1998.

Kojima, A., N. Sakurai, and J. Kishigami, Motion detection using 3D FFT spectrum, *Proceedings of International Conference on Acoustics, Speech, and Signal Processing*, V, 213-216, 1993.

Konrad, J. and E. Dubois, Bayesian estimation of motion vector fields, *IEEE Trans. Pattern Anal. Machine Intell.*, 14(9), 910-927, 1992.

Kughlin, C. D. and D. C. Hines, The phase correlation image alignment method, in *Proc. 1975 IEEE Int. Conf. on Systems, Man, and Cybernetics*, 163-165, 1975.

Li, H., P. Roivainen, and R. Forchheimer, 3-D motion estimation in model-based facial image coding, *IEEE Trans. Patt. Anal. Mach. Intell.*, 6, 545-555, 1993.

Matthies, L., T. Kanade, and R. Szeliski, Kalman filter-based algorithms for estimating depth from image sequences, *Int. J. Comput. Vision*, 3, 209-236, 1989.

Netravali, A. N. and J. D. Robbins, Motion-compensated television coding: Part I, *Bell Syst. Tech. J.*, 58(3), 631-670, 1979.

Orchard, M. T. Predictive motion-field segmentation for image sequence coding, *IEEE Transactions Aerosp. Electron. Syst.*, 3(1), 54-69, 1993.

Orchard, M. T. Visual coding standards: a research community's midlife crisis? *IEEE Signal Processing Magazine*, 43, 1998.

Pan, J. N. and Y. Q. Shi, A Kalman filter for improving optical flow accuracy along moving boundaries, *Proceedings of SPIE 1994 Visual Communication and Image Processing*, 1, 638-649, Chicago, Sept. 1994.

Pan, J. N., Y. Q. Shi, and C. Q. Shu, A Kalman filter in motion analysis from stereo image sequences, *Proceedings of IEEE 1994 International Conference on Image Processing*, 3, 63-67, Austin, TX, Nov. 1994.

Porat, B. and B. Friedlander, A frequency domain algorithm for multiframe detection and estimation of dim targets, *IEEE Transactions on Pattern Recognition and Machine Intelligence*, 12, 398-401, 1990.

Singh, A., Incremental estimation of image-flow using a Kalman filter, *Proc. 1991 IEEE Workshop on Visual Motion*, 36-43, Princeton, NJ, 1991.

Sommerfeld, A., *Mechanics of Deformable Bodies*, 1950.

Tekalp, A. M. *Digital Video Processing*, Prentice-Hall PTR, Upper Saddle River, NJ, 1995.

# Section IV

## Video Compression

# 15 Fundamentals of Digital Video Coding

In this chapter, we introduce the fundamentals of digital video coding which include digital video representation, rate distortion theory, and digital video formats. Also, we give a brief overview of image and video coding standards which will be discussed in the subsequent chapters.

## 15.1 DIGITAL VIDEO REPRESENTATION

As we discussed in previous chapters, a digital image is obtained by quantizing a continuous image both spatially and in amplitude. Digitization of the spatial coordinates is called image sampling, while digitization of the amplitude is called gray-level quantization. Suppose that a continuous image is denoted by $g(x, y)$, where the amplitude or value of $g$ at the point $(x, y)$ is the intensity or brightness of an image at that point. The transformation of a conntinuous image to a digital image can then be expressed as

$$f(m,n) = Q\big[g\big(x_o + m\Delta x, y_o + n\Delta y\big)\big], \qquad (15.1)$$

where $Q$ is a quantization operator, $x_o$ and $y_o$ are the origin of image plane, $m$ and $n$ are the discrete values 0, 1, 2, ..., and $\Delta x$ and $\Delta y$ are the sampling intervals in the horizontal and vertical directions, respectively. If the sampling process is extended to a third temporal direction (or the original signal in the temporal direction is a discrete format), a sequence, $f(m,n,t)$, is obtained as introduced in Chapter 10,

$$f(m,n,t) = Q\big[g\big(x_o + m\Delta x, y_o + n\Delta y, t_o + t\Delta t\big)\big], \qquad (15.2)$$

where $t$ is the values 0, 1, 2, ... and $\Delta t$ is the time interval.

Each point of the image or each basic element of the image is called as a pixel or pel. Each individual image is called a frame. According to the sampling theorem, the original continuous signal can be recovered exactly from its samples if the sampling frequency is higher than twice the bandwidth of the original signal (Oppenheim and Schafer, 1989). The frames are normally presented at a regular time interval so that the eye can perceive fluid motion. For example, the NTSC (National Television Systems Committee) specified a temporal sampling rate of 30 frames/second and interlace 2 to 1. Therefore, as a result of this spatio-temporal sampling, the digital signals exhibit high spatial and temporal correlation, just as the analog signals did before video data compression. In the following, we discuss the theoretical basis of video digitization. An important notion is the strong dependence between values of neighboring pixels within the same frame and between the frames themselves; this can be regarded as statistical redundancy of the image sequence. In the following section, we explain how this statistical redundancy is exploited to achieve compression of the digitized image sequence.

## 15.2 INFORMATION THEORY RESULTS (IV): RATE DISTORTION FUNCTION OF VIDEO SIGNAL

The principal goal in the design of a video-coding system is to reduce the transmission rate requirements of the video source subject to some picture quality constraint. There are only two ways to accomplish this goal: reduction of the statistical redundancy and psychophysical redundancy of the video source. The video source is normally very highly correlated, both spatially and temporally; that is, strong dependence can be regarded as statistical redundancy of the data source. If the video source to be coded in a transmission system is viewed by a human observer, the perceptual limitations of human vision can be used to reduce transmission requirements. Human observers are subject to perceptual limitations in amplitude, spatial resolution, and temporal acuity. By proper design of the coding system, it is possible to discard information without affecting perception, or at least, with only minimal degradation. In summary, we can use two factors: the statistical structure of the data source and the fidelity requirements of the end user, which make compression possible. The performance of the video compression algorithm depends on the several factors. First, and also fundamental, is the amount of redundancy contained in the video data source. In other words, if the original source contains a large amount of information, or high complexity, then more bits are needed to represent the compressed data. Second, if a lossy coding technique is used, by which some amount of loss is permitted in the reconstructed video data, then the performance of the coding technique depends on the compression algorithm and distortion measurements. In lossy coding, different distortion measurements will perceive the loss in different ways, giving different subjective results. The development of a distortion measure that can provide consistent numerical and subjective results is a very difficult task. Moreover, the majority of the video compression applications do not require lossless coding; i.e., it is not required that the reconstructed and original images be identical or reversible.

This intuitive explanation of how redundancy and lossy coding methods can be used to reduce source data is made more precise by the Shannon rate distortion theory (Berger, 1971), which addresses the problem of how to characterize both the source and the distortion measure. Let us consider the model of a typical visual communication system depicted in Figure 15.1. The source data is fed to the encoder system, which consists of two parts: source coding and channel coding. The function of the source coding is to remove the redundancy in both the spatial and temporal domains, whereas the function of channel coding is to insert the controlled redundancy, which is used to protect the transmitted data from the interference of channel noise. It should be noted that according to Shannon (1948) certain conditions allow the source and channel coding operations to be separated without any loss of optimality, such as when the sources are ergodic. However, Shannon did not indicate the complexity constraint on the coder involved. In practical systems that are limited

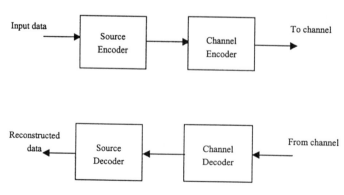

**FIGURE 15.1** A typical visual communication system.

by the complexity, this separation may not be possible (Viterbi and Omura, 1979). There is still some work on the joint optimization of the source and channel coding (Modestino et al., 1981; Sayood and Borkenhagen, 1991). Returning to rate–distortion theory, the problem addressed here is the minimizing the channel capacity requirement, while maintaining the average distortion at or below an acceptable level.

The rate distortion function $R(D)$ is the minimum average rate (bits/element), and hence minimum channel capacity, required for a given average distortion level $D$. To make this more quantitative, we suppose that the source is a sequence of pixels, and these values are encoded by successive blocks of length $N$. Each block of pixels is then described by one of a denumerable set of messages, $\{X_i\}$, with probability function, $P(X_i)$. For a given input source, $\{X_i\}$, and output, $\{Y_j\}$, the decoder system can be described mathematically by the conditional probability, $Q(Y_j/X_i)$. Therefore, the probability of the output message is

$$T(Y_j) = \sum_i P(X_i)Q(Y_j/X_i).$$ (15.3)

The information transmitted is called the average mutual information between $Y$ and $X$ and is defined for a block of length $N$ as follows:

$$I_N(X,Y) = \sum_i \sum_j P(X_i)Q(Y_j/X_i)\log_2 \frac{Q(Y_j/X_i)}{T(Y_j)}.$$ (15.4)

In the case of error-free encoding, $Y = X$ and then

$$Q(Y_j/X_i) = \begin{cases} 1, & j=i \\ 0, & j \neq i \end{cases} \quad \text{and} \quad T(Y_j) = T(Y_i).$$ (15.5)

In this case, Equation 15.4 becomes

$$I_N(X,Y) = \sum_i \sum_j P(X_i)\log_2 P(X_i) = H_N(X),$$ (15.6)

which is the $N$th-order entropy of the data source. This can also be seen as the information contained in the data source under the assumption that no correlation exists between blocks and all the correlation between elements of each $N$ length block is considered. Therefore, it requires at least $H_N(X)$ bits to code the data source without any information loss. In other words, the optimal error-free encoder requires $H_N(X)$ bits for the given data source. In the most general case, noise in the communication channel will result in error at least some of the time, causing $Y \neq X$. As a result,

$$I_N(X,Y) = H_N(X) - H_N(X/Y),$$ (15.7)

where $H_N(X/Y)$ is the entropy of the source data at the condition of decoder output $Y$. Since the entropy is a positive quantity, the source entropy is the upper bound to the mutual information; i.e.,

$$I_N(X,Y) \leq H_N(X).$$ (15.8)

Let $d\,(X,Y)$ be the average distortion between $X$ and $Y$. Then, the average distortion per pixel is defined as

$$D(Q) = \frac{1}{N}E\{d(X,Y)\} = \frac{1}{N}\sum_{i}\sum_{j}d\left(X_i,Y_j\right)P\left(X_i\right)Q\left(X_i/Y_j\right). \tag{15.9}$$

The set of all conditional probability assignments, $Q(Y/X)$, that yield average distortion less than or equal to $D*$, can be written as:

$$\{Q: D(Q) \leq D*\}. \tag{15.10}$$

The $N$-block rate distortion function is then defined as the minimum of the average mutual information, $I_N(X,Y)$, per pixel:

$$R_N\left(D^*\right) = \underset{Q:D(Q)\leq D^*}{\text{Min}} \frac{1}{N}I_N(X,Y). \tag{15.11}$$

The limiting value of the $N$-block rate distortion function is simply called the rate distortion function,

$$R\left(D^*\right) = \underset{N\to\infty}{\text{Lim}} R_N\left(D^*\right). \tag{15.12}$$

It should be clear from the above discussion that the Shannon rate distortion function is a lower bound on the transmission rate required to achieve an average distortion $D$ when the block size is infinite. In other words, when the block size is approaching infinity, the correlation between all elements within the block is considered as the information contained in the data source. Therefore, the rate obtained is the lowest rate or lower bound. Under these conditions, the rate at which a data source produces information, subject to a requirement of perfect reconstruction, is called the entropy of the data source, i.e., the information contained in the data source. It follows that the rate distortion function is a generalization of the concept of entropy. Indeed, if the distortion measure is a perfect reproduction, it is assigned zero distortion. Then, $R(0)$ is equal to the source entropy $H(X)$. Shannon's coding theorem states that one can design a coding system with rate only negligibly greater than $R(D)$ which achieves the average distortion $D$. As $D$ increases, $R(D)$ decreases monotonically and usually becomes zero at some finite value of distortion. The rate distortion function $R(D)$ specifies the minimum achievable transmission rate required to transmit a data with average distortion level $D$. The main value of this function in a practical application is that it potentially gives a measure for judging the performance of a coding system. However, this potential value has not been completely realized for video transmission. There are two reasons for this. First of all, there currently does not exist tractable and faithful mathematical models for an image source. The rate distortion function for Gaussian sources under the squared error distortion criterion can be found, but it is not a good model for images. The second reason is that a suitable distortion measure, $D$, which matches the subjective evaluation of image quality, has not been totally solved. Some results have been investigated for this task such as *JND* (just noticeable distortion) (*see* www.sarnoff.com/tech_realworld/broadcast/jnd/index.html). The issue of subjective and objective assessment of image quality has been discussed in Chapter 1. In spite of these drawbacks, the rate distortion theorem is still a mathematical basis for comparing the performance of different coding systems.

## 15.3 DIGITAL VIDEO FORMATS

In practical applications, most video signals are color signals. Various color systems have been discussed in Chapter 1. A color signal can be seen as a summation of light intensities of three primary wavelength bands. There are several color representations such as $YC_bC_r$, $RGB$, and others. It is common practice to convert one color representation to another color representation. The $YC_bC_r$ color representation is used for most video coding standards in compliance with the CCIR601 (International Radio Consultative Committee), common intermediate format (CIF), and SIF formats that are described in the following. The $Y$ component specifies the luminance information and the $C_b$ and $C_r$ components specify the color information. Conversion between the $YC_bC_r$ and $RGB$ formats can be accomplished with the following transformations, respectively.

$$\begin{bmatrix} Y \\ C_b \\ C_r \end{bmatrix} = \begin{bmatrix} 0.257 & 0.504 & 0.098 \\ -0.148 & -0.291 & 0.439 \\ 0.439 & -0.368 & -0.071 \end{bmatrix} \begin{bmatrix} R \\ G \\ B \end{bmatrix} + \begin{bmatrix} 16 \\ 128 \\ 128 \end{bmatrix}; \qquad (15.13)$$

$$\begin{bmatrix} R \\ G \\ B \end{bmatrix} = \begin{bmatrix} 1.164 & 0.000 & 1.596 \\ 1.164 & -0.392 & -0.813 \\ 1.164 & 2.017 & 0.000 \end{bmatrix} \begin{bmatrix} Y - 16 \\ C_b - 128 \\ C_r - 128 \end{bmatrix}. \qquad (15.14)$$

**Progressive and Interlaced** — Currently, most video signals that are generated by a TV camera are interlaced. These video signals are represented at 30 frames/second for an NTSC system. Each frame consists of two fields, the top field and bottom field, which are $\frac{1}{60}$ of a second apart. In the display of an interlaced frame, the top field is scanned first and the bottom field is scanned next. The top and bottom fields are composed of alternating lines of the interlaced frame. Progressive video does not consist of fields, only frames. In an NTSC system, these frames are spaced $\frac{1}{30}$ seconds apart. In contrast to interlaced video, every line within the frame is successively scanned.

**CCIR** — According to CCIR601 (*see* CCIR Recommendation 601-1) (CCIR is now known as ITU-R, International Telecommunications Union-R), a color video source has three components: a luminance component ($Y$) and two-color difference or chrominance components ($C_b$ and $C_r$ or $U$ and $V$ in some documents). The CCIR format has two options; one for the NTSC TV system and another for the PAL TV system; both are interlaced. The NTSC format uses 525 lines/frame at 30 frames/second. The luminance frames of this format have $720 \times 480$ active pixels. The chrominance frames have two kinds of formats, one has $360 \times 480$ active pixels and is referred as the 4:2:2 format, while the other has $360 \times 240$ active pixels and is referred as the 4:2:0 format. The PAL format uses 625 lines/frame at 25 frames/second. Its luminance frame has $720 \times 576$ active pixels/frame and the chrominance frame has $360 \times 576$ active pixels/frame for the 4:2:2 format and $360 \times 288$ pixels/frame for the 4:2:0 format, both at 25 frames/second.

**SIF (source input format)** — SIF has luminance resolution of $360 \times 240$ pixels/frame at 30 frames/second or $360 \times 288$ pixels/frame at 25 frames/second. For both cases, the resolution of the chrominance components is half of the luminance resolution in both horizontal and vertical dimensions. SIF can easily be obtained from a CCIR format using an appropriate antialiasing filter followed by subsampling.

**CIF (common intermediate format)** — CIF is a noninterlaced format. Its luminance resolution has $352 \times 288$ pixels/frame at 30 frames/second and the chrominance has half the luminance resolution in both vertical and horizontal dimensions. Since its line value, 288, represents half the active lines in the PAL television signal, and its picture rate, 30 frames/second, is the same as the

NTSC television signal, it is a common intermediate format for both PAL or PAL-like systems and NTSC systems. In the NTSC systems, only a line number conversion is needed, while in the PAL or PAL-like systems only a picture rate conversion is needed. For low-bit-rate applications, the quarter-SIF (QSIF) or quarter-CIF (QCIF) formats may be used since these formats have only a quarter the number of pixels of SIF and CIF formats, respectively.

**ATSC (Advanced Television Standard Committee) DTV (digital television) format** — The concept of DTV consists of SDTV (standard-definition television) and HDTV (high-definition television). Recently, in the U.S., the FCC (Federal Communication Commission) approved the ATSC-recommended DTV standard (ATSC, 1995). The DTV format is not included in the standard due to the divergent opinions of TV and computer manufacturers. Rather, it has been agreed that the picture format will be decided by the future market. The ATSC-recommended DTV formats including two kinds of formats: SDTV and HDTV. The ATSC DTV standard includes the following 18 formats:

*For HDTV*: 1920 × 1080 pixels at 23.976/24 Hz, 29.97/30 Hz, and 59.94/60 Hz progressive scan.

*For SDTV*: 704 × 480 pixels with 4:3 aspect ratio at 23.976/24 Hz, 29.97/30 Hz, 59.94/60 Hz progressive scan; 704 × 480 pixels with 16:9 aspect ratio at 23.976/24 Hz, 29.97/30 Hz, 59.94/60 Hz progressive scan; and 640 × 480 with 4:3 aspect ratio at 23.976/24 Hz, 29.97/30 Hz, 59.94/60 Hz progressive scan.

It is noted that all HDTV formats use square pixels and only part of SDTV formats uses square pixels. The number of pixels per line vs. the number of lines/frame is known as the aspect ratio.

## 15.4   CURRENT STATUS OF DIGITAL VIDEO/IMAGE CODING STANDARDS

The fast growth of digital transmission services has generated a great deal of interest in the digital transmission of video signals. Since some digitized video source signals require very high bit rates, ranging from more than 100 Mbps for broadcast-quality video to more than 1 Gbps for HDTV signals, video compression algorithms which reduce the bit rates to an affordable level on practical communication channels are required. Digital video-coding techniques have been investigated over several decades. There are two factors that make video compression possible: the statistical structure of the data in the video source and the psychophysical redundancy of human vision. Video compression algorithms can remove the spatial and temporal correlation that is normally present in the video source. In addition, human observers are subject to perceptual limitations in amplitude, spatial resolution, and temporal acuity. By proper design of the coding system it is possible to discard information without affecting perceived image quality or, at least, with only minimal degradation.

Several traditional techniques have been developed for image and video data compression. Recently, with advances in data compression and VLSI (very large scale integrated) techniques, the data compression techniques have been extensively applied to video signal compression. Video compression techniques have been under development for over 20 years and have recently emerged as the core enabling technology for a new generation of DTV (both SDTV and HDTV) and multimedia applications. Digital video systems currently being implemented (or under active consideration) include terrestrial broadcasting of digital HDTV in the U.S. (ATSC, 1993), satellite DBS (Direct Broadcasting System) (Isnardi, 1993), computer multimedia (Ada, 1993), and video via packet networks (Verbiest, 1989). In response to the needs of these emerging markets for digital video, several national and worldwide standards activities have been started over the last few years. These organizations include ISO (International Standards Organization), ITU, formally known as CCITT, International Telegraph and Telephone Consultative Committee), JPEG (Joint Photographic

Experts Group), and MPEG (Motion Picture Experts Group) as shown in Table 15.1. The related standards include JPEG standards, MPEG-1,2,4 standards, and H.261 and H.263 video teleconferencing coding standards as shown in Table 15.2. It should be noted that the JPEG standards are usually used for still image coding, but they can also be used to code video. Although the coding efficiency would be lowered, they have been shown to be useful in some applications, e.g., studio editing systems. Although they are not video-coding standards and were discussed in Chapters 7 and 8, respectively, we include them here for completeness of all international image and video coding standards.

- **JPEG Standard:** Since the mid-1980s, the ITU and ISO have been working together to develop a joint international standard for the compression of still images. Officially, JPEG (ISO/IEC, 1992a) is the ISO/IEC international standard 10918-1, "Digital Compression and Coding of Continuous-Tone Still Images," or the ITU-T recommendation T.81. JPEG became an international standard in 1992. JPEG is a DCT-based coding algorithm and continues to work on future enhancements, which may adopt wavelet-based algorithms.
- **JPEG-2000:** JPEG-2000 (*see* Joint Photographic Experts Group) is a new type of image coding system under development by JPEG for still image coding. JPEG-2000 is considering using the wavelet transform as its core technique. This is because the wavelet transform can provide not only excellent coding efficiency, but also wonderful spatial and quality scalable functionality. This standard is intended to meet the need for image compression with great flexibility and efficient interchangeability. It is also intended to offer unprecedented access into the image while still in a compressed domain. Thus, an image can be accessed, manipulated, edited, transmitted, and stored in a compressed form.
- **MPEG-1:** In 1988 ISO established the MPEG to develop standards for the coded representation of moving pictures and associated audio information for digital storage applications. MPEG completed the first phase of its work in 1991. It is known as MPEG-1 (ISO/IEC, 1992b) or ISO standard 11172, "Coding of Moving Picture and Associated Audio." The target application for this specification is digital storage media at bit-rates up to about 1.5 Mbps.
- **MPEG-2:** MPEG started its second phase of work, MPEG-2 (ISO/IEC, 1994), in 1990. MPEG-2 is an extension of MPEG-1 that allows for greater input-format flexibility, higher data rate for SDTV or HDTV applications, and better error resilience. This work resulted in the ISO standard 13818 or ITU-T Recommendation H.262, "Generic Coding of Moving Pictures and Associated Audio."
- **MPEG-4:** MPEG is now working on its fourth phase, MPEG-4 (ISO/IEC, 1998). MPEG-4 visual committee draft version 1 was approved in November 1997. The end of 1999 will define the final international standard. The MPEG-4 standard supports object-based coding technology and is aimed at providing enabling technology for a variety of functionalities and multimedia applications:
  1. Universal accessibility and robustness in error-prone environments
  2. High interactive functionality
  3. Coding of natural and synthetic data or both
  4. Compression efficiency.
- **H.261:** H.261 was adopted in 1990 and the final revision was approved in 1993 by the ITU-T. It is designed for video teleconferencing and utilizes a DCT-based motion-compensation scheme. The target bit rates are from 64 to 1920 Kbps.
- **H.263, H.263 Version 2 (H.263+), H.263++ and H.26L:** The H.263 video coding standard is specifically designed for very low bit rate applications such as video conferencing. Its technical content was completed in late 1995 and the standard was approved in early 1996. It is based on the H.261 standard with several added features: unrestricted

motion vectors, syntax-based arithmetic coding, advanced prediction, and PB-frames. The H.263 version 2 video-coding standard, also known as "H.263+," was approved in January 1998 by the ITU-T. H.263+ includes a number of new optional features based on the H.263. These new optional features are added to provide improved coding efficiency, a flexible video format, scalability, and backward-compatible supplemental enhancement information. H.263++ is the extension of H.263+ and is currently scheduled to be completed in the year 2000. H.26L is a long-term project which is looking for more efficient video-coding algorithms.

The above organizations and standards are summarized in Tables 15.1 and 15.2, respectively.

**TABLE 15.1**
**List of Some Organizations for Standardization**

| Organization | Full Name of Organization |
|---|---|
| CCITT | International Telegraph and Telephone Consultative Committee |
| ITU | International Telecommunication Union |
| JPEG | Joint Photographic Experts Group |
| MPEG | Moving Picture Experts Group |
| ISO | International Standards Organization |
| IEC | International Electrotechnical Commission |

**TABLE 15.2**
**Video/Image Coding Standards**

| Name | Completion Time | Major Features |
|---|---|---|
| JPEG | 1992 | For still image coding, DCT based |
| JPEG-2000 | 2000 | For still image coding, DWT based |
| H.261 | 1990 | For videoconferencing, 64Kbps to 1.92 Mbps |
| MPEG-1 | 1991 | For CD-ROM, 1.5 Mbps |
| MPEG-2 (H.262) | 1994 | For DTV, 2 to 15 Mbps, most extensively used |
| H.263 | 1995 | For very low bit rate coding, below 64 Kbps |
| H.263+ (version 2) | 1998 | Add new optional features to H.263 |
| MPEG-4 | 1999 | For multimedia, content-based coding |
| MPEG-4 (version 2) | 2000 | Adds more tools to MPEG-4 |
| H.263++ | 2000 | Adds more optional features to H.263+ |
| H.26L | 2000 | Functionally different, much more efficient |
| MPEG-7 | 2001 | Content description and indexing |

It should be noted that MPEG-7 in Table 15.2 is not a coding standard; it is ongoing work of MPEG. It is also interesting to note that in terms of video compression methods, there is a growing convergence toward motion-compensated, interframe DCT algorithms represented by the video coding standards. However, wavelet-based coding techniques have found recent success in the compression of still image coding in both the JPEG-2000 and MPEG-4 standards. This is because it posseses unique features in terms of high coding efficiency and excellent spatial and quality scalability. The wavelet transform has not successfully been applied to video coding due to several difficulties. For one, it is not clear how the temporal redundancy can be removed in this domain. Motion compensation is an effective technique for DCT-based video coding; however, it is not so effective for wavelet-based video coding. This is because the wavelet transform uses large block size or full frame, but motion compensation is usually performed on a limited block size. This mismatch would reduce the interframe coding efficiency. Many engineers and researchers are working on these problems.

Among these standards, MPEG-2 has had a great impact on the consumer electronics industry since the DVD (Digital Video Disk) and DTV have adopted it as core technology.

## 15.5   SUMMARY

In this chapter, several fundamental issues of digital video coding are presented. These include the representation and rate distortion function of digital video signals and the various video formats, which are widely used by the video industry. Finally, existing and emerging video coding standards are briefly introduced.

## 15.6   EXERCISES

**15-1.**   Suppose that we have 1-D digital array (it can be extended to 2-D array that may be an image), $f(i) = X_i$, ($i = 0, 1, 2, \ldots$). If we use the first-order linear predictor to predict the current component value with the previous component, such as: $X'_i = \alpha X_{i-1} + \beta$, where $\alpha$ and $\beta$ are two parameters for this linear predictor, and if we want to minimize the mean-squared error of the prediction $E\{(X_i - X'_i)^2\}$, what $\alpha$ and $\beta$ do we have to choose? Assuming that $E\{X_i\} = m$, $E\{X_i^2\} = \sigma^2$ and $E\{X_i X_{i-1}\} = \rho$, (for $i = 0, 1, 2, \ldots$), where $m$, $\sigma$, and $\rho$ are constant.

**15-2.**   To get a $128 \times 128$ or $256 \times 256$ digital image, write a program to use two $3 \times 3$ operators (Sobel operator) such as:

$$
\begin{array}{|ccc|}
\hline
-1 & -2 & -1 \\
0 & 0 & 0 \\
1 & 2 & 1 \\
\hline
\end{array}
\qquad
\begin{array}{|ccc|}
\hline
-1 & 0 & 1 \\
-2 & 0 & 2 \\
-1 & 0 & 1 \\
\hline
\end{array}
$$

to filter the image, separately. Discuss the resulting image. What will be the result if both operators are used?

**15-3.**   The convolution of two 2-D arrays is defined as:

$$y(m,n) = \sum_{k=-\infty}^{+\infty} \sum_{l=-\infty}^{+\infty} x(k,l) h(m-k, n-l)$$

and

$$\bar{x} = \begin{bmatrix} 1 & 4 & 1 \\ 2 & 5 & 3 \end{bmatrix} \qquad \bar{h} = \begin{bmatrix} 1 & 1 \\ 1 & -1 \end{bmatrix}.$$

Calculate the convolution $y(m,n)$. If $h(m,n)$ is changed to

$$\begin{bmatrix} 0 & -1 & 0 \\ -1 & 4 & -1 \\ 0 & -1 & 0 \end{bmatrix},$$

recalculate $y(m,n)$.

**15-4.** The entropy of an image source is defined as

$$H = -\sum_{k=1}^{M} p_k \log_2 p_k,$$

under the assumption that each pixel is an independent random variable. If the image is a binary image, i.e., $M = 2$, and the probability $p_1 + p_2 = 1$. If we define $p_1 = p$, then $p_2 = 1 - p$, $(0 \le p \le 1)$. The entropy can be rewritten as

$$H = -p \log_2 p - (1-p) \log_2 (1-p).$$

Find several digital binary images and compute their entropies. If one image has almost equal number of zeros and ones and another has a different number of zeros and ones, which image has larger entropy? Prove that the entropy of a binary source is maximum if the numbers of zeros and ones are equal.

**15-5.** A transformation defined as $y = f(x)$, is applied to a $256 \times 256$ digital image, where $x$ is the original pixel value and $y$ is the transformed pixel value. Obtain new images for (a) $f$ is a linear function, (b) $f$ is a logarithm, and (c) $f$ is a square function. Compare the results and indicate subjective differences of the resulting images. Repeat the experiments for different images and draw conclusions about possible use of this procedure in image processing applications.

## REFERENCES

Ada, J. A. Interactive Multimedia, *IEEE Spectrum,* 22-31, 1993.

ATSC Digital Television Standard, Doc. A/53, September 16, 1995.

Berger, T. *Rate Distortion Theory — A Mathematical Bais for Data Compression*, Prentice-Hall, Englewood Cliffs, NJ, 1971.

CCIR Recommendation 601-1, Encoding parameters for digital television for studios, 1990.

Isnardi, M. Consumers seek easy to use products, *IEEE Spectrum,* 64, 1993.

ISO/IEC JTC1 IS 11172, Coding of Moving Picture and Coding of Continuous Audio for Digital Storage Media up to 1.5 Mbps, 1992b.

ISO/IEC JTC1 IS 13818, Generic Coding of Moving Pictures and Associated Audio, 1994.

ISO/IEC JTC1 FDIS 14496-2, Information Technology — Generic Coding of Audio-Visual Objects, Nov. 19, 1998.

Just Noticeable Distortion (JND) www.sarnoff.com/tech_realworld/broadcast/jnd/index.html.

Joint Photographic Experts Group (JPEG), ISO/IEC IS 11544, ITU-T Rec. T.81, 1992a.

Modestino, J. W., D. G. Daut, and A. L. Vickers, Combined source-channel coding of image using the block cosine transform, *IEEE Trans. Commun.,* COM-29, 1262-1274, 1981.

Oppenheim, A. V. and R. W. Schafer, Discrete-Time Signal Processing, Prentice-Hall, Englewood Cliffs, NJ, 1989.

Sayood, K. and J. C. Borkenhagen, Use of residual redundancy in the design of joint source/channel coders, *IEEE Trans. Commun.,* 39(6), 838-846, 1991.

Shannon, C. E. A mathematical theory of communication, *Bell Syst. Tech. J.,* 27, 379-423, 623-656, 1948.

Verbiest, W. and L. Pinnoo, A variable bit rate video codec for asynchronous transfer mode networks, *IEEE JSAC,* 7(5), 761-770, 1989.

Viterbi, A. J. and J. K. Omura, *Principles of Digital Communication and Coding*, New York: McGraw-Hill, New York, 1979.

# 16 Digital Video Coding Standards — MPEG-1/2 Video

In this chapter, we introduce the ISO/IEC digital video coding standards, MPEG-1 (ISO/IEC, 1992) and MPEG-2 (ISO/IEC, 1995), which are extensively used in the video industry for television broadcast, visual communications, and multimedia applications.

## 16.1 INTRODUCTION

As we know, MPEG has successfully developed two standards, MPEG-1 and MPEG-2. The MPEG-1 video standard was completed in 1991 with the development of the ISO/IEC specification 11172, which is the standard for coding of moving picture and associated audio for digital storage media at up to about 1.5 Mbps. To support a wide range of application profiles the user can specify a set of input parameters including flexible picture size and frame rate. MPEG-1 was developed for multimedia CD-ROM applications. Important features provided by MPEG-1 include frame-based random access of video, fast-forward/fast-reverse searches through compressed bitstreams, reverse playback of video, and editability of the compressed bitstream. MPEG-2 is formally referred to as ISO/IEC specification 13818, which is the second phase of MPEG video coding solution for applications not originally covered by the MPEG-1 standard. Specifically, MPEG-2 was developed to provide video quality not lower than NTSC/PAL and up to HDTV quality. The MPEG-2 standard was completed in 1994. Its target bit rates for NTSC/PAL are about 2 to 15 Mbps, and it is optimized at about 4 Mbps. The bit rates used for HDTV signals are about 19 Mbps. In general, MPEG-2 can be seen as a superset of the MPEG-1 coding standard and is backward compatible to the MPEG-1 standard. In other words, every MPEG-2-compatible decoder is able to decode a compliant MPEG-1 bit stream.

In this chapter, we will briefly introduce the standard itself. Since many books and publications exist for the explanation of the standards (Haskell et al., 1997; Mitchell et al., 1997), we will pay more attention to the utility of the standard, how the standard is used, and touch on some interesting research topics that have emerged. In other words, the standards provide the knowledge for how to design the decoders that are able to decode the compliant MPEG bitstreams successfully. But the standards do not specify the means of generating these bitstreams. For instance, given some bit rate, how can one generate a bitstream that provides the best picture quality? To answer this, one needs to understand the encoding process, which is an informative part of the standard (referred to as the test model), but it is very important for the content and service providers. In this chapter, the issues related to the encoding process are described. The main contents include the following topics: preprocessing, motion compensation, rate control, statistically multiplexing multiple programs, and optimal mode decision. Some of the sections contain the authors' own research results. These research results are useful in providing examples for readers to understand how the standard is used.

## 16.2 FEATURES OF MPEG-1/2 VIDEO CODING

It should be noted that MPEG-2 video coding has the feature of being backward compatible with MPEG-1. It turns out that most of the decoders in the market are MPEG-2 compliant decoders.

For simplicity, we will start to introduce the technical detail of MPEG-1 and then describe the enhanced features of MPEG-2, which MPEG-1 does not have.

## 16.2.1  MPEG-1 Features

### 16.2.1.1  Introduction

The algorithms employed by MPEG-1 do not provide a lossless coding scheme. However, the standard can support a variety of input formats and be applied to a wide range of applications. As we know, the main purpose of MPEG-1 video is to code moving image sequences or video signals. To achieve a high compression ratio, both intraframe redundancy and interframe redundancy should be exploited. This implies that it would not be efficient to code the video signal with an intraframe-coding scheme, such as JPEG. On the other hand, to satisfy the requirement of random access, we have to use intraframe coding from time to time. Therefore, the MPEG-1 video algorithm is mainly based on discrete cosine transform (DCT) coding and interframe motion compensation. The DCT coding is used to remove the intraframe redundancy and motion compensation is used to remove the interframe redundancy. With regard to input picture format, MPEG-1 allows progressive pictures only, but offers great flexibility in the size, up to 4095 × 4095 pixels. However, the coder itself is optimized to the extensively used video SIF picture format. The SIF is a simple derivative of the CCIR601 video format for digital television applications. According to CCIR601, a color video source has three components, a luminance component ($Y$) and two chrominance components ($C_b$ and $C_r$) which are in the 4:2:0 subsampling format. Note that the 4:2:0 and 4:2:2 color formats were described in Chapter 15.

### 16.2.1.2  Layered Structure Based on Group of Pictures

The MPEG coding algorithm is a full-motion-compensated DCT and DPCM hybrid coding algorithm. In MPEG coding, the video sequence is first divided into groups of pictures or frames (GOP) as shown in Figure 16.1. Each GOP may include three types of pictures or frames: intracoded (I) picture or frame, predictive-coded (P) picture or frame, and bidirectionally predictive-coded (B) picture or frame. I-pictures are coded by intraframe techniques only, with no need for previous information. In other words, I-pictures are self-sufficient. They are used as anchors for forward and/or backward prediction. P-pictures are coded using one-directional motion-compensated prediction from a previous anchor frame, which could be either an I- or a P-picture. The distance between two nearest I-frames is denoted by $N$, which is the size of GOP. The distance between two nearest anchor frames is denoted by $M$. Parameters $N$ and $M$ both are user-selectable parameters, which are selected by user during the encoding. A larger number of $N$ and $M$ will increase the

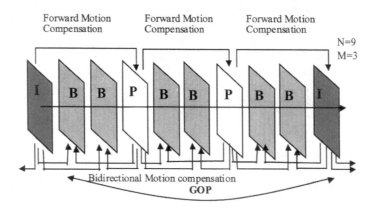

**FIGURE 16.1**  A group of pictures of video sequence in display order.

coding performance but cause error propagation or drift. Usually, $N$ is chosen from 12 to 15 and $M$ from 1 to 3. If $M$ is selected to be 1, this means no B-picture will be used. Last, B-pictures can be coded using predictions from either past or future anchor frames (I or P), or both. Regardless of the type of frame, each frame may be divided into slices; each slice consists of several macroblocks (MBs). There is no rule to decide the slice size. A slice could contain all macroblocks in a row of a frame or all macroblocks of a frame. Smaller slice size is favorable for the purpose of error resilience, but will decrease coding performance due to higher overhead. A macroblock contains a $16 \times 16$ $Y$ component and spatially corresponding $8 \times 8$ $C_b$ and $C_r$ components. A macroblock has four luminance blocks and two chrominance blocks (for 4:2:0 sampling format) and the macroblock is also the basic unit of adaptive quantization and motion compensation. Each block contains $8 \times 8$ pixels over which the DCT operation is performed.

To exploit the temporal redundancy in the video sequence, the motion vector for each macroblock is estimated from two original luminance pictures using a block-matching algorithm. The criterion for the best match between the current macroblock and a macroblock in the anchor frame is the minimum mean absolute error. Once the motion vector for each macroblock is estimated, pixel values for the target macroblock can be predicted from the previously decoded frame. All macroblocks in the I-frame are coded in intramode with no motion compensation. Macroblocks in P- and B-frames can be coded in several modes. Among the modes are intracoded and intercoded with motion compensation. This decision is made by mode selection. Most encoders depend on the values of predicted differences to make this decision. Within each slice, the values of motion vectors and DC values of each macroblock are coded using DPCM. The detailed specifications of this coding can be found in the document proposed by the MPEG video committee (ISO/IEC, 1995). The structure of MPEG implies that if an error occurs within I-frame data, it will be propagated through all frames in the GOP. Similarly, an error in a P-frame will affect the related P- and B-frames, while B-frame errors will be isolated.

### 16.2.1.3   Encoder Structure

The typical MPEG-1 video encoder structure is shown in Figure 16.2. Since the encoding order is different from the display order, the input sequence has to be reordered for encoding. For example, if we choose the GOP size ($N$) to be 12, and the distance between two nearest anchor frames ($M$) to be 3, the display order and encoding order are as shown in Table 16.1.

It should be noted that in the encoding order or in the bitstream the first frame in a GOP is always an I-picture. In the display order the first frame can be either an I-picture or the first B-picture of the consecutive series of B-pictures which immediately precedes the first I-picture, and the last

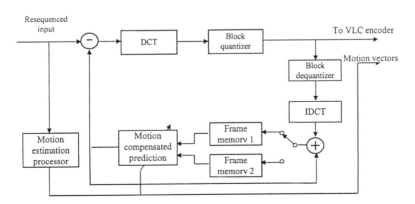

**FIGURE 16.2**   Typical MPEG-1 encoder structure. (From ISO/IEC, MPEG-2, Test Model 5, ISO-IEC/STCI/ SC29/WGII, April, 1993. With permission.)

**TABLE 16.1**
**Display Order and Encoding Order**

| Display Order | 0 | 1 | 2 | 3 | 4 | 5 | 6 | 7 | 8 | 9 | 10 | 11 | 12 |
|---|---|---|---|---|---|---|---|---|---|---|---|---|---|
| Encoding order | 0 | 3 | 1 | 2 | 6 | 4 | 5 | 9 | 7 | 8 | 12 | 10 | 11 |
| Coding type | I | P | B | B | P | B | B | P | B | B | I | B | B |

picture in a GOP is an anchor picture, either an I- or P-picture. The first GOP always starts with an I-picture and, as a consequence, this GOP will have fewer B-pictures than the other GOPs.

The MPEG-1 video compression technique uses motion compensation to remove the interframe redundancy. The concept of motion compensation is based on the estimation of motion between video frames. The fundamental model that is used assumes that a translational motion can approximate the motion of a block. If all elements in a video scene are approximately spatially displaced, the motion between frames can be described by a limited number of motion parameters. In other words, the motion can be described by motion vectors for translatory motion of pixels. Since the spatial correlation between adjacent pixels is usually very high, it is not necessary to transmit motion information for each coded image pixel. This would be too expensive and the coder would never be able to reach a high compression ratio. The MPEG video uses the macroblock structure for motion compensation; i.e., for each $16 \times 16$ macroblock only one or sometimes two motion vectors are transmitted. The motion vectors for any block are found within a search window that can be up to 512 pixels in each direction. Also, the matching can be done at half-pixel accuracy, where the half-pixel values are computed by averaging the full-pixel values (Figure 16.3).

For interframe coding, the prediction differences or error images are coded and transmitted with motion information. A 2-D DCT is used for coding both the intraframe pixels and the predictive error pixels. The image to be coded is first partitioned into $8 \times 8$ blocks. Each $8 \times 8$ pixel block is then subject to an $8 \times 8$ DCT, resulting in a frequency domain representation of the block as shown in Figure 16.4.

The goal of the transformation is to decorrelate the block data so that the resulting transform coefficients can be coded more efficiently. The transform coefficients are then quantized. During

**FIGURE 16.3**   Half-pixel locations in motion compensation.

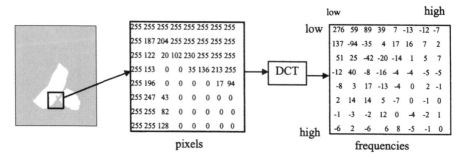

**FIGURE 16.4**   Example of $8 \times 8$ DCT.

the process of quantization a weighted quantization matrix is used. The function of quantization matrix is to quantize high frequencies with coarser quantization steps that will suppress high frequencies with no subjective degradation, thus taking advantage of human visual perception characteristics. The bits saved for coding high frequencies are used for lower frequencies to obtain better subjective coded images. There are two quantizer weighting matrices in Test Model 5 (TM5) (ISO/IEC, 1993), an intraquantizer weighting matrix and a nonintraquantizer weighting matrix; the latter is flatter since the energy of coefficients in interframe coding is more uniformly distributed than in intraframe coding.

In intra macroblocks, the DC value, $dc$, is an 11-bit value before quantization and it will be quantized to 8, 9, or 10 bits according to the setting of parameter. Thus, the quantized DC value, QDC, is calculated as

$$\text{8-bit: QDC} = dc//8, \text{ 9-bit: QDC} = dc//4, \text{ or 10-bit: QDC} = dc//2, \tag{16.1}$$

where symbol $//$ means integer division with rounding to the nearest integer and the half-integer values are rounded away for zero unless otherwise specified. The AC coefficients, $ac(i,j)$, are first quantized by individual quantization factors to the value of $ac \sim (i,j)$:

$$ac \sim (i,j) = \left(16 * ac(i,j)\right)//W_I(i,j), \tag{16.2}$$

where $W_I(i,j)$ is the element at the $(i,j)$ position in the intraquantizer weighting matrix shown in Figure 16.5.

The quantized level $QAC(i,j)$ is given by

$$QAC(i,j) = \left[ac \sim (i,j) + \text{sign}\left(ac \sim (i,j) * \left((p * mquant)//q\right)\right)\right]/(2 * mquant), \tag{16.3}$$

where $mquant$ is the quantizer scale or step which is derived for each macroblock by rate control algorithm, and $p = 3$ and $q = 4$ in TM5 (ISO/IEC, 1993). For nonintra macroblocks,

$$ac \sim (i,j) = \left(16 * ac(i,j)\right) // W_N(i,j), \tag{16.4}$$

where $W_N(i,j)$ is the nonintraquantizer weighting matrix in Figure 16.5 and

$$QAC(i,j) = ac \sim (i,j)/(2 * mquant). \tag{16.5}$$

An example of encoding an intrablock is shown in Figure 16.6.

| | | | | | | |
|---|---|---|---|---|---|---|
| 8 | 16 | 19 | 22 | 26 | 27 | 29 | 34 |
| 16 | 16 | 22 | 24 | 27 | 29 | 34 | 37 |
| 19 | 22 | 26 | 27 | 29 | 34 | 34 | 38 |
| 22 | 22 | 26 | 27 | 29 | 34 | 37 | 40 |
| 22 | 26 | 27 | 29 | 32 | 35 | 40 | 48 |
| 26 | 27 | 29 | 32 | 35 | 40 | 48 | 58 |
| 26 | 27 | 29 | 34 | 38 | 46 | 56 | 69 |
| 27 | 29 | 35 | 38 | 46 | 56 | 69 | 83 |

Intra quantizer weighting matrix

| | | | | | | |
|---|---|---|---|---|---|---|
| 16 | 17 | 18 | 19 | 20 | 21 | 22 | 23 |
| 17 | 18 | 19 | 20 | 21 | 22 | 23 | 24 |
| 18 | 19 | 20 | 21 | 22 | 23 | 24 | 25 |
| 19 | 20 | 21 | 22 | 23 | 24 | 26 | 27 |
| 20 | 21 | 22 | 23 | 25 | 26 | 27 | 28 |
| 21 | 22 | 23 | 24 | 26 | 27 | 28 | 30 |
| 22 | 23 | 24 | 26 | 27 | 28 | 30 | 31 |
| 23 | 24 | 25 | 27 | 28 | 30 | 31 | 33 |

Nonintra quantizer weighting matrix

**FIGURE 16.5**  Quantizer matrices for intra- and nonintracoding.

Intra quantizer weighting matrix

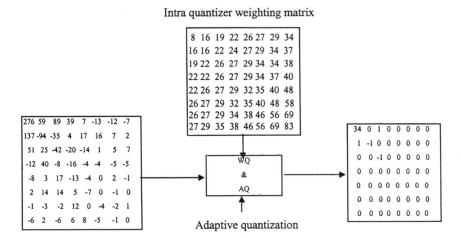

**FIGURE 16.6**   An example of coding an intrablock.

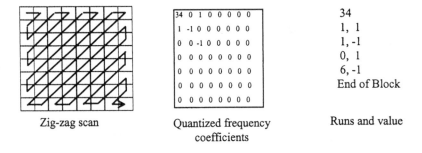

| Zig-zag scan | Quantized frequency coefficients | Runs and value |

**FIGURE 16.7**   Zigzag scans to get pairs of zero-runs and value.

The coefficients are processed in zigzag order since the most energy is usually concentrated in the lower-order coefficients. The zigzag ordering of elements in an $8 \times 8$ matrix allows for a more efficient run-length coder. This is illustrated in Figure 16.7.

With the zigzag order, the run-length coder converts the quantized frequency coefficients to pairs of zero runs and nonzero coefficients:

$$34 \ 0 \ 1 \ 0 \ -1 \ 1 \ 0 \ 0 \ 0 \ 0 \ 0 \ 0 \ -1 \ 0 \ 0 \ 0 \ 0....$$

After parsing we obtain the pairs of zero runs and values:

$$34 \ | \ 0 \ 1 \ | \ 0 \ -1 \ | \ 1 \ | \ 0 \ 0 \ 0 \ 0 \ 0 \ 0 \ -1 \ | \ 0 \ 0 \ 0 \ 0....$$

These pairs of runs and values are then coded by a Huffman-type entropy coder. For example, for the above run/value, pairs are

| Run/Value 34 | VLC (Variable Length Code) |
| --- | --- |
| 1, 1 | 0110 |
| 1, −1 | 0111 |
| 0, 1 | 110 |
| 6, −1 | 0001011 |
| End of block | 10 |

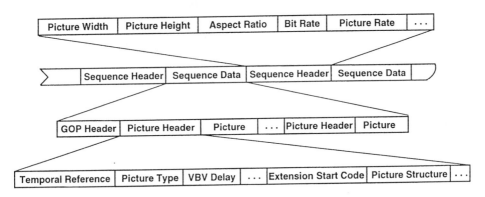

**FIGURE 16.8**  Description of layered structure of compressed bitstream.

The VLC tables are obtained by statistically optimizing a large number of training video sequences and are included in the MPEG-2 specification. The same idea is applied to code the DC values, motion vectors, and other information. Therefore, the MPEG video standard contains a number of VLC tables.

### 16.2.1.4   Structure of the Compressed Bitstream

After coding, all the information is converted to binary bits. The MPEG video bitstream consists of several well-defined layers with headers and data fields. These layers include sequence, GOP, picture, slice, macroblock, and block. The important syntax elements contained in each layer can be summarized in Table 16.2. The typical structure of the MPEG-1 video-compressed bitstream is shown in Figure 16.8. The syntax elements contained in the headers and the amount of bits defined for each element can be found in the standard.

For picture layer, a frame of picture is first partitioned into macroblocks (16 × 16 for luminance and 8 × 8 for chrominance in the 4:2:0 color representation). The compressed bitstream structure at this layer is shown in Figure 16.9. It is important to note that most elements in the syntax are coded by VLC. The tables of these variable run-length codes are obtained through the simulation of a large number of training video sequences.

**TABLE 16.2**
**Summary of Important Syntax of Each Layer**

| Name of Layer | Important Syntax Elements |
|---|---|
| Sequence | Picture size and frame rate |
|  | Bit rate and buffering requirement |
|  | Programmable coding parameters |
| GOP | Random access unit |
|  | Time code |
| Picture | Timing information (buffer fullness, temporal reference) |
|  | Coding type (I, P, or B) |
| Slice | Intraframe addressing information |
|  | Coding reinitialization (error resilience) |
| MB | Basic coding structure |
|  | Coding mode |
|  | Motion vectors |
|  | Quantization |
| Block | DCT coefficients |

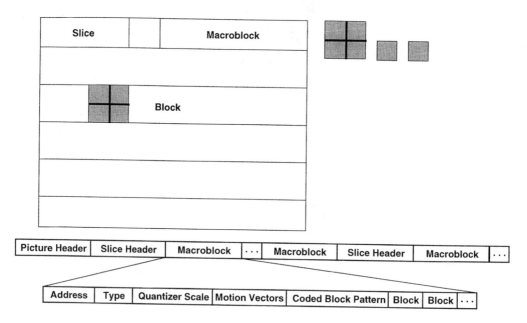

**FIGURE 16.9**   Picture layer data structure.

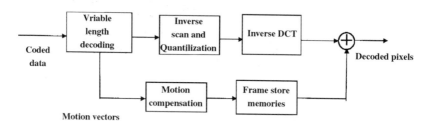

**FIGURE 16.10**   Simplified MPEG video decoder. (From ISO/IEC, MPEG-2 Test Model 5, April, 1993. With permission.)

### 16.2.1.5   Decoding Process

The decoding process is an inverse procedure of encoding. The block diagram of a typical decoder is shown in Figure 16.10

The variable-length decoder (VLD) first decodes the coded data or video bitstream. This process yields the quantized DCT coefficients and motion vector data for each macroblock. The coefficients are inversely scanned and dequantized. The decoded DCT coefficients are then inverse-transformed to obtain the spatial-domain pixels. If the macroblock was intracoded, these pixels represent the reconstructed values, without any further processing. However, if the macroblock is intercoded, then motion compensation is performed to add the prediction from the corresponding reference frame or frames.

### 16.2.2   MPEG-2 Enhancements

The basic coding structure of MPEG-2 video is the same as that of MPEG-1 video, that is, intraframe and interframe DCT with I-, P-, and B-pictures is used. The most important features of MPEG-2 video coding include:

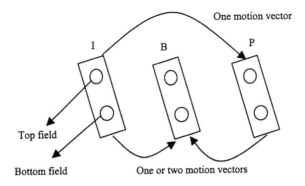

**FIGURE 16.11**  Frame-based prediction of MPEG-1 video coding.

- Field/frame prediction modes for supporting the interlaced video input;
- Field/frame DCT coding syntax;
- Downloadable quantization matrix and alternative scan order;
- Scalability extension.

The above enhancement items are all coding performance improvements that are related to the support of interlaced material. There are also several noncompression enhancements, which include:

- Syntax to facilitate 3:2 pull-down in the decoder;
- Pan and scan codes with $1/16$ pixel resolution;
- Display flags indicating chromaticity, subcarrier amplitude, and phase (for NTSC/PAL/ SECAM source material).

In the following, each of these enhancements is introduced.

### 16.2.2.1  Field/Frame Prediction Mode

In MPEG-1 video, we always code each picture as a frame structure, whether the original material is progressive or interlaced. If the original sequence is interlaced, each frame consists of two fields: top field and bottom field as shown in Figure 16.11. We still can use frame-based prediction if we consider the two fields as a frame, such as that shown in Figure 16.11.

In Figure 16.11, three frames are coded as I-, B-, and P-frames and each frame consists of two fields. The P-frame is predicted with the I-frame with one motion vector. The B-frame can be predicted only with I-frame (forward prediction) or only with P-frame (backward prediction) or from both I- and P-picture (bidirectional prediction), the forward and backward prediction needs only one motion vector and the bidirectional prediction needs two motion vectors.

MPEG-2 video provides an enhanced prediction mode to support interlaced material, which uses the adaptive field/frame selection, based on the best match criteria. Each frame consists of two fields: top field and bottom field. Each field can be predicted from either field of the previous anchor frame. The possible prediction modes are shown in Figure 16.12.

In a field-based prediction, the top field of the current frame can be either predicted from the top field or the bottom field of an anchor frame as shown in Figure 16.12. The solid arrow represents the prediction from the top field, and the dashed arrow represents the prediction from the bottom field. The same is also true for bottom field of the current frame. If the current frame is a P-frame, there could be up to two motion vectors used to make the prediction (one for top field and one for bottom field); if the current frame is a B-frame, there could be up to four motion vectors (each field could be bidirectional prediction which needs two motion vectors). At the macroblock level

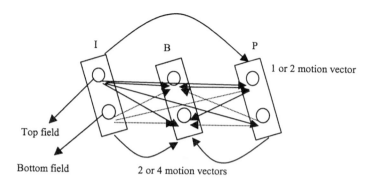

**FIGURE 16.12**   Field-based prediction of enhanced option of MPEG-2 video coding.

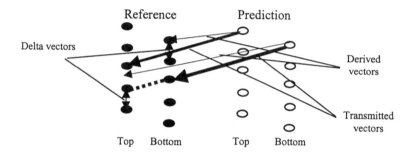

**FIGURE 16.13**   Dual prime prediction in MPEG-2 video coding.

of MPEG-2, several coding modes are added to support these new field-based predictions. Additionally, there is another new prediction mode supported by the MPEG-2 syntax. This is the special prediction mode referred to as dual prime prediction. The basic idea of dual prime prediction is to code a set of field motion vectors with a scaling to a near or far field, plus a transmitted delta value. Due to the correlation of adjacent pixels, the dual prime coding of field vectors can save the number of bits used for field motion vectors. The dual prime prediction is shown in Figure 16.13. In Figure 16.13, the value of one field motion vector and the value of the delta motion vector are transmitted; the motion vectors for other field are derived from the above two values.

It should be noted that only the P-picture is allowed to use dual prime prediction. In other words, if the dual prime prediction is used in the encoder, there will be no B-pictures. The reason for this restriction is to limit the required memory bandwidth for a real system implementation.

### 16.2.2.2   Field/Frame DCT Coding Syntax

Another important feature to support interlaced material is to allow adaptive selection of the field/frame DCT coding as shown in Figure 16.14.

In Figure 16.14, the middle is a luminance macroblock of $16 \times 16$ pixels, the black rectangular represents the 8 pixels in the top field and the white rectangular represents the 8 pixels in the bottom field. The left is the field DCT in which each $8 \times 8$ block contains only the pixels from the same field. The eight in the frame DCT, each $8 \times 8$ block contains the pixels from both top field and bottom field.

At the macroblock level for interlaced video, the field-type DCT may be selected when the video scene contains less detail and experiences large motion. Since the difference between adjacent fields may be large when there is large motion between fields, it may be more efficient to group the fields together, rather than the frames. In this way, the possibility that there exists more correlation among the fields can be exploited. Ultimately, this can provide much more efficient

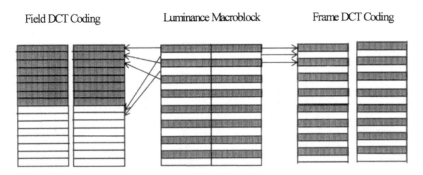

**FIGURE 16.14**   Frame and field DCT for interlaced video.

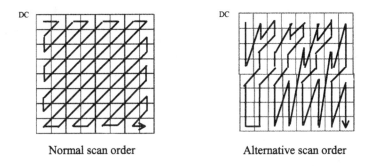

**FIGURE 16.15**   Two zigzag scan methods for MPEG-2 video coding.

coding since the block data are represented with fewer coefficients, especially if there is not much detail contained in the scene.

### 16.2.2.3   Downloadable Quantization Matrix and Alternative Scan Order

A new feature in MPEG-2 regarding the quantization matrix is that it can be downloaded for every frame. This may be helpful if the input video characteristics are very dynamic. In general, the quantizer matrices are different for intracoding and nonintracoding. With 4:2:0 format, only two matrices are used, one for the intrablocks and another for the nonintrablocks. With 4:2:2 or 4:4:4 formats four matrices are used, both an intra- and a nonintramatrix are used for the luminance and chrominance blocks. If the matrix load flags are not set, the decoder will use default matrices. The formats 4:2:0, 4:2:2 are defined in Chapter 15. In the 4:4:4 format, the luminance and two chrominance pictures have the same picture size.

In the picture layer, there is a flag that can be set for an alternative scan of DCT blocks, instead of using the zigzag scan discussed earlier. Depending on the spectral distribution, the alternative scan can yield run lengths that better exploit the multitude of zero coefficients. The zigzag scan and alternative scan are shown in Figure 16.15.

The normal zigzag scan is used for MPEG-1 and as an option for MPEG-2. The alternative scan is not supported by MPEG-1 and is an option for MPEG-2. For frame-type DCT of interlaced video, more energy may exist at the bottom part of the block; hence the run-length coding may be better off with the alternative scan.

### 16.2.2.4   Pan and Scan

In MPEG-2 there are several parameters defined in the sequence display extension and picture display extension. These parameters are used to display a specified rectangle within a reconstructed

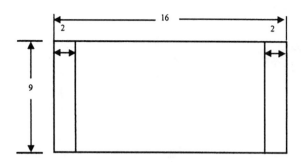

**FIGURE 16.16**    An example of pan-scan.

frame. They include display horizontal size and display vertical size in the sequence display extension, and frame center horizontal offset and frame center vertical offset in the picture display extension. A typical example using pan-scan parameters is the conversion of a 16:9 frame to a 4:3 frame. The 4:3 region is defined by display horizontal size and display vertical size, and the 16:9 frame is defined by horizontal size and vertical size. If we choose the display horizontal size to be 4 pixels less than the horizontal size, and keep the display vertical size as the same as the vertical size, then we can obtain a 4:3 pictures on the display. Figure 16.16 shows the conversion of 16:9 to the 4:3 frame using the pan-scan parameter, but there is no center offset involved in this example.

### 16.2.2.5   Concealment Motion Vector

The concealment motion vector is a new tool supported by MPEG-2. This tool is useful in concealing errors in the noisy channel environment where the transmitted data may be lost or corrupted. The basic idea of a concealment motion vector is that the motion vectors are sent for the intracoded macroblock. These motion vectors are referred to as concealment motion vectors (CMV) which should be used in macroblocks immediately below the one in which the CMV occurs. The details are described in the section about error concealment.

### 16.2.2.6   Scalability

MPEG-2 video has several scalable modes, which include spatial scalability, temporal scalability, SNR (signal-to-noise ratio) scalability, and data partitioning. These scalability tools allow a subset of any bitstream to be decoded into meaningful imagery. Moreover, scalability is a useful tool for error resilience on prioritized transmission media. The drawback of scalability is that some coding efficiency is lost as a result of extra overhead. Here, we briefly introduce the basic notions of the above scalability features.

Spatial scalability allows multiresolution coding, which is suitable for video service internet-working applications. In spatial scalability, a single video source is split into a base layer (lower spatial resolution) and enhancement layers (higher spatial resolution). For example, a CCIR601 video can be down-sampled to SIF format with spatial filtering, which can serve as the base layer video. The base layer or low-resolution video can be coded with MPEG-1 or MPEG-2, and the higher-resolution layer must be coded by MPEG-2-supported syntax. For the up-sampled lower layer, an additional prediction mode is available in the MPEG-2 encoder. This is a flexible technique in terms of bit rate ratios, and the enhancement layer can be used in high-quality service. The problem with spatial scalability is that there exists some bit rate penalty due to overhead and there is also a moderate increase in complexity. A block diagram that illustrates encoding with spatial scalability is shown in Figure 16.17. In Figure 16.17, the output of decoding and spatial up-sampling block provides an additional choice of prediction for the MPEG-2 compatible coder, but not the only choice of prediction. The prediction can be obtained from HDTV input itself, also depending on the prediction select criterion such as the minimum prediction difference.

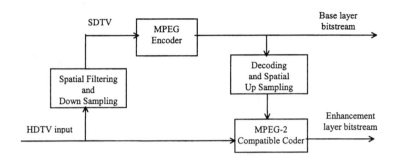

**FIGURE 16.17**   Block diagram of spatial scalability encoder.

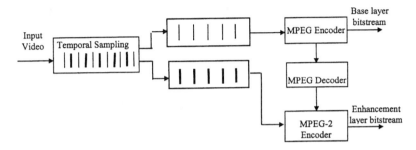

**FIGURE 16.18**   Block diagram of temporal scalability.

It should be noted that the spatial scalability coding allows the base layer to be coded independently from the enhancement layer. In other words, the base layer or lower layer bitstream is generated without regard for the enhancement layer and can be decoded independently. The enhancement layer bitstream is additional information, which can be seen as the prediction error based on the base layer data. This implies that the enhancement layer is useless without the base. However, this type of structure can find a lot of applications such as error concealment, which will be discussed in the following section.

Temporal scalability is a scalable coding technique in the temporal domain. An example of a two-layer temporal scalable coder is shown in Figure 16.18. The example uses temporal scalability to decompose the progressive image sequence to two interlaced image sequences; then one is coded as the base layer and one as the enhancement layer. Of course, the decomposition could be different. For the enhancement layer, there are two choices in making predictions. One choice for prediction is available between frames of base layer and enhancement layer, and the other is between frames from the enhancement layer itself. It should be noted that the spatial resolution of two layers is the same and the combined temporal rate of two layers is the full temporal rate of the source. Again, it should be noted that the decoding output of the base layer bitstream by the MPEG decoder provides an additional choice of prediction but not the only choice of predictions.

The SNR scalability provides a mechanism for transmitting two-layer service with the same spatial resolution but different quality levels. The lower layer is coded at a coarse quantization step at 3 to 5 Mbps to provide NTSC/PAL/SECAM-quality video for low-capacity channels. In the enhancement layer, the difference between original and the coarse-quantized signals is then coded with a finer quantizer to generate an enhancement bitstream for high-quality video applications.

The above three scalability schemes all generate at least two bitstreams, one for the base layer and the other for the enhancement layer, and the lower-layer bitstream can be independently decoded to provide low spatial resolution, low quality, or low frame rate video, respectively. There is another scalability scheme, data partitioning, in which the base layer bitstream cannot be independently decoded. In data partitioning, a single video source is split into a high-priority portion, which can be better protected, and low-priority portion, which is less important with regard to the reconstructed video quality. The priority breakpoint in the syntax specifies which syntax elements are coded as low priority (for example, the higher-order DCT coefficients in the intercoded blocks).

## 16.3  MPEG-2 VIDEO ENCODING

### 16.3.1  INTRODUCTION

MPEG video compression is a generic standard that is essential for the growth of the digital video industry, as mentioned previously. Although the MPEG video coding standard recommended a general coding methodology and syntax for the creation of a legitimate MPEG bitstream, there are many areas of research left open regarding how to generate high-quality MPEG bitstreams. This allows the designers of an MPEG encoder great flexibility in developing and implementing their own MPEG-specific algorithms, leading to product differentiation on the marketplace. To design a performance-optimized MPEG-2 encoder system, several major areas of research have to be considered. These include image preprocessing, motion estimation, coding mode decisions, and rate control. Algorithms for all of these areas in an encoder should aim to minimize subjective distortion for a prescribed bit rate and operating delay constraint. The preprocessing includes the noise reduction and the removal of redundant fields, which are contained in the detelecine material. The telecine material is used for the movie industry, which contains 24 progressive frames/second. The TV signal is 30 frames/second. The detelecine process converts the 24-frames/second film signal to the 30-frames/second TV signal. This is also referred to as 3:2 pull-down process. Since the 30-frames/second detelecine material only contains 24 frames/second of unique pictures, the encoder has to detect and remove the redundant fields for obtaining better coding performance. The procession of noise reduction can reduce the bits wasted for coding random noise. Motion compensation is used to remove the temporal redundancy in the video signals. The motion vectors between the anchor picture and the current picture are obtained with motion estimation algorithms. Except for I-pictures each macroblock can be inter- or intracoded, which is determined by the mode decision. The investigation of motion estimation algorithms is an important research topic since different motion estimation schemes may result in different coding efficiency. Rate control is always applied for non-variable-bit rate (non-VBR) coding. The purpose of rate control is to assign the bits for each macroblock properly under the constraints of total bit rate budget and buffer size. This is also an important topic since the optimized bit assignment scheme will result in better coding performance and better subjective reconstruct quality at a given bit rate. In this section, areas of preprocessing and motion estimation are covered. The topics of rate control and optimum mode decision are discussed in later sections.

### 16.3.2  PREPROCESSING

For low-bit-rate video coding, preprocessing is sometimes applied to the video signals before coding to increase the coding efficiency. Usually, preprocessing implies a filtering of the video signals that are corrupted by random and burst noise for various reasons, such as imperfections of the scanner, transmission, or recording medium. Noise reduction not only improves the visual quality but also increases the performance of video coding. Noise reduction can be achieved by filtering each frame independently. There are a variety of spatial filters which have been developed for image noise filtering and restoration that can be used for noise reduction task (Cano and Benard, 1983; Katsaggelos et al., 1991). On the other hand, it is also possible to filter the video sequence temporally along the motion trajectories using motion compensation (Sezan et al., 1991). However, it was shown that among the recursive stationary methods the motion-compensated spatiotemporal filtering performed better than spatial or motion-compensated temporal filtering alone (Ozkan et al., 1993).

Another important type of preprocessing is detelecine processing. Since movie material is originally shot at 24 progressive frames/second, standard conversion to television at 30 frames/second is made by a 3:2 pull-down process, which periodically inserts a repeated field, giving 30-frames/second telecine source material. The 3:2 pull-down has been described in Chapter 10, and will not be repeated here. Since the 30-frames/second detelecine material only contains 24 frames/second of unique pictures, it is necessary to detect and remove the redundant fields before or during encoding. Rather than directly encoding the 30-frames/second detelecine material, one can remove the redundant fields first and then encode 24 frames/second of unique material, thereby realizing higher coding quality at the same bit rate. The decoder can simply reconstruct the redundant fields before presenting them.

Television broadcast programmers frequently switch between telecine material and natural 30-frames/second material, such as when splicing to and from various sources of movies, ordinary television programs, and commercials. An MPEG-2 encoder should be able to cope with these transitions and consistently produce decent pictures. During movie segments, the encoder should realize the gains from coding at the lower frame rate after detelecine. Ideally, the process of source transition from the lower 24-frames/second rate to the higher 30-frames/second rate should not cause any quality drop of every encoded frame. The quality of encoded frames should maintain the same as the case where the detelecine process is ignored and all material, regardless of source type, is coded at 30 frames/second.

### 16.3.3  MOTION ESTIMATION AND MOTION COMPENSATION

In principle, for coding video signals if the motion trajectory of each pixel could be measured, then only the initial or anchor reference frame and the motion vector information need to be coded. In such a way the interframe redundancy will be removed. To reproduce the pictures, one can simply propagate each pixel along its motion trajectory. Since there is also a cost for transmitting motion vector information, in practice one can only measure the motion vectors of a group of pixels, which will share the cost for transmission of the motion information. Of course, at the same time the pixels in the same group are assumed to have the same motion information. This is not always true since the pixels in the block may move in different directions, or some of them may belong to the background. Therefore, both motion vectors and the prediction difference have to be transmitted. Usually, the block matching can be considered as the most practical method for motion estimation because of less hardware complexity. In the block-matching method, the image frame is divided into fixed-size small rectangular blocks such as $16 \times 16$ or $16 \times 8$ in MPEG video coding. Each block is assumed to undergo a linear translation and the displacement vector of each block and the predictive errors are coded and transmitted. The related issues for motion estimation and compensation include a motion vector searching algorithm, searching range, matching criteria and coding method. Although the matching criteria, and searching algorithms have been discussed in Chapter 11, we will briefly introduce them here for the sake of completeness.

### 16.3.3.1  Matching Criterion

The matching of the blocks can be determined according to the various criteria including the maximum cross-correlation, the minimum mean square error (MSE), the minimum mean absolute difference (MAD) and maximum matching pixel count (MPC). For MSE and MAD, the best matching block is reached if the MSE or MAD is minimized at that location. In practice, we use MAD instead of MSE as the matching criterion because of its computational simplicity. The minimum MSE criterion is not commonly used in hardware implementations because it is difficult to realize the square operation. However, the performance of the MAD criterion deteriorates as the search area becomes larger as a result of the presence of several local minima. In the maximum MPC criterion, each pixel in the block is classified as either a matching pixel or a mismatching pixel according to the prediction difference whether which is smaller than a preset threshold. The best matching is then determined by the maximum number of matching pixels. However, the MPC criterion requires a threshold comparator and a counter.

### 16.3.3.2  Searching Algorithm

Finding the best-matching block requires optimizing the matching criterion over all possible candidate displacement vectors at each pixel. The so-called full-search logarithmic search, and hierarchical searching algorithms can accomplish this.

*Full search*: The full-search algorithm evaluates the matching criterion for all possible values within the predefined searching window. If the search window is restricted to a $[-p, p]$ square, for each motion vector there are $(2p + 1)^2$ search locations. For a block size of $M \times N$ pixels, at each search location we compare $N \times M$ pixels. If we know the matching criterion and how many operations are needed for each comparison, then we can calculate the computation complexity of the full-search algorithm. Full search is computationally expensive, but guarantees finding the global optimal matching within a defined searching range.

*Logarithmic search*: Actually, the expected accuracy of motion estimation algorithms varies according to the applications. In motion-compensated video coding, all one seeks is a matching block in terms of some metric, even if the match does not correlate well with the actual projected motion. Therefore, in most cases, search strategies faster than full searches are used, although they lead to suboptimal solutions. These faster search algorithms evaluate the criterion function only at a predetermined subset of the candidate motion vector locations instead of all possible locations. One of these faster search algorithms is the logarithmic search. Its more popular form is referred to as the three-step search. We explain the three-step search algorithm with the help of Figure 16.19, where only the search frame is depicted. Search locations corresponding to each of the steps in the three-step search procedure are labeled 1, 2, and 3. In the first step, starting from pixel 0 we compute MAD for the nine search locations labeled 1. The spacing between these search locations here is 4. Assume that MAD is minimum for the search location (4,4) which is circled 1. In the second step, the criterion function is evaluated at eight locations around the circled 1 which are labeled 2. The spacing between locations is now 2 pixels. Assume now the minimum MAD is at the location (6,2), which is also circled. Thus, the new search origin is the circled 2, which is located at (6.2). For the third step, the spacing is now set to 1 and the eight locations labeled 3 are searched. The search procedure is terminated at this point and the output of the motion vector is (7,1). Additional steps may be incorporated into the procedure if we wish to obtain subpixel accuracy in the motion estimations. Then, the search frame needs to be interpolated to evaluate the criterion function at subpixel locations.

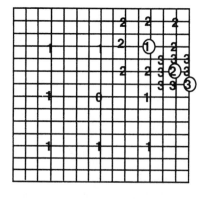

**FIGURE 16.19**  Three-step search.

*Hierarchical motion estimation*: Hierarchical representations of images in the form of a Laplacian pyramid or wavelet transform are also quite often used with the block-matching method for improved motion estimation. The basic idea of hierarchical block matching is to perform motion estimation at each level successively, starting with the lowest resolution level. The lower resolution levels serve to determine a rough estimate of the motion information using relatively larger blocks. The estimate of the motion vector at a lower resolution level is then passed onto the next higher resolution level as an initial estimate. The higher resolution levels are used to fine-tune the motion vector estimate. At higher resolution levels, relatively smaller window sizes can be used since we start with a good initial estimate. The hierarchical motion estimate can significantly reduce the implementation complexity since its search method is very efficient. However, such a method requires increased storage because of the need to keep pictures at different resolutions. Furthermore, this scheme may yield inaccurate motion vectors for regions containing small objects. Since the search starts at the lowest resolution of the hierarchy, regions containing small objects may be eliminated and thus fail to be tracked. On the other hand, the creation of low-resolution pictures provides some immunity to noise. Results of experiments performed by one of the authors have shown that, compared with full-search, the two-layer hierarchical motion estimation reduces the search complexity of factor 10 at the price of degrading reconstruction quality from about 0.2 to 0.6 dB for frame-mode coding, from 0.26 to 0.38 dB for field-mode coding, and only 0.16 to 0.37 dB for frame/field adaptive coding, for different video sequences in the case of a fixed bit rate of 4 Mbps. In the case of VBR coding, similar results can be observed from the rate distortion curves.

In the above discussion, we have restricted the motion vector estimation to integer pixel grids, or pixel accuracy. Actually, the motion vectors can be estimated with fractional or subpixel accuracy. In MPEG-2 video coding the half-pixel accuracy motion estimation can be used. Half-pixel accuracy can easily be achieved by interpolating the current and reference pictures by a factor of two and then using any of the motion estimation methods described previously.

### 16.3.3.3  Advanced Motion Estimation

Progress has recently been made in several aspects of motion estimation, which are described as follows.

*Motion estimation using a reduced set of image data*: The methods to reduce search complexity with subsampling and pyramid processing are well known and can be found in the literatures (Sun, 1994). However, the reduction by lowering the precision of each sample does not appear to have been extensively studied. Some experimental results have shown that performance degradation of the hierarchical motion estimation algorithm is not serious when each layer up to a four-layer pyramid is limited to 6 bits/sample. At 4 to 5 bits/sample the performance is degraded 0.2 dB over full precision.

*Overlapped motion estimation* (Katto et al., 1994): A limitation of block matching is that it generates a significant proportion of motion vectors that do not represent the true motion present in the scene. One possible reason is that the motion vectors are estimated without reference to any picture data outside of the nonoverlapping blocks. This problem has been addressed by overlapped motion estimation. In the case of the overlapped motion compensation, motion-compensated regions translated by the motion vectors are overlapped with each other. Then, a window function is used to determine the weighting factors for each vector. This technique has been adopted into the H.263 video coding standard. Some improvements have been clearly identified for low-bit-rate coding.

*Frequency domain motion estimation*: An alternative to spatial-domain block-matching methods is to estimate motion vectors in the frequency domain through calculating the cross-correlation (Young and Kingsbury, 1993). Most international standards, such as MPEG, H.263, as well as the proposed HDTV standard, use the DCT and block-based motion estimation as essential elements to achieve spatial and temporal compression, respectively. The new motion estimation approach is proposed in the DCT domain (Koc and Liu, 1998). This method of motion estimation has certain merits over conventional methods. It has very low computational complexity and is robust even in a noisy environment. Moreover, the motion-compensation loop in the encoder is much simplified due to replacing the IDCT out of the loop (Koc and Liu, 1998).

*Generalized block matching*:  In generalized block matching, the encoded frame is divided into triangular, rectangular, or arbitrary quadrilateral patches. We then search for the best-matching triangular or quadrilateral patch in the search frame under a given spatial transformation. The choice of patch shape and the spatial transform are mutual related. For example, triangular patches offer sufficient degrees of freedom with affine transformation, which has only six independent parameters. The bilinear transform has eight free parameters. Hence, it is suitable for use with rectangular or quadrilateral patches. Generalized block matching is usually only adaptively used for those blocks where standard block matching is not satisfactory for avoiding imposed computational load.

## 16.4   RATE CONTROL

### 16.4.1   INTRODUCTION OF RATE CONTROL

The purpose of rate control is to optimize the perceived picture quality and to achieve a given constant average bit rate by controlling the allocation of the bits. From the viewpoint of rate control, the encoding can be classified into VBR coding and constant bit rate (CBR) coding. The VBR coding can provide a constant picture quality with variable coding bit rate, while the CBR will provide a constant bit rate with a nonuniform picture quality. Rate control and buffer regulation is an important issue for both VBR and CBR applications. In the case of VBR encoding, the rate controller attempts to achieve optimum quality for a given target rate. In the case of CBR encoding and real-time application, the rate control scheme has to satisfy the low-latency and VBV (video buffering verifier) buffer constraints. The VBV is a hypothetical decoder, which is conceptually connected to the output of an encoder (see Appendix C of ISO/IEC, 1995). The bitstream generated by the encoder is placed into the VBV buffer at the CBR rate that is being used. The rate control has to assure that the VBV will not be overflow or underflow. In addition, the rate control scheme has to be applicable to a wide variety of sequences and bit rates. At the GOP level, the total number of available bits is allocated among the various picture types, taking into account the constraints of the decoder buffer, so that the perceived quality is balanced. Within each picture, the available bits are allocated among the macroblocks to maximize the visual quality and to achieve the desired target of encoded bits for the whole picture.

### 16.4.2   RATE CONTROL OF TEST MODEL 5 FOR MPEG-2

As we described before, the standard only defines the syntax for decoding. The test model is an example of the encoder, which may not be optimal; however, it can provide a compliant compressed bitstream. Also, the test model served as a reference during the development of the standard. The TM5 rate control algorithm consists of three steps to adapting the macroblock quantization parameter for controlling the bit rate.

#### 16.4.2.1   Step 1: Target Bit Allocation

The target bit allocation is the first step of rate control. Before coding a picture, we need to estimate the number of bits available for coding this picture. The estimation is based on several factors. These include the picture type, buffer fullness, and picture complexity. The estimation of picture complexity is based on the number of bits and quantization parameter used for coding the same type of previous picture in the GOP. The initial complexity values are given according to the type of picture:

$$X_i = 160 * \text{bit-rate}/115$$

$$X_p = 60 * \text{bit-rate}/115 \qquad (16.6)$$

$$X_b = 42 * \text{bit-rate}/115,$$

where the subscripts $i$, $p$, and $b$ stand for picture types I, P, and B (this will be applied to the formulas in this section). After a picture of a certain type ($I$, $P$, or $B$) is encoded, the respective "global complexity measure" ($X_i$, $X_p$, and $X_b$) is updated as

$$X_i = S_i Q_i, X_p = S_p Q_p, \text{ and } X_b = S_b Q_b, \qquad (16.7)$$

where $S_i$, $S_p$, $S_b$ are the number of bits generated by encoding this picture and $Q_i$, $Q_p$, $Q_b$ are the average quantization parameters computed the actual quantization values used during the encoding of all the macroblocks including the skipped macroblocks. This estimation is very intuitive since, if the picture is more complicated, more bits are needed to encode it. The quantization parameter (step or interval) is used to normalize this measure because the number of bits generated by the encoder is inversely proportional to the quantization step. The quantization step can also be considered as a measure of coded picture quality. The target number of bits for the next picture in the GOP ($T_i$, $T_p$, and $T_b$) is computed as follows:

$$T_i = \max \left\{ \frac{R}{1 + \dfrac{N_p X_p}{X_i K_p} + \dfrac{N_b X_b}{X_i K_b}}, \text{bit-rate}/8 * \text{picture-rate} \right\}$$

$$T_p = \max \left\{ \frac{R}{N_p + \dfrac{N_b K_p X_b}{X_b K_p}}, \text{bit-rate}/8 * \text{picture-rate} \right\} \qquad (16.8)$$

$$T_b = \max \left\{ \frac{R}{N_b + \dfrac{N_p K_p X_p}{X_p K_p}}, \text{bit-rate}/8 * \text{picture-rate} \right\},$$

where $K_p$ and $K_b$ are "universal" constants dependent on the quantization matrices. For the matrices of TM5, $K_p = 1.0$ and $K_b = 1.4$. The $R$ is the remaining number of bits assigned to the GOP and after coding the picture this number is updated by subtracting the bit used for the picture. $N_p$ and $N_b$ are the number of P-pictures and B-pictures remaining in the current GOP in the encoding order. The problem of the above target bit assignment algorithm is that it does not handle scene changes efficiently.

### 16.4.2.2   Step 2: Rate Control

Within a picture, the bits used for each macroblock is determined by the rate control algorithm. Then a quantizer step is derived from the number of bits available for the macroblock to be coded. The following is an example of rate control for P-picture.

In Figure 16.20, $d_0^p$ is initial virtual buffer fullness, the $T_p$ is the target bits for P-picture. $B_j$ is the number of bits generated by encoding all macroblocks in the picture up to and including $j$th macroblock. $MB\_cnt$ is the number of macroblocks in the picture. Before encoding the $j$th macroblock the virtual buffer fullness is adjusted during the encoding according to the following equation for the P-picture:

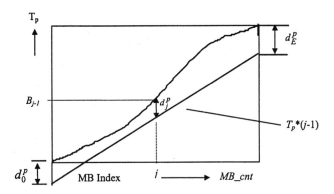

**FIGURE 16.20**   Rate control for P-picture. (From ISO/IEC, MPEG-2, Test Model 5, April 1993. With permission.)

$$d_j^p = d_0^p + B_{j-1} - \frac{T_p(j-1)}{MB\_cnt}. \tag{16.9}$$

Then the quantization step is computed with the equation:

$$Q_j^p = \frac{d_j^p}{r}, \tag{16.10}$$

where the "reaction parameter" $r$ is given by $r = 2 * bit\text{-}rate/picture\text{-}rate$ and $d_j^p$ is the fullness of the appropriate virtual buffer. This procedure is shown in Figure 16.20. The fullness of the virtual buffer for the last macroblock is used for encoding the next picture of the same type as the initial fullness.

The above example can be extended to the general case for all I-, P-, and B-pictures. Before encoding the $j$th macroblock, we compute the fullness of the appropriate virtual buffer:

$$d_j^i = d_0^i + B_{j-1} - \frac{T_i(j-1)}{MB\_cnt} \quad \text{or}$$

$$d_j^p = d_0^p + B_{j-1} - \frac{T_p(j-1)}{MB\_cnt} \quad \text{or} \tag{16.11}$$

$$d_j^b = d_0^b + B_{j-1} - \frac{T_b(j-1)}{MB\_cnt}.$$

Depending on the picture type, where $d_0^i$, $d_0^p$, $d_0^b$ are initial fullness of the virtual buffers and $d_j^i$, $d_j^p$, $d_j^b$ are the fullness of virtual buffer at $j$th macroblock — one for each picture type. From the number of bits of the virtual buffer fullness, we compute the quantization step $Q_j$ for macroblock $j$ according to the buffer fullness:

$$Q_j = \frac{d_j * 31}{r}. \tag{16.12}$$

The initial values of the virtual buffer fullness are

$$d_0^i = 10 \cdot r/31$$

$$d_0^p = K_p \cdot d_0^i \qquad (16.13)$$

$$d_0^b = K_b \cdot d_0^i$$

$K_p$ and $K_b$ are constants which are defined in Equation 16.8.

### 16.4.2.3 Step 3: Adaptive quantization

Adaptive quantization is the last step of the TM5 rate control. It is noted that for active areas or busy areas, the human eyes are not so sensitive to the quantization noise, while the smooth areas are more sensitive to the quantization noise as discussed in Chapter 1. Based on this observation we modulate the quantization step obtained from the previous step in such a way to increase the quantization step for active areas and reduce the quantization step for the smooth areas. In other words, we use more bits in the smooth areas and fewer bits for the active areas. The experiment results have shown that the subjective quality is higher with the adaptive quantization step than without this step. The procedure of adaptive quantization in TM5 is as follows. First, the spatial activity measure for the *j*th macroblock is calculated from the four luminance frame–organized subblocks and the four luminance field–organized blocks using the intrapixel values:

$$act_j = 1 + \operatorname*{Min}_{sblk=1,8} (var\_sblk), \qquad (16.14)$$

where *var_sblk* is the variance of each spatial $8 \times 8$ block, which value is calculated as

$$var\_sblk = \frac{1}{64} \sum_{k=1}^{64} (P_k - P_{mean})^2 \qquad (16.15)$$

and $P_k$ is the pixel value in the original $8 \times 8$ block and $P_{mean}$ is the mean value of the block which is calculated as

$$P_{\text{mean}} = \frac{1}{64} \sum_{k=1}^{64} P_k. \qquad (16.16)$$

The normalized activity factor $N\_act_j$ is

$$N\_act_j = \frac{2 \cdot act_j + avg\_act}{act_j + 2 \cdot avg\_act} \qquad (16.17)$$

where *avg_act* is the average value of $act_j$ the last picture to be encoded. Therefore, this value will not give good results when a scene change occurs. On the first picture, this parameter takes the value of 400. Finally, we can obtain the modulated quantization step for *j*th macroblock:

$$mquant_j = Q_j \cdot N\_act_j \qquad (16.18)$$

where $Q_j$ is the reference quantization step value obtained in the last step. The final value of *mquant_j* is clipped to the range of [1,31] and is used and coded as described in the MPEG standard.

As we indicated before, the TM5 rate control provides only a reference model. It is not optimized in many aspects. Therefore, there is still a lot of room for improving the rate control algorithm, such as to provide more precise estimation of average activity by preprocessing. In the following section, we will investigate the optimization problem for mode decision combined with rate control, which can provide a significant quality improvement as shown by experimental results.

## 16.5  OPTIMUM MODE DECISION

### 16.5.1  PROBLEM FORMATION

This section addresses the problem of determining the optimal MPEG (ISO/IEC, 1995) coding strategy in terms of the selection of macroblock coding modes and quantizer scales. In the TM5, the rate control operates independently from the coding mode selection for each macroblock. The coding mode is decided based only upon the energy of predictive residues. Actually, the two processes, coding mode decision and rate control, are intimately related to each other and should be determined jointly in order to achieve optimal coding performance. A constrained optimization problem can be formulated based on the rate–distortion characteristics, or $R(D)$ curves, for all the macroblocks that compose the picture being coded. Distortion for the entire picture is assumed to be decomposable and expressible as a function of individual macroblock distortions, with this being the objective function to minimize. The determination of the optimal solution is complicated by the MPEG differential encoding of motion vectors and dc coefficients, which introduce dependencies that carry over from macroblock to macroblock for a duration equal to the slice length. As an approximation, a near-optimum greedy algorithm can be developed. Once the upper bound in performance is calculated, it can be used to assess how well practical suboptimum methods perform.

Prior related work dealing with dependent quantization for MPEG include the work done by Ramchandran et al. (1994) and Lee and Dickerson (1994). Those works treated the problem of bit allocation where there is temporal dependency in coding complexity across I-, P-, and B-frames. While these techniques represent the most proper bit allocation strategies across frames from a theoretical viewpoint, no practical real-time MPEG encoding system will use even those proposed simplified techniques because they require an unwieldy number of preanalysis encoding passes over the window of dependent frames (one MPEG GOP). To overcome these computational burdens, more pragmatic solutions that can realistically be implemented have been considered by Sun et al. (1997). In this work, the major emphasis is not on the problem of bit allocation among I-, P-, and B-frames; rather, the authors choose to utilize the frame-level allocation method provided by the TM5. In this way, frame-level coding complexities are estimated from past frames without any forward preanalysis knowledge of future frames. This type of analysis forms the most reasonable set of assumptions for a practical real-time encoding system. Another method that extends the basic TM5 idea to alter frame budgets heuristically in the case of scene changes, use of dynamic GOP size, and temporal masking effects can be found in Wang (1995). These techniques also offer very effective and practical solutions for implementation. Given the chosen method for frame-level bit budget allocation, the focus of this section is to optimize macroblock coding modes and quantizers jointly within each frame.

There exists many choices for the macroblock coding mode under the MPEG-2 standard for P- and B-pictures, including intramode, no-motion-compensation mode, frame/field/dual-prime motion compensation intermode, forward/backward/average intermode, and field/frame DCT mode. In the standard TM5 reference (ISO/IEC, 1993), the coding mode for each macroblock is selected by comparing the energy of predictive residuals. For example, the intra/inter decision is determined by a comparison of the variance of the macroblock pixels against the variance of the predictive

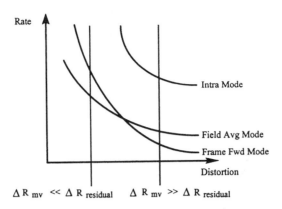

**FIGURE 16.21**   R(D) curves for different macroblock coding modes.

residuals; the interprediction mode is selected to be the intermode that has the least predictive residual MSE. The coding mode selected by the TM5 criteria does not result in the optimal coding performance.

In attempting to achieve optimal coding performance, it is important to realize that coding modes should be determined jointly with rate control because the best coding mode depends upon the operating point for rate. In deciding which of the various coding modes is best, one should consider what the operating point is for distortion, and also consider the trade-off between spending bits for coding the prediction residuals and bits for coding motion vectors.

The number of bits used for coding the macroblock is the sum of bits used for coding motion vectors and bits used for coding residuals:

$$R_{MB} = R_{mv} + R_{\text{residual}} \tag{16.19}$$

For example, in Figure 16.21, consider the decision between (1) frame-mode forward prediction and (2) field-mode bidirectional prediction. Mode (2) will almost always produce a prediction that has lower MSE than mode (1). However, mode (1) requires coding of fewer motion vectors than mode (2). Which mode is best? The answer depends on the operating point for distortion. When coding at a very coarse quant scale, mode (1) can perform better than mode (2) because the difference in bits required for coding motion vectors between the two modes may be much greater than the difference in bits required for coding residuals between the two modes. However, when coding at a fine quant scale, mode (2) can perform better than mode (1) because mode (2) provides a better prediction and the bits required for motion vectors would become negligible compared with bits for coding residuals.

Coding mode decisions and rate control can be determined jointly and optimally starting from the basics of constrained optimization using $R(D)$ curves. This optimal solution would be an *a posteriori* solution that assumes complete knowledge of $R(D)$. We investigate an optimal solution for objective functions of the form:

$$D_{PICT} = \sum_{i=1} D_{MBi}, \tag{16.20}$$

which states that the distortion for the picture, $D_{PICT}$, can be measured as an accumulation of individual macroblock distortions, $D_{MB}$, for all *NMB* number of macroblocks in the picture. We minimize this objective function subject to having individual macroblock distortions being uniform over the picture:

$$D_1 = D_2 = \cdots = D_{NMB} \qquad (16.21)$$

and having the bits generated from coding each macroblock, $R_{MB}$, sum to a target bit allocation for the entire picture, $R_{PICT}$:

$$\sum_{i=1} R_{MBi} = R_{PICT} \qquad (16.22)$$

The choice for the macroblock distortion measure, $D_{MB}$, can be the MSE computed over the pixels in the macroblock, or it can be a measure that reflects subjective distortion more accurately, such as luminance- and frequency-weighted MSE. Other choices for $D_{MB}$ may be the quantizer scale used for coding the macroblock, or, better yet, the quantizer scale weighted by an activity-masking factor. In this chapter, we select distortion for each macroblock $i$ to be a spatial-masking-activity-weighted quantizer scale:

$$D_{MBi} = qscale_i / N\_act_i, \qquad (16.23)$$

where $N\_act_i \in [0.5, 2.0]$ is the normalized spatial masking activity quantizer weighting factor, as defined in the TM5:

$$N\_act_i = \frac{2 * act_i + avg\_act}{act_i + 2 * avg\_act}, \qquad (16.24)$$

where $act_i$ is the minimum luma block spatial variance for macroblock $i$ and $avg\_act$ is the average value of $act_i$ over the last picture to be coded. $N\_act_i$ reflects the relative amount of quantization error that can be tolerated for macroblock $i$ as compared with the rest of the macroblocks that compose the picture. $N\_act_i$ depends strongly on whether the macroblock belongs to a smooth, edge, or textured region of the picture. Hence, the macroblock distortion metric is space variant and depends on the context of the local picture characteristics surrounding each macroblock. We assume that maintaining the same $D_{MBi}$ for all macroblocks, or selecting the quantizer scales directly proportional to $N\_act_i$ in such a manner, corresponds to maintaining uniform subjective quality throughout the picture. The masking-activity-weighted quantizer scale is a somewhat coarse measure for image quality, but it reflects subjective image quality better than MSE or PSNR (peak signal-to-noise ratio), and it is a practical metric to compute that lends itself to an additive form for distortion.

It is important to note that the resulting distortion measure for the picture $D_{PICT}$ is really only meaningful as a relative comparison figure for the same identical picture (thus having the same masking activities) quantized different ways. It is not useful comparing two different images. PSNR is only useful in this sense too, although with poorer subjective accuracy.

In the following, a procedure for obtaining the optimal coding performance with the joint optimization of coding mode selection and rate control is discussed. Since this method would be too complex to implement, a practical suboptimal heuristic algorithm is presented. Some simulation results and comparisons between the different algorithms — TM5 algorithm, near-optimum algorithm, and the practical suboptimum algorithm are also provided to assist the reader in understanding the differences in performance.

## 16.5.2    PROCEDURE FOR OBTAINING THE OPTIMAL MODE

### 16.5.2.1    Optimal Solution

The solution to the optimization problem is unique because the objective function is monotonic and the individual macroblock $R(D)$ functions are also monotonic. In order to solve for the optimal set of macroblock modes and quant-scales for the picture ($\overrightarrow{mode}$ and $\overrightarrow{qscale}$), the differential encoding of motion vectors and intra-dc coefficients as done in MPEG should be accounted for. According to MPEG, each slice has its own differential encoding chain. At the start of each slice, prediction motion vectors are reset to zero. As each macroblock is encoded in raster scan order, the macroblock motion vectors are encoded differentially with respect to prediction motion vectors that depend on the coding mode of the previous macroblock. These prediction motion vectors may be reset to zero in the case that the previous macroblock was coded as intra or skipped. Similarly, dc coefficients in continuous runs of intramacroblocks are encoded differentially with respect to the previous intramacroblock. The intra dc predictors are reset at the start of every slice, and at inter or skipped macroblocks. Slice boundaries delimit independent self-contained decodable units. Finding the optimal set of coding modes for the macroblocks in each slice entails a search through a trellis of dimensions $S$ stages by $M$ states per stage, with $S$ being the slice size and $M$ being the number of coding modes being considered (Figure 16.22). This trellis structure arises because there are $M^2$ distinct rate distortion, $R_{mode|previous\text{-}mode}(D)$, characteristic curves corresponding to each of $M$ coding modes, with each in turn having a different dependency for each of $M$ coding modes of the previous macroblock. We now consider populating the trellis links with values by sampling the set of these $M^2 S$ rate–distortion curves at a specific distortion level. For a given fixed macroblock distortion level, $D_{MB}$, each link on the trellis is assigned a cost equal to the number of bits to code a macroblock in a certain mode given the mode from which the preceding macroblock was coded. For any group of links entering a node, the cost of these links differs only because of the difference in bits caused by the motion vector and dc coefficient coding dependency upon the prior macroblock.

The computational requirements per slice involve:

- To determine link costs in the trellis, the number of "code the macroblock" operations (i.e., DCT + Quantization + RLC/VLC) is equal to $M^2 S$.
- After determining all trellis link costs, the number of path searches is equal to $M^S$.

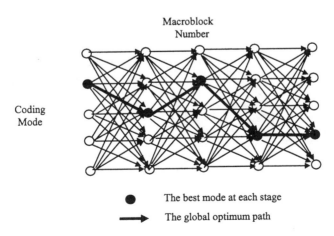

**FIGURE 16.22**   Full-search trellis, $M^S$ ($M$ is number of modes at each stage and $S$ is the length of slice) searches needed to obtain the best path.

A general iterative procedure for obtaining the optimal solution is as follows:

1. Initialize a guess for $D_{MB} = D_{MB0}$. Since $D_{MB}$ is the same for every macroblock in the picture, this sets an initial guess for the operating distortion level of the picture.
2. Perform for each slice in the picture:
   - For each macroblock in the slice and the mode considered, determine the quantizer scale which yields the distortion level $D_{MB}$, i.e., $q_s = f(D_{MB})$, where $f$ is the function that describes the relationship between quantizer scale $q_s$ and distortion $D_{MB}$. If we use the spatial-masking-activity-weighted quantizer scale as a measure of distortion (as from Equation 16.4), then $q_s$ equals $N\_act * D_{MB}$.
   - Compute all the link costs in the trellis representing the slice. The link costs, $R_{MBi}$ (mode $k$ | mode $j$), represents the number of resulting bits (total bits for coding residual, motion vectors, and macroblock header) for coding macroblock $i$ in mode $k$ given that the preceding macroblock was coded in mode $j$.
   - Search through the trellis to find the path that has the lowest $\Sigma R_{MBi}$ over the slice.
3. Compute $\Sigma R_{MBi}$ for all macroblocks in the picture and compare to target $R_{PICT}$.
   - If $|\Sigma R_{MBi} - R_{PICT}| < \varepsilon$, then the optimal $\overrightarrow{\text{mode}}$ and $\overrightarrow{\text{qscale}}$ has been found for picture. Repeat the process for the next picture.
   - If $\Sigma R_{MBi} < R_{PICT}$, then decrement $D_{MB} = D_{MB} - \Delta D_{MB}$ and go to step 2.
   - If $\Sigma R_{MBi} > R_{PICT}$, then increment $D_{MB} = D_{MB} + \Delta D_{MB}$ and go to step 2.

### 16.5.2.2  Near-Optimal Greedy Solution

The solution from the full exponential-order search requires an unwieldy amount of computations. To avoid the heavy computational burden, we can use a greedy approach (Lee and Dickerson, 1994) to simplify and sidestep the dependency problems of the full-search method. In the greedy algorithm, the best coding mode selection for the current macroblock depends only upon the best mode of the previous coded macroblock. Therefore, the upper bound we obtain is a near-optimum solution instead of a global optimum. Figure 16.23 illustrates the greedy algorithm. After coding a macroblock in each of the $M$ modes, the mode resulting in the least number of bits is chosen to be "best." The very next macroblock is coded with dependencies to that chosen "best" mode. The computations per slice are reduced to $M \times S$ "code the macroblock" operations and $M \times S$ comparisons. A general iterative procedure for obtaining the greedy solution is as follows:

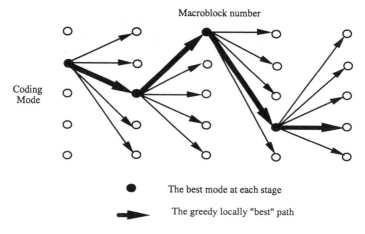

FIGURE 16.23   Greedy approach, $M \times S$ comparisons needed to obtain the locally "best" path.

1. Initialize a guess for $D_{MB} = D_{MBO}$.
2. Perform for each macroblock:
   - For each mode considered, determine the quantizer scale that yields the distortion level $D_{MB}$, i.e., $q_s = f(D_{MB})$, where $f$ is the function we mentioned previously.
   - For each mode, code the macroblock in that mode with that $q_s$ value and record the resulting number of generated bits, $R_{MBi(\text{mode } il \text{ mode } j)}$. The macroblock is coded based on the already determined mode of the preceding macroblock.
   - The "best" mode for macroblock $i$ is the mode for which $R_{MBi(\text{mode } il \text{ mode } j)\,\text{mode}}$ is smallest. This yields $R_{MBi}$ bits for macroblock $i$.
3. Compute $\Sigma R_{MBi}$ for all macroblocks in the picture and compare to target $R_{PICT}$.
   - If $|\,\Sigma R_{MBi} - R_{PICT}\,| < \varepsilon$, then the optimal $\overrightarrow{\text{mode}}$ and $\overrightarrow{\text{qscale}}$ has been found for the picture. Repeat the process for the next picture.
   - If $\Sigma R_{MBi} < R_{PICT}$, then decrement $D_{MB} = D_{MB} - \Delta D_{MB}$ and go to step 2.
   - If $\Sigma R_{MBi} > R_{PICT}$, then increment $D_{MB} = D_{MB} + \Delta D_{MB}$ and go to step 2.

### 16.5.3  PRACTICAL SOLUTION WITH NEW CRITERIA FOR THE SELECTION OF CODING MODE

It is obvious that the near-optimal solution discussed in the previous section is not a practical method because of its complexity. To determine the best mode, we have to know how many bits it takes to code each macroblock in every mode with the same distortion level. The total number of bits for each macroblock, $R_{MB}$, consists of three parts, bits for coding motion vectors, $R_{mv}$, bits for coding the predictive residue, $R_{res}$, and bits for coding macroblock header information, $R_{header}$, such as macroblock type, quantizer scale, and coded-block pattern.

$$R_{MB} = R_{mv} + R_{res} + R_{header}. \qquad (16.25)$$

The number of bits for motion vectors, $R_{mv}$, can be easily obtained by VLC table lookup. But to obtain the number of bits for coding the predictive residue, one has to go through the three step coding procedure: (1) DCT, (2) quantization, and (3) VLC as shown in Figure 16.24. At step 3, $R_{res}$ is obtained with a lookup table according to the run length of zeros and the level of quantized coefficients, i.e., $R_{res}$ depends on the pair of values of run and level:

$$R_{res} = f(run, level). \qquad (16.26)$$

As stated above, to obtain the upper-bound coding performance, all three steps are needed for each coding mode, and then the coding mode resulting in the least number of bits is selected as the best mode.

To obtain a much less computationally intensive method, it is preferred to use a statistical model of DCT coefficient bit usage vs. variance of the prediction residual and quantizer step size. This will provide an approximation of the number of residual bits, $R_{res}$. For this purpose we assume that the run and level pair in Equation 16.26 is strongly dependent on values of the quantizer scale, $q_s$, and the variance of the residue, $V_{res}$, for each macroblock. Intuitively, we would expect the number of bits to encode a macroblock is proportional to the variance of the residual and inversely

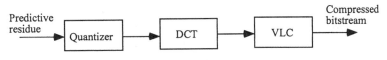

**FIGURE 16.24**  Coding stages to find bit count.

proportional to the value of quantizer step size. Therefore, a statistical model can be constructed by plotting $R_{res}$ vs. the independent variables $V_{res}$ and $q_s$ over a large set of representative macroblock pixels from images typical of natural video material. This results in a scatter plot showing tight correlation, and hence a surface can be fit through the data points. It was found that Equation 16.26 can be approximately expressed as:

$$R_{res} \approx f\left(q_s, V_{res}\right) = \left(K/\left(Cq_s + q_s^2\right)\right) V_{res}, \tag{16.27}$$

where $K$ and $C$ are constants found through surface-fitting regression. If we assume $R_{header}$ is a relatively fixed component that does not vary much with macroblock coding mode and can be ignored, then Equation 16.25 can be approximately replaced by:

$$R_{MB'} = R_{mv} + \left(K/\left(Cq_s + q_s^2\right)\right) V_{res}. \tag{16.28}$$

The value of $R_{MB'}$ reflects the variable portion of bit usage that is dependent on coding mode, and can be used as the measure for selecting the coding mode in our encoder. For a given quantizer step size, the mode resulting in the smallest value of $R_{MB'}$ is chosen as the "best" mode. It is obvious that, in the use of this new measurement to select the coding mode, the computational complexity increase over the TM5 is very slight (the same identical calculation for $V_{res}$ is made in the TM5).

## 16.6 STATISTICAL MULTIPLEXING OPERATIONS ON MULTIPLE PROGRAM ENCODING

In this section, the strategies for statistical multiplexing operation on the multiple program encoding will be introduced. This topic is an extension of rate control into the case of multiple program encoding. First, a background survey of general encoding and multiplexing modes is reviewed. Second, the specific algorithm used in some current systems is introduced, its shortcomings are addressed, and possible amendments to the basic algorithm are described. Some potential research topics such as modeling strategies and methods for solving the problem are proposed for investigation. These topics may be good research topics for the interested graduate student.

### 16.6.1 BACKGROUND OF STATISTICAL MULTIPLEXING OPERATION

In many applications, several video sources may often be combined, or multiplexed, onto a single link for transmission. At the receiving end, the individual sources of data from the multiplexed data are demultiplexed and supplied to the intended receivers. For example, in an ATM network scenario many video sources originating from a local area are multiplexed onto a wide-area backbone trunk. In a satellite-broadcasting scenario, several video sources are multiplexed for transmission through a transponder. In a cable TV scenario, hundreds of video programs are broadcast onto a cable bus. Since the transmission channel, such as a trunk, a transponder, or a cable, is always an expensive resource, the limited channel capacity should be exploited as much as possible. The goal of statistical multiplexing encoding is to make the best use of the limited channel capacity possible. There are several approaches to encoding and multiplexing a plurality of video sources. In the following, we will compare the methods and describe the situation where each method is applicable. The qualitative comparisons are made in terms of trade-offs among factors of computation, implementation complexity, encoded picture quality, buffering delay, and channel utilization. To understand the statistical multiplexing method, we first introduce a simple case of deterministic multiplexing function of a CBR encoder. The standard method for performing the encoding and multiplexing function is to encode the source independently with a CBR. The

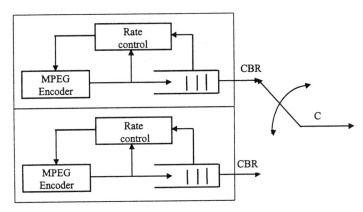

**FIGURE 16.25** Independent encoding/muxing of CBR sources.

CBR encoder produces an encoded bit steam, representing the video supplied to it, at a predetermined CBR. To produce a CBR, the CBR encoder utilizes a rate buffer and feedback control mechanism that continually modifies the amount of quantization applied to the video signal, as shown in Figure 16.25.

The CBR encoder provides a CBR with varying encoded picture quality. This means that the degree of quantization applied depends upon the coding complexity of the current frame offered to the MPEG compression algorithm. Fine quantization is then applied to those frames that have low spatial and/or temporal coding complexity, and conversely coarse quantization is applied to frames that possess high spatial and temporal coding complexity in order to meet the bit rate. However, varying the quantization level corresponds to varying the video quality. Thus, in a CBR encoder, spatial and temporal complexity tends to be encoded in such a manner that the subjective quality of the reproduced image is lower than that of less complex images. This makes any form of rate control inherently bad in the sense that control is always imposed in a direction contrary to the goal of achieving uniform image quality. Usually, bit rates for CBR encoders are chosen so that the moderately difficult scenes can be coded to an acceptable quality level. Given that moderately difficult scenes give good results, then all simpler scenes will yield even better results with the given rate, while very difficult scenes will result in noticeable degradation. Since CBR encoders produce CBR, the multiplexing of a plurality of sources is very simple. The required channel capacity would simply be the sum of all the individual CBRs. Deterministic time or frequency division multiplexing of the individual CBR bitstreams onto the channel is a well-known and simple process. So with CBR encoding, uniformly consistent image quality is impossible for the video sequence with varying scene complexity, but the reward is the ease of multiplexing. The penalty of CBR coding with easy multiplexing may not only be nonuniform picture quality, also result in lower efficiency of channel bandwidth employment. Better efficiency can be gained by statistical multiplexing, whereby each source is encoded at a VBR coding approach. The VBR coding will result in uniform or consistent coded image quality by fixing the quantization scale or by modulating the quantization scale to a limited extent according to activity-masking attributes of the human visual system. Then, the bit rates generated by VBR coding vary with the coding complexity of the incoming video source material. Statistical multiplexing is referred to as StatMux. The coding gain of StatMux is possible through sharing of the channel resource jointly among the encoders. For example, two MPEG encoders may assign the appearance of their I-pictures at different time; this may reduce the limitation of the maximum channel bandwidth requirement since coding an I-picture may generate a large number of bits. This may not be a good example for practical applications. However, this explains that the process of StatMux is not a zero-sum game whereby one encoder's gain must be exactly another encoder's loss. In the process of StatMux, one encoder's gain is obtained by using the channel bandwidth that another encoder does not need at that time or that would bring a very marginal gain for another encoder at that time. More exactly, this concept of gains through sharing arises when the limited amount of bits is dynamically appropriated toward encoders that can best utilize those bits in substantially improving their image quality during complex segments and eschewed from encoders that can improve their image quality only marginally during easy segments. It is obvious that the CBR-encoded sources do not need statistical multiplexing since the bandwidth for each encoded source is well defined. The gain of statistical multiplexing is only possible with VBR-encoded sources. In the following section, we discuss two kinds of multiplexing with multiple VBR-encoded sources.

### 16.6.2 VBR Encoders in StatMux

There are two multiplexing methods for encoding multiple sources with VBR encoders, open loop and closed loop. Each VBR encoder in open-loop-multiplexing mode produces the most consistently uniform predefined image quality level regardless of the coding complexity of the incoming video sources. The image quality is decided by fixing the quantization scale. When the quantization scale is fixed, the SNR is fixed under assumption of white Gaussian quantization noise. Sometimes, the quantization scale is slightly modulated according to the image activity to match the human visual system, for example, in the method in MPEG-2 TM5. The resulting VBR bit rate process is generated by allowing the encoder to use freely however many bits needed to meet the predetermined quality level. Usually, each video source encoded by a VBR encoder in the open-loop mode is not geographically colocated and cannot be encoded jointly. However, the resulting VBR processes do share the channel "jointly," in the sense that the total channel bandwidth is not rigidly allocated among the sources in a fixed manner such as is done in CBR operation mode, where each source has the fixed portion of channel bandwidth. The instantaneous combined rates of all the VBR encoders may exceed the channel capacity, especially in the case when all the encoders generate bursts of bits at the same time. Then, the joint buffer will overflow, thereby leading to loss of data. However, there always still exists a possibility to utilize the channel capacity more efficiently by carefully allocating the loading conditions without loss of data. But totally open-loop VBR coding is not stationary and it is hard to achieve both good channel utilization and very limited data loss. A practical method of VBR transmission for use in the ATM environment involves placing limitations on the degree of variability allowed in VBR processes. Figure 16.26 illustrates the idea of self-regulating VBR encoders.

The difference between the proposed VBR encoder and a totally open-loop VBR encoder is that a looser form of rate control is imposed to the VBR encoder in order to avoid violating transmission constraints that are agreed to by the user and the network as part of the contract

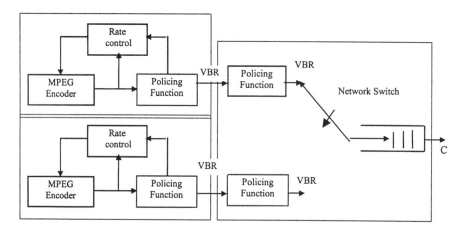

**FIGURE 16.26**   Independent encoding/muxing of geographically dispersed VBR sources.

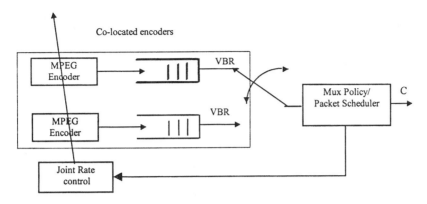

**FIGURE 16.27**   Method of joint rate control and multiplexing.

negotiated during the call setup stage. The rate control will match the policing function, which is enforced by the network. Looser rate control means that the rate control is not so strict as the one in the CBR case because it allows for the encoder to vary its output bit rate according to the coding complexity up to a certain degree as decided by the policing function.

In some applications such as the TV broadcasting or cable TV, the video sources may be geographically colocated at the same site. In such scenarios, additional gains can be realized by the StatMux in which the sources are jointly encoded and jointly multiplexed. By using a common rate controller, all encoders operate in VBR mode but without contending and stepping over one another as in independent VBR encoding and multiplexing. The joint rate controller assigns the total available channel capacity to each encoder so that a certain common quality level is maintained. The bit rates assigned to each individual encoder by joint rate control dynamically change based on the coding complexities of each video source to achieve the most uniform quality among the encoders and along the time for each encoder. In such a joint rate control method, although each encoder produces its own variable rate bits, the sum of bits produced by all encoders combined together is a CBR to fit the channel capacity. Such an idea is shown in Figure 16.27.

### 16.6.3   RESEARCH TOPICS OF STATMUX

The major problem of StatMux is how to allocate the bit rate resource among the video sources that share the common channel bit rate and are jointly encoded by a joint rate controller. This allocation should be based on the coding complexity of each source. The bit rate, $R_i(t)$, for encoder $i$ at time $t$ according to the normalized coding complexity of all encoders for the GOP period ending at time $t$, such as

$$R_i(t) = \frac{X_i(t)}{\sum_{j=1}^{N} X_j(t)} C,$$
(16.29)

where $X_i(t)$ is the coding complexity of source for encoder $i$ at the time $t$ over a GOP period and $C$ is the total channel capacity. Also the bit rate assignment has to be updated from time to time to trace the variation of source complexity. In the following, we will discuss several topics which may be research topics for graduate students.

**Forward Analysis** — Without forward analysis, scene transitions are unanticipated and lead to incorrect bit allocation for a brief, transient period following the scene changes. If the bit allocation of a current video segment is based on the complexity of previous video segments and is adjusted by the available bit rate resource, those video segments which change from easy coding complexity to difficult coding complexity suffer the greatest degradation without preanalysis of upcoming increased complexity. Preanalysis could be performed with a dual set of encoders operating with a certain preprocessing delay ahead of the actual encoding process. As a simple example, we start to assign the equal portion of bit rate for each encoder; then we can obtain the average quantization scale for this GOP that can be considered as the forward analysis results of coding complexity. The real coding process can operate on the coding complexity obtained by the preanalysis. If we choose one or two GOPs according to the synchronous status of the input video sources to perform the preanalysis, it will result in small buffering delay.

**Potential Modeling Strategies and Methods** — Several modeling strategies and methods have been investigated to find a suitable procedure for classifying sources and determining what groups of sources can appropriately be jointly encoded together for transmission over a common channel to meet a specified image quality level. These modeling strategies and methods include modeling of video encoding, modeling of source coding complexity, and source classification. The modeling of a video-encoding algorithm involves measuring the operating performance of the individual encoders or characterizing its rate distortion function for a variety of scenes. Embodied into this model are the MPEG algorithms implemented for motion estimation, mode decision, rate control, and their joint optimization issues. It has been speculated that a hyperbolic functional form of

$$Rate = X/Distortion \qquad (16.30)$$

would be appropriate over the normal operating bit rate range of 3 to 7 Mbps for MPEG-2 encoded CCIR601-sized videos. The hyperbolic shape of rate distortion curves would be also suitable for all video scenes. Actually, we can use a set of collected rate distortion data pairs with an encoder to fit a hyperbola through the points as shown in Figure 16.28 and estimate the shape parameter $X$. The value of $X$ will be used to present the coding complexity offered to an encoder. For modeling at the GOP level, the rate would be the number of bits used to code that GOP and the distortion can be chosen as the averaging quantization scale over the GOP. In some of the literature, the distortion is taken as the average PSNR over the GOP or overall sequence. If it is assumed that the quantization noise is modeled by white Gaussian noise, then both distortion measures are equivalent.

After obtaining the correct coding complexity parameters, we can improve the StatMux algorithm by assigning an encoding bit budget to each encoder based on the GOP level normalized complexity measure $X$ that each encoder is encoding. The GOP level normalized complexity measure $X(n)$ is defined as

$$X(n) = \sum_{i \in GOP} T(i)Q(i), \qquad (16.31)$$

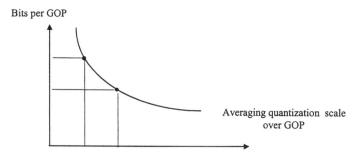

**FIGURE 16.28**   Rate distortion modeling of encoding algorithm and video source.

where $n$ is the GOP number, $T(i)$ is the total number of bits used for encoding picture $i$, and $Q(i)$ is the average quantization scale used for encoding picture $i$. Some research results have shown that the $X(n)$ is insensitive to the operating bit rate, therefore, $X(n)$ is a reliable measure of the loading characteristics of a video source. Therefore, the study of accurate model of the random process of $x(n)$ is very important for improving the operations of the StatMux algorithm. The accurate model of $X(n)$ reflects the loading characteristics of the video source which dictates the share of total bit budget that an encoder expects to get. Several statistical models have been proposed to describe the complexity measure, $X(n)$. For example, an autoregressive process model is proposed for the intrascene $X(n)$ process. This proposed model is based on the following observations; the complexity measure within a single scene has a skewed distribution by the Gamma function, and furthermore, the complexity measure within a scene displays a strong temporal correlation and the form of the correlation is essentially exponential. The definition for the $M$th order autoregressive model is

$$X(n) = \sum_{m=1}^{M} a(m) \bullet X(n-m) + e(n),$$  (16.32)

where $e(n)$ is the white noise process and $a(m)$ terms are the innovation filter coefficients. The statistics of the model such as the mean value, the variance, the correlation, and the marginal distribution are used to match those of actual signals by adjusting the $a(m)$ terms, $e(n)$ and $M$. Other cases, such as scene transition model, intercoded scene models, we leave as project topics for graduate students.

## 16.7  SUMMARY

In this chapter, the technical details of MPEG video are introduced. The technical detail of MPEG standards includes the decoding process of MPEG-1 and MPEG-2 video. Although the encoding process is not a standard part, it is very important for content and service providers. We discuss the most important parts of encoding techniques. Some examples such as the joint optimizing of mode decision and rate control are good examples to understand how the standard is used.

## 16.8  EXERCISES

16-1.  According to your understanding, give several reasons to explain why the MPEG standards specify only decoding as a normative part and define encoding as an informative part (TM5).

16-2.  Can an MPEG-2 video decoder decode a bitstream generated by an MPEG-1 video encoder? Summarize the main difference between the MPEG-1 and MPEG-2 video standards.

16-3.  Prefiltering may reduce the noise of original video source and increase the coding efficiency. But at the same time prefiltering will result in a certain information loss. Conduct a project to investigate at what bit rate range prefiltering may benefit the coding efficiency for some video sources.

16-4.  Use TM5 rate control to encode several video sequences (such as Flower Garden sequence) in two ways: (a) with adaptive quantization step, (b) without adaptive quantization step (Equation 6.16). Compare and discuss the numerical results and subjective results (observe the smooth areas carefully).

**16-5.** Why does MPEG-2 use different quantizer matrices for intra- and intercoding? Conduct a project to use different quantization matrices to encode several video sequences and report the results.

**16-6.** Conduct a project to encode several video sequences (a) with B-picture; (b) without B-picture. Compare the numerical and subjective results. Observe what difference exists between the sequences with fast motion and the sequence with slow motion. (Typical bit rates for CCIR601 sequences are 4 to 6 Mbps).

## REFERENCES

Cano, D. and M. Benard, 3-D Kalman filtering of image sequences, in *Image Sequence Processing and Dynamic Scene Analysis*, T. S. Huang, Ed., Springer, Berlin, 1983, pp. 563-579.

Haskell, B. G., A. Puri, and A. N. Netravali, *Digital Video: Introduction to MPEG-2*, Chapman Hall, New York, 1997.

Isnardi, M. A. Consumers seek easy to use products, *IEEE Spectrum*, 64, 1993.

ISO/IEC 11172, International Standard, 1992.

ISO/IEC, MPEG-2 Test Model 5, ISO-IEC/JTC1/SC29/WG11, April, 1993.

ISO/IEC 13818 MPEG-2 International Standard, Video Recommendation ITU-T H.262, Jan. 10, 1995.

Katsaggelos, A. K., R. P. Kleihorst, S. N. Efstratiadis, and R. L. Lagendijk, Adaptive image sequence noise filtering methods, *Proceeding of SPIE Visual Communication and Image Processing*, Boston, Nov. 10-13, 1991.

Katto, J., J. Ohki, S. Nogaki, and M. Ohta, A wavelet codec with overlapped motion compensation for very low bit rate enviroment, *IEEE Trans. Circuits Syst. Video Technol.*, 4(3), 328-338, 1994.

Koc, U.-V. and K. J. R. Liu, DCT-based motion estimation, *IEEE Trans. Image Process.*, 7, 948-965, 1998.

Lee, J. and B.W. Dickerson, Temporally adaptive motion interpolation exploiting temporal masking in visual perception, *IEEE Trans. on Image Proc.*, 3(5), 513-526, 1994.

Mitchell, J. L., W. B. Pennebaker, C. E. Fogg, and D. J. LeGall, *MPEG Video Compression Standard*, Chapman Hall, New York, 1997.

Ozkan, M. K., M. I. Sezan, and A. M. Tekalp, Adaptive motion-compensated filtering of noisy image sequences, *IEEE Trans on Circuits and Systems for Video Technology*, 3(4), 277-290, 1993.

Ramchandran, K., A. Ortega, and M. Vetterli, Bit allocation for dependent quantization with application to MPEG video coders, *IEEE Trans. on Image Proc.*, 3(5), 533-545, 1994.

Sezan, M. I., M. K. Ozkan, and S. V. Fogel, Temporal adaptive filtering of noisy image sequences using a robust motion estimation algorithm, *IEEE ICASSP*, 2429-2432, 1991.

Sun, H. Sarnoff Internal technical report, May 1994.

Sun, H., W. Kwok, M. Chien, and C. H. John Ju, MPEG coding performance improvement by jointly optimization coding mode decision and rate control, *IEEE Trans. Circuits Syst. Video Technol.*, 7(3), 449-458, 1997.

Wang, L. Rate control for MPEG-2 video coding, *SPIE on Visual Communications and Image Processing*, pp. 53-64, Taipei, Taiwan, May 1995.

Young, R. W. and N. G. Kingsbury, Frequency-domain motion estimation using a complex lapped transform, *IEEE Trans. Image Process.*, 2(1), 2-17, 1993.

# 17 Application Issues of MPEG-1/2 Video Coding

This chapter is an extension of the previous chapter. We introduce several important application issues of MPEG-1/2 video which include the ATSC (Advanced Television Standard Committee) DTV standard which has been adopted by the FCC (Federal Communications Commission) as the TV standard in the United States, transcoding, down-conversion decoder, and error concealment.

## 17.1 INTRODUCTION

Digital video signal processing is an area of science and engineering that has developed rapidly over the past decade. The maturity of the moving picture expert group (MPEG) video-coding standard is a very important achievement for the video industry and provides strong support for digital transmission and storage of video signals. The MPEG coding standard is now being deployed for a variety of applications, which include high-definition television (HDTV), teleconferencing, direct broadcasting by satellite (DBS), interactive multimedia terminals, and digital video disk (DVD). The common feature of these applications is that the different source information such as video, audio, and data are all converted to the digital format and then mixed together to a new format which is referred to as the bitstream. This new format of information is a revolutionary change in the multimedia industry, since the digitized information format, i.e., the bitstream, can be decoded not only by traditional consumer electronic products such as television but also by the digital computer. In this chapter, we will present several application examples of MPEG-1/2 video standards, which include the ATSC DTV standard, transcoding, down-conversion decoder, and error concealment. The DTV standard is the application extension of the MPEG video standard. The transcoding and down-conversion decoders are the practical application issues which increase the features of compression-related products. The error concealment algorithms provide the tool for transmitting the compressed bitstream over noisy channels.

## 17.2 ATSC DTV STANDARDS

### 17.2.1 A Brief History

The birth of digital television (DTV) in the U.S. has undergone several stages: the initial stage, the competition stage, the collaboration stage, and the approval stage (Reitmeier, 1996). The concept of high-definition television (HDTV) was proposed in Japan in the late 1970s and early 1980s. During that period, Japan and Europe continued to make efforts in the development of analog television transmission systems, such as MUSE and HD-MAC systems. In early 1987, U.S. broadcasters fell behind in this field and felt they should take action to catch up with the new HDTV technology and petitioned the FCC to reserve a spectrum for terrestrial broadcasting of HDTV. As a result, the Advisory Committee on Advanced Television Service (ACATS) was founded in August 1987. This committee takes the role of recommending a standard to the FCC for approval. Thus, the process of selecting an appropriate HDTV system for the U.S. started. At the initial stage between 1987 and 1990, there were over 23 different analog systems proposed; among these systems two typical approaches were extended definition television (EDTV) which fits into a single 6-MHz channel and the high definition television (HDTV) approach which requires two 6-MHz channels. By 1990, ACATS had established the Advanced Television Test Center (ATTC), an official testing laboratory sponsored by broadcasters to conduct extensive laboratory tests in Virginia and field tests in Charlotte, NC. Also, the industry had formed the Advanced Television Standards Committee (ATSC) to perform the task of drafting the official standard documents of the selected winning system.

As we know, the current ATSC-proposed television standard is a digital system. In early 1990, the FCC issued a very difficult request to industry about the DTV standard. The FCC required the industry to provide full-quality HDTV service in a single 6-MHz channel. Having recognized the technical difficulty of this requirement at that time, the FCC also stated that this service could be provided by a simulcast service in which programs would be simultaneously broadcasted in both NTSC and the new television system. However, the FCC decided not to assign new spectrum bands for television. This means that simulcasting would occur in the already crowded VHF and UHF spectrum. The new television system had to use low-power transmission to avoid excessive inter-ference into the existing NTSC services. Also, the new television system had to use a very aggressive compression approach to squeeze a full HDTV signal into the 6-MHz spectrum. One good thing was that backward compatibility with NTSC was not required. Actually, under these constraints the backward compatibility had already become impossible. Also, this goal could not be achieved by any of the previously proposed systems and it caused most of the competing proponents to reconsider their approaches. Engineers realized that it was almost impossible to use the traditional analog approaches to reach this goal and that the solution may be in digital approaches. After a few months of consideration, General Instrument announced its first digital system proposal for HDTV, DigiCigher, in June 1990. In the following half year, three other digital systems were proposed: the Advanced Digital HDTV by the Advanced Television Research Consortium, which included Thomson, Philips, Sarnoff, and NBC in November 1990; Digital Spectrum Compatible HDTV by Zenith and AT&T in December 1990; and Channel Compatible Digicipher by General Instrument and the Massachusetts Institute of Technology in January 1991. Thus, the competition stage started. The prototypes of four competing digital systems and the analog system, Narrow MUSE, proposed by NHK (Nippon Houson Kyokai, the Japan Broadcasting Corporation), were officially tested and extensively analyzed during 1992. After a first round of tests, it was concluded that the digital systems would be continued for further improvement and would be adopted. In February 1992, the ACATS recommended digital HDTV for the U.S. standard. It also recommended that the competing systems be either further improved and retested, or be combined into a new system. In the middle of 1993, the former competitors joined in a Grand Alliance. Then the DTV development entered the collaboration stage. The Grand Alliance began a collaborative effort to create the best system which combines the best features and capabilities of the formerly competing systems into a single "best of the best" system. After 1 year of joint effort by the seven Grand Alliance members, the Grand Alliance provided a new system that was prototyped and extensively tested in the laboratory and field. The test results showed that the system is indeed the best of the best compared with the formerly competing systems (Grand Alliance, 1994). The ATSC then recommended this system to the FCC as the candidate HDTV standard in the United States. During the following period, the computer industry realized that DTV provides the signals that can now be used for computer applications and the TV industry was invading its terrain. It presented different opinions about the signal format and was especially opposed to the interlaced format. This reaction delayed the approval of the ATSC standard. After a long debate, the FCC finally approved the ATSC standard in early 1997. But, the FCC did not specify the picture formats and leaves this issue to be decided by the market.

## 17.2.2  TECHNICAL OVERVIEW OF ATSC SYSTEMS

The ATSC DTV system has been designed to satisfy the FCC requirements. The basic requirement is that no additional frequency spectrum will be assigned for DTV broadcasting. In other words, during a transition period, both NTSC and DTV service will be simultaneously broadcast on different channels and DTV can only use the taboo channels. This approach allows a smooth transition to DTV, such that the services of the existing NTSC receivers will remain and gradually be phased out of existence in the year 2006. The simulcasting requirement causes some technical difficulties in DTV design. First, the high-quality HDTV program must be delivered in a 6-MHz channel to make efficient use of spectrum and to fit allocation plans for the spectrum assigned to television broadcasting. Second, a low-power and low-interference signal must be used so that simulcasting in the same frequency allocations as current NTSC service does not cause excessive interference with the existing NTSC receiving, since the taboo channels are generally unsuitable for broadcasting an NTSC signal due to high interference. In addition to satisfying the frequency spectrum requirement, the DTV standard has several important features, which allow DTV to achieve interoperability with computers and data communications. The first feature is the adoption of a layered digital system architecture. Each individual layer of the system is designed to be interoperable with other systems at the corresponding layers. For example, the square pixel and progressive scan picture format should be provided to allow computers access to the compression layer or picture layer depending on the capacity of the computers and the ATM-like packet format for the ATM network to access the transport layer. Second, the DTV standard uses a header/descriptor approach to provide maximum flexible operating characteristics. Therefore, the layered architecture is the most important feature of DTV standards. The additional advantage of layering is that the elements of the system can be combined with other technologies to create new applications. The system of DTV standard includes four layers: the picture layer, the compression layer, the transport layer, and the transmission layer.

### 17.2.2.1  Picture Layer

At the picture layer, the input video formats have been defined. The Executive Committee of the ATSC has approved release of statement regarding the identification of the HDTV and Standard Definition Television (SDTV) transmission formats within the ATSC DTV standards. There are six video formats in the ATSC DTV standard, which are "High Definition Television." These formats are listed in Table 17.1.

The remaining 12 video formats are not HDTV format. These formats represent some improvements over analog NTSC and are referred to as "SDTV." These are listed in Table 17.2.

These definitions are fully supported by the technical specifications for the various formats as measured against the internationally accepted definition of HDTV established in 1989 by the ITU and the definitions cited by the FCC during the DTV standard development process. These formats cover a wide variety of applications, which include motion picture film, currently available HDTV production equipment, the NTSC television standard, and computers such as personal computers and workstations. However, there is no simple technique which can convert images from one pixel

**TABLE 17.1**
**HDTV Formats**

| Spatial Format (X × Y active pixels) | Aspect Ratio | Temporal Rate (Hz progressive scan) |
|---|---|---|
| 1920 × 1080 (square pixel) | 16:9 | 23.976/24 |
| | | 29.97/30 |
| | | 59.94/60 |
| 1280 × 720 (square pixel) | 16:9 | 23.976/24 |
| | | 29.97/30 |
| | | 59.94/60 |

**TABLE 17.2**
**SDTV Formats**

| Spatial Format (X × Y active pixels) | Aspect Ratio | Temporal Rate (Hz progressive scan) |
|---|---|---|
| 704 × 480 (CCIR601) | 16:9 or 4:3 | 23.976/24 |
|  |  | 29.97/30 |
|  |  | 59.94/60 |
| 640 × 480 (VGA, square pixel) | 4:3 | 23.976/24 |
|  |  | 29.97/30 |
|  |  | 59.94/60 |

format and frame rate to another that achieve interoperability among film and the various worldwide television standards. For example, all low-cost computers use square pixels and progressive scanning, while current television uses rectangular pixels and interlaced scanning. The video industry has paid a lot of attention to developing format-converting techniques. Some techniques such as deinterlacing, down/up-conversion for format conversion have already been developed. It should be noted that the broadcasters, content providers, and service providers can use any one of these DTV format. This results in a difficult problem for DTV receiver manufacturers who have to provide all kinds of DTV receivers to decode all these formats and then to convert the decoded signal to its particular display format. On the other hand, this requirement also gives receiver manufacturers the flexibility to produce a wide variety of products that have different functionality and cost, and the consumers freedom to choose among them.

### 17.2.2.2  Compression Layer

The raw data rate of HDTV of 1920 × 1080 × 30 × 16 (16 bits per pixel corresponds to 4:2:2 color format) is about 1 Gbps. The function of the compression layer is to compress the raw data from about 1 Gbps to the data rate of approximately 19 Mbps to satisfy the 6-MHz spectrum requirement. This goal is achieved by using the main profile and high level of the MPEG-2 video standard. Actually, during the development of the Grand Alliance HDTV system, many research results were adopted by the MPEG-2 standard at the same time; for example, the support for interlaced video format and the syntax for data partitioning and scalability. The ATSC DTV standard is the first and most important application example of the MPEG-2 standard. The use of MPEG-2 video compression fundamentally enables ATSC DTV devices to interoperate with MPEG-1/2 computer multimedia applications directly at the compressed bitstream level.

### 17.2.2.3  Transport Layer

The transport layer is another important issue for interoperability. The ATSC DTV transport layer uses the MPEG-2 system transport stream syntax. It is a fully compatible subset of the MPEG-2 transport protocol. The basic function of the transport layer is to define the basic format of data packets. The purposes of packetization include:

- Packaging the data into the fixed-size cells or packets for forward error correction (FEC) encoding to protect the bit error due to the communication channel noise;
- Multiplexing the video, audio, and data of a program into a bitstream;
- Providing time synchronization for different media elements;
- Providing flexibility and extensibility with backward compatibility.

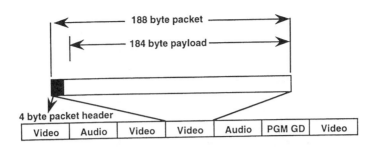

**FIGURE 17.1**   Packet structure of ATSC DTV transport layer.

The transport layer of ATSC DTV uses a fixed-length packet. The packet size is 188 bytes consisting of 184 bytes of payload and 4 bytes of header. Within the packet header, the 13-bit packet identifier (PID) is used to provide the important capacity to combine the video, audio, and ancillary data stream into a single bitstream as shown in Figure 17.1. Each packet contains only a single type of data (video, audio, data, program guide, etc.) identified by the PID.

This type of packet structure packetizes the video, audio, and auxiliary data separately. It also provides the basic multiplexing function that produces a bitstream including video, five-channel surround-sound audio, and an auxiliary data capacity. This kind of transport layer approach also provides complete flexibility to allocate channel capacity to achieve any mix among video, audio, and other data services. It should be noted that the selection of 188-packet length is a trade-off between reducing the overhead due to the transport header and increasing the efficiency of error correction. Also, one ATSC DTV packet can be completely encapsulated with its header within four ATM packets by using 1 AAL byte per ATM header leaving 47 usable payload bytes times 4, for 188 bytes. The details of the transport layer is discussed in the chapter on MPEG systems.

**Transmission Layer** — The function of the transmission layer is to modulate the transport bitstream into a signal that can be transmitted over the 6-MHz analog channel. The ATSC DTV system uses a trellis-coded eight-level vestigial sideband (8-VSB) modulation technique to deliver approximately 19.3 Mbps in the 6-MHz terrestrial simulcast channel. VSB modulation inherently requires only processing the in-phase signal sampled at the symbol rate, thus reducing the complexity of the receiver, and ultimately the cost of implementation. The VSB signal is organized in a data frame that provides a training signal to facilitate channel equalization for removing multipath distortion. However, from several field-test results, the multipath distortion is still a serious problem of terrestrial simulcast receiving. The frame is organized into segments each with 832 symbols. Each transmitted segment consists of one synchronization byte (four symbols), 187 data bytes, and 20 R-S parity bytes. This corresponds to a 188-byte packet, which is protected by 20-byte R-S code. Interoperability at the transmission layer is required by different transmission media applications. The different media use different modulation techniques now, such as QAM for cable and QPSK for satellite. Even for terrestrial transmission, European DVB systems use OFDM transmission. The ATV receivers will not only be designed to receive terrestrial broadcasts, but also the programs from cable, satellite, and other media.

## 17.3   TRANSCODING WITH BITSTREAM SCALING

### 17.3.1   BACKGROUND

As indicated in the previous chapters, digital video signals exist everywhere in the format of compressed bitstreams. The compressed bitstreams of video signals are used for transmission and storage through different media such as terrestrial TV, satellite, cable, the ATM network, and the

Internet. The decoding of a bitstream can be implemented in either hardware or software. However, for high-bit-rate compressed video bitstreams, specially designed hardware is still the major decoding approach due to the speed limitation of current computer processors. The compressed bitstream as a new format of video signal is a revolutionary change to video industry since it enables many applications. On the other hand, there is a problem of bitstream conversion. Bitstream conversion or transcoding can be classified as bit rate conversion, resolution conversion, and syntax conversion. Bit rate conversion includes bit rate scaling and the conversion between constant bit rate (CBR) and variable bit rate (VBR). Resolution conversion includes spatial resolution conversion and temporal resolution conversion. Syntax conversion is needed between different compression standards such as JPEG, MPEG-1, MPEG-2, H.261, and H.263. In this section, we will focus on the topic of bit rate conversion, especially on bit rate scaling since it finds wide application and readers can extend the idea to other kinds of transcoding. Also, we limit ourselves to focus on the problem of scaling an MPEG CBR-encoded bitstream down to a lower CBR. The other kind of transcoding, down-conversion decoder, will be presented in a separate section.

The basic function of bitstream scaling may be thought of as a black box, which passively accepts a precoded MPEG bitstream at the input and produces a scaled bitstream, which meets new constraints that are not known *a priori* during the creation of the original precoded bitstream. The bitstream scaler is a transcoder, or filter, that provides a match between an MPEG source bitstream and the receiving load. The receiving load consists of the transmission channel, the destination decoder, and perhaps a destination storage device. The constraint on the new bitstream may be bound by a variety of conditions. Among them are the peak or average bit rate imposed by the communications channel, the total number of bits imposed by the storage device, and/or the variation of bit usage across pictures due to the amount of buffering available at the receiving decoder.

While the idea of bitstream scaling has many concepts similar to those provided by the various MPEG-2 scalability profiles, the intended applications and goals differ. The MPEG-2 scalability methods (data partitioning, SNR scalability, spatial scalability, and temporal scalability) are aimed at providing encoding of source video into multiple service grades (that are predefined at the time of encoding) and multitiered transmission for increased signal robustness. The multiple bitstreams created by MPEG-2 scalability are hierarchically dependent in such a way that by decoding an increasing number of bitstreams, higher service grades are reconstructed. Bitstream scaling methods, in contrast, are primarily decoder/transcoder techniques for converting an existing precoded bitstream to another one that meets new rate constraints. Several applications that motivate bitstream scaling include the following:

1. Video-On-Demand — Consider a video-on-demand (VOD) scenario wherein a video file server includes a storage device containing a library of precoded MPEG bitstreams. These bitstreams in the library are originally coded at high quality (e.g., studio quality). A number of clients may request retrieval of these video programs at one particular time. The number of users and the quality of video delivered to the users are constrained by the outgoing channel capacity. This outgoing channel, which may be a cable bus or an ATM trunk, for example, must be shared among the users who are admitted to the service. Different users may require different levels of video quality, and the quality of a respective program will be based on the fraction of the total channel capacity allocated to each user. To accommodate a plurality of users simultaneously, the video file server must scale the stored precoded bitstreams to a reduced rate before it is delivered over the channel to respective users. The quality of the resulting scaled bitstream should not be significantly degraded compared with the quality of a hypothetical bitstream so obtained by coding the original source material at the reduced rate. Complexity cost is not such a critical factor because only the file server has to be equipped with the bitstream scaling hardware, not every user. Presumably, video service providers would be willing to pay a high cost for delivering the possible highest-quality video at a prescribed bit rate.

As an option, a sophisticated video file server may also perform scaling of multiple original precoded bitstreams jointly and statistically multiplex the resulting scaled VBR bitstreams into the channel. By scaling the group of bitstreams jointly, statistical gains can be achieved. These statistical gains can be realized in the form of higher and more uniform picture quality for the same channel capacity. Statistical multiplexing over a DirecTv transponder (Isnardi, 1993) is one example of an application of video statistical multiplexing.

2. Trick-play Track on Digital VTRs — In this application, the video bitstream is scaled to create a sidetrack on video tape recorders (VTRs). This sidetrack contains very coarse quality video sufficient to facilitate trick-modes on the VTR (e.g., FF and REW at different speeds). Complexity cost for the bitstream scaling hardware is of significant concern in this application since the VTR is a mass consumer item subject to mass production.

3. Extended-Play Recording on Digital VTRs — In this application, video is broadcast to users' homes at a certain broadcast quality (~6 Mbps for standard-definition video and ~24 Mbps for high-definition video). With a bitstream scaling feature in their VTRs, users may record the video at a reduced rate, akin to extended-play (EP) mode on today's VHS recorders, thereby recording a greater duration of video programs onto a tape at lower quality. Again, hardware complexity costs would be a major factor here.

## 17.3.2   BASIC PRINCIPLES OF BITSTREAM SCALING

As described previously, the idea of scaling an MPEG-2-compressed bitstream down to a lower bit rate is initiated by several applications. One problem is the criteria that should be used to judge the performance of an architecture that can reduce the size or rate of an MPEG-compressed bitstream. Two basic principles of bitstream scaling are (1) the information in the original bitstream should be exploited as much as possible, and (2) the resulting image quality of the new bitstream with a lower bit rate should be as close as possible to a bitstream created by coding the original source video at the reduced rate. Here, we assume that for a given rate the original source is encoded in an optimal way. Of course, the implementation of hardware complexity also has to be considered. Figure 17.2 shows a simplified encoding structure of MPEG encoding in which the rate control mechanism is not shown.

In this structure, a block of image data is first transformed to a set of coefficients; the coefficients are then quantized with a quantizer step which is decided by the given bit rate budget, or number of bits assigned to this block. Finally, the quantized coefficients are coded in variable-length coding to the binary format, which is called the bitstream or bits.

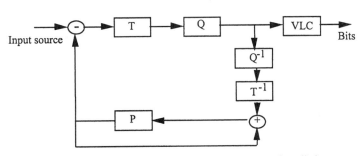

T-- transform, Q-- quantizer, P--motion-compensated prediction
VLC-- variable length

**FIGURE 17.2**   Simplified encoder structure. T = transform, Q = quantizer, P = motion-compensated prediction, VLC = variable length.

From this structure it is obvious that the performance of changing the quantizer step will be better than cutting higher frequencies when the same amount of rate needs to be reduced. In the original bitstream the coefficients are quantized with finer quantization steps which are optimized at the original high rate. After cutting the coefficients of higher frequencies, the rest of the coefficients are not quantized with an optimal quantizer. In the method of requantization all coefficients are requantized with an optimal quantizer which is determined by the reduced rate; the performance of the requantization method must be better than the method of cutting high frequencies to reach the reduced rate. The theoretical analysis is given in Section 17.3.4.

In the following, several different architectures that accomplish the bitstream scaling are discussed. The different methods have varying hardware implementation complexities; each has its own degree of trade-off between required hardware and resulting image quality.

## 17.3.3   ARCHITECTURES OF BITSTREAM SCALING

Four architectures for bitstream scaling are discussed. Each of the scaling architectures described has its own particular benefits that are suitable for a particular application.

Architecture 1:   The bitstream is scaled by cutting high frequencies.
Architecture 2:   The bitstream is scaled by requantization.
Architecture 3:   The bitstream is scaled by reencoding the reconstructed pictures with motion vectors and coding decision modes extracted from the original high-quality bitstream.
Architecture 4:   The bitstream is scaled by reencoding the reconstructed pictures with motion vectors extracted from the original high-quality bitstream, but new coding decisions are computed based on reconstructed pictures.

Architectures 1 and 2 are considered for VTR applications such as trick-play modes and EP recording. Architectures 3 and 4 are considered for and other applicable StatMux scenarios.

### 17.3.3.1   Architecture 1: Cutting AC Coefficients

A block diagram illustrating architecture 1 is shown in Figure 17.3a. The method of reducing the bit rate in architecture 1 is based on cutting the higher-frequency coefficients. The incoming precoded CBR stream enters a decoder rate buffer. Following the top branch leading from the rate buffer, a VLD is used to parse the bits for the next frame in the buffer to identify all the variable-length codewords that correspond to ac coefficients used in that frame. No bits are removed from the rate buffer. The codewords are not decoded, but just simply parsed by the VLD parser to determine codeword lengths. The bit allocation analyzer accumulates these ac bit counts for every macroblock in the frame and creates an ac bit usage profile as shown in Figure 17.3(b). That is, the analyzer generates a running sum of ac DCT coefficient bits on a macroblock basis:

$$PV_N = \sum AC\_BITS, \tag{17.1}$$

where $PV_N$ is the profile value of a running sum of $AC$ codeword bits until the macroblock $N$. In addition, the analyzer counts the sum of all coded bits for the frame, TB (total bits). After all macroblocks for the frame have been analyzed, a target value $TV_{AC}$, of ac DCT coefficient bits per frame is calculated as

$$TV_{AC} = PV_{LS} - \alpha * TB - B_{EX}, \tag{17.2}$$

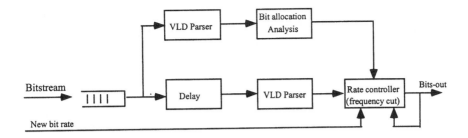

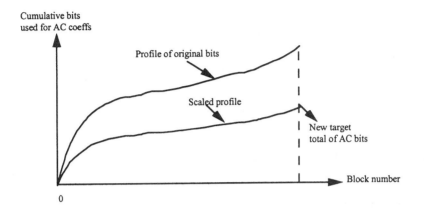

**FIGURE 17.3** (a) Architecture 1: cutting high frequencies. (b) Profile map.

where $TV_{AC}$ is the target value of $AC$ codeword bits per frame, $PV_{LS}$ is the profile value at the last macroblock, $\alpha$ is the percentage by which the preencoded bitstream is to be reduced, $TB$ is the total bits, and $B_{EX}$ is the amount of bits by which the previous frame missed its desired target. The profile value of $AC$ coefficient bits is scaled by the factor $TV_{AC}/PV_{LS}$. Multiplying each $PV_N$ performs scaling by that factor to generate the linearly scaled profile shown in Figure 17.3(b). Following the bottom branch from the rate buffer, a delay is inserted equal to the amount of time required for the top branch analysis processing to be completed for the current frame. A second VLD parser accesses and removes all codeword bits from the buffer and delivers them to a rate controller. The rate controller receives the scaled target bit usage profile for the amount of ac bits to be used within the frame. The rate controller has memory to store all coefficients associated with the current macroblock it is operating on. All original codeword bits at a higher level than ac coefficients (i.e., all fixed-length header codes, motion vector codes, macroblock type codes, etc.) are held in memory and will be remultiplexed with all $AC$ codewords in that macroblock that have not been excised to form the outgoing scaled bitstream. The rate controller determines and flags in the macroblock codeword memory which $AC$ codewords to keep and which to excise. $AC$ codewords are accessed from the macroblock codeword memory in the order $AC11$, $AC12$, $AC13$, $AC14$, $AC15$, $AC16$, $AC21$, $AC22$, $AC23$, $AC24$, $AC25$, $AC26$, $AC31$, $AC32$, $AC33$, etc., where $ACij$ denotes the $i$th $AC$ codewords from $j$th block in the macroblock if it is present. As the $AC$ codewords are accessed from memory, the respective codeword bits are summed and continuously compared with the scaled profile value to the current macroblock, less the number of bits for insertion of $EOB$ (end-of-block) codewords. Respective $AC$ codewords are flagged as kept until the running sum of $AC$ codewords bits exceeds the scaled profile value less $EOB$ bits. When this condition is met, all remaining $AC$ codewords are marked for being excised. This process continues until all macroblocks have their kept codewords reassembled to form the scaled bitstream.

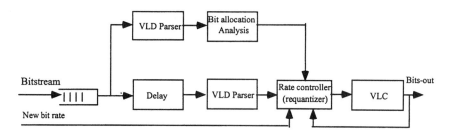

**FIGURE 17.4**   Architecture 2: increasing quantization step.

### 17.3.3.2   Architecture 2: Increasing Quantization Step

Architecture 2 is shown in Figure 17.4. The method of bitstream scaling in architecture 2 is based on increasing the quantization step. This method requires additional dequantizer/quantizer and variable-length coding (VLC) hardware over the first method. Like the first method, it also makes a first VLD pass on the bitstream and obtains a similar scaled profile of target cumulative codeword bits vs. macroblock count to be used for rate control.

The rate control mechanism differs from this point on. After the second-pass VLD is made on the bitstream, quantized DCT coefficients are dequantized. A block of finely quantized DCT coefficients is obtained as a result of this. This block of DCT coefficients is requantized with a coarser quantizer scale. The value used for the coarser quantizer scale is determined adaptively by making adjustments after every macroblock so that the scaled target profile is tracked as we progress through the macroblocks in the frame:

$$Q_N = Q_{NOM} + G * \left( \sum_{N-1} \left( BU - PV_{N-1} \right) \right), \tag{17.3}$$

where $Q_N$ is the quantization factor for macroblock $N$, $Q_{NOM}$ is an estimate of the new nominal quantization factor for the frame, $\sum_{N-1} BU$ is the cumulative amount of coded bits up to macroblock $N - 1$, and $G$ is a gain factor which controls how tightly the profile curve is tracked through the picture. $Q_{NOM}$ is initialized to an average guess value before the very first frame, and updated for the next frame by setting it to $Q_{LS}$ (the quantization factor for the last macroblock) from the frame just completed. The coarsely requantized block of DCT coefficients is variable-length-coded to generate the scaled bitstream. The rate controller also has provisions for changing some macroblock-layer codewords, such as the macroblock-type and coded-block pattern to ensure a legitimate scaled bitstream that conforms to MPEG-2 syntax.

### 17.3.3.3   Architecture 3: Reencoding with Old Motion Vectors
###                and Old Decisions

The third architecture for bitstream scaling is shown in Figure 17.5. In this architecture, the motion vectors and macroblock coding decision modes are first extracted from the original bitstream, and at the same time the reconstructed pictures are obtained from the normal decoding procedure. Then the scaled bitstream is obtained by reencoding the reconstructed pictures using the old motion vectors and macroblock decisions from the original bitstream. The benefits obtained from this architecture compared with full decoding and reencoding is that no motion estimation and decision computation is needed.

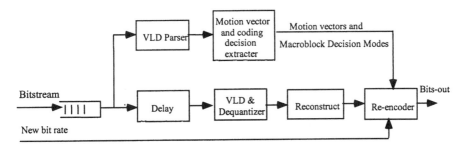

**FIGURE 17.5**   Architecture 3.

### 17.3.3.4   Architecture 4: Reencoding with Old Motion Vectors and New Decisions

Architecture 4 is a modified version of architecture 3 in which new macroblock decision modes are computed during reencoding based on reconstructed pictures. The scaled bitstream created this way is expected to yield an improvement in picture quality because the decision modes obtained from the high-quality original bitstream are not optimal for reencoding at the reduced rate. For example, at higher rates the optimal mode decision for a macroblock is more likely to favor bidirectional field motion compensation over forward frame motion compensation. But at lower rates, only the opposite decision may be true. In order for the reencoder to have the possibility of deciding on new macroblock coding modes, the entire pool of motion vectors of every type must be available. This can be supplied by augmenting the original high-quality bitstream with ancillary data containing the entire pool of motion vectors during the time it was originally encoded. It could be inserted into the user data every frame. For the same original bit rate, the quality of an original bitstream obtained this way is degraded compared with an original bitstream obtained from architecture 3 because the additional overhead required for the extra motion vectors steals away bits for actual encoding. However, the resulting scaled bitstream is expected to show quality improvement over the scaled bitstream from architecture 3 if the gains from computing new and more accurate decision modes can overcome the loss in original picture quality. Table 17.3 outlines the hardware complexity savings of each of the three proposed architectures as compared with full decoding and reencoding.

### 17.3.3.5   Comparison of Bitstream Scaling Methods

We have described four architectures for bitstream scaling which are useful for various applications as described in the introduction. Among the four architectures, architectures 1 and 2 do not require

---

**TABLE 17.3**
**Hardware Complexity Savings over Full Decoding/Reencoding**

| Coding Method | Hardware Complexity Savings |
|---|---|
| Architecture 1 | No decoding loop, no DCT/IDCT, no frame store memory, no encoding loop, no quantizer/dequantizer, no motion compensation, no VLC, simplified rate control |
| Architecture 2 | No decoding loop, no DCT/IDCT, no frame store memory, no encoding loop, no motion compensation, simplified rate control |
| Architecture 3 | No motion estimation, no macroblock coding decisions |
| Architecture 4 | No motion estimation |

entire decoding and encoding loops or frame store memory for reconstructed pictures, thereby saving significant hardware complexity. However, video quality tends to degrade through the group of pictures (GOP) until the next I-picture due to drift in the absence of decoder/encoder loops. For large scaling, say, for rate reduction greater than 25%, architecture 1 produces poor-quality blocky pictures, primarily because many bits were spent in the original high-quality bitstream on finely quantizing the dc and other very low-order ac coefficients. Architecture 2 is a particularly good choice for VTR applications since it is a good compromise between hardware complexity and reconstructed image quality. Architectures 3 and 4 are suitable for VOD server applications and other StatMux applications.

### 17.3.4 ANALYSIS

In this analysis, we assume that the optimal quantizer is obtained by assigning the number of bits according to the variance or energy of the coefficients. It is slightly different from MPEG standard which will be explained later, but the principal concept is the same and the results will hold for the MPEG standard. We first analyze the errors caused by cutting high coefficients and increasing the quantizer step. The optimal bit assignment is given by Jayant and Noll (1984):

$$R_{k0} = R_{av0} + \frac{1}{2} \log_2 \frac{\sigma_k^2}{\left( \prod_{i=0}^{N-1} \sigma_i^2 \right)^{1/N}}, \quad k = 0, 1, \ldots, N-1, \tag{17.4}$$

where $N$ is the number of coefficients in the block, $R_{k0}$ is the number of bits assigned to the $k$th coefficient, $R_{av0}$ is the average number of bits assigned to each coefficient in the block, i.e., $R_{T0} = N \cdot R_{av0}$, is the total bits for this block under a certain bit rate, and $\sigma_k^2$ is the variance of $k$th coefficient. Under the optimal bit assignment (17.4), the minimized average quantizer error, $\sigma_{q0}^2$, is

$$\sigma_{q0}^2 = \frac{1}{N} \sum_{k=1}^{N-1} \sigma_{qk}^2 = \frac{1}{N} \sum_{k=1}^{N-1} 2^{-2R_{k0}} \cdot \sigma_k^2, \tag{17.5}$$

where $\sigma_{qk}^2$ is the quantizer error of $k$th coefficient. According to Equation 17.4, we have two major methods to reduce the bit rate, cutting high coefficients or decreasing the $R_{av}$, i.e., increasing the quantizer step. We are now analyzing the effects on the reconstructed errors caused by the method of cutting high-order coefficients. Assume that the number of the bits assigned to the block is reduced from $R_{T0}$ to $R_{T1}$. Then the bits to be reduced, $\Delta R_1$, are equal to $R_{T0} - R_{T1}$.

In the case of cutting high frequencies, say, the number of coefficients is reduced from $N$ to $M$, then

$$R_{k0} = 0 \text{ for } K < M, \text{ and } \Delta R_1 = R_{T0} - R_{T1} = \sum_{k=M}^{N-1} R_{k0}. \tag{17.6}$$

the quantizer error increased due to the cutting is

$$\Delta \sigma_{q1}^2 = \sigma_{q1}^2 - \sigma_{q0}^2 = \frac{1}{N} \left( \sum_{k=0}^{M-1} 2^{-2R_{k0}} \cdot \sigma_k^2 + \sum_{k=M}^{N-1} \sigma_k^2 - \sum_{k=0}^{N-1} 2^{-2R_{k0}} \cdot \sigma_k^2 \right) \tag{17.7}$$

$$= \frac{1}{N}\left(\sum_{k=M}^{N-1}\sigma_k^2 - \sum_{k=M}^{N-1}2^{-2R_{k0}}\cdot\sigma_k^2\right)$$

$$= \frac{1}{N}\sum_{k=M}^{N-1}\left(1-2^{-2R_{k0}}\right)\cdot\sigma_k^2,$$

where $\sigma_{q1}^2$ is the quantizer error after cutting the high frequencies.

In the method of increasing quantizer step, or decreasing the average bits, from $R_{av0}$ to $R_{av2}$, assigned to each coefficient, the number of bits reduced for the block is

$$\Delta R_2 = R_{T0} - R_{T2} = N\cdot\left(R_{av0} - R_{av2}\right) \tag{17.8}$$

and the bits assigned to each coefficient become now

$$R_{k2} = R_{av2} + \frac{1}{2}\log_2\frac{\sigma_k^2}{\left(\displaystyle\prod_{i=0}^{N-1}\sigma_i^2\right)^{1/N}}, \quad k=0, 1, ..., N-1, \tag{17.9}$$

The corresponding quantizer error increased by the cutting bits is

$$\Delta\sigma_{q2}^2 = \sigma_{q2}^2 - \sigma_{q0}^2 = \frac{1}{N}\left(\sum_{k=0}^{N-1}2^{-2R_{k2}}\cdot\sigma_k^2 - \sum_{k=0}^{N-1}2^{-2R_{k0}}\cdot\sigma_k^2\right)$$

$$= \frac{1}{N}\sum_{k=0}^{N-1}\left(2^{-2R_{k2}} - 2^{-2R_{k0}}\right)\cdot\sigma_k^2, \tag{17.10}$$

where $\sigma_{q2}^2$ is the quantizer error at the reduced bit rate.

If the same number of bits is reduced, i.e., $\Delta R_1 = \Delta R_2$, it is obvious that $\Delta\sigma_{q2}^2$ is smaller than $\Delta\sigma_{q1}^2$ since $\sigma_{q2}^2$ is the minimized value at the reduced rate. This implies that the performance of changing the quantizer step will be better than cutting higher frequencies when the same amount of rate needs to be reduced. It should be noted that in the MPEG video coding, more sophisticated bit assignment algorithms are used. First, different quantizer matrices are used to improve the visual perceptual performance. Second, different VLC tables are used to code the DC values and the AC transform coefficients and the run-length coding is used to code the pairs of the zero-run length and the values of amplitudes. However, in general, the bits are still assigned according to the statistical model that indicates the energy distribution of the transform coefficients. Therefore, the above theoretical analysis will hold for the MPEG video coding.

## 17.4 DOWN-CONVERSION DECODER

### 17.4.1 BACKGROUND

Digital video broadcasting has had a major impact in both academic and industrial communities. A great deal of effort has been made to improve the coding efficiency at the transmission side and

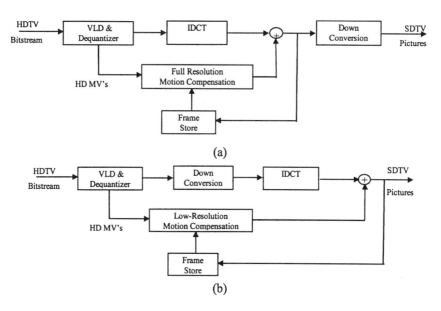

**FIGURE 17.6** Decoder structures. (a) Block diagram of full-resolution decoder with down-conversion in the spatial domain. The quality of this output will serve as a drift-free reference. (b) Block diagram of low-resolution decoder. Down-conversion is performed within the decoding loop and is a frequency domain process. Motion compensation is performed from a low-resolution reference using motion vectors that are derived from the full-resolution encoder. Motion compensation is a spatial domain process.

offer cost-effective implementations in the overall end-to-end system. Along these lines, the notion of format conversion is becoming increasingly popular. On the transmission side, there are a number of different formats that are likely candidates for digital video broadcast. These formats vary in horizontal, vertical, and temporal resolution. Similarly, on the receiving side, there are a variety of display devices that the receiver should account for. In this section, we are interested in the specific problem of how to receive an HDTV bitstream and display it at a lower spatial resolution. In the conventional method of obtaining a low-resolution image sequence, the HD bitstream is fully decoded; then it is simply prefiltered and subsampled (ISO/IEC, 1993). The block diagram of this system is shown in Figure 17.6(a); it will be referred to as a full-resolution decoder (FRD) with spatial down-conversion. Although the quality is very good, the cost is quite high due to the large memory requirements. As a result, low-resolution decoders (LRDs) have been proposed to reduce some of the costs (Ng, 1993; Sun, 1993; Boyce et al., 1995; Bao et al., 1996). Although the quality of the picture will be compromised, significant reductions in the amount of memory can be realized; the block diagram for this system is shown in Figure 17.6(b). Here, incoming blocks are subject to down-conversion filters within the decoding loop. In this way, the down-converted blocks are stored into memory rather than the full-resolution blocks. To achieve a high-quality output with the low-resolution decoder, it is important to take special care in the algorithms for down-conversion and motion compensation (MC). These two processes are of major importance to the decoder as they have significant impact on the final quality. Although a moderate amount of complexity within the decoding loop is added, the reductions in external memory are expected to provide significant cost savings, provided that these algorithms can be incorporated into the typical decoder structure in a seamless way.

As stated above, the filters used to perform the down-conversion are an integral part of the low-resolution decoder. In Figure 17.6(b), the down-conversion is shown to take place before the IDCT. Although the filtering is not required to take place in the DCT domain, we initially assume that it takes place before the adder. In any case, it is usually more intuitive to derive a down-conversion filter in the frequency domain rather than in the spatial domain; this has been described by Pang et al. (1996), Merhav and Bhaskaran (1997), and Mokry and Anastassiou (1994). The major drawback of these approaches is that high frequency data is lost or not preserved very well. To overcome this, a method of down-conversion, which better preserves high-frequency data within the macroblock has been reported by Bao et al. (1996), Vetro and Sun (1998a); this method is referred to as frequency synthesis.

Although the above statement of the problem has only mentioned filtering-based approaches to memory reduction within the decoding loop, readers should be aware that other techniques have also been proposed. For the most part, these approaches rely on methods of embedded compression. For instance, de With et al. (1998) quantized the data being written to memory adaptively using a block predictive coding scheme; then a segment of macroblocks is fit into a fixed length packet. Similarly, Yu et al. (1999) proposed an adaptive min-max quantizer and edge detector. With this method, each macroblock is compressed to a fixed size to simplify memory access. Another, simpler approach may be to truncate the 8-bit data to 7 or 6 bits. However, in this case, it is expected the drift would accumulate very fast and result in poor reconstruction quality. Bruni et al. (1998) used a vectors quantization method, and Lei (1999) described a wavelet-based approach. Overall, these approaches offer exceptional techniques to reduce the memory requirements, but in most cases the reconstructed video would still be a high-resolution signal. The reason is that compressed high-resolution data are stored in memory rather than the raw, low-resolution data. For this reason, the remainder of this section emphasizes the filtering-based approach, in which the data stored in memory represent the actual low-resolution picture data.

The main novelty of the system that we describe is the filtering which is used to perform motion compensation from low-resolution anchor frames. It is well known that prediction drift has been difficult to avoid. It is partly due to the loss of high-frequency data from the down-conversion and partly due to the inability to recover the lost information. Although prediction drift cannot be totally avoided in a low-resolution decoder, it is possible to reduce the effects of drift significantly in contrast to simple interpolation methods. The solution that we described is optimal in the least-squares sense and is dependent on the method of down-conversion that is used (Vetro and Sun, 1998b). In its direct form, the solution cannot be readily applied to a practical decoding scheme. However, it is shown that a cascaded realization is easily implemented into the FRD-type structure (Vetro et al., 1998).

## 17.4.2 FREQUENCY SYNTHESIS DOWN-CONVERSION

The concept of frequency synthesis was first reported by Bao et al. (1996) and later expanded upon by Vetro and Sun (1998b). The basic premise is to better preserve the frequency characteristics of a macroblock in comparison to simpler methods which extract or cut specified frequency components of an $8 \times 8$ block. To accomplish this, the four blocks of a macroblock are subject to a global transformation — this transformation is referred to as frequency synthesis. Essentially, a single-frequency domain block can be realized using the information in the entire macroblock. From this, lower-resolution blocks can be achieved by cutting out the low-order frequency components of the synthesized block — this action represents the down-conversion process and is generally represented in the following way:

$$\tilde{\underline{A}} = X \underline{A}, \tag{17.11}$$

where $\underline{A}$ denotes the original DCT macroblock, $\tilde{\underline{A}}$ denotes the down-converted DCT block, and $X$ is a matrix which contains the frequency synthesis coefficients. The original idea for frequency synthesis down-conversion was to extract an $8 \times 8$ block directly from the $16 \times 16$ synthesized block in the DCT domain as shown in Figure 17.7(a). The advantage of doing this is that the down-converted DCT block is directly applicable to an $8 \times 8$ IDCT (for which fast algorithms exist). The major drawback with regard to computation is that each frequency component in the synthesized

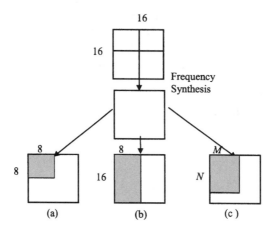

**FIGURE 17.7**   Concept of frequency synthesis down-conversion; (a) 256-tap filter applied to every frequency component to achieve vertical and horizontal down-conversion by a factor of two frame-based filtering; (b) 16-tap filter applied to frequency components in the same row to achieve horizontal down-conversion by two, picture structure is irrelevant; (c) illustration that the amount of synthesized frequency components which are retained is arbitrary.

block is dependent on all of the frequency components in each of the $8 \times 8$ blocks, i.e., each synthesized frequency component is the result of a 256-tap filter. The major drawback with regard to quality is that interlaced video with field-based predictions should not be subject to frame-based filtering (Vetro and Sun, 1998b). If frame-based filtering is used, it becomes impossible to recover the appropriate field-based data that is required to make field-based predictions. In areas of large motion, severe blocking artifacts will result.

Obviously, the original approach would incur too much computation and quality degradation, so, instead, the operations are performed separately and vertical down-conversion is performed on a field basis. In Figure 17.7(b), it is shown that a horizontal-only down-conversion can be performed. To perform this operation, a 16-tap filter is ultimately required. In this way, only the relevant row information is applied as the input to the horizontal filtering operation and the structure of the incoming video has no bearing on the down-conversion process. The reason is that the data in each row of a macroblock belong to the same field; hence the format of the output block will be unchanged. It is noteworthy that the set of filter coefficients is dependent on the particular output frequency index. For 1-D filtering, this means that the filters used to compute the second output index, for example, are different from those used to compute the fifth output index. Similar to the horizontal down-conversion, vertical down-conversion can also be applied as a separate process. As reasoned earlier, field-based filtering is necessary for interlaced video with field-based predictions.

However, since a macroblock consists of eight lines for the even field and eight lines for the odd field, and the vertical block unit is 8, frequency synthesis cannot be applied. Frequency synthesis is a global transformation and is only applicable when one wishes to observe the frequency characteristics over a larger range of data than the basic unit. Therefore, to perform the vertical down-conversion, we can simply cut the low-order frequency components in the vertical direction. This loss that we accept in the vertical direction is justified by the ability to perform accurate low-resolution MC that is free from severe blocking artifacts.

In the above, we have explained how the original idea to extract an $8 \times 8$ DCT block is broken down into separable operations. However, since frequency synthesis provides an expression for every frequency component in the new $16 \times 16$ block, it makes sense to generalize the down-conversion process so that decimation, which are multiples of $1/16$ can be performed. In Figure 17.7(c), an $M \times N$ block is extracted. Although this type of down-conversion filtering may not be appropriate before the IDCT operation and may not be appropriate for a bitstream containing

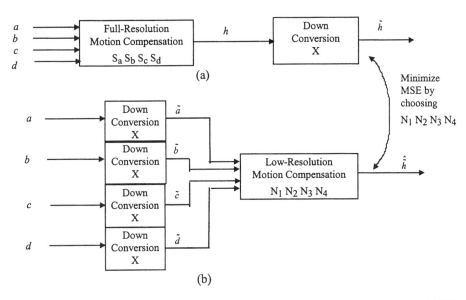

**FIGURE 17.8**  Comparison of decoding methods to achieve low-resolution image sequence. (a) FRD with spatial down-conversion; (b) LRD. The objective is to minimize the MSE between the two outputs by choosing $N_1$, $N_2$, $N_3$, and $N_4$ for a fixed down-conversion. (From Vetro, A. et al., *IEEE Trans. Consumer Elec.,* 44(3), 1998. With permission.)

field-based predictions, it may be applicable elsewhere, e.g., as a spatial domain filter somewhere else in the system and/or for progressive material. To obtain a set of spatial domain filters, an appropriate transformation can be applied. In this way, Equation 17.8 is expressed as

$$\tilde{a} = x\underline{a}, \tag{17.12}$$

where the lowercase counterparts denote spatial equivalents. The expression which transforms $X$ to $x$ is derived in Appendix A, Section 17.4.6.

### 17.4.3  LOW-RESOLUTION MOTION COMPENSATION

The focus of this section is to provide an expression for the optimal set of low-resolution MC filters given a set of down-conversion filters. The resulting filters are optimal in the least-squares sense as they minimize the mean squared error (MSE) between a reference block and a block obtained through low-resolution MC. The results that have been derived by Vetro and Sun (1998a) assume that a spatial domain filter, $x$, is applied to incoming macroblocks to achieve the down-conversion. The scheme shown in Figure 17.8(a) illustrates the process by which reference blocks are obtained. First, full-resolution motion compensation is performed on macroblocks $\underline{a}$, $\underline{b}$, $\underline{c}$, and $\underline{d}$ to yield $\underline{h}$. To execute this process, the filters $S_a^{(r)}, S_b^{(r)}, S_c^{(r)}$, and $S_d^{(r)}$ are used. Basically, these filters represent the masking/averaging operations of the motion compensation in a matrix form. More on the composition of these filters can be found in Appendix B, Section 17.4.7. Once $\underline{h}$ is obtained, it is down-converted to $\tilde{\underline{h}}$ via the spatial filter, $x$:

$$\tilde{\underline{h}} = x\underline{h}. \tag{17.13}$$

The above block is considered to be the drift-free reference. On the other hand, in the scheme of Figure 17.8(b), the blocks $\underline{a}$, $\underline{b}$, $\underline{c}$, and $\underline{d}$ are first subject to the down-conversion filter, $x$, to yield the down-converted blocks, $\tilde{a}$, $\tilde{b}$, $\tilde{c}$, and $\tilde{d}$, respectively. By using these down-converted blocks as input to the low-resolution motion compensation process, the following expression can be assumed:

$$\hat{\underline{h}} = \begin{bmatrix} N_1 & N_2 & N_3 & N_4 \end{bmatrix} \begin{bmatrix} \tilde{a} \\ \tilde{b} \\ \tilde{c} \\ \tilde{d} \end{bmatrix},$$
(17.14)

where $N_k$, $k = 1,2,3,4$ are the unknown filters which are assumed to perform the low-resolution motion compensation, and $\underline{h}$ is the low-resolution prediction. These filters are solved by differentiating the following objective function (Vetra and Sun, 1998a):

$$J\{N_k\} = \left\| \underline{h} - \hat{\underline{h}} \right\|^2,$$
(17.15)

with respect to each unknown filter and setting each result equal to zero. It can be verified that the optimal least-squares solution for these filters is given by

$$N_1^{(r)} = x S_a^{(r)} x^+; \quad N_2^{(r)} = x S_b^{(r)} x^+$$
$$N_3^{(r)} = x S_c^{(r)} x^+; \quad N_4^{(r)} = x S_d^{(r)} x^+,$$
(17.16)

where

$$x^+ = x^T \left( x x^T \right)^{-1}$$
(17.17)

is the Moore–Penrose inverse (Lancaster and Tismenetsky, 1985) for an $m \times n$ matrix with $m \leq n$. In the solution of Equation 17.16, the superscript $r$ is added to the filters, $N_k$, due to their dependency on the full-resolution motion compensation filters. In using these filters to perform the low-resolution motion compensation, the MSE between $\underline{h}$ and $\hat{\underline{h}}$ is minimized. It should be emphasized that Equation 17.16 represents a generalized set of MC filters which are applicable to any $x$, which operates on a single macroblock. For the special case of the $4 \times 4$ cut, these filters are equivalent to the ones that were determined by Morky and Anastassiou (1994) to minimize the drift.

In Figure 17.9, two equivalent MC schemes are shown. However, for implementation purposes, the optimal MC scheme is realized in a cascade form rather than a direct form. The reason is that the direct-form filters are dependent on the matrices, which perform full-resolution MC. Although, these matrices were very useful in analytically expressing the full-resolution MC process, they require a huge amount of storage due to their dependency on the prediction mode, motion vector, and half-pixel-accuracy. Instead, the three linear processes in Equation 17.13 are separated, so that an up-conversion, full-resolution MC, and down-conversion can be performed. Although one may be able to guess such a scheme, we have proved here that it is an optimal scheme provided the up-conversion filter is a Moore–Penrose inverse of the down-conversion filter. Vetro and Sun (1998b), the optimal MC scheme, which employs frequency synthesis, to a nonoptimal MC scheme, which employs bilinear interpolation, and an optimal MC scheme, which employs the $4 \times 4$ cut down-conversion. Significant reductions in the amount of drift were realized by both optimal MC schemes over the method, which used bilinear interpolation as the method of up-conversion. But more importantly, a 35% reduction in the amount of drift was realized by the optimal MC scheme using frequency synthesis over the optimal MC scheme using the $4 \times 4$ cut.

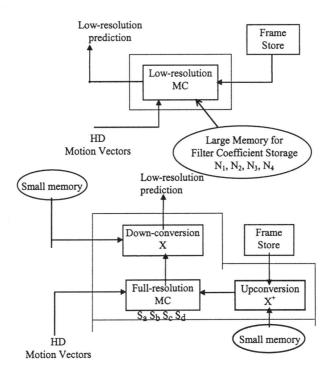

**FIGURE 17.9** Optimal low-resolution MC scheme: direct form (top) vs. cascade form (bottom). Both forms yield equivalent quality, but vary significantly in the amount of internal memory. (From Vetra, A., et al., *IEEE Trans. Consumer Elec.,* 44(3), 1998. With permission.)

## 17.4.4 THREE-LAYER SCALABLE DECODER

In this section, we show how the key algorithms for down-conversion and motion compensation are integrated into a three-layer scalable decoder. The central concept of this decoder is that three layers of resolution can be decoded using a decreased amount of memory for the lower resolution layers. Also, regardless of which layer is being decoded, much of the logic can be shared. Three possible decoder configurations are considered: full-memory decoder (FMD), half-memory decoder (HMD), and quarter-memory decoder (QMD). The low-resolution decoder configurations are based on the key algorithms, which were described for down-conversion and motion compensation. In the following, three possible architectures are discussed that provide equal quality, but vary in system-level complexity. The first (ARCH1) is based on the low-resolution decoder modeled in Figure 17.6(b), the second (ARCH2) is very similar, but attempts to reduce the IDCT computation, while the third (ARCH3) is concerned with the amount of interface with an existing high-level decoder.

With regard to functionality, all of the architectures share similar characteristics. For one, an efficient implementation is achieved by arranging the logic in a hierarchical manner, i.e., employing separable processing. In this way, the FMD configuration is the simplest and serves as the logic core on which other decoder configurations are built. In the HMD configuration, an additional horizontal down-conversion and up-conversion are performed. In the QMD configuration, all of the logic components from the HMD are utilized, such that an additional vertical down-conversion is performed after a horizontal down-conversion, and an additional vertical up-conversion is performed after a horizontal up-conversion. In summary, the logic for the HMD is built on the logic for the FMD, and the logic for the QMD is built on the logic of the HMD. The total system contains a moderate increase in logic, but HD bitstreams may be decoded to a lower resolution with a smaller amount of external memory. By simply removing external memory, lower layers can be achieved at a reduced cost.

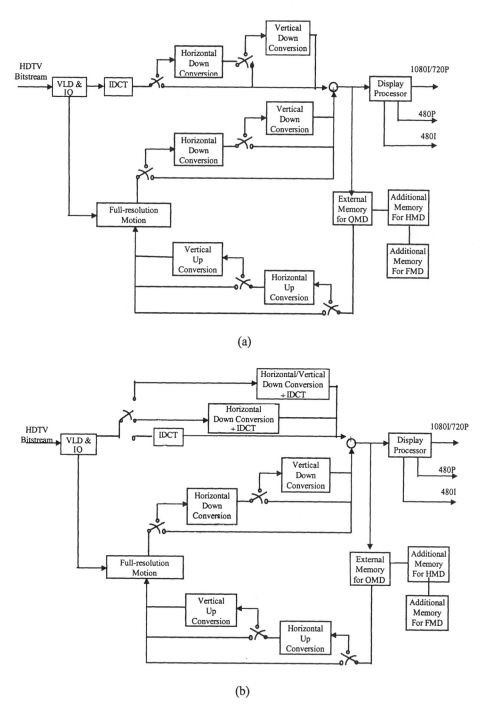

**FIGURE 17.10**  Block diagram of various three-layer scalable decoder architectures; all architectures provide equal quality with varying system complexity: (a) ARCH1, derived directly from block diagram of assumed low-resolution decoder; (b) ARCH2, reduce computation of IDCT by combining down-conversion and IDCT filters; (c) ARCH3, minimize interface with existing HL decoder by moving linear filtering for down-conversion outside of the adder. (From Vetro, A. et al., *IEEE Trans. Consumer Elec.,* 44(3), 1998. With permission.)

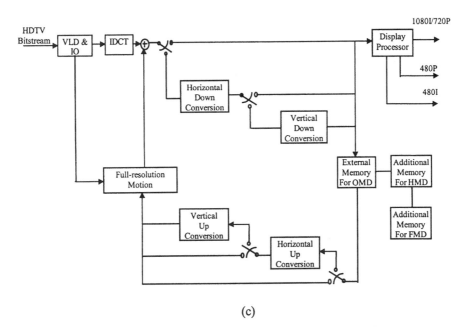

(c)

**FIGURE 17.10** (continued)

The complete block diagram of ARCH1 is shown in Figure 17.10(a). The diagram shown here assumes two things: (1) the initial system model of a low-resolution decoder from Figure 17.6(b) is assumed, and (2) the down-conversions in the incoming branch are performed after the IDCT to avoid any confusion regarding macroblock format conversions in the DCT domain (Vetro and Sun, 1998b). In looking at the resulting system, it is evident that full computation of the IDCT is required, and that two independent down-conversion operations must be performed. The latter is necessary so that low-resolution predictions are added to low-resolution residuals. Overall, the increase in logic for the added feature of memory savings is quite small. However, it is evident that ARCH1 is not the most cost-effective implementation, but it represents the foundation of previous assumptions, and allows us to analyze better the impact of the two modified architectures to follow.

In Figure 17.10(b), the block diagram of ARCH2 is shown. In this system, realizing that the IDCT operation is simply a linear filter reduces the combined computation for the IDCT and down-conversion. In the FMD, we know that a fast IDCT is applied separately to the rows and columns of an $8 \times 8$ block. For the HMD, our goal is to combine the horizontal down-conversion with the horizontal IDCT. In the 1-D case, an $8 \times 16$ matrix can represent the horizontal down-conversion, and an $8 \times 8$ matrix can represent the horizontal IDCT. Combining these processes, such that the down-conversion operates on the incoming DCT rows first, results in a combined $8 \times 16$ matrix. To complete the transformation, the remaining columns can then be applied to the fast IDCT. In the above description, computational savings are achieved in two places: first, the horizontal IDCT is fully absorbed into the down-conversion computation which must take place anyway, and, second, the fast IDCT is utilized for a smaller amount of columns. In the case of the QMD, these same principles can be used to combine the vertical down-conversion with the vertical IDCT. In this case, one must be aware of the macroblock type (field-DCT or frame-DCT) so that an appropriate filter can be applied. In contrast to the previous two architectures, ARCH3 assumes that the entire front-end processing of the decoder is used; it is shown in Figure 17.5. In this way, the adder is always a full-resolution adder, whereas in ARCH1 and ARCH2, the adder needed to handle all three layers of resolution. The major benefit of ARCH3 is that it does not require many interfaces with the existing decoder structure. The memory is really the only place where a new interface needs to be defined. Essentially, a down-conversion filtering may be applied before storing the data, and an up-conversion filtering may be applied, as the data is needed for full-resolution MC. This final architecture is similar in principle to the embedded compression schemes that were mentioned in the beginning of this section. The main difference is that the resolution of the data is decreased rather than compressed. This allows a simpler means of low-resolution display.

### 17.4.5 SUMMARY OF DOWN-CONVERSION DECODER

A number of integrated solutions for a scalable decoder have been presented. Each decoder is capable of decoding directly to a lower resolution using a reduced amount of memory in comparison with the memory required by the high-level decoder. The method of frequency synthesis is successful in better preserving the high-frequency data within a macroblock, and the filtering that is used to perform optimal low-resolution MC is capable of minimizing the drift. It has been shown that a realizable implementation can be achieved, such that the filters for optimal low-resolution MC are equivalent to an up-conversion, full-resolution MC, and for down-conversion, where the up-conversion filters are determined by a Moore–Penrose inverse of the down-conversion. The amount of logic required by these processes is kept minimal since they are realized in a hierarchical structure. Since the down-conversion and up-conversion processes are linear, the architecture design is flexible in that equal quality can be achieved with varying levels of system complexity. The first architecture that we examined came from the initial assumptions that were made on the low-resolution decoder, i.e., a down-conversion is performed before the adder. It was noted that a full IDCT computation was required and that a down-conversion must be performed in two places. As a result, a second architecture was presented to reduce the IDCT computation, and a third was presented to minimize the amount of interface with the existing high-level decoder. The major point here is that the advantages of ARCH2 and ARCH3 cannot be realized by a single architecture. The reason is that performing a down-conversion in the incoming branch reduces the IDCT computation; therefore, a down-conversion must be performed after the full-resolution MC as well. In any case, equal quality is offered by each architecture and the quality is of commercial grade.

### 17.4.6 DCT-TO-SPATIAL TRANSFORMATION

Our objective in this section is to express the following DCT domain relationship:

$$\tilde{A}(k,l) = \sum_{p=0}^{M-1} \sum_{q=0}^{N-1} \left[ X_{k,l}(p,q) A(p,q) \right] \tag{17.18}$$

as

$$\tilde{a}(i,j) = \sum_{s=0}^{M-1} \sum_{t=0}^{N-1} \left[ X_{i,j}(s,t) a(s,t) \right], \tag{17.19}$$

where $\tilde{A}$ and $\tilde{a}$ are the DCT and spatial output, $A$ and $a$ are the DCT and spatial input, and $X$ and $x$ are the DCT and spatial filters, respectively. By definition, the $M \times N$ DCT transform is defined by

$$A(k,l) = \sum_{i=0}^{M-1} \sum_{j=0}^{N-1} a(i,j) \psi_k^M(i) \psi_l^N(j) \tag{17.20}$$

and its inverse, the $M \times N$ IDCT by

$$a(i,j) = \sum_{k=0}^{M-1} \sum_{l=0}^{N-1} A(k,l) \psi_k^M(i) \psi_l^N(j), \tag{17.21}$$

where the basis function is given by

$$\psi_k^N = \sqrt{\frac{2}{N}}\ \alpha(k)\cos\left(\frac{2i+1}{2N}k\pi\right) \tag{17.22}$$

and

$$\alpha(k) = \begin{cases} \dfrac{1}{\sqrt{2}} & \text{for} \quad k = 0; \\ 1 & \text{for} \quad k \neq 0. \end{cases} \tag{17.23}$$

By substituting Equation 17.22 into the expression for the IDCT yields

$$\tilde{a}(i,j) = \sum_{k=0}^{M-1}\sum_{l=0}^{N-1}\psi_k^M(i)\psi_l^N(j)\cdot\left[\sum_{p=0}^{M-1}\sum_{q=0}^{N-1}X_{k,l}(p.q)A(p.q)\right]$$

$$= \sum_{p=0}^{M-1}\sum_{q=0}^{N-1}A(p,q)\cdot\left[\sum_{k=0}^{M-1}\sum_{l=0}^{N-1}X_{k,l}(p,q)\psi_k^M(i)\psi_l^N(j)\right]. \tag{17.24}$$

Substituting the DCT definition into the above gives the following,

$$\tilde{a}(i,j) = \sum_{p=0}^{M-1}\sum_{q=0}^{N-1}\left[\sum_{s=0}^{M-1}\sum_{t=0}^{N-1}a(s,t)\psi_p^M(s)\psi_q^N(t)\right]\sum_{k=0}^{M-1}\sum_{l=0}^{N-1}\left[X_{k,l}(p,q)\cdot\psi_p^M(i)\psi_q^N(j)\right]. \tag{17.25}$$

Finally, Equation 17.17 can be formed with

$$x_{i,j}(s,t) = \sum_{k=0}^{M-1}\sum_{l=0}^{N-1}\psi_k^M(i)\cdot\psi_l^N(j)\left[\sum_{k=0}^{M-1}\sum_{l=0}^{N-1}\left(X_{k,l}(p,q)\cdot\psi_p^M(s)\psi_q^N(t)\right)\right] \tag{17.26}$$

and the transformation is fully defined.

### 17.4.7 Full-Resolution Motion Compensation in Matrix Form

In 2-D, a motion compensated macroblock may have contributions from at most four macroblocks per motion vector. As noted in Figure 17.11, macroblocks $a$, $b$, $c$, and $d$ include four $8 \times 8$ blocks each. These subblocks are raster-scanned so that each macroblock can be represented as a vector. According to the motion vector, $(dx, dy)$, a local reference, $(y_1, y_2)$, is computed to indicate where the origin of the motion compensated block is located; the local reference is determined by

$$y_1 = dy - 16\cdot\left[Integer(dy/16) - \gamma(dy)\right]$$

$$y_2 = dx - 16\cdot\left[Integer(dx/16) - \gamma(dx)\right], \tag{17.27}$$

where

$$\gamma(d) = \begin{cases} 1, & \text{if} \quad d < 0 \text{ and } d\bmod 16 = 0 \\ 0, & \text{otherwise}. \end{cases} \tag{17.28}$$

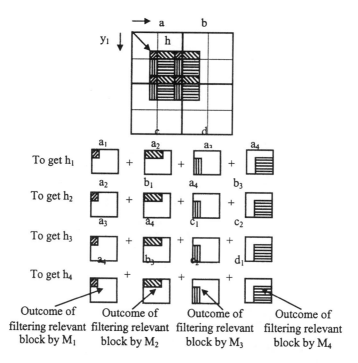

**FIGURE 17.11**   Relationship between the input and output blocks of the motion compensation process in the FRD. (From Vetro, A. et al., *IEEE Trans. Consumer Elec.,* 44(3), 1998. With permission.)

The reference point for this value is the origin of the upper-left-most input macroblock. With this, the motion-compensated prediction may be expressed as

$$
\underline{h} = \begin{bmatrix} h_1 \\ \hline h_2 \\ \hline h_3 \\ \hline h_4 \end{bmatrix} = \begin{bmatrix} S_a^{(r)} & S_b^{(r)} & S_c^{(r)} & S_d^{(r)} \end{bmatrix} \cdot \begin{bmatrix} a \\ \hline b \\ \hline c \\ \hline d \end{bmatrix}; \quad r = 1, 2, 3, 4. \tag{17.29}
$$

As an example, Figure 17.11 considers $(y_1, y_2) \in [0,7]$, which implies that $r = 1$. In this case the motion compensation filters are given by

$$
S_a^{(1)} = \begin{bmatrix} M_1 & M_2 & M_3 & M_4 \\ 0 & M_1 & 0 & M_3 \\ 0 & 0 & M_1 & M_2 \\ 0 & 0 & 0 & M_1 \end{bmatrix}, \quad
S_b^{(1)} = \begin{bmatrix} 0 & 0 & 0 & 0 \\ M_2 & 0 & M_4 & 0 \\ 0 & 0 & 0 & 0 \\ 0 & 0 & M_2 & 0 \end{bmatrix},
$$

$$
S_c^{(1)} = \begin{bmatrix} 0 & 0 & 0 & 0 \\ 0 & 0 & 0 & 0 \\ M_3 & M_4 & 0 & 0 \\ 0 & M_3 & 0 & 0 \end{bmatrix}, \quad
S_c^{(1)} = \begin{bmatrix} 0 & 0 & 0 & 0 \\ 0 & 0 & 0 & 0 \\ 0 & 0 & 0 & 0 \\ M_4 & 0 & 0 & 0 \end{bmatrix}. \tag{17.30}
$$

In the above equations, the $M_1$, $M_2$, $M_3$, and $M_4$ matrices operate on the relevant $8 \times 8$ blocks of $a$, $b$, $c$, and $d$. Their elements will vary according to the amount of overlap as indicated by $(y_1, y_2)$ and the type of prediction. The type of prediction may be frame based or field based and is predicted with half-pixel accuracy. As a result, the matrices $S_a^{(r)}$, $S_b^{(r)}$, $S_c^{(r)}$, and $S_d^{(r)}$, are extremely sparse and may only contain nonzero values of 1, ½, and ¼. For different values of $(y_1,y_2)$ the configuration of the above matrices will change: $y_1 \in [0,7]$ and $y_2 \in [8,15]$ implies $r = 2$; $y_1 \in [8,15]$ and $y_2 \in [0,7]$ implies $r = 3$; $y_1, y_2 \in [8,15]$ implies $r = 4$. The resulting matrices can easily be formed using the concepts illustrated in Figure 17.11.

## 17.5   ERROR CONCEALMENT

### 17.5.1   BACKGROUND

Practical communications channels available for delivery of compressed digital video are characterized by occasional bit error and/or packet loss, although the actual impairment mechanism varies widely with the specific medium under consideration. The class of decoder error concealment schemes described here is based on identification and predictive replacement of picture regions affected by bit error or data loss. It is noted that this approach is based on conversion (via appropriate error/loss detection mechanisms) of the transmission medium into an erasure channel in which all error or loss events can be identified in the received bit-stream. In a block-structured compression algorithm such as MPEG, all channel impairments are manifested as erasures of video units (such as MPEG macroblocks or slices). Concealment at the decoder is then based on exploiting temporal and spatial picture redundancy to obtain an estimate of erased picture areas. The efficiency of error concealment depends on redundancies in pictures and on redundancies in the compressed bitstream that are not removed by source coding. Block compression algorithms do not remove a considerable amount of inter-block redundancies, such as structure, texture, and motion information about objects in the scene.

To be more specific, error resilience for compressed video can be achieved through the addition of suitable transport and error concealment methods, as outlined in the system block diagram shown in Figure 17.12.

The key elements of such a robust video delivery system are outlined below:

- The video signal is encoded using an appropriate video compression syntax such as MPEG. Note that we have restricted consideration primarily to the practical case in which the video compression process itself is not modified, and robustness is achieved through additive transport and decoder concealment mechanisms (except for I-frame motion described in Section 17.4.3). This approach simplifies encoder design, since it separates

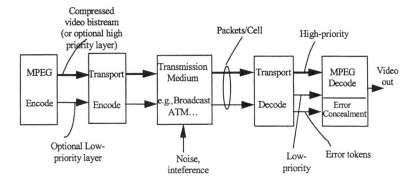

**FIGURE 17.12**   System block diagram of visual communication system.

media-independent video compression functions from media-dependent transport operations. On the receiver side, although a similar separation is substantially maintained, the video decoder must be modified to support an "error token" interface and error concealment functionality.

- Compressed video data is organized into a systematic data structure with appropriate headers for identification of the temporal and spatial pixel-domain location of encoded data (Joseph et al., 1992b). When an erroneous/lost packet is detected, these video units serve as resynchronization points for resumption of normal decoding, while the headers provide a means for precisely locating regions of the picture that were not correctly received. Note that two-tier systems may require additional transport-level support for high- and low-priority (HP/LP) resynchronization (Siracusa et al., 1993).
- The video bitstream may optionally be segregated into two layers for prioritized transport (Ghanbari, 1989; Kishno et al., 1989; Karlsson and Vetterli, 1989; Zdepski et al., 1989; Joseph et al., 1992a,b; Siracusa, 1993) when a high degree of error resilience is required. Note that separation into high and low priorities may be achieved either by using a hierarchical (layered) compression algorithm (Ghanbari, 1989; Siracusa, 1993) or by direct codeword parsing (Zdepski et al., 1989; 1990). Note that both these layering mechanisms have been accepted for standardization by MPEG-2 (ISO/IEC, 1995).
- Once the temporal and spatial location(s) corresponding to lost or incorrectly received packets is determined by the decoder, it will execute an error-concealment procedure for replacement of lost picture areas with subjectively acceptable material estimated from available picture regions (Harthanck et al., 1986; Jeng and Lee, 1991; Wang and Zhu, 1991). Generally, this error concealment procedure will be applied to all erased blocks in one-tier (single-priority) transmission systems, while for two-tier (HP/LP) channels the concealment process may optionally ignore loss of LP data.
- In the following subsections, the technical detail of some commonly used error concealment algorithms is provided. Specifically, we focus on the recovery of codeword errors and errors that affect the pixels within a macroblock.

### 17.5.2  ERROR CONCEALMENT ALGORITHMS

In general, design of specific error-concealment strategies depends on the system design. For example, if two-layered transmission is used, the receiver should be designed to conceal high-priority error and low-priority error with different strategies. Moreover, if some redundancy ("steering information") could be added to the encoder the concealment could be more efficient. However, we first assume that the encoder is defined for maximum compression efficiency, and that concealment is only performed in the receiver. It should be noted that some exemptions exist for this assumption. These exemptions include the use of I-frame motion vectors, scalability concealment, and limitation of slice length (to perform acceptable concealment in the pixel domain the limitation of slice length exists, i.e., the length of slices cannot be longer than one row of picture). Figure 17.13 shows a block diagram of a generic one/two-tier video decoder with error concealment.

Note that the figure shows two stages of decoder concealment in the codeword domain and pixel domain, respectively. Codeword domain concealment, in which locally generated decodable codewords (e.g., B-picture motion vectors, end-of-block code, etc.) are inserted into the bitstream, is convenient for implementation of simple temporal replacement functions (which in principle can also be performed in the pixel domain). The second stage of pixel domain processing is for temporal and spatial operations not conveniently done in the codeword domain. Advanced spatial processing will generally have to be performed in the pixel domain, although limited codeword domain options can also be identified.

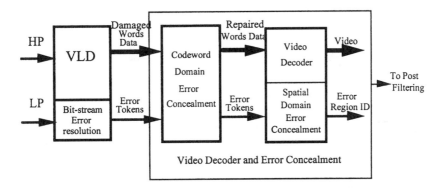

**FIGURE 17.13**    MPEG video decoder with error concealment.

### 17.5.2.1    Codeword Domain Error Concealment

The codeword domain concealment receives video data and error tokens from the transport processor/VLD. Under normal conditions, no action is taken and the data are passed along to the video decoder. When an error token is received, damaged data are repaired to the extent possible by insertion of locally generated codewords and resynchronization codes. An error region ID is also created to indicate the image region to be concealed by subsequent pixel domain processing. Two mechanisms have been used in codeword domain error concealment: neglect the effect of lost data by declaring an end of block (EOB), or replace the lost data with a pseudo-code to handle the macroblock-types or other VLC codes. If high-level data such as dc or macroblock header is lost, the codeword domain concealment with pseudo-codes can only provide signal resynchronization (decodability) and replaces the image scene with a fixed gray level in the error region. Obviously, further improvement is needed in the video decoder. This task is implemented with the error concealment in the video decoder. It is desirable to replace erased I- or P-picture regions with a reasonably accurate estimate to minimize the impact of frame-to-frame propagation.

### 17.5.2.2    Spatiotemporal Error Concealment

In general, two basic approaches are used for spatial domain error concealment: temporal replacement and spatial interpolation. In temporal replacement, as shown in Figure 17.14, the spatially corresponding ones in the previously decoded data with motion compensation replace the damaged blocks in the current frame if motion information is available. This method exploits temporal redundancy in the reconstructed video signals and provides satisfactory results in areas with small

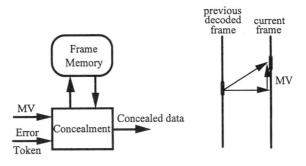

**FIGURE 17.14**    Error concealment uses temporal replenishment with motion compensation.

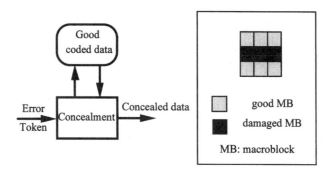

**FIGURE 17.15**   Error concealment uses spatial interpolation with the data from good neighbors. (From Sun, H. and Kwok, W. *IEEE Trans. Image Proc.,* 4(4), 470–477, 1995. With permission.)

motion and for which motion vectors are provided. If motion information is lost, this method will fail in the moving areas. In the method of spatial interpolation as shown in Figure 17.15, the lost blocks are interpolated by the data from the adjacent nonerror blocks with maximally smooth reconstruction criteria or other techniques.

In this method, the correlation between adjacent blocks in the received and reconstructed video signals is exploited. However, severe blurring will result from this method if data in adjacent blocks are also lost. In an MPEG decoder, temporal replacement outlined above is based on previously decoded anchor (I, P) pictures that are available in the frame memory. If motion vectors corresponding to pixels in the erasure region can also be estimated, this temporal replacement operation can be improved via motion compensation. Also, in the MPEG decoder, groups of video pixels (blocks, macroblocks, or slices) are separately decoded, so that pixel values and motion information corresponding to adjacent picture regions are generally available for spatial concealment. However, estimation from horizontally adjacent blocks may not always be useful since cell loss tends to affect a number of adjacent blocks (due to the MPEG and ATM data structures); also differential encoding between horizontally adjacent blocks tends to limit the utility of data obtained from such neighbors. Therefore, most of the usable spatial information will be located in blocks above or below the damaged region. That is, vertical processing/concealment is found to be most useful due to the transmission order of the data.

For I-pictures, the damaged data can be reconstructed by either temporal replacement from the previously decoded anchor frame or by spatial interpolation from good neighbors. These two methods will be discussed later. For P- and B-pictures, the main strategy to conceal the lost data is to replace the region with pixels from the corresponding (and possibly motion-compensated) location in the previously decoded anchor. In this replacement the motion vectors play a very important role. In other words, if "good" estimates of motion information can be obtained, its use may be the least noticeable correction. Since DPCM coding for motion vectors only exploited the correlations between the horizontally neighboring macroblocks, the redundancy between the vertical neighborhood still exists after encoding. Therefore, the lost motion information can be estimated from the vertical neighbors. In the following, three algorithms that have been developed for error concealment in the video decoder are described.

**Algorithm 1**: Spatial interpolation of missing I-picture data and temporal replacement for P- and B-pictures with motion compensation (Sun et al., 1992a):

For I-pictures, dc values of damaged block are replaced by the interpolation from the closest top and bottom good neighbors; the ac coefficients of those blocks are synthesized from the dc values of the surrounding neighboring blocks.

For P-pictures, the previously decoded anchor frames with motion compensation replace the lost blocks. The lost motion vectors are estimated by interpolation of the ones from the top and bottom macroblocks. If motion vectors in both top and bottom macroblocks are not available, zero motion vectors are used. The same strategy is used for B-pictures; the only difference is that the closest anchor frame is used. In other words, the damaged part of the B-picture could be replaced by either the forward or backward anchor frame, depending on its temporal position.

**Algorithm 2:** Temporal replacement of missing I-picture data and temporal replacement for P- and B-pictures with top motion compensation:

For I-pictures, the damaged blocks are replaced with the colocated ones in the previously decoded anchor frame.

For P- and B-pictures, the closest previously decoded anchor frame replaces the damaged part with motion compensation as in the Algorithm 1. The only difference is that the motion vectors are estimated only from the closest top macroblock instead of interpolation of top and bottom motion vectors. This makes the implementation of this scheme much easier. If these motion vectors are not available, then zero motion vectors are used.

In the above two algorithms, the damaged blocks in an I-picture (anchor frame) are concealed by two methods: temporal replacement and spatial interpolation. Temporal replacement is able to provide high-resolution image data to substitute for lost data; however, in motion areas, a big difference might exist between the current intracoded frame and the previously decoded frame. In this case, temporal replacement will produce large shearing distortion unless some motion-based processing can be applied at the decoder. However, this type of processing is not generally available since it is a computationally demanding task to compute motion trajectories locally at the decoder. In contrast, the spatial interpolation approach synthesizes lost data from the adjacent blocks in the same frame. Therefore, the intraframe redundancy between blocks is exploited, while the potential problem of severe blurring due to insufficient high-order ac coefficients for active areas. To alleviate this problem, an adaptive concealment strategy can be used as a compromise; this is described in Algorithm 3.

**Algorithm 3:** Adaptive spatiotemporal replacement of missing I-picture data and temporal replacement with motion compensation for P- and B-pictures:

For I-pictures, the damaged blocks are concealed with temporal replacement or spatial interpolation according to the decision made by the top and bottom macroblocks, which is shown in Figure 17.16. The decision of which concealment method to use will be based on the more cheaply obtained measures of image activity from the neighboring top and bottom macroblocks. One candidate for the decision processor is to make the decision based on prediction error statistics measured in the neighborhood. The decision region is shown in Figure 17.16, where

$$VAR = E\left[(x - \hat{x})^2\right],$$
$$VAROR = E\left[x^2\right] - \mu^2,$$
(17.31)

and $x$ is the neighboring good macroblock data, $\hat{x}$ is the data of the corresponding macroblock in the previously decoded frame at the colocated position, and $\mu$ is the average value of the neighboring good macroblock data in the current frame. One can appreciate that VAR is indicative of the local motion and VAROR of the local spatial detail. If VAR > VAROR and VAR > $T$, where $T$ is a preset threshold value which is set to 5 in the experiments, the concealment method is spatial interpolation; if VAR < VAROR or VAR < T, the concealment method is temporal replacement.

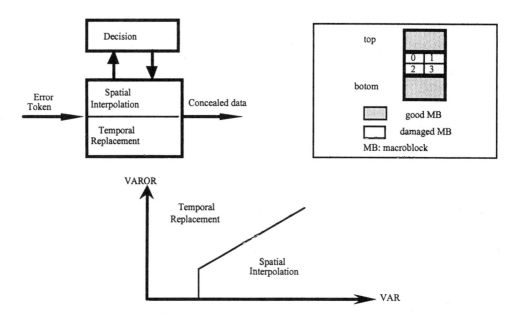

**FIGURE 17.16**  Adaptive error concealment strategy. (From Sun, H. and Kwok, W., *IEEE Trans. Image Proc.*, 4(4), 470–477, 1995. With permission.)

It should be noted that the concealment for luminance is performed on a block basis instead of macroblock basis, while the chrominance is still on the macroblock basis. The detailed decisions for the luminance blocks are described as follows:

- If both top and bottom are temporally replaced, then four blocks (0, 1, 2, and 3) are replaced by the colocated ones (colocated means no motion compensation) in the previously decoded frame.
- If top is temporally replaced and bottom is spatially interpolated, then blocks 0 and 1 are replaced by the colocated ones in the previously decoded anchor frame and blocks 2 and 3 are interpolated from the block boundaries.
- If top is spatially interpolated and bottom is temporally replaced, then blocks 0 and 1 are interpolated from the boundaries, and blocks 2 and 3 are replaced by the colocated ones in the previously decoded anchor frame.
- If both top and bottom are not temporally replaced, all four blocks are spatially interpolated.

In spatial interpolation, a maximal smoothing technique with boundary conditions under certain smoothness measures is used. The spatial interpolation process is carried out with two steps: the mean value of the damaged block is first bilinearly interpolated with ones from the neighboring blocks; then spatial interpolation for each pixel is performed with a Laplacian operator. Minimizing the Laplacian on the boundary pixels using the iterative process (Wang and Zhu, 1991) enforces the process of maximum smoothness.

For P- and B-pictures a similar concealment method is used as in Algorithm 2 except motion vectors from top and bottom neighboring macroblocks are used for top two blocks and bottom two blocks, respectively.

A schematic block diagram for implementation of adaptive error concealment for intracoded frames is given in Figure 17.17. Corrupted macroblocks are first indicated by error tokens obtained via the transport interface. Then, a decision regarding which concealment method (temporal replacement or spatial interpolation) should be used is based on easily obtained measures of image activity

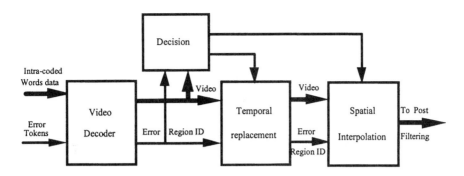

**FIGURE 17.17**  Two-stage error concealment strategy. (From Sun, H. and Kwok, W., *IEEE Trans. Image Proc.*, 4(4), 470–477, 1995. With permission.)

from the neighboring top and bottom macroblocks. The corrupted macroblocks are first classified into two classes according to the local activities. If local motion is smaller than spatial detail, the corrupted macroblocks are defined as the first class and will be concealed by temporal replacement; when local motion is greater than local spatial detail, the corrupted macroblocks are defined as the second class and will be concealed by spatial interpolation. The overall concealment procedure consists of two stages. First, temporal replacement is applied to all corrupted macroblocks of the first class throughout the whole frame. After the temporal replacement stage, the remaining unconcealed damaged macroblocks of the second class are more likely to be surrounded by valid image macroblocks. A stage of spatial interpolation is then performed on them. This will now result in less blurring, or the blurring will be limited to smaller areas. Therefore, a good compromise between shearing (discontinuity or shift of edge or line) and blurring can be obtained.

### 17.5.3  ALGORITHM ENHANCEMENTS

As discussed above, I-picture errors, which are imperfectly concealed, will tend to propagate through all frames in the group of pictures (GOP). Therefore, it is desirable to develop enhancements for the basic spatiotemporal error concealment technique to improve further the accuracy with which missing I-picture pixels are replaced. Three new algorithms have been developed for this purpose. The first is an extension of the spatial restoration technique outlined earlier, and is based on processing of edge information in a large local neighborhood to obtain better restoration of the missing data. The second and third are variations which involve encoder modifications aimed at improved error concealment performance. Specifically, information such as I-picture pseudo-motion vectors, or low-resolution data in a hierarchical compression system are added in the encoder. These redundancies can significantly benefit error concealment in the decoders that must operate under higher cell loss/error conditions, while having a relatively modest impact on nominal image quality.

#### 17.5.3.1  Directional Interpolation

Improvements in spatial interpolation algorithms (for use with MPEG I-pictures) have been proposed (Kwok and Sun, 1993; Sun and Kwok, 1995). In these studies, additional smoothness criteria and/or directional filtering are used for estimating the picture area to be replaced. The new algorithms utilize spatially correlated edge information from a large local neighborhood of surrounding pixels and perform directional or multidirectional interpolation to restore the missing block. The block diagram illustrating the general principle of the restoration process is shown in Figure 17.18.

Three parts are included in the restoration processing: edge classification, spatial interpolation, and pattern mixing. The function of the classifier is to select the top one, two, or three directions that strongly characterize edge orientations in the surrounding neighborhood. Spatial interpolation

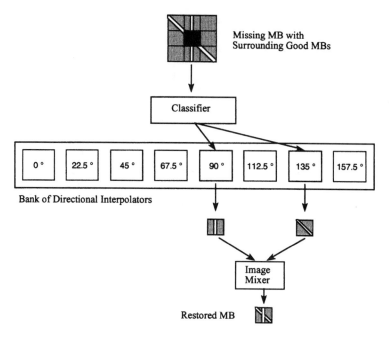

**FIGURE 17.18**   The multidirectional edge restoration process. (From Sun, H. and Kwok, W., *IEEE Trans. Image Proc.*, 4(4), 470–477, 1995. With permission.)

is performed for each of the directions determined by the classifier. For a given direction, a series of 1-D interpolations are carried out along that direction. All of the missing pixels are interpolated from a weighted average of good neighborhood pixels. The weights depend inversely on the distance from the missing pixel to the good neighborhood pixels. The purpose of pattern mixing is to extract strong characteristic features of two or more images and merge them into one image, which is then used to replace the corrupted one. Results show that these algorithms are capable of providing subjectively better edge restoration in missing areas, and may thus be useful for I-picture processing in high-error-rate scenarios. However, the computational practicality of these edge-filtering techniques needs further investigation for given application scenarios.

### 17.5.3.2  I-Picture Motion Vectors

Motion information is very useful in concealing losses in P- and B-frames, but is not available for I-pictures. This limits the concealment algorithm to spatial or direct temporal replacement options described above, which may not always be successful in moving areas of the picture. If motion vectors are made available for all MPEG frames (including intracoded ones) as an aid for error concealment (Sun et al., 1992a), good error concealment performance can be obtained without the complexity of adaptive spatial processing. Therefore, a syntax extension has been adopted by the MPEG-2 where motion vectors can be transmitted in an I-picture as the redundancy for error-concealment purposes (Sun et al., 1992b). The macroblock syntax is unchanged, however, motion vectors are interpreted in the following way: the decoded forward motion vectors belong to the macroblock spatially below the current macroblock, and describe how that macroblock can be replaced from the previous anchor frame in the event that the macroblock cannot be recovered. Simulation results have shown that subjective picture quality with I-picture motion vectors is noticeably superior to conventional temporal replacement, and that the overhead for transmitting the additional motion vectors is less than 0.7% of the total bit rate at a bit rate of about 6 to 7 Mbps.

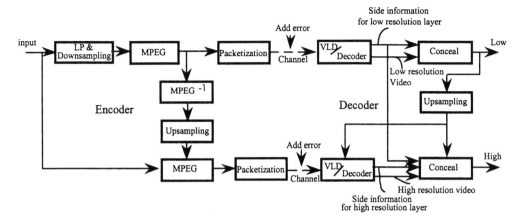

**FIGURE 17.19**   Block diagram of spatial scalability with error concealment.

### 17.5.3.3   Spatial Scalable Error Concealment

This approach for error concealment of MPEG video is based on the scalability (or hierarchy) feature of MPEG-2 (ISO/IEC, 1995). Hierarchical transmission provides more possibilities for error concealment, when a corresponding two-tier transmission media is available. A block diagram illustrating the general principle of coding system with spatial scalability and error concealment is shown in Figure 17.19.

It should be noted that the concept of scalable error concealment is different from the two-tier concept with data partitioning. Scalable concealment uses the spatial scalability feature in MPEG-2, while the two-tier case uses the data partitioning feature of MPEG-2, in which the data corresponds to the same spatial resolution layer but is partitioned to two parts with a breakpoint. In spatial scalability, the encoder produces two separate bitstreams: one for the low-resolution base layer and another for the high-resolution enhancement. The high-resolution layer is encoded with an adaptive choice of temporal prediction from previous anchor frames and compatible spatial prediction (obtained from the up-sampled low-resolution layer) corresponding to the current temporal reference. In the decoder, redundancies that exist in the scaling data greatly benefit the error concealment processing. In a simple experiment with spatially scalable MPEG-2, we consider a scenario in which losses in the high-resolution MPEG-2 video are concealed with information from the low-resolution layer. Actually, there are two kinds of information in the lower layer that can be used to conceal the data loss in the high-resolution layer: up-sampled picture data and scaled motion information. Therefore, three error concealment approaches are possible:

1. *Up-sampled substitution*: Lost data are replaced by colocated up-sampled data in the low-resolution decoded frame. The up-sampled picture is obtained from the low-resolution picture with proper up-sampling filter.
2. *Mixed substitution*: Lost macroblocks in I-picture are replaced by colocated up-sampled macroblocks in the low-resolution decoded frame, while lost macroblocks in P- and B-picture are temporally replaced by the previously decoded anchor frame with the motion vectors for the low-resolution layer.
3. *Motion vector substitution*: The previously decoded anchor frame with the motion vectors replaces lost macroblocks for the low-resolution layer appropriately scaled.

Since motion vectors are not available for I-pictures, obviously, method 3 does not work for I-pictures (unless I-picture motion vectors, concealment motion vectors, of MPEG-2 are generated

**TABLE 17.4**
**Subjective Quality Comparison**

| Picture Material | Items | Alg 1 | Alg 2 | Alg 3 | Comments |
|---|---|---|---|---|---|
| Still | Blurring | High | None | Low | Temporal replacement works very well in no- |
| | Shearing | None | None | None | motion areas |
| | Artifact blocking | Medium | None | Low | |
| Slow motion | Blurring | High | None | Low | Temporal replacement works well in slow- |
| | Shearing | None | Low | Low | motion areas |
| | Artifact blocking | Medium | None | Low | |
| Fast motion | Blurring | High | None | Medium | Temporal replacement causes more shearing; |
| | Shearing | None | High | Low | spatial interpolation results in blurring; adaptive |
| | Artifact blocking | High | Low | Medium | strategy limits blurring in smaller areas |
| Overall | The adaptive strategy of steering the temporal replacement and spatial interpolation according to the | | | | |
| | measures of local activity and local motion gives a good compromise between shearing and blurring | | | | |

in encoder). Simulation results have shown that, on average, the up-sampled substitution outperforms the other two, and mixed substitution also provides acceptable results in the case of video
with smooth motion.

### 17.5.4 SUMMARY OF ERROR CONCEALMENT

In this section, a general class of error-concealment algorithms for MPEG video has been discussed.
The error-concealment approaches that have been described are practical for current MPEG decoder
implementations, and have been demonstrated to provide significant robustness. Specifically, it has
been shown that the adaptive spatiotemporal algorithm can provide reasonable picture quality at
cell loss ratios (CLR) as high as $10^{-3}$ when used in conjunction. These results confirm that
compressed video is far less fragile than originally believed when appropriate transport and concealment techniques are employed. The results can be summarized as in Table 17.4.

Several concealment algorithm extensions based on directional filtering, I-picture pseudo-
motion vectors, and MPEG-2 scalability were also considered and shown to provide performance
gains that may be useful in certain application scenarios. In view of the practical benefits of robust
video delivery, it is recommended that such error resilience functions (along with associated
transport structures) be important for implementation in emerging TV, HDTV, teleconferencing,
and multimedia systems if the cell loss rates on these transmission systems are significant. Particularly for terrestrial broadcasting and ATM network scenarios, we believe that robust video delivery
based on decoder error concealment is an essential element of a viable system design.

## 17.6 SUMMARY

In this chapter, several application issues of MPEG-2 are discussed. The most successful application
of MPEG-2 is the U.S. HDTV standard. The other application issues include transcoding with
bitstream scaling, down-conversion decoding, and error concealment. Transcoding is a very interesting topic that converts the bitstreams between different standards. The error concealment is very
useful in the noisy communication channels such as terrestrial television broadcasting. The down-
conversion decoder responds to the market requirement during the DTV transition-period and long-
term need for displaying DTV signals on computer monitors.

## 17.7 EXERCISES

**17-1.** In DTV applications, describe the advantages and disadvantages of interlaced format and progressive format. Explain why the computer industry favors progressive format and TV manufacturers like interlaced format.

**17-2.** Do all DTV formats have square pixel format? Why is square pixel format important for digital television?

**17-3.** The bitstream scaling is one kind of transcoding; according to your knowledge, describe several other kinds of transcoding (such as MPEG-1 to JPEG) and propose a feasible solution to achieve the transcoding requirements.

**17-4.** What type of MPEG-2 frames will cause a higher degree of error propagation if errors occur? What technique of error concealment is allowed by the MPEG-2 syntax? Using this technique, perform simulations with several images to determine the penalty in the case of no errors.

**17-5.** To reduce the drift in a down-conversion decoder, what coding parameters can be chosen at the encoder? Will these actions affect the coding performance?

**17-6.** What are the advantages and disadvantages of a down-conversion decoder in the frequency domain and spatial domain?

## REFERENCES

Bao, J., H. Sun, and T. Poon, HDTV down-conversion decoder, *IEEE Trans. Consumer Elec.,* 42(3), 402-410, 1996.

Boyce, J., J. Henderson, and L. Pearlestien, An SDTV decoder with HDTV capability: an all-format ATV decoder, presented at SMPTE Fall Conference, New Orleans, 1995.

Bruni, R., A. Chimienti, M. Lucenteforte, D. Pau, and R. Sannino, A novel adaptive vector quantization method for memory reduction in MPEG-2 HDTV receivers, *IEEE Trans. Consumer Elec.,* 44(3), 537-544, 1998.

de With, P. H. N., P. H. Frencken, and M. v.d. Schaar-Mitrea, An MPEG decoder with embedded compression for memory reduction, *IEEE Trans. Consumer Elec.,* 44(3), 545-555, 1998.

Ghanbari, M. Two-layer coding of video signals for VBR networks, *IEEE J. Selected Areas Comm.,* 7(5), 771-781, 1989.

Gonzalez, R. C. and P. Wintz, *Digital Image Processing,* 2nd ed., Addison-Wesley, Reading, MA, 1987, 232-233.

Grand Alliance, HDTV System Specification Version 2.0, December 7, 1994.

Harthanck, W., W. Keesen, and D. Westerkamp, Concealment techniques for block encoded TV-signals, presented at Picture Coding Symposium, 1986.

Isnardi, M. A. Consumers seek easy-to-use products, *IEEE Spect.,* 64, Jan. 1993.

ISO/IEC, MPEG Test Model 5, ISO/IEC JTC/SC29/WG11 Document. April, 1993.

ISO/IEC, MPEG-2 International Standard. Video Recommendation ITU-T H.262, ISO/IEC 13818-2, Jan. 10, 1995.

Jayant, N. N. and P. Noll, *Digital Coding of Waveforms to Speech and Video,* Prentice-Hall, Englewood Cliffs, NJ, 1984.

Jeng, F.-C. and S. H. Lee, Concealment of bit error and cell loss in inter-frame coded video transmission, *ICC Proceeding,* ICC'91, 496-500, 1991.

Joseph, K., S. Ng, D. Raychaudhuri, R. Saint Girons, T. Savatier, R. Siracusa, and J. Zdepski, MPEG++: A robust compression and transport system for digital HDTV, *Signal Proc. Image Comm.,* 4, 307-323, 1992a.

Joseph, K., S. Ng, D. Raychaudhuri, R. Saint Girons, R. Siracusa, and J. Zdepski, Prioritization and transport in the ADTV digital simulcast system, *Proceedings ICCE '92,* 1992b.

Karlsson, G. and M. Vetterli, Packet video and its integration into the network architecture, *IEEE J. Selected Areas Communication,* 739-751, 1989.

Kishino, F., K. Manabe, Y. Hayashi, and H. Yasuda, Variable bit-rate coding of video signals for ATM networks, *IEEE J. Selected Areas Comm.,* 7(5), 801-806, 1989.

Kwok, W. and H. Sun, Multi-directional interpolation for spatial error concealment, *IEEE Trans. Consumer Elec.*, 455-460, 1993.

Lancaster, P. and M. Tismenetsky, *The Theory of Matrices with Application*, Academic Press, Boston, 1985.

Lei, S., A quadtree embedded compression algorithm for memory saving DTV decoders, *Proceedings International Conference on Consumer Electronics*, Los Angeles, CA, June 1999.

Merhav, N. and V. Bhaskaran, Fast algorithms for DCT-domain image down-sampling and for inverse motion compensation, *IEEE Trans. Circ. Syst. Video Technol.*, 7(3), 468-476, 1997.

Mokry, R. and D. Anastassiou, Minimal error drift in frequency scalability for motion-compensated DCT coding, *IEEE Trans. Circ. Syst. Video Technol.*, 4(4), 392-406, 1994.

Ng, S. Thompson Consumer Electronics, Low Resolution HDTV Receivers, U.S. patent 5,262,854, Nov. 16, 1993.

Pang, K. K., H. G. Lim, S. Dunstan, and J. M. Badcock, Frequency domain decimation and interpolation techniques, presented at Picture Coding Symposium, Melbourne, Australia, March 1996.

Perkins, M. and D. Arnstein. Statistical multiplexing of multiple MPEG-2 video programs in a single channel, *SMPTE J.*, September 1995.

Reitmeier, G. A. The U.S. advanced television standard and its impact on VLSI, *VLSI and Signal Processing, Systems for Signal, Image, and Video Technology*, Kluwer Academic Press, 1997.

Siracusa, R., K. Joseph, J. Zdepski, and D. Raychaudhuri, Flexible and robust packet transport for digital HDTV, *IEEE J. Selected Areas Comm.*, 11(1), ISACEM, 88-98, 1993.

Sun, H. Hierarchical decoder for MPEG compressed video data *IEEE Trans. Consumer Elec.*, 39(3), 559-562, 1993.

Sun, H., K. Challapali, and J. Zdepski, Error concealment in simulcast AD-HDTV decoder, *IEEE Trans. Consumer Elec.*, 38(3), 108-118, 1992.

Sun, H., M. Uz, J. Zdepski, and R. Saint Girons, A Proposal for Increased Error Resilience, ISO-IEC/JTC1?SC29/WG11, MPEG92, Sept. 30, 1992b.

Sun, H. and W. Kwok, Restoration of damaged block transform coded image using projection onto convex sets, *IEEE Trans. Image Process.*, 4(4), IIPRE4, 470-477, 1995.

Vetro, A. and H. Sun, On the motion compensation within a down-conversion decoder, *J. Elec. Imag.*, 7(3), 1998a.

Vetro, A. and H. Sun, Frequency domain down-conversion using an optimal motion compensation scheme, *J. Imag. Sci. Technol.*, 9(4), 1998b.

Vetro, A., H. Sun, P. DaGraca, and T. Poon, Minimum drift architectures for three-layer scalable DTV decoding, *IEEE Trans. Consumer Elec.*, 44(3), 1998.

Wang, Y. and Q.-F. Zhu, Signal loss recovery in DCT-based image and video codecs, *Proceedings of SPIE on Visual Communication and Image Processing*, Boston, 667-678, Nov. 1991.

Yu, H., W.-M. Lam, B. Canfield, and B. Beyers, Block-based image processor for memory efficient MPEG video decoding, *Proceedings International Conference on Consumer Electronics*, Los Angeles, CA, June 1999.

Zdepski, J. et al., Packet transport of rate-free interframe DCT compressed digital video on a CSMA/CD LAN, *Proceedings IEEE Global Conference on Communications*, Dallas TX, Nov. 1989.

Zdepski, J. et al., Prioritized Packet Transport of VBR CCITT H.261 Format Compressed Video on a CSMA/CD LAN, Third International Workshop on Packet Video, Morristown, NJ, March 22-23, 1990.

# 18 MPEG-4 Video Standard: Content-Based Video Coding

This chapter provides an overview of the ISO MPEG-4 standard. The MPEG-4 work includes natural video, synthetic video, audio and systems. Both natural and synthetic video have been combined into a single part of the standard, which is referred to as MPEG-4 visual (ISO/IEC, 1998a). It should be emphasized that neither MPEG-1 nor MPEG-2 considers synthetic video (or computer graphics) and the MPEG-4 is also the first standard to consider the problem of content-based coding. Here, we focus on the video parts of the MPEG-4 standard.

## 18.1 INTRODUCTION

As we discussed in the previous chapters, MPEG has completed two standards: MPEG-1 that was mainly targeted for CD-ROM applications up to 1.5 Mbps and MPEG-2 for digital TV and HDTV applications at bit rates between 2 and 30 Mbps. In July 1993, MPEG started its new project, MPEG-4, which was targeted at providing technology for multimedia applications. The first working draft (WD) was completed in November 1996, and the committee draft (CD) of version 1 was completed in November 1997. The draft international standard (DIS) of MPEG-4 was completed in November of 1998, and the international standard (IS) of MPEG-4 version 1 was completed in February of 1999. The goal of the MPEG-4 standard is to provide the core technology that allows efficient content-based storage, transmission, and manipulation of video, graphics, audio, and other data within a multimedia environment. As we mentioned before, there exist several video-coding standards such as MPEG-1/2, H.261, and H.263. Why do we need a new standard for multimedia applications? In other words, are there any new attractive features of MPEG-4 that the current standards do not have or cannot provide? The answer is yes. The MPEG-4 has many interesting features that will be described later in this chapter. Some of these features are focused on improving coding efficiency; some are used to provide robustness of transmission and interactivity with the end user. However, among these features the most important one is the content-based coding. MPEG-4 is the first standard that supports content-based coding of audio visual objects. For content providers or authors, the MPEG-4 standard can provide greater reusability, flexibility, and manageability of the content that is produced. For network providers, MPEG-4 will offer transparent information, which can be interpreted and translated into the appropriate native signaling messages of each network. This can be accomplished with the help of relevant standards bodies that have the jurisdiction. For end users, MPEG-4 can provide much functionality to make the user terminal have more capabilities of interaction with the content. To reach these goals, MPEG-4 has the following important features:

The contents such as audio, video, or data are represented in the form of primitive audio visual objects (AVOs). These AVOs can be natural scenes or sounds, which are recorded by video camera or synthetically generated by computers.

The AVOs can be composed together to create compound AVOs or scenes.

The data associated with AVOs can be multiplexed and synchronized so that they can be transported through network channels with certain quality requirements.

## 18.2   MPEG-4 REQUIREMENTS AND FUNCTIONALITIES

Since the MPEG-4 standard is mainly targeted at multimedia applications, there are many requirements to ensure that several important features and functionalities are offered. These features include
the allowance of interactivity, high compression, universal accessibility, and portability of audio
and video content. From the MPEG-4 video requirement document, the main functionalities can
be summarized by the following three aspects: content-based interactivity, content-based efficient
compression, and universal access.

### 18.2.1   Content-Based Interactivity

In addition to provisions for efficient coding of conventional video sequences, MPEG-4 video has
the following features of content-based interactivity.

#### 18.2.1.1   Content-Based Manipulation and Bitstream Editing

The MPEG-4 supports the content-based manipulation and bitstream coding without the need for
transcoding. In MPEG-1 and MPEG-2, there is no syntax and no semantics for supporting true
manipulation and editing in the compressed domain. MPEG-4 provides the syntax and techniques
to support content-based manipulation and bitstream editing. The level of access, editing, and
manipulation can be done at the object level in connection with the features of content-based
scalability.

#### 18.2.1.2   Synthetic and Natural Hybrid Coding (SNHC)

The MPEG-4 supports combining synthetic scenes or objects with natural scenes or objects. This
is for "compositing" synthetic data with ordinary video, allowing for interactivity. The related
techniques in MPEG-4 for supporting this feature include sprite coding, efficient coding of 2-D
and 3-D surfaces, and wavelet coding for still textures.

#### 18.2.1.3   Improved Temporal Random Access

The MPEG-4 provides and efficient method to access randomly, within a limited time, and with
the fine resolution parts, e.g., video frames or arbitrarily shaped image objects from an audiovisual
sequence. This includes conventional random access at very low bit rate. This feature is also
important for content-based bitstream manipulation and editing.

### 18.2.2   Content-Based Efficient Compression

One initial goal of MPEG-4 is to provide a highly efficient coding tool with high compression at
very low rates. But this goal has now extended to a large range of bit rates from 10 Kbps to
5 Mbps, which covers QSIF to CCIR601 video formats. Two important items are included in this
requirement.

#### 18.2.2.1   Improved Coding Efficiency

The MPEG-4 video standard provides subjectively better visual quality at comparable bit rates
compared with the existing or emerging standards, including MPEG-1/2 and H.263. MPEG-4 video
contains many new tools, which optimize the code in different bit rate ranges. Some experimental
results have shown that it outperforms MPEG-2 and H.263 at the low bit rates. Also, the content-
based coding reaches the similar performance of the frame-based coding.

### 18.2.2.2 Coding of Multiple Concurrent Data Streams

The MPEG-4 provides the capability of coding multiple views of a scene efficiently. For stereo-scopic video applications, MPEG-4 allows the ability to exploit redundancy in multiple viewing points of the same scene, permitting joint coding solutions that allow compatibility with normal video as well as the ones without compatibility constraints.

## 18.2.3 UNIVERSAL ACCESS

The another important feature of the MPEG-4 video is the feature of universal access.

### 18.2.3.1 Robustness in Error-Prone Environments

The MPEG-4 video provides strong error robustness capabilities to allow access to applications over a variety of wireless and wired networks and storage media. Sufficient error robustness is provided for low-bit-rate applications under severe error conditions (e.g., long error bursts).

### 18.2.3.2 Content-Based Scalability

The MPEG-4 video provides the ability to achieve scalability with fine granularity in content, quality (e.g., spatial and temporal resolution), and complexity. These scalabilities are especially intended to result in content-based scaling of visual information.

## 18.2.4 SUMMARY OF MPEG-4 FEATURES

From above description of MPEG-4 features, it is obvious that the most important application of MPEG-4 will be in a multimedia environment. The media that can use the coding tools of MPEG-4 include computer networks, wireless communication networks, and the Internet. Although it can also be used for satellite, terrestrial broadcasting, and cable TV, these are still the territories of MPEG-2 video since MPEG-2 already has made such a large impact in the market. A large number of silicon solutions exist and its technology is more mature compared with the current MPEG-4 standard. From the viewpoint of coding theory, we can say there is no significant breakthrough in MPEG-4 video compared with MPEG-2 video. Therefore, we cannot expect to have a significant improvement of coding efficiency when using MPEG-4 video over MPEG-2. Even though MPEG-4 optimizes its performance in a certain range of bit rates, its major strength is that it provides more functionality than MPEG-2. Recently, MPEG-4 added the necessary tools to support interlaced material. With this addition, MPEG-4 video does support all functionalities already provided by MPEG-1 and MPEG-2, including the provision to compress efficiently standard rectangular-sized video at different levels of input formats, frame rates, and bit rates.

Overall, the incorporation of an object- or content-based coding structure is the feature that allows MPEG-4 to provide more functionality. It enables MPEG-4 to provide the most elementary mechanism for interactivity and manipulation with objects of images or video in the compressed domain without the need for further segmentation or transcoding at the receiver, since the receiver can receive separate bitstreams for different objects contained in the video. To achieve content-based coding, the MPEG-4 uses the concept of a video object plane (VOP). It is assumed that each frame of an input video is first segmented into a set of arbitrarily shaped regions or VOPs. Each such region could cover a particular image or video object in the scene. Therefore, the input to the MPEG-4 encoder can be a VOP, and the shape and the location of the VOP can vary from frame to frame. A sequence of VOPs is referred to as a video object (VO). The different VOs may be encoded into separate bitstreams. MPEG-4 specifies demultiplexing and composition syntax which provide the tools for the receiver to decode the separate VO bitstreams and composite them into a

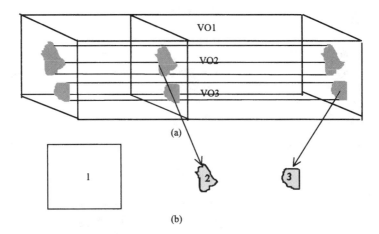

**FIGURE 18.1**    Video object definition and format: (a) video object, (b) VOPs.

frame. In this way, the decoders have more flexibility to edit or rearrange the decoded video objects. The detailed technical issues will be addressed in the following sections.

## 18.3    TECHNICAL DESCRIPTION OF MPEG-4 VIDEO

### 18.3.1    OVERVIEW OF MPEG-4 VIDEO

The major feature of MPEG-4 is to provide the technology for object-based compression, which is capable of separately encoding and decoding video objects. To explain the idea of object-based coding clearly, we should review the set of video object-related definitions. An image scene may contain several objects. In the example of Figure 18.1, the scene contains the background and two objects. The time instant of each video object is referred to as the VOP. The concept of a VO provides a number of functionalities of MPEG-4, which are either impossible or very difficult in MPEG-1 or MPEG-2 video coding. Each video object is described by the information of texture, shape, and motion vectors. The video sequence can be encoded in a way that will allow the separate decoding and reconstruction of the objects and allow the editing and manipulation of the original scene by simple operation on the compressed bitstream domain. The feature of object-based coding is also able to support functionality such as warping of synthetic or natural text, textures, image, and video overlays on reconstructed video objects.

Since MPEG-4 aims at providing coding tools for multimedia environments, these tools not only allow one to compress natural video objects efficiently, but also to compress synthetic objects, which are a subset of the larger class of computer graphics. The tools of MPEG-4 video includes the following:

- Motion estimation and compensation
- Texture coding
- Shape coding
- Sprite coding
- Interlaced video coding
- Wavelet-based texture coding
- Generalized temporal and spatial as well as hybrid scalability
- Error resilience.

The technical details of these tools will be explained in the following sections.

## 18.3.2 Motion Estimation and Compensation

For object-based coding, the coding task includes two parts: texture coding and shape coding. The current MPEG-4 video texture coding is still based on the combination of motion-compensated prediction and transform coding. Motion-compensated predictive coding is a well-known approach for video coding. Motion compensation is used to remove interframe redundancy, and transform coding is used to remove intraframe redundancy, as in the MPEG-2 video-coding scheme. However, there are lots of modifications and technical details in MPEG-4 for coding a very wide range of bit rates. Moreover, MPEG-4 coding has been optimized for low-bit-rate applications with a number of new tools. In other words, MPEG-4 video coding uses the most common coding technologies, such as motion compensation and transform coding, but at the same time, it modifies some traditional methods such as advanced motion compensation and also creates some new features, such as sprite coding.

The basic technique to perform motion-compensated predictive coding for coding a video sequence is motion estimation (ME). The basic ME method used in the MPEG-4 video coding is still the block-matching technique. The basic principle of block matching for motion estimation is to find the best-matched block in the previous frame for every block in the current frame. The displacement of the best-matched block relative to the current block is referred to as the motion vector (MV). Positive values for both motion vector components indicate that the best-matched block is on the bottom right of the current block. The motion-compensated prediction difference block is formed by subtracting the pixel values of the best-matched block from the current block, pixel by pixel. The difference block is then coded by a texture-coding method. In MPEG-4 video coding, the basic technique of texture coding is a discrete cosine transformation (DCT). The coded motion vector information and difference block information is contained in the compressed bitstream, which is transmitted to the decoder. The major issues in the motion estimation and compensation are the same as in the MPEG-1 and MPEG-2 which include the matching criterion, the size of search window (searching range), the size of matching block, the accuracy of motion vectors (one pixel or half-pixel), and inter/intramode decision. We are not going to repeat these topics and will focus on the new features in the MPEG-4 video coding. The feature of the advanced motion prediction is a new tool of MPEG-4 video. This feature includes two aspects: adaptive selection of $16 \times 16$ block or four $8 \times 8$ blocks to match the current $16 \times 16$ block and overlapped motion compensation for luminance block.

### 18.3.2.1 Adaptive Selection of $16 \times 16$ Block or Four $8 \times 8$ Blocks

The purpose of the adaptive selection of the matching block size is to enhance coding efficiency further. The coding performance may be improved at low bit rate since the bits for coding prediction difference could be greatly reduced at the limited extra cost for increasing motion vectors. Of course, if the cost of coding motion vectors is too high, this method will not work. The decision in the encoder should be very careful. For explaining the procedure of how to make decisions, we define $\{C(i,j), i,j = 0, 1,..., N-1\}$ to be the pixels of the current block and $\{P(i,j), i,j = 0, 1, ..., N-1\}$ to be the pixels in the search window in the previous frame. The sum of absolute difference (SAD) is calculated as

$$SAD_N(x,y) = \begin{cases} \displaystyle\sum_{i=0}^{N-1}\sum_{j=0}^{N-1}|C(i,j) - P(i,j)| - T & \text{if}(x,y) = (0,0) \\ \displaystyle\sum_{i=0}^{N-1}\sum_{j=0}^{N-1}|C(i,j) - P(i+x,j+y)| & \text{otherwise,} \end{cases} \tag{18.1}$$

where $(x, y)$ is the pixel within the range of searching window, and $T$ is a positive constant. The following steps then make the decision:

Step 1:  To find $SAD_{16}(MV_x, MV_y)$;
Step 2:  To find $SAD_8(MV1_x, MV1_y)$, $SAD_8(MV2_x, MV2_y)$, $SAD_8(MV3_x, MV3_y)$, and $SAD_8(MV4_x, MV4_y)$;
Step 3:  If

$$\sum_{i=1}^{4} SAD_8\left(MV_{ix}, MV_{iy}\right) < SAD_{16}\left(MV_x, MV_y\right) - 128,$$

then choose $8 \times 8$ prediction; otherwise, choose $16 \times 16$ prediction.

If the $8 \times 8$ prediction is chosen, there are four motion vectors for the four $8 \times 8$ luminance blocks that will be transmitted. The motion vector for the two chrominance blocks is then obtained by taking an average of these four motion vectors and dividing the average value by a factor of two. Since each motion vector for the $8 \times 8$ luminance block has half-pixel accuracy, the motion vector for the chrominance block may have a sixteenth pixel accuracy.

### 18.3.2.2  Overlapped Motion Compensation

This kind of motion compensation is always used for the case of four $8 \times 8$ blocks. The case of one motion vector for a $16 \times 16$ block can be considered as having four identical $8 \times 8$ motion vectors, each for an $8 \times 8$ block. Each pixel in an $8 \times 8$ of the best-matched luminance block is a weighted sum of three prediction values specified in the following equation:

$$p'(i,j) = \left(H_0(i,j) \cdot q(i,j) + H_1(i,j) \cdot r(i,j) + H_2(i,j) \cdot s(i,j)\right)/8, \tag{18.2}$$

where division is with round-off. The weighting matrices are specified as:

$$H_0 = \begin{bmatrix} 4 & 5 & 5 & 5 & 5 & 5 & 5 & 4 \\ 5 & 5 & 5 & 5 & 5 & 5 & 5 & 5 \\ 5 & 5 & 6 & 6 & 6 & 6 & 5 & 5 \\ 5 & 5 & 6 & 6 & 6 & 6 & 5 & 5 \\ 5 & 5 & 6 & 6 & 6 & 6 & 5 & 5 \\ 5 & 5 & 6 & 6 & 6 & 6 & 6 & 6 \\ 5 & 5 & 5 & 5 & 5 & 5 & 5 & 5 \\ 4 & 5 & 5 & 5 & 5 & 5 & 5 & 4 \end{bmatrix}, \quad H_1 = \begin{bmatrix} 2 & 2 & 2 & 2 & 2 & 2 & 2 & 2 \\ 1 & 1 & 2 & 2 & 2 & 2 & 1 & 1 \\ 1 & 1 & 1 & 1 & 1 & 1 & 1 & 1 \\ 1 & 1 & 1 & 1 & 1 & 1 & 1 & 1 \\ 1 & 1 & 1 & 1 & 1 & 1 & 1 & 1 \\ 1 & 1 & 1 & 1 & 1 & 1 & 1 & 1 \\ 1 & 1 & 2 & 2 & 2 & 2 & 1 & 1 \\ 2 & 2 & 2 & 2 & 2 & 2 & 2 & 2 \end{bmatrix}, \text{ and}$$

$$H_2 = \begin{bmatrix} 2 & 1 & 1 & 1 & 1 & 1 & 1 & 2 \\ 2 & 2 & 1 & 1 & 1 & 1 & 2 & 2 \\ 2 & 2 & 1 & 1 & 1 & 1 & 2 & 2 \\ 2 & 2 & 1 & 1 & 1 & 1 & 2 & 2 \\ 2 & 2 & 1 & 1 & 1 & 1 & 2 & 2 \\ 2 & 2 & 1 & 1 & 1 & 1 & 2 & 2 \\ 2 & 2 & 1 & 1 & 1 & 1 & 2 & 2 \\ 2 & 1 & 1 & 1 & 1 & 1 & 1 & 2 \end{bmatrix}$$

It is noted that $H_0(i,j) + H_1(i,j) + H_2(i,j) = 8$ for all possible $(i,j)$. The value of $q(i,j)$, $r(i,j)$, and $s(i,j)$ are the values of the pixels in the previous frame at the locations,

$$q(i,j) = p(i + MV_x^0, j + MV_y^0),$$

$$r(i,j) = p(i + MV_x^1, j + MV_y^1), \tag{18.3}$$

$$s(i,j) = p(i + MV_x^2, j + MV_y^2),$$

where $(MV_x^0, MV_y^0)$ is the motion vector of the current $8 \times 8$ luminance block $p(i,j)$, $(MV_x^1, MV_y^1)$ is the motion vector of the block either above (for $j = 0,1,2,3$) or below (for $j = 4,5,6,7$) the current block and $(MV_x^2, MV_y^2)$ is the motion vector of the block either to the left (for $i = 0,1,2,3$) or right (for $i = 4,5,6,7$) of the current block. The overlapped motion compensation can reduce the prediction noise at a certain level.

### 18.3.3   TEXTURE CODING

Texture coding is used to code the intra-VOPs and the prediction residual data after motion compensation. The algorithm for video texture coding is based on the conventional $8 \times 8$ DCT with motion compensation. DCT is performed for each luminance and chrominance block, where the motion compensation is performed only on the luminance blocks. This algorithm is similar to those in H.263 and MPEG-1 as well as MPEG-2. However, MPEG-4 video texture coding has to deal with the requirement of object-based coding, which is not included in the other video-coding standards. In the following we will focus on the new features of the MPEG-4 video coding. These new features include the intra-DC and AC prediction for I-VOP and P-VOP, the algorithm of motion estimation and compensation for arbitrary shape VOP, and the strategy of arbitrary shape texture coding. The definitions of I-VOP, P-VOP, and B-VOP are similar to the I-picture, P-picture, and B-picture in Chapter 16 for MPEG-1 and MPEG-2.

#### 18.3.3.1   Intra-DC and AC Prediction

In the intramode coding, the predictive coding is not only applied on the DC coefficients but also the AC coefficients to increase the coding efficiency. The adaptive DC prediction involves the selection of the quantized DC (QDC) value of the immediately left block or the immediately above block. The selection criterion is based on comparison of the horizontal and vertical DC gradients around the block to be coded. Figure 18.2 shows the three surrounding blocks "A," "B," and "C" to the current block "X" whose QDC is to be coded where block "A", "B," and "C" are the immediately left, immediately left and above, and immediately above block to the "X," respectively. The QDC value of block "X," $QDC_X$, is predicted by either the QDC value of block "A," $QDC_A$,

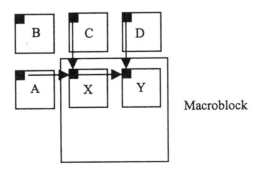

Macroblock

**FIGURE 18.2**   Previous neighboring blocks used in DC prediction. (From ISO/IEC 14496-2 Video Verification Model V.12, N2552, Dec. 1998. With permission.)

or the QDC value of block "C," $QDC_C$, based on the comparison of horizontal and vertical gradients as follows:

$$\text{If} \quad \left|QDC_A - QDC_B\right| < \left|QDC_B - QDC_C\right|, \quad QDC_P = QDC_C;$$
$$\text{Otherwise} \quad QDC_P = QDC_A. \tag{18.4}$$

The differential DC is then obtained by subtracting the DC prediction, $QDC_P$, from $QDC_X$. If any of block "A", "B," or "C" are outside of the VOP boundary, or they do not belong to an intracoded block, their QDC value are assumed to take a value of 128 (if the pixel is quantized to 8 bits) for computing the prediction. The DC prediction is performed similarly for the luminance and each or the two chrominance blocks.

For AC coefficient prediction, either coefficients from the first row or the first column of a previous coded block are used to predict the cosited (same position in the block) coefficients in the current block. On a block basis, the same rule for selecting the best predictive direction (vertical or horizontal direction) for DC coefficients is also used for the AC coefficient prediction. A difference between DC prediction and AC prediction is the issue of quantization scale. All DC values are quantized to the 8 bits for all blocks. However, the AC coefficients may be quantized by the different quantization scales for the different blocks. To compensate for differences in the quantization of the blocks used for prediction, scaling of prediction coefficients becomes necessary. The prediction is scaled by the ratio of the current quantization step size and the quantization step size of the block used for prediction. In the cases when AC coefficient prediction results in a larger range of prediction errors as compared with the original signal, it is desirable to disable the AC prediction. The decision of AC prediction switched on or off is performed on a macroblock basis instead of a block basis to avoid excessive overhead. The decision for switching on or off AC prediction is based on a comparison of the sum of the absolute values of all AC coefficients to be predicted in a macroblock and that of their predicted differences. It should be noted that the same DC and AC prediction algorithm is used for the intrablocks in the intercoded VOP. If any blocks used for prediction are not intrablocks, the QDC and QAC values used for prediction are set to 128 and 0 for DC and AC prediction, respectively.

### 18.3.3.2  Motion Estimation/Compensation of Arbitrarily Shaped VOP

In previous sections we discussed the general issues of motion estimation (ME) and motion compensation (MC). Here we are going to discuss the ME and MC for coding the texture in the arbitrarily shaped VOP. In an arbitrarily shaped VOP, the shape information is given by either binary shape information or alpha components of a gray-level shape information. If the shape information is available to both encoder and decoder, three important modifications have to be considered for the arbitrarily shaped VOP. The first is for the blocks, which are located in the border of VOP. For these boundary blocks, the block-matching criterion should be modified. Second, a special padding technique is required for the reference VOP. Finally, since the VOPs have arbitrary shapes rather than rectangular shapes, and the shapes change from time to time, an agreement on a coordinate system is necessary to ensure the consistency of motion compensation. At the MPEG-4 video, the absolute frame coordinate system is used for referencing all of the VOPs. At each particular time instance, a bounding rectangle that includes the shape of that VOP is defined. The position of upper-left corner in the absolute coordinate in the VOP spatial reference is transmitted to the decoder. Thus, the motion vector for a particular block inside a VOP is referred to as the displacement of the block in absolute coordinates.

Actually, the first and second modifications are related since the padding of boundary blocks will affect the matching of motion estimation. The purpose of padding aims at more accurate block matching. In the current algorithm, the repetitive padding is applied to the reference VOP for performing motion estimation and compensation. The repetitive padding process is performed as the following steps:

Define any pixel outside the object boundary as a zero pixel.

Scan each horizontal line of a block (one 16 × 16 for luminance and two 8 × 8 for chrominance). Each scan line is possibly composed of two kinds of line segments: zero segments and nonzero segment. It is obvious that our task is to pad zero segments. There are two kinds of zero segments: (1) between an end point of the scan line and the end point of a nonzero segment, and (2) between the end points of two different nonzero segments. In the first case, all zero pixels are replaced by the pixel value of the end pixel of nonzero segment; for the second kind of zero segment, all zero pixels take the averaged value of the two end pixels of the nonzero segments.

Scan each vertical line of the block and perform the identical procedure as described for the horizontal line.

If a zero pixel is located at the intersection of horizontal and vertical scan lines, this zero pixel takes the average of two possible values.

For the rest of zero pixels, find the closest nonzero pixel on the same horizontal scan line and the same vertical scan line (if there is a tie, the nonzero pixel on the left or the top of the current pixel is selected). Replace the zero pixel by the average of these two nonzero pixels.

For a fast-moving VOP, padding is further extended to the blocks outside the VOP but immediately next to the boundary blocks. These blocks are padded by replacing the pixel values of adjacent boundary blocks. This extended padding is performed in both horizontal and vertical directions. Since block matching is replaced by polygon matching for the boundary blocks of the current VOP, the SAD values are calculated by the modified formula:

$$SAD_N(x,y) = \begin{cases} \sum_{i=0}^{N-1}\sum_{j=0}^{N-1} |c(i,j) - p(i,j)| \cdot \alpha(i,j) - C & \text{if } (x,y) = (0,0); \\ \sum_{i=0}^{N-1}\sum_{j=0}^{N-1} |c(i,j) - p(i+x, j+y)| \cdot \alpha(i,j) - C & \text{otherwise,} \end{cases} \tag{18.5}$$

where $C = N_B/2 + 1$ and $N_B$ is the number of pixels inside the VOP and in this block and $\alpha(i,j)$ is the alpha component specifying the shape information, and it is not equal to zero here.

### 18.3.3.3 Texture Coding of Arbitrarily Shaped VOP

During encoding the VOP is represented by a bounding rectangle that is formed to contain the video object completely but with minimum number of macroblocks in it, as shown in Figure 18.3. The detailed procedure of VOP rectangle formation is given in MPEG-4 video VM (ISO/IEC, 1998b).

There are three types of macroblocks in the VOP with arbitrary shape: the macroblocks that are completely located inside of the VOP, the macroblocks that are located along the boundary of the VOP, and the macroblocks outside of the boundary. For the first kind of macroblock, there is no need for any particular modified technique to code them and just use of normal DCT with entropy coding of quantized DCT coefficients such as coding algorithm in H.263 is sufficient. The second kind of macroblocks, which are located along the boundary, contains two kinds of 8 × 8 blocks: the blocks lie along the boundary of VOP and the blocks do not belong to the arbitrary shape but lie inside the rectangular bounding box of the VOP. The second kind of blocks are referred

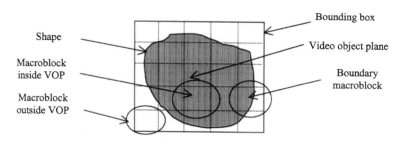

**FIGURE 18.3**  A VOP is represented by a bounding rectangular box.

to as transparent blocks. For those $8 \times 8$ blocks that do lie along the boundary of VOP, there are two different methods that have been proposed: low-pass extrapolation (LPE) padding and shape-adaptive DCT (SA-DCT). All blocks in the macroblock outside of boundary are also referred to as transparent blocks. The transparent blocks are skipped and not coded at all.

1. Low-pass extrapolation padding technique: This block-padding technique is applied to intracoded blocks, which are not located completely within the object boundary. To perform this padding technique we first assign the mean value of those pixels that are located in the object boundary (both inside and outside) to each pixel outside the object boundary. Then an average operation is applied to each pixel $p(i,j)$ outside the object boundary starting from the upper-left corner of the block and proceeding row by row to the lower-right corner pixel:

$$p(i,j) = \left[ p(i,j-1) + p(i-1,j) + p(i,j+1) + p(i+1,j) \right] / 4. \tag{18.6}$$

If one or more of the four pixels used for filtering are outside of the block, the corresponding pixels are not considered for the average operation and the factor ¼ is modified accordingly.

2. SA-DCT: The shape-adaptive DCT is only applied to those $8 \times 8$ blocks that are located on the object boundary of an arbitrarily shaped VOP. The idea of the SA-DCT is to apply 1-D DCT transformation vertically and horizontally according to the number of active pixels in the row and column of the block, respectively. The size of each vertical DCT is the same as the number of active pixels in each column. After vertical DCT is performed for all columns with at least one active pixel, the coefficients of the vertical DCTs with the same frequency index are lined up in a row. The DC coefficients of all vertical DCTs are lined up in the first row, the first-order vertical DCT coefficients are lined up in the second row, and so on. After that, horizontal DCT is applied to each row. As the same as for the vertical DCT, the size of each horizontal DCT is the same as the number of vertical DCT coefficients lined up in the particular row. The final coefficients of SA-DCT are concentrated into the upper-left corner of the block. This procedure is shown in the Figure 18.4.

The final number of the SA-DCT coefficients is identical to the number of active pixels of the image. Since the shape information is transmitted to the decoder, the decoder can perform the inverse shape-adapted DCT to reconstruct the pixels. The regular zigzag scan is modified so that the nonactive coefficient locations are neglected when counting the runs for the run-length coding of the SA-DCT coefficients. It is obvious that for a block with all $8 \times 8$ active pixels, the SA-DCT becomes a regular $8 \times 8$ DCT and the scanning of the coefficients is identical to the zigzag scan. All SA-DCT coefficients are quantized and coded in the same way as the regular DCT coefficients

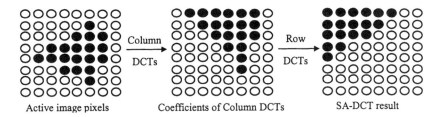

Active image pixels      Coefficients of Column DCTs      SA-DCT result

**FIGURE 18.4** Illustration of SA-DCT. (From ISO/IEC 14496-2 Video Verification Model V.12, N2552, Dec. 1998. With permission.)

employing the same quantizers and VLC code tables. The SA-DCT is not included in MPEG-4 video version 1, but it is being considered for inclusion into version 2.

### 18.3.4 SHAPE CODING

Shape information of the arbitrarily shaped objects is very useful not only in the field of image analysis, computer vision, and graphics, but also in object-based video coding. MPEG-4 video coding is the first to make an effort to provide a standardized approach to compress the shape information of objects and contain the compressed results within a video bitstream. In the current MPEG-4 video coding standard, the video data can be coded on an object basis. The information in the video signal is decomposed to shape, texture, and motion. This information is then coded and transmitted within the bitstream. The shape information is provided in binary format or gray scale format. The binary format of shape information consists of a pixel map, which is generally the same size as the bounding box of the corresponding VOP. Each pixel takes on one of two possible values indicating whether it is located within the video object or not. The gray scale format is similar to the binary format with the additional feature that each pixel can take on a range of values, i.e., times an alpha value. Alpha typically has a normalized value of 0 to 1. The alpha value can be used to blend two images on a pixel-by-pixel basis in this way: new pixel = (alpha)(pixel A color) + (1 − alpha)(pixel B color).

Now let us discuss how to code the shape information. As we mentioned, the shape information is classified as binary shape or gray scale shape. Both binary and gray scale shapes are referred to as an alpha plane. The alpha plane defines the transparency of an object. Multilevel alpha maps are frequently used to blend different images. A binary alpha map defines whether or not a pixel belongs to an object. The binary alpha planes are encoded by modified content-based arithmetic encoding (CAE), while the gray scale alpha planes are encoded by motion-compensated DCT coding, which is similar to texture coding. For binary shape coding, a rectangular box enclosing the arbitrarily shaped VOP is formed as shown in Figure 18.3. The bounded rectangle box is then extended in both vertical and horizontal directions on the right-bottom side to the multiple of $16 \times 16$ blocks. Each $16 \times 16$ block within the rectangular box is referred to as binary alpha block (BAB). Each BAB is associated with colocated macroblock. The BAB can be classified as three types: transparent block, opaque block, and alpha or shape block. The transparent block does not contain any information about an object. The opaque block is entirely located inside the object. The alpha or shape block is located in the area of the object boundary; i.e., a part of block is inside of object and the rest of block is in background. The value of pixels in the transparent region is zero. For shape coding, the type information will be included in the bitstream and signaled to the decoder as a macroblock type. But only the alpha blocks need to be processed by the encoder and decoder. The methods used for each shape format contain several encoding modes. For example, the binary shape information can be encoded using either an intra- or intermode. Each of these modes can be further divided into lossy and lossless options. Gray scale shape information also contains intra- and intermodes; however, only a lossy option is used.

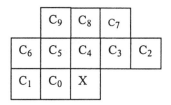

**FIGURE 18.5**   Template for defining the context of the pixel, X, to be coded in intramode. (From ISO/IEC 14496-2 Video Verification Model V.12, N2552, Dec. 1998. With permission.)

### 18.3.4.1   Binary Shape Coding with CAE Algorithm

As mentioned previously, the CAE is used to code each binary pixel of the BAB. For a P-VOP, the BAB may be encoded in intra- or intermode. Pixels are coded in scan-line order, i.e., row by row for both modes. The process for coding a given pixel includes three steps: (1) compute a context number, (2) index a probability table using the context number, and (3) use the indexed probability to drive an arithmetic encoder. In intramode, a template of 10 pixels is used to define the causal context for predicting the shape value of the current pixel as shown in Figure 18.5. For the pixels in the top and left boundary of the current macroblock, the template of causal context will contain the pixels of the already transmitted macroblocks on the top and on the left side of the current macroblock. For the two rightmost columns of the VOP, each undefined pixel such as $C_7$, $C_3$, and $C_2$, of the context is set to the value of its closest neighbor inside the macroblock, i.e., $C_7$ will take the value of $C_8$ and $C_3$ and $C_2$ will take the value of $C_4$.

A 10-bit context is calculated for each pixel, $X$ as

$$C = \sum_{k=0}^{9} C_k \cdot 2^k . \tag{18.7}$$

This causal context is used to predict the shape value of the current pixel. For encoding the state transition, a context-based arithmetic encoder is used. The probability table of the arithmetic encoder for the 1024 contexts was derived from sequences that are outside of the test set. Two bytes are allocated to describe the symbol probability for each context; the table size is 2048 bytes. To increase coding efficiency and rate control, the algorithm allows lossy shape coding. In lossy shape coding a macroblock can be down-sampled by a factor of two or four resulting in a subblock of size $8 \times 8$ pixels or $4 \times 4$ pixels, respectively. The subblock is then encoded using the same method as for full-size block. The down-sampling factor is included in the encoded bitstream and then transmitted to the decoder. The decoder decodes the shape data and then up-samples the decoded subblock to full macroblock size according to the down-sampling factor. Obviously, it is more efficient to code shape using a high down-sampling factor, but the coding errors may occur in the decoded shape after up-sampling. However, in the case of low-bit-rate coding, lossy shape coding may be necessary since the bit budget may not be enough for lossless shape coding. Depending on the up-sampling filter, the decoded shape can look somewhat blocky. Several up-sampling filters were investigated. The best-performing filter in terms of subjective picture quality is an adaptive nonlinear up-sampling filter. It should be noted that the coding efficiency of shape coding also depends on the orientation of the shape data. Therefore, the encoder can choose to code the block as described above or transpose the macroblock prior to arithmetic coding. Of course, the transpose information has to be signaled to the decoder.

For shape coding in a P-VOP or B-VOP, the intermode may be used to exploit the temporal redundancy in the shape information with motion compensation. For motion compensation, a 2-D integer pixel motion vector is estimated using full search for each macroblock in order to minimize

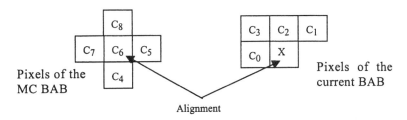

**FIGURE 18.6** Template for defining the context of the pixel, X, to be coded in intermode. (From ISO/IEC 14496-2 Video Verification Model, N2552, Dec. 1998. With permission.)

the prediction error between the previously coded VOP shape and the current VOP shape. The shape motion vectors are predictively encoded with respect to the shape motion vectors of neighboring macroblocks. If no shape motion vector is available, texture motion vectors are used as predictors. The template for intermode differs from the one used for intramode. The intermode template contains 9 pixels among which 5 pixels are located in the previous frame and 4 are the current neighbors as shown in Figure 18.6.

The intermode template defines a context of 9 pixels. Accordingly, a 9-bit context or 512 contexts, can be computed in a similar way to Equation 18.7:

$$C = \sum_{k=0}^{8} C_k \cdot 2^k. \tag{18.8}$$

The probability for one symbol is also described by 2 bytes giving a probability table size of 1024 bytes. The idea of lossy coding can also be applied to the intermode shape coding by downsampling the original BABs. For intermode shape coding, the total bits for coding the shape consist of two parts, one part for coding motion vectors and another for prediction residue. The encoder may decide that the shape representation achieved by just using motion vectors is sufficient; thus bits for coding the prediction error can be saved. Actually, there are seven modes to code the shape information of each macroblock: (1) transparent, (2) opaque, (3) intra, inter (4) with and (5) without shape motion vectors, and inter (6) with and (7) without shape motion vectors and prediction error coding. These different options with optional down-sampling and transposition allow for encoder implementations of different coding efficiency and implementation complexity. Again, this is a problem of encoder optimization, which does not belong to the standard.

### 18.3.4.2 Gray Scale Shape Coding

The gray scale shape information is encoded by separately encoding the shape and transparency information as shown in Figure 18.7. For a transparent object, the shape information is referred to as the support function and is encoded using the binary shape-coding method. The transparency or alpha values are treated as the texture of luminance and encoded using padding, motion compensation, and the same $8 \times 8$ block DCT approach for the texture coding. For an object with varying alpha maps, shape information is encoded in two steps. The boundary of the object is first losslessly encoded as a binary shape, and then the actual alpha map is encoded as texture coding.

The binary shape coding allows one to describe objects with constant transparency, while gray scale shape coding can be used to describe objects with arbitrary transparency, providing for more flexibility for image composition. One application example is a gray scale alpha shape that consists of a binary alpha shape with the value around the edges tapered from 255 to 0 to provide for a smooth composition with the background. The description of each video object layer includes the information to give instruction for selecting one of six modes for feathering. These six modes

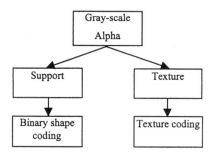

**FIGURE 18.7**　Gray scale shape coding.

include (1) no effects, (2) linear feathering, (3) constant alpha, (4) linear feathering and constant alpha, (5) feathering filter, and (6) feathering filter and constant alpha. The detailed description of the function of these modes are given in the reference of version 12 (ISO/IEC, 1998b).

## 18.3.5　Sprite Coding

As mentioned previously, MPEG-4 video has investigated a number of new tools, which attempt to improve the coding efficiency at low bit rates compared with MPEG-1/2 video coding. Among these tools, sprite coding is an efficient technology to reach this goal. A sprite is a specially composed video object that is visible throughout an entire piece of video sequence. For example, the sprite generated from a panning sequence contains all the visible pixels of the background throughout the video sequence. Portions of the background may not be seen in certain frames due to the occlusion of the foreground objects or the camera motion. This particular example is one of the static sprites. In other words, a static sprite is a possible still image. Since the sprite contains all visible background scenes of a segment video sequence where the changes within the background content are mainly caused by camera parameters, the sprite can be used for direct reconstruction of the background VOPs or as the prediction of the background VOPs within the video segment. The sprite-coding technology first efficiently transmits this background to the receiver and then stores it in a frame at both encoder and decoder. The camera parameters are then transmitted to the decoder for each frame so that the appropriate part of the background scene can be either used as the direct reconstruction or as the prediction of the background VOP. Both cases can significantly save the coding bits and increase the coding efficiency. There are two types of sprites, static sprite and dynamic sprite, which are being considered as coding tools for MPEG-4 video. A static sprite is used for a video sequence in which the objects in a scene can be separated into foreground objects and a static background. A static sprite is a special VOP, which is generated by copying the background from a video sequence. This copying includes the appropriate warping and cropping. Therefore, a static sprite is always built off-line. In contrast, a dynamic sprite is dynamically built during the predictive coding. It can be built either online or off-line. The static sprite has shown significant coding gain over existing compression technology for certain video sequences. The dynamic sprite is more complicated in the real-time application due to the difficulty of updating the sprite during the coding. Therefore, the dynamic sprite has not been adopted by version 1 of the standard. Additionally, both sprites are not easily applied to the generic scene content. Also, there is another kind of classification of sprite coding according to the method of sprite generation, namely, off-line and online sprites. Off-line is always used for static sprite generation. Off-line sprites are well suited for synthetic objects and objects that mostly undergo rigid motion. Online sprites are only used for dynamic sprites. Online sprites provide a no-latency solution in the case of natural sprite objects. Off-line dynamic sprites provide an enhanced predictive coding environment. The sprite is built with a similar way in both off-line and online methods. In particular, the same global motion estimation algorithm is exploited. The difference is that the off-line sprite is

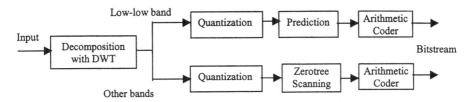

**FIGURE 18.8** Block diagram of encoder of wavelet-based texture coding, DWT stands for discrete wavelet transform.

built before starting the encoding process while, in the online sprite case, both the encoder and the decoder build the same sprite from reconstructed VOPs. This is why the online dynamic sprites are more complicated in the implementation. The online sprite is not included in version 1, and will most likely not be considered for version 2 either. In sprite coding, the chrominance components are processed in the same way as the luminance components, with the properly scaled parameters according to the video format.

### 18.3.6 Interlaced Video Coding

Since June of 1997, MPEG-4 has extended its application to support interlaced video. Interlaced video consists of two fields per frame, which are referred to as the even field and the odd field. MPEG-2 has a number of tools, which are used to deal with field structure of video signals. These tools include frame/field-adaptive DCT coding and frame/field-adaptive motion compensation. However, the field issue in MPEG-4 has to be considered on a VOP basis instead of the conventional frame basis. When field-based motion compensation is specified, two field motion vectors and the corresponding reference fields are used to generate the prediction from each reference VOP. Shape information has to be considered in the interlaced video for MPEG-4.

### 18.3.7 Wavelet-Based Texture Coding

In MPEG-4 there is a texture-coding mode which is used to code the texture or still image such as in JPEG. The basic technique used in this mode is wavelet-based transform coding. The reason for adopting wavelet transform instead of DCT for still texture coding is not only its high coding efficiency, but also because the wavelet can provide excellent scalability, both spatial scalability and SNR scalability. Since the principle of wavelet-based transform coding for image compression has been explained in Chapter 8, we just briefly describe the coding procedure of this mode. The block diagram of the encoder is shown in Figure 18.8.

#### 18.3.7.1 Decomposition of the Texture Information

The texture or still image is first decomposed into bands using a bank of analysis filters. This decomposition can be applied recursively on the obtained bands to yield a decomposition tree of subbands. An example of decomposition to depth 2 is shown in Figure 18.9.

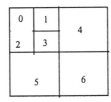
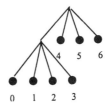

**FIGURE 18.9** An example of wavelet decomposition of depth 2.

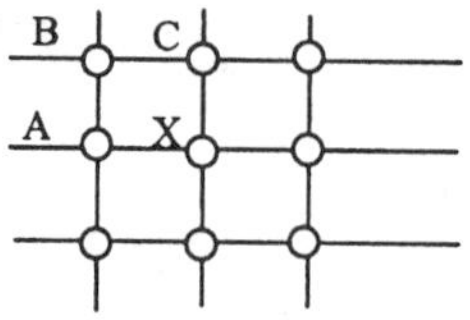

If $|W_A - |W_B| < |W_B - W_C|$, $W_{Xp} = W_C$, else $W_{Xp} = W_A$.

**FIGURE 18.10**   Adaptive DPCM coding of the coefficients in the lowest band.

### 18.3.7.2  Quantization of Wavelet Coefficients

After decomposition, the coefficients of the lowest band are coded independently of the other bands. These coefficients are quantized using a uniform midriser quantizer. The coefficients of high bands are quantized with a multilevel quantization. The multilevel quantization provides a very flexible approach to support the correct trade-off between levels and type of scalability, complexity, and coding efficiency for any application. All quantizers for the higher bands are uniform midrise quantizers with a dead zone that is twice the quantizer step size. The levels and quantization steps are determined by the encoder and specified in the bitstream. To achieve scalability, a bi-level quantization scheme is used for all multiple quantizers. This quantizer is also uniform and midrise with a dead zone that is twice the quantization step. The coefficients outside of the dead zone are quantized to 1 bit. The number of quantizers is equal to the maximum number of bit planes in the wavelet transform representation. In this bi-level quantizer, the maximum number of bit planes instead of a quantization step size is specified in the bitstream.

### 18.3.7.3  Coding of Wavelet Coefficients of Low–Low Band and Other Bands

The quantized coefficients at the lowest band are DPCM coded. Each of the current coefficients is predicted from three other quantized coefficients in its neighborhood in a way shown in Figure 18.10.

The coefficients in high bands are coded with the zerotree algorithm (Shapiro, 1993), which has been discussed in Chapter 8.

### 18.3.7.4  Adaptive Arithmetic Coder

The quantized coefficients and the symbols generated by the zerotree are coded using an adaptive arithmetic coder. In the arithmetic coder three different tables which correspond to the different statistical models have been utilized. The method used here is very similar to one in Chapter 8. Further detail can be found in MPEG-4 (ISO/IEC, 1998a).

### 18.3.8  Generalized Spatial and Temporal Scalability

The scalability framework is referred to as generalized scalability that includes the spatial and the temporal scalability similar to MPEG-2. The major difference is that MPEG-4 extends the concept of scalability to be content based. This unique functionality allows MPEG-4 to be able to resolve objects into different VOPs. By using the multiple VOP structure, different resolution enhancement can be applied to different portions of a video scene. Therefore, the enhancement layer may only be applied to a particular object or region of the base layer instead of to the entire base layer. This is a feature that MPEG-2 does not have.

In spatial scalability, the base layer and the enhancement layer can have different spatial resolutions. The base-layer VOPs are encoded in the same way as the nonscalable encoding technique described previously. The VOPs in the enhancement layer are encoded as P-VOPs or B-VOPs, as shown in Figure 18.11. The current VOP in the enhancement layer can be predicted

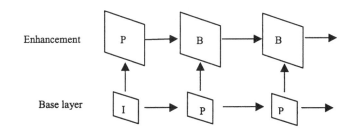

**FIGURE 18.11** Illustration of spatial scalability.

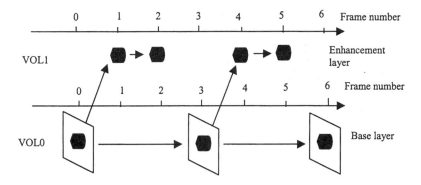

**FIGURE 18.12** An example of temporal scalability. (From ISO/IEC 14496-2 Video Verification Model V.12, N2552, Dec. 1998. With permission.)

from either the up-sampled base layer VOP or the previously decoded VOP at the same layer as well as both of them. The down-sampling and up-sampling processing in spatial scalability is not a part of the standard and can be defined by the user.

In temporal scalability, a subsequence of subsampled VOP in the time domain is coded as a base layer. The remaining VOPs can be coded as enhancement layers. In this way, the frame rate of a selected object can be enhanced so that it has a smoother motion than other objects. An example of temporal scalability is illustrated in Figure 18.12. In Figure 18.12, the $VOL_0$ is the entire frame with both an object and a background, while $VOL_1$ is a particular object in $VOL_0$. $VOL_0$ is encoded with a low frame rate and $VOL_1$ is the enhancement layer. The high frame rate can be reached for the particular object by combining the decoded data from both the base layer and the enhancement layer. Of course, the B-VOP is also used in temporal scalability for coding the enhancement layer, which is another type of temporal scalability. As in spatial scalability, the enhancement layer can be used to improve either the entire base layer frame resolution or only a portion of the base layer resolution.

### 18.3.9 ERROR RESILIENCE

The MPEG-4 visual coding standard provides error robustness and resilience to allow access of image and video data over a wide range of storage and transmission media. The error resilience tool development effort is divided into three major areas, which include resynchronization, data recovery, and error concealment. As with other coding standards, MPEG-4 makes heavy use of variable-length coding to reach high coding performance. However, if even 1 bit is lost or damaged, the entire bitstream becomes undecodable due to loss of synchronization. The resynchronization tools attempt to enable resynchronization between the decoder and the bitstream after a transmission error or errors have been detected. Generally, the data between the synchronization point prior to the error and the first point, where synchronization is reestablished, are discarded. The purpose of resynchronization is to localize effectively the amount of data discarded by the decoder; then the other methods such as error concealment can be used to conceal the damaged areas of a decoded picture. Currently, the resynchronization approach adopted by MPEG-4 is referred to as a packet approach. This approach is similar to the group of block (GOB) structure used in H.261 and H.263. In the GOB structure, the GOB contains a start code, which provides the location information of the GOB. MPEG-4 adopted a similar approach in which a resynchronization marker is periodically inserted into the bitstream at the particular macroblock locations. The resynchronization marker is used to indicate the start of new video packet. This marker is distinguished from all possible VLC codewords as well as the VOP start code. The packet header information is then provided at the start of a video packet. The header contains the information necessary to restart the decoding process. This includes the macroblock number of the first macroblock contained in this packet and the quantization parameter necessary to decode the first macroblock. The macroblock number provides the necessary spatial resynchronization while the quantization parameter allows the differential decoding process to be resynchronized. It should be noted that when error resilience is used within MPEG-4, some of the compression efficiency tools need to be modified. For example, all predictively encoded information must be contained within a video packet to avoid error propagation. In conjunction with the video packet approach to resynchronization, MPEG-4 has also adopted a fixed-interval synchronization method which requires that VOP start-codes and resynchronization markers appear only at legal fixed-interval locations in the bitstream. This will help to avoid the problems associated with start-code emulation. In this case, when fixed-interval synchronization is utilized, the decoder is only required to search for a VOP start-code at the beginning of each fixed interval. The fixed-interval synchronization method extends this approach to any predetermined interval.

After resynchronization is reestablished, the major problem is recovering lost data. A new tool called reversible variable-length codes (RVLC) is developed for the purpose of data recovery. In this approach, the variable-length codes are designed such that the codes can be read both in the forward and the reverse direction. An example of such a code includes codewords like 111, 101, 010. All these codewords can be read reversibly. However, it is obvious that this approach will reduce the coding efficiency that is achieved by the entropy coder. Therefore, this approach is used only in the cases where error resilience is important.

Error concealment is an important component of any error-robust video coding. The error-concealment strategy is highly dependent on the performance of the resynchronization technique. Basically, if the resynchronization method can efficiently localize the damaged data area, the error concealment strategy becomes much more tractable. Error concealment is actually a decoder issue if there is no additional information provided by the encoder. There are many approaches to error concealment, which are referred to in Chapter 17.

## 18.4  MPEG-4 VISUAL BITSTREAM SYNTAX AND SEMANTICS

The common feature of MPEG-4 and MPEG-1/MPEG-2 is the layered structure of the bitstream. MPEG-4 defines a syntactic description language to describe the exact binary syntax of an audio-visual object bitstream, as well as that of the scene description information. This provides a consistent and uniform way to describe the syntax in a very precise form, while at the same time simplifying bitstream compliance testing. The visual syntax hierarchy includes the following layers:

- Video session (VS)
- Video object (VO)
- Video object layer (VOL) or texture object layer (TOL)
- Group of video object plane (GOV)
- Video object plane (VOP)

A typical video syntax hierarchy is shown in Figure 18.13.

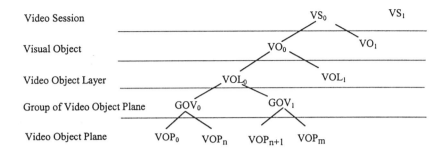

**FIGURE 18.13** MPEG-4 video syntax hierarchy.

The video session (VS) is the highest syntactic structure of the coded video bitstream. A VS is a collection of one or more VOs. A VO can consist of one or more layers. Since MPEG-4 is extended from video coding to visual coding, the type of visual objects not only includes video objects, but also still texture objects, mesh objects, and face objects. These layers can be either video or texture. Still texture coding is designed for high-visual-quality applications in transmission and rendering of texture. The still coding algorithm supports a scalable representation of image or synthetic scene data such as luminance, color, and shape. This is very useful for progressive transmission of images or 2-D/3-D synthetic scenes. The images can be gradually built up in the terminal monitor as they are received. The bitstreams for coded mesh objects are nonscalable; they define the structure and motion of a 2-D mesh. The texture of the mesh has to be coded as a separate video object. The bitstreams for face objects are also nonscalable; these bitstreams contain the face animation parameters. VOs are coded with different types of scalability. The base layer can be decoded independently and the enhancement layers can only be decoded with the base layer. In the special case of a single rectangular VO, all of the MPEG-4 layers can be related to MPEG-2 layers. That is, VS is the same as VO since in this case a single VO is a video sequence, VOL or TOL is the same as the sequence scalable extension, GOV is like the GOP, and VOP is a video frame. VO sequence may contain one or more VOs coded concurrently. The VO header information contains the start code followed by profile and level identification and a VO identification to indicate the type of object, which may be a VO, a still texture object, a mesh object, or a face object. The VO may contain one or more VOLs. In the VOL, the VO can be coded with spatial or temporal scalability. Also, the VO may be encoded in several layers from coarse to fine resolution. Depending on the application need, the decoder can choose the number of layers to decode. A VO at a specified time is called a video object plane (VOP). Thus, a VO contains many VOPs. A scene may contain many VOs. Each VO can be encoded to an independent bitstream. A collection of VOPs in a VOL is called a group of VOPs (GOV). This concept corresponds to the group of pictures (GOP) in MPEG-1 and MPEG-2. A VOP is then coded by shape coding and texture coding, which is specified at lower layers of syntax, such as the macroblock and block layer. The VOP or higher-than-VOP layer always commences with a start code and is followed by the data of lower layers, which is similar to the MPEG-1 and MPEG-2 syntax.

## 18.5   MPEG-4 VIDEO VERIFICATION MODEL

Since all video-coding standards define only the bitstream syntax and decoding process, the use of test models to verify and optimize the algorithms is needed during the development process. For this purpose a common platform with a precise definition of encoding and decoding algorithms has to be provided. The test model (TM) of MPEG-2 took the above-mentioned role. The TM of MPEG-2 was updated continually from version 1.0 to version 5.0, until the MPEG-2 Video IS (International Standard) was completed. MPEG-4 video uses a similar tool during the development process; this tool in MPEG-4 is called the Verification Model (VM). So far, the MPEG-4 video VM has gradually evolved from version 1.0 to version 12.0 and in the process has addressed an increasing number of desired functionalities such as content-based scalability, error resilience, coding efficiency, and so on. The material presented in this section is different from Section 18.3. Section 18.3 presented the technologies adopted or that will be adopted by MPEG-4, while this section provides an example of how to use the standard, for example, how to encode or generate the MPEG-4-compliant bitstream. Of course, the decoder is also included in the VM.

### 18.5.1   VOP-Based Encoding and Decoding Process

Since the most important feature of MPEG-4 is an object-based coding method, the input video sequence is first decomposed into separate VOs, these VOs are then encoded into separate bitstreams so that the user can access and manipulate (cut, paste, etc.) the video sequence in the bitstream domain. Instances of VOs in a given time are called a video object plane (VOP). The bitstream also contains the composition information to indicate where and when each VOP is to be displayed. At the decoder, the user may be allowed to change the composition of the scene displayed by interactively changing the composition information.

### 18.5.2   Video Encoder

For object-based coding, the encoder consists mainly of two parts: the shape coding and the texture coding of the input VOP. The texture coding is based on the DCT coding with traditional motion-compensated predictive coding. The VOP is represented by means of a bounding rectangular as described previously. The phase between luminance and chrominance pixels of the bounding rectangular has to be correctly set to the 4:2:0 format as in MPEG-1/2. The block diagram of encoding structure is shown in Figure 18.14.

The core technologies used in VOP coding of MPEG-4 have been described previously. Here we are going to discuss several encoding issues. Although these issues are essential to the performance and application, they are not dependent on the syntax. As a result, they are not included as normative parts of the standard, but are included as informative annexes.

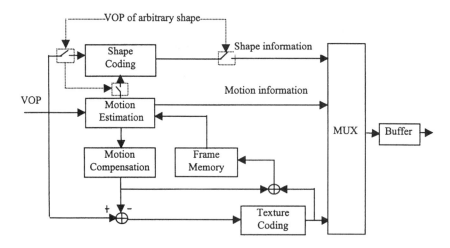

**FIGURE 18.14**   Block diagram of MPEG-4 video encoder structure.

### 18.5.2.1   Video Segmentation

Object-based coding is the most important feature of MPEG-4. Therefore, the tool for object boundary detection or segmentation is a key issue in efficiently performing the object-based coding scheme. But the method of decomposing a natural scene to several separate objects is not specified by the standard since it is a preprocessing issue. There are currently two kinds of algorithms for segmentation of video objects. One kind of algorithm is an automatic segmentation algorithm. In the case of real-time applications, the segmentation must be done automatically. Real-time automatic segmentation algorithms are currently not mature. An automatic segmentation algorithm has been proposed in MPEG96/M960 (Colonnese and Russo, 1996). This algorithm separates regions corresponding to moving objects from regions belonging to a static background for each frame of a video sequence. The algorithm is based on a motion analysis for each frame. The motion analysis is performed along several frames to track each pixel of the current frame and to detect whether the pixel belongs to the moving objects.

Another kind of segmentation algorithms is one that is user assisted or "semiautomatic." In non-real-time applications, the semiautomatic segmentation may be used effectively and give better results than the automatic segmentation. In the core experiments of MPEG-4, a semiautomatic segmentation algorithm was proposed in MPEG97/M3147 (Choi et al., 1997). The block diagram of the semiautomatic segmentation is shown in Figure 18.15.

This technique consists of two steps. First, the intraframe segmentation is applied to the first frame, which is considered as a frame that either contains newly appeared objects or a reset frame. Then the interframe segmentation is applied to the consecutive frames. For intraframe, the segmentation is processed by a user manually or semiautomatically. The user uses a graphical user interface (GUI) to draw the boundaries of objects of interest. The user can mask the entire objects all the way around objects using a mouse with a predefined thickness of the line (number of pixels). A marked swath is then achieved by the mouse, and this marked area is assumed to contain the object boundaries. A boundary-detection algorithm is applied to the marked area to create the real object boundaries. For interframe segmentation, an object boundary-tracking algorithm is proposed to obtain the object boundaries of the consecutive frames. At first, the boundary of the previous object is extracted and the motion estimation is performed on the object boundary. The object boundary

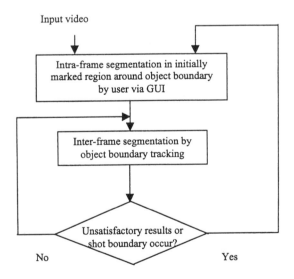

**FIGURE 18.15**   Block diagram of a user-assisted VO segmentation method.

of the current frame is initially obtained by motion compensation and then refined by using temporal information and spatial information all the way around the object boundary. Finally, the refined object boundary can be obtained. As mentioned previously, the segmentation technique is an important tool for object-based processing in MPEG-4, but it is not defined by the standard. The method described here is just an example provided by the core experiments of MPEG-4. There are many other algorithms under investigation, such as the circular Viterbi algorithm described by Lin et al. (1998).

### 18.5.2.2  Intra/Intermode Decision

For inter-VOP coding, a macroblock can be coded in one of four modes. These four modes include direct coding mode, forward coding, backward coding, and bidirectional coding. In the encoder we have to decide which mode is the best. The mode decision is an important part of encoding optimization. An example of the selection of an optimized mode decision has been given in Chapter 17 for an MPEG-2 encoder. The same technique can be extended to an MPEG-4 encoder. The basic idea of mode decision is to choose the coding mode that results in the best operation point on the rate–distortion curve. For obtaining the best operation point on the rate–distortion curve, the encoder has to compare all possible coding modes and choose the best one. This is a very complicated procedure. In the MPEG-2 case, we used a quadratic model to unify the measures of bits used to code prediction residues and the motion vectors. A simplified mode but near-optimized mode decision method has resulted. Here, the VM.12 proposes the following steps to make coding mode decisions. First, the motion-compensated prediction error is calculated by each of the four modes. Next, the SAD of each of the motion-compensated prediction macroblocks is calculated and compared with the variance of the macroblock to be coded. Then, a mode of generating the smallest SAD (for direct mode, a bias is applied) is selected. For the interlaced video, more coding modes are involved. This method of mode decision is simple, but it is not optimal since the cost for coding motion vectors is not considered. Consequently, the mode may not lie on the best operation point on the distortion curve. But again, this is an encoding issue; the encoding designers have the freedom to use their own algorithm. The VM just provides an example of an encoder that can generate the compliant bitstream.

### 18.5.2.3  Off-Line Sprite Generation

The sprite is a useful tool in MPEG-4 for coding a certain kind of video sequences at very low bit rates. The method of generating a sprite for a video sequence is an encoder issue. The VM gives an example of off-line sprite generation. For a natural VO, a sprite is referred to as a representative view collected from a video sequence. Before decoding, the sprite is transmitted to the decoder. Then the motion compensation can be performed by using the sprite from which the video can be reconstructed. The effectiveness of video reconstruction depends on whether the motion of the object can be effectively represented by a global motion model such as translation, zooming, affine, and perspective. The key technology of the sprite generation is the motion estimation to find perspective motion parameters. This can be implemented by many algorithms described in this book such as the three-step matching technique. The block diagram of sprite generation using the perspective motion estimation is shown as in Figure 18.16.

The sprite is generated from the input video sequence by the following steps. First, the first frame is used as the initial value of sprite. From the second frame, the motion estimation is applied to find the perspective motion parameters between two frames. The current frame is wrapped toward the initial sprite using the perspective motion vectors to get wrapped image. Then the wrapped image is blended with initial sprite to obtain a updated sprite. This procedure is continued to the entire video sequence. The final sprite is then generated.

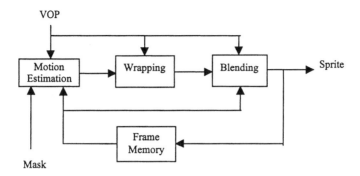

**FIGURE 18.16**   Block diagram of sprite generation.

## 18.5.2.4   Multiple VO Rate Control

As we know, the purpose of rate control is to obtain the best coding performance for a given bit rate in constant-bit-rate video coding. In MPEG-4 video coding, there is an additional objective for rate control: how to assign the bits among multiple VOs. In the multiple VO video coding rate control algorithm, the total target is first adjusted based on the buffer fullness, and then distributed proportional to the size of the object, the motion which the object is experiencing, and its maximum absolute differences. Based on the new individual targets and second-order model parameters (Lee et al., 1997), appropriate quantization parameters can be calculated for each VO. To compromise the trade-offs in spatial and temporal coding, two modes of operation have been introduced. With these modes, suitable decisions can be made to differentiate between low- and high-bit-rate coding. In addition, a shape rate control algorithm has been included. The algorithm for performing the joint rate control can be decomposed into a preencoding stage and a postencoding stage. The preencoding stage consists of (1) the target bit estimation, (2) joint buffer control, (3) pre-frame-skip control, and (4) the quantization level and alpha threshold calculation. The postencoding stage consists of (1) updating the rate–distortion model, (2) post-frame-skip control, and (3) determining the mode of operation. The initialization process is very similar to the single VOP initialization process. Since a single buffer is used, the buffer drain rate and initializations remain the same, but many of the parameters are extended to vector quantities. As a means of regulating the trade-offs between spatial and temporal coding, two modes of operation are introduced: low mode and high mode. When encoding at high bit rates, the availability of bits allows the algorithm to be flexible in its target assignment to each VO. Under these circumstances, it is reasonable to impose homogeneous quality among each VO. Therefore, the inclusion of $MAD^2[i]$ is essential to the target distribution and should carry the highest weighting. On the other hand, when the availability of bits is limited, it is very difficult (if not impossible) to achieve homogeneous quality among the VOs. Under these conditions, it is desirable to spend fewer bits on the background and more bits on the foreground. Consequently, the significance of the variance has decreased and the significance of the motion has increased. Besides regulating the quality within each frame, it is also important to regulate the temporal quality as well, i.e., to keep the frame skipping to a minimum. In high mode, this is very easy to do since the availability of bits is plentiful. However, in low mode, frame-skipping occurs much more often. In fact, the number of frames being skipped is a good indication in which mode the algorithm should be operating. Overall, this particular algorithm is able to achieve the target bit rate successfully, effectively code arbitrarily shaped objects, and maintain a stable buffer (Vetro et al., 1999).

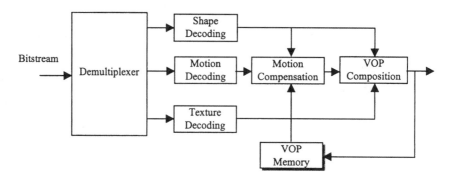

**FIGURE 18.17**   VOP decoder structure.

### 18.5.3  VIDEO DECODER

The decoder mainly consists of three parts: shape, motion, and texture decoding. The decoder block diagram is shown in Figure 18.17. At the decoder the bitstream is first demultiplexed into shape information and motion information as well as texture information. The reconstructed VOP is obtained by the right combination of the shape, texture, and motion information. The shape decoding is a unique feature of the MPEG-4 decoder. The basic technology of shape decoding is context-based arithmetic decoding and block-based motion compensation.

The primary data structure is denoted is the binary alpha block (BAB). The BAB is a square block of binary pixels representing the opacity or transparency for the pixels in a specified block-shaped spatial region of size $16 \times 16$ pixels which is colocated with each texture macroblock. The block diagram of a texture decoder is shown in Figure 18.18.

Texture decoding is similar to the video decoder in MPEG-1/2 except for inverse DC/AC prediction and more quantization methods. The DC prediction is different from the one used in MPEG-1/2. In MPEG-4 the DC coefficient is adaptively predicted from the above block or left block. The AC prediction is similar to the one used in H.263 but is not used in the MPEG-1/2. For motion compensation, the motion vectors must be decoded. The horizontal and vertical motion vector components are decoded differentially by using a prediction from the spatial neighborhood consisting of three motion vectors already decoded. The final motion vector is obtained by adding the prediction motion vector values to the decoded differential motion values. Also, in MPEG-4 video coding the several advanced motion compensation modes, such as four $8 \times 8$ motion vector compensation and overlapped motion compensation, have to be handled. Another issue of motion compensation in MPEG-4 is raised by VOP-based coding. To perform motion-compensated prediction on a VOP basis, a special padding technique is used for each of macroblock that lies on the shape boundary of the VOP. The padding process defines the values of pixels, which are located outside the VOP for prediction of arbitrarily shaped objects. Padding for luminance pixels and chrominance pixels is defined in the standard (ISO/IEC, 1998a). The additional decoding issues

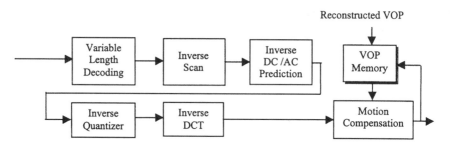

**FIGURE 18.18**   Block diagram of texture decoding.

which are special for MPEG-4 include sprite decoding, generalized scalable decoding, and still texture decoding. We do not go into further detail for these topics. Interested readers can get detail from the standard documents. The outputs of decoded results are the reconstructed VOPs that are finally sent to the compositor. In the compositor, the VOPs are recursively blended in the order specified by the VOP composition order. It should be noted that the decoders could take advantage of object-based decoding. They are able to be flexible in the composition of the reconstructed VOPs such as reallocating, rotation, or other editing actions.

## 18.6  SUMMARY

In this chapter, the new video-coding standard, MPEG-4 is introduced. The unique feature of MPEG-4 video is content-based coding. This feature allows the MPEG-4 to provide much functionality, which other video-coding standards do not have. The key technologies used in MPEG-4 video are described. These technologies provide basic tools for MPEG-4 video to provide object-based coding functionality. Finally, the video verification model, a platform of MPEG-4 development and an encoding and decoding example, is described.

## 18.7  EXERCISES

**18-1.** Why is object- or content-based coding the most important feature of MPEG-4 visual coding standard? Describe several applications for this feature.

**18-2.** What are the new coding tools in MPEG-4 visual coding that are different from MPEG-2 video coding? Is MPEG-4 backward compatible with MPEG-2?

**18-3.** MPEG-4 video coding has the feature of using either a $16 \times 16$ block motion vector or an $8 \times 8$ block motion vector. For what kind of video sequences will the $8 \times 8$ block motion increase coding efficiency? For what kind of video sequences will the $8 \times 8$ block motion compensation decrease the coding efficiency?

**18-4.** What approaches for error resilience are supported by the MPEG-4 syntax? Make a comparison with the error resilience method adopted in MPEG-2 (supported by MPEG-2 syntax), and indicate their relative advantages and disadvantages.

**18-5.** Design an arithmetic coder for zerotree coding and write a program to test it with several images.

**18-6.** The Sprite is a new feature of MPEG-4 video coding. MPEG-4 specifies the syntax for sprite coding, but does not give any detail about how to generate a sprite. Conduct a project to generate an off-line sprite for a video sequence and use it for coding the video sequence. Do you observe any increased coding efficiency? When do you expect to see such an increase?

**18-7.** Shape coding (binary-shape coding) is an important part of MPEG-4 due to object-based coding. Besides the shape coding method used in MPEG-4, name another shape coding method. Conduct a project to compare the method you know with the method proposed in MPEG-4. (Do not expect to get better performance, but expect to reduce the complexity.)

## REFERENCES

Choi, J. G., M. Kim, H. Lee, and C. Ahn, Partial experiments on a user-assisted segmentation technique for video object plane generation, MPEG97/M3147, San Jose Meeting of ISO/IEC JTC1/SC29/WG11, Feb. 1998.

Colonnese, S. and G. Russo, FUB results on core experiment N2: comparison of automatic segmentation techniques, MPEG96/M960, Tempere, July 1996.

ISO/IEC 14496-2, Coding of audio-visual objects, part 2, Dec. 18, 1998a.

ISO/IEC 14496-2 Video Verification Model V.12, N2552, December, 1998b.

Lee, H. J., T. Chiang, and Y. Q. Zhang, Scalable rate control for very low bit-rate coding, *Proceedings Internatinoal Conference on Image Processing (ICIP'97)*, II, 768-771, Santa Barbara, CA, Oct. 1997.

Lin, I. J., S. Y. Kung, A. Vetro, and H. Sun, *Circular Viterbi: Boundary Detection with Dynamic Programming*, MPEG, 1998.

Shapiro, J. Embedded image coding using zerotrees of wavelet coefficients, *IEEE Trans. Signal Proc.*, 3445-3462, 1993.

Vetro, A., H. Sun, and Y. Wang, MPEG-4 rate control for multiple video objects, *IEEE Trans. Circ. Syst. Video Technol.*, 9(1), 186-199, 1999.

# 19 ITU-T Video Coding Standards H.261 and H.263

This chapter introduces ITU-T video coding standards H.261 and H.263, which are established mainly for videophony and videoconferencing. The basic technical detail of H.261 is presented. The technical improvements with which H.263 achieves high coding efficiency are discussed. Features of H.263+, H.263++, and H.26L are presented.

## 19.1 INTRODUCTION

Very low bit rate video coding has found many industry applications such as wireless and network communications. The rapid convergence of standardization of digital video-coding standards is the reflection of several factors: the maturity of technologies in terms of algorithmic performance, hardware implementation with VLSI technology, and the market need for rapid advances in wireless and network communications. As stated in the previous chapters, these standards include JPEG for still image coding and MPEG-1/2 for CD-ROM storage and digital television applications. In parallel with the ISO/IEC development of the MPEG-1/2 standards, the ITU-T has developed H.261 (ITU-T, 1993) for videotelephony and videoconferencing applications in an ISDN environment.

## 19.2 H.261 VIDEO-CODING STANDARD

The H.261 video-coding standard was developed by ITU-T study group XV during 1988 to 1993. It was adopted in 1990 and the final revision approved in 1993. This is also referred to as the $P \times 64$ standard because it encodes the digital video signals at the bit rates of $P \times 64$ Kbps, where $P$ is an integer from 1 to 30, i.e., at the bit rates 64 Kbps to 1.92 Mbps.

### 19.2.1 OVERVIEW OF H.261 VIDEO-CODING STANDARD

The H.261 video-coding standard has many features in common with the MPEG-1 video-coding standard. However, since they target different applications, there exist many differences between the two standards, such as data rates, picture quality, end-to-end delay, and others. Before indicating the differences between the two coding standards, we describe the major similarity between H.261 and MPEG-1/2. First, both standards are used to code similar video format. H.261 is mainly used to code the video with the common intermediate format (CIF) or quarter-CIF (QCIF) spatial resolution for teleconferencing application. MPEG-1 uses CIF, SIF, or higher spatial resolution for CD-ROM applications. The original motivation for developing the H.261 video-coding standard was to provide a standard that can be used for both PAL and NTSC television signals. But later, the H.261 was mainly used for videoconferencing and the MPEG-1/2 was used for digital television (DTV), VCD (video CD), and DVD (digital video disk). The two TV systems, PAL and NTSC, use different line and picture rates. The NTSC, which is used in North America and Japan, uses 525 lines per interlaced picture at 30 frames/second. The PAL system is used for most other countries, and it uses 625 lines per interlaced picture at 25 frames/second. For this purpose, the CIF was adopted as the source video format for the H.261 video coder. The CIF format consists of 352 pixels/line, 288 lines/frame, and 30 frames/second. This format represents half the active lines of the PAL signal and the same picture rate of the NTSC signal. The PAL systems need only perform a picture rate conversion and NTSC systems need only perform a line number conversion. Color pictures consist of one luminance and two color-difference components (referred to as $Y C_b$, $C_r$ format) as specified by the CCIR601 standard. The $C_b$ and $C_r$ components are the half-size on both horizontal and vertical directions and have 176 pixels/line and 144 lines/frame. The other format, QCIF, is used for very low bit rate applications. The QCIF has half the number of pixels and half the number of lines of CIF format. Second, the key coding algorithms of H.261 and MPEG-1 are very similar. Both H.261 and MPEG-1 use DCT-based coding to remove intraframe redundancy and motion compensation to remove interframe redundancy.

Now let us describe the main differences between the two coding standards with respect to coding algorithms. The main differences include:

- H.261 uses only I- and P-macroblocks but no B-macroblocks, while MPEG-1 uses three macroblock types, I-, P-, and B-macroblocks (I-macroblock is in intraframe-coded macroblock, P-macroblock is a predictive-coded macroblock, and B-macroblock is a bidirectionally coded macroblock), as well as three picture types, I-, P-, and B-pictures as defined in Chapter 16 for the MPEG-1 standard.

- There is a constraint of H.261 that for every 132 interframe-coded macroblocks, which corresponds to 4 GOBs (group of blocks) or to one-third of the CIF pictures, it requires at least one intraframe-coded macroblock. To obtain better coding performance at low-bit-rate applications, most encoding schemes of H.261 prefer not to use intraframe coding on all the macroblocks of a picture, but only on a few macroblocks in every picture with a rotational scheme. MPEG-1 uses the GOP (group of pictures) structure, where the size of GOP (the distance between two I-pictures) is not specified.

- The end-to-end delay is not a critical issue for MPEG-1, but is critical for H.261. The video encoder and video decoder delays of H.261 need to be known to allow audio compensation delays to be fixed when H.261 is used in interactive applications. This will allow lip synchronization to be maintained.

- The accuracy of motion compensation in MPEG-1 is up to a half-pixel, but is only a full-pixel in H.261. However, H.261 uses a loop filter to smooth the previous frame. This filter attempts to minimize the prediction error.

- In H.261, a fixed picture aspect ratio of 4:3 is used. In MPEG-1, several picture aspect ratios can be used and the picture aspect ratio is defined in the picture header.

- Finally, in H.261, the encoded picture rate is restricted to allow up to three skipped frames. This would allow the control mechanism in the encoder some flexibility to control the encoded picture quality and satisfy the buffer regulation. Although MPEG-1 has no restriction on skipped frames, the encoder usually does not perform frame skipping. Rather, the syntax for B-frames is exploited, as B-frames require much fewer bits than P-pictures.

## 19.2.2 TECHNICAL DETAIL OF H.261

The key technologies used in the H.261 video-coding standard are the DCT and motion compensation. The main components in the encoder include DCT, prediction, quantization (Q), inverse DCT (IDCT), inverse quantization (IQ), loop filter, frame memory, variable-length coding, and coding control unit. A typical encoder structure is shown in Figure 19.1.

The input video source is first converted to the CIF frame and then is stored in the frame memory. The CIF frame is then partitioned into GOBs. The GOB contains 33 macroblocks, which are $1/12$ of a CIF picture or $1/3$ of a QCIF picture. Each macroblock consists of six $8 \times 8$ blocks among which four are luminance ($Y$) blocks and two are chrominance blocks (one of $C_b$ and one of $C_r$).

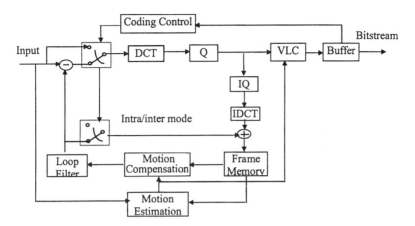

**FIGURE 19.1**  Block diagram of a typical H.261 video encoder. (From ITU-T Recommendation H.261, March 1993. With permission.)

For the intraframe mode, each $8 \times 8$ block is first transformed with DCT and then quantized. The variable-length coding (VLC) is applied to the quantized DCT coefficients with a zigzag scanning order such as in MPEG-1. The resulting bits are sent to the encoder buffer to form a bitstream.

For the interframe-coding mode, frame prediction is performed with motion estimation in a similar manner to that in MPEG-1, but only P-macroblocks and P-pictures, no B-macroblocks and B-pictures, are used. Each $8 \times 8$ block of differences or prediction residues is coded by the same DCT coding path as for intraframe coding. In the motion-compensated predictive coding, the encoder should perform the motion estimation with the reconstructed pictures instead of the original video data, as it will be done in the decoder. Therefore, the IQ and IDCT blocks are included in the motion compensation loop to reduce the error propagation drift. Since the VLC operation is lossless, there is no need to include the VLC block in the motion compensation loop. The role of the spatial filter is to minimize the prediction error by smoothing the previous frame that is used for motion compensation.

The loop filter is a separable 2-D spatial filter that operates on an $8 \times 8$ block. The corresponding 1-D filters are nonrecursive with coefficients ¼, ½, ¼. At block boundaries, the coefficients are 0, 1, 0 to avoid the taps falling outside the block. It should be noted that MPEG-1 uses subpixel accurate motion vectors instead of a loop filter to smooth the anchor frame. The performance comparison of two methods should be interesting.

The role of coding control includes the rate control, the buffer control, the quantization control, and the frame rate control. These parameters are intimately related. The coding control is not the part of the standard; however, it is an important part of the encoding process. For a given target bit rate, the encoder has to control several parameters to reach the rate target and at the same time provide reasonable coded picture quality.

Since H.261 is a predictive coder and the VLCs are used everywhere, such as coding quantized DCT coefficients and motion vectors, a single transmission error may cause a loss of synchronization and consequently cause problems for the reconstruction. To enhance the performance of the H.261 video coder in noisy environments, the transmitted bitstream of H.261 can optionally contain a BCH (Bose, Chaudhuri, and Hocquengham) (511,493) forward error-correction code.

The H.261 video decoder performs the inverse operations of the encoder. After optional error correction decoding, the compressed bitstream enters the decoder buffer and then is parsed by the variable-length decoder (VLD). The output of the VLD is applied to the IQ and IDCT where the data are converted to the values in the spatial domain. For the interframe-coding mode, the motion

| 1 | 2 | 3 | 4 | 5 | 6 | 7 | 8 | 9 | 10 | 11 |
|----|----|----|----|----|----|----|----|----|----|----|
| 12 | 13 | 14 | 15 | 16 | 17 | 18 | 19 | 20 | 21 | 22 |
| 23 | 24 | 25 | 26 | 27 | 28 | 29 | 30 | 31 | 32 | 33 |

**FIGURE 19.2**    Arrangement of macroblocks in a GOB. (From ITU-T Recommendation H.261, March 1993. With permission.)

compensation is performed and the data from the macroblocks in the anchor frame are added to the current data to form the reconstructed data.

### 19.2.3    SYNTAX DESCRIPTION

The syntax of H.261 video coding has a hierarchical layered structure. From the top to the bottom the layers are picture layer, GOB layer, macroblock layer, and block layer.

#### 19.2.3.1    Picture Layer

The picture layer begins with a 20-bit picture start code (PSC). Following the PSC, there are temporal reference (5-bit), picture type information (PTYPE, 6-bit), extra insertion information (PEI, 1-bit), and spare information (PSPARE). Then the data for GOBs are followed.

#### 19.2.3.2    GOB Layer

A GOB corresponds to 176 pixels by 48 lines of $Y$ and 88 pixels by 24 lines of $C_b$ and $C_r$. The GOB layer contains the following data in order: 16-bit GOB start code (GBSC), 4-bit group number (GN), 5-bit quantization information (GQUANT), 1-bit extra insertion information (GEI), and spare information (GSPARE). The number of bits for GSPARE is variable depending on the set of GEI bits. If GEI is set to "1," then 9 bits follow, consisting of 8 bits of data and another GEI bit to indicate whether a further 9 bits follow, and so on. Data of the GOB header are then followed by data for macroblocks.

#### 19.2.3.3    Macroblock Layer

Each GOB contains 33 macroblocks, which are arranged as in Figure 19.2. A macroblock consists of 16 pixels by 16 lines of $Y$ that spatially correspond to 8 pixels by 8 lines each of $C_b$ and $C_r$. Data in the bitstream for a macroblock consist of a macroblock header followed by data for blocks. The macroblock header may include macroblock address (MBA) (variable length), type information (MTYPE) (variable length), quantizer (MQUANT) (5 bits), motion vector data (MVD) (variable length), and coded block pattern (CBP) (variable length). The MBA information is always present and is coded by VLC. The VLC table for macroblock addressing is shown in Table 19.1. The presence of other items depends on macroblock type information, which is shown in the VLC Table 19.2.

#### 19.2.3.4    Block Layer

Data in the block layer consists of the transformed coefficients followed by an end of block (EOB) marker (10 bits). The data of transform coefficients (TCOEFF) is first converted to the pairs of RUN and LEVEL according to the zigzag scanning order. The RUN represents the number of successive zeros and the LEVEL represents the value of nonzero coefficients. The pairs of RUN and LEVEL are then encoded with VLCs. The DC coefficient of an intrablock is coded by a fixed-length code with 8 bits. All VLC tables can be found in the standard document (ITU-T, 1993).

**TABLE 19.1**
**VLC Table for Macroblock Addressing**

| MBA | Code | MBA | Code | MBA | Code |
|---|---|---|---|---|---|
| 1 | 1 | 13 | 0000 1000 | 25 | 0000 0100 000 |
| 2 | 011 | 14 | 0000 0111 | 26 | 0000 0011 111 |
| 3 | 010 | 15 | 0000 0110 | 27 | 0000 0011 110 |
| 4 | 0011 | 16 | 0000 0101 11 | 28 | 0000 0011 101 |
| 5 | 0010 | 17 | 0000 0101 10 | 29 | 0000 0011 100 |
| 6 | 0001 1 | 18 | 0000 0101 01 | 30 | 0000 0011 011 |
| 7 | 0001 0 | 19 | 0000 0101 00 | 31 | 0000 0011 010 |
| 8 | 0000 111 | 20 | 0000 0100 11 | 32 | 0000 0011 001 |
| 9 | 0000 110 | 21 | 0000 0100 10 | 33 | 0000 0011 000 |
| 10 | 0000 1011 | 22 | 0000 0100 011 | MBA stuffing | 0000 0001 111 |
| 11 | 0000 1010 | 23 | 0000 0100 010 | Start code | 0000 0000 0000 0001 |
| 12 | 0000 1001 | 24 | 0000 0100 001 | | |

**TABLE 19.2**
**VLC Table for Macroblock Type**

| Prediction | MQUANT | MVD | CBP | TCOEFF | VLC |
|---|---|---|---|---|---|
| Intra | | | | x | 0001 |
| Intra | x | | | x | 0000 001 |
| Inter | | | x | x | 1 |
| Inter | x | | x | x | 0000 1 |
| Inter+MC | | x | | | 0000 0000 1 |
| Inter+MC | | x | x | x | 0000 0001 |
| Inter+MC | x | x | x | x | 0000 0000 01 |
| Inter+MC+FIL | | x | | | 001 |
| Inter+MC+FIL | | x | x | x | 01 |
| Inter+MC+FIL | x | x | x | x | 0000 01 |

*Notes:*
1. "x" means that the item is present in the macroblock,
2. It is possible to apply the filter in a non-motion-compensated macroblock by declaring it as MC+FIL but with a zero vector.

## 19.3   H.263 VIDEO-CODING STANDARD

The H.263 video-coding standard (ITU-T, 1996) is specifically designed for very low bit rate applications such as practical video telecommunication. Its technical content was completed in late 1995 and the standard was approved in early 1996.

### 19.3.1   OVERVIEW OF H.263 VIDEO CODING

The basic configuration of the video source coding algorithm of H.263 is based on the H.261. Several important features that are different from H.261 include the following new options: unrestricted motion vectors, syntax-based arithmetic coding, advanced prediction, and PB-frames. All these features can be used together or separately for improving the coding efficiency. The H.263

**TABLE 19.3**
**Number of Pixels per Line and the Number of Lines for Each Picture Format**

| Picture Format | Number of Pixels for Luminance ($dx$) | Number of Lines for Luminance ($dy$) | Number of Pixels for Chrominance ($dx/2$) | Number of Lines for Chrominance ($dy/2$) |
|---|---|---|---|---|
| Sub-QCIF | 128 | 96 | 64 | 48 |
| QCIF | 176 | 144 | 88 | 72 |
| CIF | 352 | 288 | 176 | 144 |
| 4CIF | 704 | 576 | 352 | 288 |
| 16CIF | 1408 | 1152 | 704 | 576 |

video standard can be used for both 625-line and 525-line television standards. The source coder operates on the noninterlaced pictures at picture rate about 30 pictures/second. The pictures are coded as luminance and two color difference components ($Y$, $C_b$, and $C_r$). The source coder is based on a CIF. Actually, there are five standardized formats which include sub-QCIF, QCIF, CIF, 4CIF, and 16CIF. The detail of formats is shown in Table 19.3.

It is noted that for each format, the chrominance is a quarter the size of the luminance picture, i.e., the chrominance pictures are half the size of the luminance picture in both horizontal and vertical directions. This is defined by the ITU-R 601 format. For CIF format, the number of pixels/line is compatible with sampling the active portion of the luminance and color difference signals from a 525- or 626-line source at 6.75 and 3.375 MHz, respectively. These frequencies have a simple relationship to those defined by the ITU-R 601 format.

### 19.3.2 TECHNICAL FEATURES OF H.263

The H.263 encoder structure is similar to the H.261 encoder with the exception that there is no loop filter in H.263 encoder. The main components of the encoder include block transform, motion-compensated prediction, block quantization, and VLC. Each picture is partitioned into groups of blocks, which are referred to as GOBs. A GOB contains a multiple number of 16 lines, $k * 16$ lines, depending on the picture format ($k = 1$ for sub-QCIF, QCIF; $k = 2$ for 4CIF; $k = 4$ for 16CIF). Each GOB is divided into macroblocks that are the same as in H.261 and each macroblock consists of four $8 \times 8$ luminance blocks and two $8 \times 8$ chrominance blocks. Compared with H.261, H.263 has several new technical features for the enhancement of coding efficiency for very low bit rate applications. These new features include picture-extrapolating motion vectors (or unrestricted motion vector mode), motion compensation with half-pixel accuracy, advanced prediction (which includes variable-block-size motion compensation and overlapped block motion compensation), syntax-based arithmetic coding, and PB-frame mode.

#### 19.3.2.1 Half-Pixel Accuracy

In H.263 video coding, half-pixel accuracy motion compensation is used. The half-pixel values are found using bilinear interpolation as shown in Figure 19.3.

Note that H.263 uses subpixel accuracy for motion compensation instead of using a loop filter to smooth the anchor frames as in H.261. This is also done in other coding standards, such as MPEG-1 and MPEG-2, which also use half-pixel accuracy for motion compensation. In MPEG-4 video, quarter-pixel accuracy for motion compensation has been adopted as a tool for version 2.

#### 19.3.2.2 Unrestricted Motion Vector Mode

Usually motion vectors are limited within the coded picture area of anchor frames. In the unrestricted motion vector mode, the motion vectors are allowed to point outside the pictures. When the values

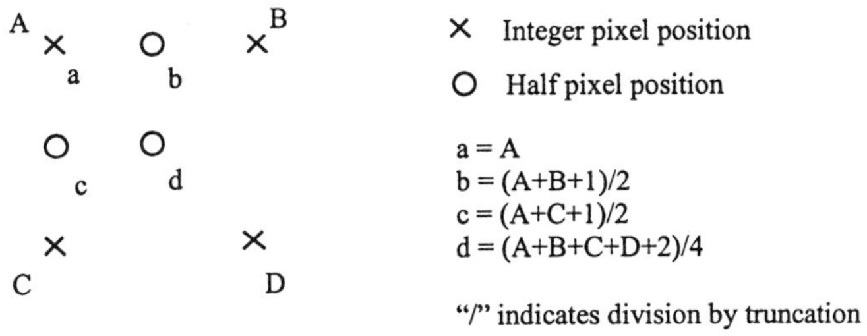

**FIGURE 19.3**   Half-pixel prediction by bilinear interpolation.

of the motion vectors exceed the boundary of the anchor frame in the unrestricted motion vector mode, the picture-extrapolating method is used. The values of reference pixels outside the picture boundary will take the values of boundary pixels. The extension of the motion vector range is also applied to the unrestricted motion vector mode. In the default prediction mode, the motion vectors are restricted to the range of [−16, 15.5]. In the unrestricted mode, the maximum range for motion vectors is extended to [−31.5, 31.5] under certain conditions.

### 19.3.2.3  Advanced Prediction Mode

Generally, the decoder will accept no more than one motion vector per macroblock for baseline algorithm of H.263 video-coding standard. However, in the advanced prediction mode, the syntax allows up to four motion vectors to be used per macroblock. The decision to use one or four vectors is indicated by the macroblock type and coded block pattern for chrominance (MCBPC) codeword for each macroblock. How to make this decision is the task of the encoding process.

The following example gives the steps of motion estimation and coding mode selection for the advanced prediction mode in the encoder.

Step 1.   Integer pixel motion estimation:

$$SAD_N(x,y) = \sum_{i=0}^{N-1} \sum_{j=0}^{N-1} |original - previous|, \tag{19.1}$$

where $SAD$ is the sum of absolute difference, values of $(x, y)$ are within the search range, $N$ is equal to 16 for $16 \times 16$ block, and $N$ is equal to 8 for $8 \times 8$ block.

$$SAD_{4 \times 8} = \sum SAD_8(x,y) \tag{19.2}$$

$$SAD_{inter} = \min\left(SAD_{16}(x,y), SAD_{4 \times 8}\right). \tag{19.3}$$

Step 2.   Intra/intermode decision:
If $A < (SAD_{inter} - 500)$, this macroblock is coded as intra-MB; otherwise, it is coded as inter-MB, where $SAD_{inter}$ is determined in step 1, and

$$A = \sum_{i=0}^{15} \sum_{j=0}^{15} |original = MB_{mean}| \tag{19.4}$$

$$MB_{mean} = \frac{1}{256} \sum_{i=0}^{15} \sum_{j=(\ )15} original.$$

If this macroblock is determined to be coded as inter-MB, go to step 3.

Step 3.   Half-pixel search:

In this step, half-pixel search is performed for both $16 \times 16$ blocks and $8 \times 8$ blocks as shown in Figure 19.3.

Step 4.   Decision on $16 \times 16$ or four $8 \times 8$ (one motion vector or four motion vectors per macroblock):

If $SAD_{4x8} < SAD_{16} - 100$, four motion vectors per macroblock will be used, one of the motion vectors is used for all pixels in one of the four luminance blocks in the macroblock, otherwise, one motion vector will be used for all pixels in the macroblock.

Step 5.   Differential coding of motion vectors for each of $8 \times 8$ luminance block is performed as in Figure 19.4.

When it has been decided to use four motion vectors, the $MVD_{CHR}$ motion vector for both chrominance blocks is derived by calculating the sum of the four luminance vectors and dividing by 8. The component values of the resulting $1/16$ pixel resolution vectors are modified toward the position as indicated in the Table 19.4.

Another advanced prediction mode is overlapped motion compensation for luminance. Actually, this idea is also used by MPEG-4, which has been described in Chapter 18. In the overlapped motion compensation mode, each pixel in an $8 \times 8$ luminance block is a weighted sum of three values divided by 8 with rounding. The three values are obtained by the motion compensation with three motion vectors: the motion vector of the current luminance block and two of four "remote"

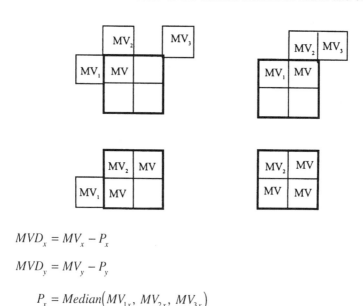

$$MVD_x = MV_x - P_x$$

$$MVD_y = MV_y - P_y$$

$$P_x = Median\left(MV_{1x}, MV_{2x}, MV_{3x}\right)$$

$$P_y = Median\left(MV_{1y}, MV_{2y}, MV_{3y}\right)$$

$P_x = P_y = 0$, if MB is intracoded or block is outside of picture boundary

**FIGURE 19.4**   Differential coding of motion vectors.

**TABLE 19.4**
**Modification of 1/16 Pixel Resolution Chrominance Vector Components**

| 1/16 Pixel Position | 0 | 1 | 2 | 3 | 4 | 5 | 6 | 7 | 8 | 9 | 10 | 11 | 12 | 13 | 14 | 15 | /16 |
|---|---|---|---|---|---|---|---|---|---|---|---|---|---|---|---|---|---|
| Resulting Position | 0 | 0 | 0 | 1 | 1 | 1 | 1 | 1 | 1 | 1 | 1 | 1 | 1 | 1 | 2 | 2 | /2 |

vectors. These remote vectors include the motion vector of the block to the left or right of the current block and the motion vector of the block above or below the current block. The remote motion vectors from other GOBs are used in the same way as remote motion vectors inside the current GOB. For each pixel to be coded in the current block, the remote motion vectors of the blocks at the two nearest block borders are used, i.e., for the upper half of the block the motion vector corresponding to the block above the current block is used while for the lower half of the block the motion vector corresponding to the block below the current block is used. Similarly, the left half of the block uses the motion vector of the block at the left side of the current block and the right half uses the one at the right side of the current block. To make this clearer, let ($MV_x^0$, $MV_y^0$) be the motion vector for the current block, ($MV_x^1$, $MV_y^1$) be the motion vector for the block either above or below, and ($MV_x^2$, $MV_y^2$) be the motion vector of the block either to the left or right of the current block. Then the value of each pixel, $p(x, y)$ in the current $8 \times 8$ luminance block is given by

$$p(x,y) = \left(q(x,y)\cdot H_0 + r(x,y)\cdot H_1 + s(x,y)\cdot H_2(x,y) + 4\right)/8, \tag{19.5}$$

where

$$q(x,y) = p\left(x + MV_x^0, y + MV_y^0\right), \quad r(x,y) = p\left(x + MV_x^1, y + MV_y^1\right),$$

and

$$s(x,y) = p\left(x + MV_x^2,\ y + MV_y^2\right), \tag{19.6}$$

$H_0$ is the weighting matrix for prediction with the current block motion vector, $H_1$ is the weighting matrix for prediction with the top or bottom block motion vector and $H_2$ is the weighting matrix for prediction with the left or right block motion vector. This applies to the luminance block only. The values of $H_0$, $H_1$, and $H_2$ are shown in Figure 19.5.

$H_0$

| | | | | | | | |
|---|---|---|---|---|---|---|---|
| 4 | 5 | 5 | 5 | 5 | 5 | 5 | 4 |
| 5 | 5 | 5 | 5 | 5 | 5 | 5 | 5 |
| 5 | 5 | 6 | 6 | 6 | 6 | 5 | 5 |
| 5 | 5 | 6 | 6 | 6 | 6 | 5 | 5 |
| 5 | 5 | 6 | 6 | 6 | 6 | 5 | 5 |
| 5 | 5 | 6 | 6 | 6 | 6 | 5 | 5 |
| 5 | 5 | 5 | 5 | 5 | 5 | 5 | 5 |
| 4 | 5 | 5 | 5 | 5 | 5 | 5 | 4 |

$H_1$

| | | | | | | | |
|---|---|---|---|---|---|---|---|
| 2 | 2 | 2 | 2 | 2 | 2 | 2 | 2 |
| 1 | 1 | 2 | 2 | 2 | 2 | 1 | 1 |
| 1 | 1 | 1 | 1 | 1 | 1 | 1 | 1 |
| 1 | 1 | 1 | 1 | 1 | 1 | 1 | 1 |
| 1 | 1 | 1 | 1 | 1 | 1 | 1 | 1 |
| 1 | 1 | 1 | 1 | 1 | 1 | 1 | 1 |
| 1 | 1 | 2 | 2 | 2 | 2 | 1 | 1 |
| 2 | 2 | 2 | 2 | 2 | 2 | 2 | 2 |

$H_2$

| | | | | | | | |
|---|---|---|---|---|---|---|---|
| 2 | 1 | 1 | 1 | 1 | 1 | 1 | 2 |
| 2 | 2 | 1 | 1 | 1 | 1 | 2 | 2 |
| 2 | 2 | 1 | 1 | 1 | 1 | 2 | 2 |
| 2 | 2 | 1 | 1 | 1 | 1 | 2 | 2 |
| 2 | 2 | 1 | 1 | 1 | 1 | 2 | 2 |
| 2 | 2 | 1 | 1 | 1 | 1 | 2 | 2 |
| 2 | 2 | 1 | 1 | 1 | 1 | 2 | 2 |
| 2 | 1 | 1 | 1 | 1 | 1 | 1 | 2 |

**FIGURE 19.5** Weighting matrices for overlapped motion compensation.

It should be noted that the above coding scheme is not optimized in the selection of mode decision since the decision depends only on the values of predictive residues. Optimized mode decision techniques that include the above possibilities for prediction have been considered by Weigand (1996).

### 19.3.2.4 Syntax-Based Arithmetic Coding

As in other video-coding standards, H.263 uses VLC and variable-length decoding (VLC/VLD) to remove the redundancy in the video data. The basic principle of VLC is to encode a symbol with a specific table based on the syntax of the coder. The symbol is mapped to an entry of the table in a table lookup operation, then the binary codeword specified by the entry is sent to a bitstream buffer for transmitting to the decoder. In the decoder, an inverse operation, VLD, is performed to reconstruct the symbol by the table lookup operation based on the same syntax of the coder. The tables in the decoder must be the same as the one used in the encoder for encoding the current symbol. To obtain better performance, the tables are generated in a statistically optimized way (such as a Huffman coder) with a large number of training sequences. This VLC/VLD process implies that each symbol be encoded into a fixed-integral number of bits. An optional feature of H.263 is to use arithmetic coding to remove the restriction of fixed-integral number bits for symbols. This syntax-based arithmetic coding mode may result in bit rate reductions.

### 19.3.2.5 PB-Frames

The PB-frame is a new feature of H.263 video coding. A PB-frame consists of two pictures, one P-picture and one B-picture, being coded as one unit, as shown in Figure 19.6. Since H.261 does not have B-pictures, the concept of a B-picture comes from the MPEG video-coding standards. In a PB-frame, the P-picture is predicted from the previous decoded I- or P-picture and the B-picture is bidirectionally predicted both from the previous decoded I- or P-picture and the P-picture in the PB-frame unit, which is currently being decoded.

Several detailed issues have to be addressed at macroblock level in PB-frame mode:

- If a macroblock in the PB-frame is intracoded, the P-macroblock in the PB-unit is intracoded and the B-macroblock in the PB-unit is intercoded. The motion vector of intercoded PB-macroblock is used for the B-macroblock only.
- A macroblock in PB-frame contains 12 blocks for 4:2:0 format, six (four luminance blocks and two chrominance blocks) from the P-frame and six from the B-frame. The data for the six P-blocks are transmitted first and then for the six B-blocks.
- Different parts of a B-block in a PB-frame can be predicted with different modes. For pixels where the backward vector points inside of coded P-macroblock, bidirectional prediction is used. For all other pixels, forward prediction is used.

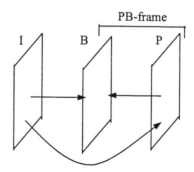

**FIGURE 19.6**   Prediction in PB-frames mode. (From ITU-T Recommendation H.263, May 1996. With permission.)

## 19.4 H.263 VIDEO CODING STANDARD VERSION 2

### 19.4.1 OVERVIEW OF H.263 VERSION 2

The H.263 version 2 (ITU-T, 1998) video-coding standard, also known as H.263+, was approved in January 1998 by the ITU-T. H.263 version 2 includes a number of new optional features based on the H.263 video-coding standard. These new optional features are added to broaden the application range of H.263 and to improve its coding efficiency. The main features are flexible video format, scalability, and backward-compatible supplemental enhancement information. Among these new optional features, five of them are intended to improve the coding efficiency and three of them are proposed to address the needs of mobile video and other noisy transmission environments. The features of scalability provide the capability of generating layered bitstreams, which are spatial scalability, temporal scalability, and signal-to-noise ratio (SNR) scalability similar to those defined by the MPEG-2 video-coding standard. There are also other modes of H.263 version 2 that provide some enhancement functions. We will describe these features in the following section.

### 19.4.2 NEW FEATURES OF H.263 VERSION 2

The H.263 version 2 includes a number of new features. In the following we briefly describe the key techniques used for these features.

#### 19.4.2.1 Scalability

The scalability function allows for encoding the video sequences in a hierarchical way that partitions the pictures into one basic layer and one or more enhancement layers. The decoders have the option of decoding only the base layer bitstream to obtain lower-quality reconstructed pictures or further decode the enhancement layers to obtain higher-quality decoded pictures. There are three types of scalability in H.263: temporal scalability, SNR scalability, and spatial scalability.

Temporal scalability (Figure 19.7) is achieved by using B-pictures as the enhancement layer. All three types of scalability are similar to the ones in the MPEG-2 video-coding standard. The B-pictures are predicted from either or both a previous and subsequent decoded picture in the base layer.

In SNR scalability (Figure 19.8), the pictures are first encoded with coarse quantization in the base layer. The differences or coding error pictures between a reconstructed picture and its original in the base layer encoder are then encoded in the enhancement layer and sent to the decoder providing an enhancement of SNR. In the enhancement layer there are two types of pictures. If a picture in the enhancement layer is only predicted from the base layer, it is referred to as an EI picture. It is a bidirectionally predicted picture if it uses both a prior enhancement layer picture and a temporally simultaneous base layer reference picture for prediction. Note that the prediction

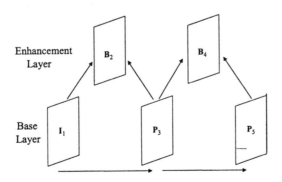

**FIGURE 19.7**   Temporal scalability. (From ITU-T Recommendation H.263, May 1996. With permission.)

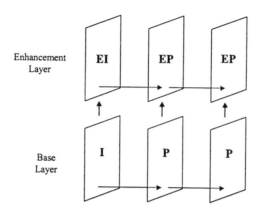

**FIGURE 19.8**   SNR scalability. (From ITU-T Recommendation H.263, May 1996. With permission.)

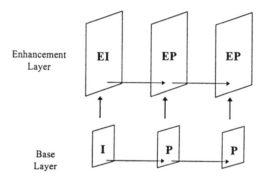

**FIGURE 19.9**   Spatial scalability. (From ITU-T Recommendation H.263, May 1996. With permission.)

from the reference layer uses no motion vectors. However, EP (enhancement P) pictures use motion vectors when predicted from their temporally prior reference picture in the same layer. Also, if more than two layers are used, the reference may be the lower layer instead of the base layer.

In spatial scalability (Figure 19.9), lower-resolution pictures are encoded in the base layer or lower layer. The differences or error pictures between up-sampled decoded base layer pictures and their original picture are encoded in the enhancement layer and sent to the decoder providing the spatial enhancement pictures. As in MPEG-2, spatial interpolation filters are used for the spatial scalability. There are also two types of pictures in the enhancement layer: EI and EP. If a decoder is able to perform spatial scalability, it may also need to be able to use a custom picture format. For example, if the base layer is sub-QCIF ($128 \times 96$), the enhancement layer picture would be $256 \times 192$, which does not belong to a standard picture format.

Scalability in H.263 can be performed with multilayers. In the case of multilayer scalability, the picture layer used for upward prediction in an EI or EP picture may be an I, P, EI, or EP picture, or may be the P part of a PB or improved PB frame in the base layer as shown in Figure 19.10.

### 19.4.2.2   Improved PB-Frames

The difference between the PB-frame and the improved PB-frame is that bidirectional prediction is used for B-macroblocks in the PB-frame, while in the improved PB-frame, B-macroblocks can be coded in three prediction modes: bidirectional prediction, forward prediction, and backward prediction. This means that in forward prediction or backward prediction only one motion vector is used for a $16 \times 16$ macroblock instead of using two motion vectors for a $16 \times 16$ macroblock in

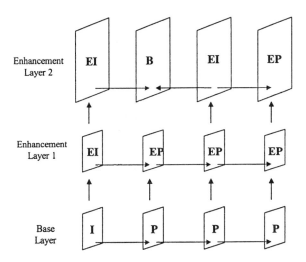

**FIGURE 19.10**   Multilayer scalability. (From ITU-T Recommendation H.263, May 1996. With permission.)

bidirectional prediction. In the very low bit rate case, this mode can improve the coding efficiency by saving bits for coding motion vectors.

### 19.4.2.3   Advanced Intracoding

The advantage of intracoding is to protect the error propagation since intracoding does not depend on the previous decoded picture data. However, the problem of intracoding is that more bits are needed since the temporal correlation between frames is not exploited. The idea of advanced intracoding (AIC) is used to address this problem. The coding efficiency of intracoding is improved by the use of following three methods:

1. Intrablock prediction using neighboring intrablocks for the same color component ($Y$, $C_b$, or $C_r$): A particular intracoded block may be predicted from the block above or left to the current block being decoded, or from both. The main purpose of these predictions is to use the correlation between neighboring blocks. For example, the first row of AC coefficients may be predicted from those in the block above, the first column of AC coefficients may be predicted from those in the left, and the DC value may be predicted as an average from the block above and left.
2. Modified inverse quantization for intracoefficients: Inverse quantization of the intra-DC coefficient is modified to allow a varying quantization step size. Inverse quantization of all intra-AC coefficients is performed without a "dead-zone" in the quantizer reconstruction spacing.
3. A separate VLC for intracoefficients: To improve intracoding a separate VLC table is used for all intra-DC and intra-AC coefficients. The price paid for this modification is the use of more tables.

### 19.4.2.4   Deblocking Filter

The deblocking filter (DF) is used to improve the decoded picture quality further by smoothing the block artifacts. Its function in improving picture quality is similar to overlapped block motion compensation. The filter operations are performed across $8 \times 8$ block edges using a set of four pixels on both horizontal and vertical directions at the block boundaries, such as shown in Figure 19.11. In the figure, the filtering process is applied to the edges. The edge pixels, $A$, $B$, $C$, and $D$, are replaced by $A_1$, $B_1$, $C_1$, and $D_1$ by the following operations:

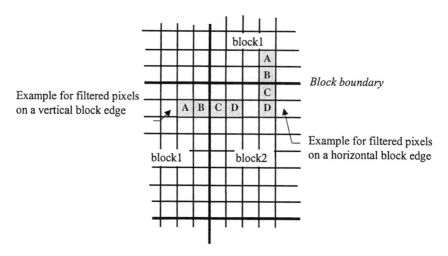

**FIGURE 19.11** Positions of filtered pixels. (From ITU-T Recommendation H.263, May 1996. With permission.)

$$B_1 = \mathrm{clip}(B + d_1) \tag{19.7a}$$

$$C_1 = \mathrm{clip}(C - d_1) \tag{19.7b}$$

$$A_1 = A - d_2 \tag{19.7c}$$

$$D_1 = D + d_2 \tag{19.7d}$$

$$d = (A - 4B + 4C - D)/8 \tag{19.7e}$$

$$d_1 = f(d, S) \tag{19.7f}$$

$$d_2 = \mathrm{clip}\,d_1\big((A - D)/4, d_1/2\big), \tag{19.7g}$$

where clip is a function of clipping the value to the range of 0 to 255, clip $d(x, d)$ is a function that clips $x$ to the range of from $-d$ to $+d$, and the value $S$ is a function of quantization step QUANT that is defined in Table 19.5.

**TABLE 19.5**
**The Value $S$ as a Function of Quantization Step (QUANT)**

| QUANT | 1 | 2 | 3 | 4 | 5 | 6 | 7 | 8 | 9 | 10 | 11 | 12 | 13 | 14 | 15 | 16 |
|-------|---|---|---|---|---|---|---|---|---|----|----|----|----|----|----|----|
| S | 1 | 1 | 2 | 2 | 3 | 3 | 4 | 4 | 4 | 5 | 5 | 6 | 6 | 7 | 7 | 7 |
| QUANT | 17 | 18 | 19 | 20 | 21 | 22 | 23 | 24 | 25 | 26 | 27 | 28 | 29 | 30 | 31 | |
| S | 8 | 8 | 8 | 9 | 9 | 9 | 10 | 10 | 10 | 11 | 11 | 11 | 12 | 12 | 12 | |

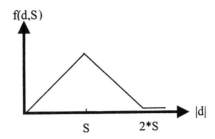

**FIGURE 19.12** The plot of function of $f(d,S)$. (From ITU-T Recommendation H.263, May 1996. With permission.)

The function $f(d, S)$ is defined as

$$f(d,S) = \text{sign}(d) * \left(\max\left(0,\ \text{abs}(d)\right) - \left(\max\left(0,\ 2 * \text{abs}(d) - S\right)\right)\right). \tag{19.8}$$

This function can be described by Figure 19.12. From the figure, it can be seen that this function is used to control the amount of distortion introduced by filtering. The filter has an effect only if $d$ is smaller than $2S$. Therefore, some features such an isolated pixel, corner, etc. would be reserved during the nonlinear filtering since for those features the value $d$ may exceed the $2S$. The function $f(d,S)$ is also designed to ensure that small mismatch between encoder and decoder will remain small and will not allow the mismatch to be propagated over multiple pictures. For example, if the filter is simply switched on or off with a mismatch of only +1 or –1 for $d$, then this will cause the filter to be switched on at the encoder and off at the decoder, or vice versa. It should be noted that the deblocking filter proposed here is an optional selection. It is a result of a large number of simulations; it may be effective for some sequences, but may be not effective for all kinds of video sequences.

### 19.4.2.5 Slice Structured Mode

A slice contains a video picture segment. In the coding syntax, a slice is defined as a slice header followed by consecutive macroblocks in scanning order. The slice structured (SS) mode is designed to address the needs of mobile video and other unreliable transmission environments. This mode contains two submodes: the rectangular slice (RS) submode and the arbitrarily slice ordering (ASO) submode. In the rectangular submode, a slice contains a rectangular region of a picture, such that the slice header specifies the width. The macroblocks in this slice are in scan order within the rectangular region. In the arbitrarily slice ordering submode, the slices may appear in any order within the bitstream. The arbitrarily arrangement of slices in the picture may provide an environment for obtaining better error concealment. The reason is that the damaged areas caused by packet loss may be isolated from each other and can be easily concealed by the well-decoded neighboring blocks. In this submode, there is usually no data dependency that can cross the slice boundaries, except for the deblocking filter mode since the slices may not be decoded in the normal scan order.

### 19.4.2.6 Reference Picture Selection

With optional mode of the reference picture selection (RPS), the encoder is allowed to use a modified interframe prediction method. In this method, additional picture memories are used. The encoder may select one of the picture memories to suppress the temporal error propagation due to the interframe coding. The information to indicate which picture is selected for prediction is included in the encoded bitstream that is allowed by syntax. The strategy used by the encoder to select the picture to be used for prediction is open for algorithm design. This mode can use the backward channel message that is sent from a decoder to an encoder to inform the encoder which part of which pictures have been correctly decoded. The encoder can use the message from the backward channel to decide which picture will provide better prediction. From the above description of reference picture selection mode, it becomes evident that this mode is useful for improving the performance over unreliable channels.

### 19.4.2.7 Independent Segmentation Decoding

The independent segmentation decoding (ISD) mode is another option of H.263 video coding which can be used for unreliable transmission environments. In this mode, each video picture segment is decoded without the presence of any data dependencies across slice boundaries or across GOB boundaries, i.e., with complete independence from all other video picture segments and all data outside the same video picture segment location in the reference pictures. This independence includes no use of motion vectors outside of the current video picture segment for motion prediction or remote motion vectors for overlapped motion compensation in the advanced prediction mode, no deblocking filter operation, and no linear interpolation across the boundaries of current video picture segment.

### 19.4.2.8 Reference Picture Resampling

The reference picture resampling (RPR) mode allows a prior-coded picture to be resampled, or wrapped, before it is used as a reference picture. The idea of using this mode is similar to the idea of global motion, which is expected to obtain better performance of motion estimation and compensation. The wrapping is defined by four motion vectors for the corners of the reference picture as shown in Figure 19.13.

For the current picture with horizontal size $H$ and vertical size $V$, four conceptual motion vectors, $MV_{OO}$, $MV_{OV}$, $MV_{HO}$, and $MV_{HV}$ are defined for the upper-left, lower-left, upper-right, and lower-right corners of the picture, respectively. These motion vectors as wrapping parameters have to be coded with VLC and included in the bitstream. These vectors are used to describe how to move the corners of the current picture to map them onto the corresponding corners of the previous decoded pictures as shown in Figure 19.13. The motion compensation is performed using bilinear interpolation in the decoder with the wrapping parameters.

### 19.4.2.9 Reduced-Resolution Update

When encoding a video sequence with highly active scenes, the encoder may have a problem providing sufficient subjective picture quality at low-bit-rate coding. The reduced-resolution update (RRU) mode is expected to be used in this case for improving the coding performance. This mode allows the encoder to send update information for a picture that is encoded at a reduced resolution to create a final image at the higher resolution. At the encoder, the pictures in the sequence are

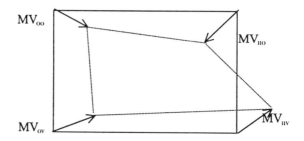

**FIGURE 19.13** Reference picture resampling.

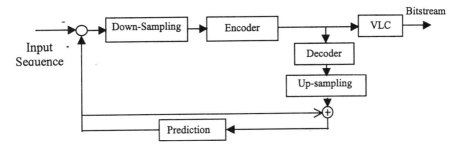

**FIGURE 19.14**   Block diagram of encoder with RRU mode.

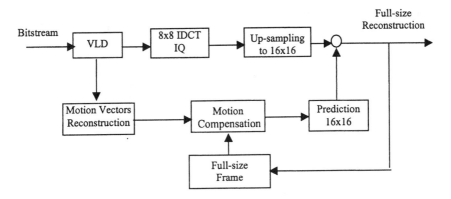

**FIGURE 19.15**   Block diagram of decoder with RRU mode.

first down-sampled to a quarter-size (half in both horizontal and vertical directions) and then the resulting low-resolution pictures are encoded as shown in Figure 19.14.

A decoder with this mode is more complicated than one without this mode. The block diagram of decoding process with the RRU mode is shown in Figure 19.15.

The decoder with RRU mode has to deal with several new issues. First, the reconstructed pictures are up-sampled to the full size for display. However, the reference pictures have to be extended to the integer times of $32 \times 32$ macroblocks if it is necessary. The pixel values in the extended areas take the values of the original border pixels. Second, the motion vectors for $16 \times 16$ macroblocks in the encoder are used for the up-sampled $32 \times 32$ macroblock in the decoder. Therefore, an additional procedure is needed to reconstruct the motion vectors for each up-sampled $16 \times 16$ macroblock including chrominance macroblocks. Third, bilinear interpolation is used for up-sampling in the decoder loop. Finally, in the boundary of a reconstructed picture, a block boundary filter is used along the edges of the $16 \times 16$ reconstructed blocks at the encoder as well as on the decoder. There are two kinds of block boundary filters that have been proposed. One is the previously described deblocking filter. The other one is defined as follows. If two pixels, $A$ and $B$, are neighboring pixels and $A$ is in block 1 and $B$ is in block 2, respectively, then the filter is designed as

$$A_1 = (3*A + B + 2)/4 \tag{19.9a}$$

$$B_1 = (A + 3*B + 2)/2, \tag{19.9b}$$

where $A_1$ and $B_1$ are the pixels after filtering and "/" is division with truncation.

### 19.4.2.10  Alternative Inter-VLC and Modified Quantization

The alternative inter-VLC (AIV) mode is developed for improving coding efficiency of interpicture coding for the pictures containing significant scene changes. This efficiency improvement is obtained by allowing some VLC codes originally designed for intrapicture to be used for interpicture coefficients. The idea is very intuitive and simple. When a rapid scene change occurs in the video sequence, interpicture prediction becomes difficult. This results in large prediction differences, which are similar to the intrapicture data. Therefore, the use of intrapicture VLC tables instead of using interpicture tables may obtain better results. However there is no syntax definition for this mode. In other words, the encoder may use the intra-VLC table for encoding an interblock without informing the decoder. After receiving all coefficient codes of a block, the decoder will first decode these codewords with the inter-VLC tables. If the addressing of coefficients stays inside the 64 coefficients of a block, the VLC will accept the results even if some coding mismatch exists. Only if coefficients outside the block are addressed, will the codewords be interpreted according to the intra-VLC table. The modified quantization mode is designed for providing several features, which can improve the coding efficiency. First, with this mode more flexible control of the quantizer step can be specified in the dequantization field. The dequantization field is no longer a 2-bit fixed-length field; it is a variable-length field which can either be 2 or 6 bits depending on the first bit. Second, in this mode the quantization parameter of the chrominance coefficients is different from the quantization parameter of the luminance coefficients. The chrominance fidelity can be improved by specifying a smaller quantization step for chrominance than that for luminance. Finally, this mode allows the extension of a range of coefficient values. This provides more accuracy representation of any possible true coefficient value with the accuracy allowed by the quantization step. However, the range of quantized coefficient levels is restricted to those, which can reasonably occur, to improve the detectability or errors and minimize decoding complexity.

### 19.4.2.11  Supplemental Enhancement Information

The usage of supplemental information may be included in the bitstream in the picture layer to signal enhanced display capabilities or to provide tagging information for external usage. This supplemental enhancement information includes full-picture freeze/freeze-release request, partial-picture freeze/freeze-release request, resizing partial-picture freeze request, full-picture snapshot tag, partial-picture snapshot tag, video time segment start/end tag, progressive refinement segment start/end tag, and chroma key information. The full-picture freeze request is used to indicate that the contents of the entire prior displayed video picture will be kept and not updated by the contents in the current decoded picture. The picture freeze will be kept under this request until the full-picture freeze-release request occurs in the current or subsequent picture-type information. The partial-picture freeze request indicates that the contents of a specified rectangular area of the prior displayed video picture are frozen until the release request is received or time-out occurs. The resizing partial-picture freeze request is used to change the specified rectangular area for the partial picture. One use of this information is to keep the contents of a picture in the corner of display unchanged for a time period for commercial use or some other purpose. All information given by the tags indicates that the current picture is labeled as either a still image snapshot or a subsequence of video data for external usage. The progressive refinement segment tag is used to indicate the display period of the pictures with better quality. The chroma keying information is used to request "transparent" and "semitransparent" pixels in the decoded video pictures (Chen et al., 1997). One application of the chroma key is to describe simply the shape information of objects in a video sequence.

## 19.5  H.263++ VIDEO CODING AND H.26L

H.263++ is the next version of H.263. It considers adding more optional enhancements to H.263 and is the extension of H.263 version 2. It is currently scheduled be completed late in the year

2000. H.26L, the L standards for long term, is a project to seek more-efficient video-coding algorithms that will be much better than the current H.261 and H.263 standards. The algorithms for H.26L can be fundamentally different from the current DCT with motion compensation framework that is used for H.261, H.262 (MPEG-2), and H.263. The expected improvements on the current standards include several aspects: higher coding efficiency, more functionality, low complexity permitting software implementation, and enhanced error robustness. H.26L addresses very low bit rate, real-time, low end-to-end delay applications. The potential application targets can be Internet videophones, sign-language or lip-reading communications, video storage and retrieval service, multipoint communication, and other visual communication systems. H.263L is currently scheduled for approval in the year 2002.

## 19.6  SUMMARY

In this chapter, the video-coding standards for low-bit-rate applications are introduced. These standards include H.261, H.263, H.263 version 2, and the versions under development, H.263++ and H.263L. H.261 and H.263 are extensively used for videoconferencing and other multimedia applications at low bit rates. In H.263 version 2, all new negotiable coding options are developed for special applications. Among these options, five options, which include advanced intracoding mode, alternative inter-VLC mode, modified quantization mode, de-blocking filter mode, and improved PB-frame mode, are intended to improve coding efficiency. Three modes, including slice structured mode, reference picture selection mode, and independent segment decoding mode, are used to meet the need of mobile video applications. The others provide the functionality of scalability such as spatial, temporal, and SNR scalability. H.26L is a future standard to meet the requirements of very low bit rate, real-time, low end-to-end delay, and other advanced performance needs.

## 19.7  EXERCISES

**19-1.** What is the enhancement of H.263 over H.261? Describe the applications of each enhanced tool of H.263.

**19-2.** Compared with MPEG-1 and MPEG-2, which features of H.261 and H.263 are used to improve coding performance at low bit rates? Explain the reasons.

**19-3.** What is the difference between spatial scalability and reduced resolution update mode in H.263 video coding?

**19-4.** Conduct a project to compare the results by using deblocking filters in the coding loop and out of the coding loop. Which method will cause less drift if a large number of pictures are contained between two consecutive I-pictures?

## REFERENCES

Chen, T., C. T. Swain, and B. G. Haskell, Coding of sub-regions for content-based scalable video, *IEEE Trans. Circuits Syst. Video Technol.,* 7(1), 256-260, 1997.

ITU-T Recommendation H.261, Video Codec for Audiovisual Services at px64 kbit/s, March 1993.

ITU-T Recommendation H.263, Video Coding for Low Bit Rate Communication, Draft H.263, May 2, 1996.

ITU-T Recommendation H.263, Video Coding for Low Bit Rate Communication, Draft H.263, January 27, 1998.

Weigand, T. et al., Rate-distortion optimized mode selection for very low bit-rate video coding and the emerging H.263 standard, *IEEE Trans. Circ. Syst. Video Technol.,* 6(2), 182-190, 1996.

# 20 MPEG System — Video, Audio, and Data Multiplexing

In this chapter, we present the methods and standards requiring how to multiplex and synchronize the MPEG-coded video, audio, and other data into a single bitstream or multiple bitstreams for storage and transmission.

## 20.1 INTRODUCTION

ISO/IEC MPEG has completed work on the ISO/IEC 11172 and 13818 standards known as MPEG-1 and MPEG-2, respectively, which deal with the coding of digital audio and video signals. Currently, ISO/IEC is working on ISO/IEC 14496 known as MPEG-4 that is object-based generic coding for multimedia applications. As mentioned in the previous chapters, the MPEG-1, 2, and 4 standards are designed as generic standards and as such are suitable for use in a wide range of audiovisual applications. The coding part of the standards convert the digital visual, audio, and data signals to the compressed formats that are represented as binary bits. The task of the MPEG system is focused on multiplexing and synchronizing the coded audio, video, and data into a single bitstream or multiple bitstreams. In other words, the digital compressed video, audio, and data are all first represented as binary formats which are referred to as bitstreams, and then the function of system is to mix the bitstreams from video, audio, and data together. For this purpose, several issues have to be addressed by the system part of the standard:

- Distinguishing different data, such as audio, video, or other data;
- Allocating bandwidth during muxing;
- Reallocating or decoding the different data during demuxing;
- Protecting the bitstreams in error-prone media and detecting the errors;
- Dynamically multiplexing several bitstreams.

Additional requirements for the system should include extensibility issues, such as:

- New service extensions should be possible;
- Existing decoders should recognize and ignore data they cannot understand;
- The syntax should have extension capacity.

It should also be noted that all system-timing signals are included in the bitstream. This is the big difference between digital systems and traditional analog systems in which the timing signals are transmitted separately. In this chapter, we will introduce the concept of systems and give detailed explanations for existing standards such as MPEG-2. However, we will not go through the standards page by page to explain the syntax, we will pay more attention to the core parts of the standard and the parts which always cause confusion during implementation. One of the key issues is system timing. For MPEG-4, we will give a presentation of the current status of the system part of the standards.

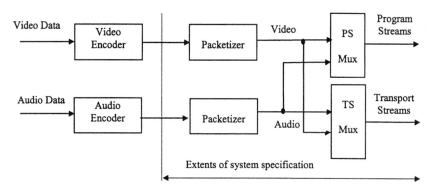

**FIGURE 20.1**  Simplified overview of system layer scope. (From ISO/IEC 13818-1, 1996. With permission.)

## 20.2   MPEG-2 SYSTEM

The MPEG-2 system standard is also referred to as ITU-T Rec. H.222.0/ISO/IEC 13818-1 (ISO/IEC, 1996). The ISO document gives a very detailed description of this standard. A simplified overview of this system is shown in Figure 20.1.

The MPEG-2 system coding is specified in two forms: the transport stream and the program stream. Each is optimized for a different set of applications. The audio and video data are first encoded by an audio and a video encoder, respectively. The coded data are the compressed bitstreams, which follow the syntax rules specified by the video-coding standard 13818-2 and audio-coding standard 13818-3. The compressed audio and video bitstreams are then packetized to the packetized elementary streams (PES). The video PES and audio PES are coded by system coding to the transport stream or program stream according to the requirements of the application.

The system coding provides a coding syntax which is necessary and sufficient to synchronize the decoding and presentation of the video and audio information; at the same time it also has to ensure that data buffers in the decoders do not overflow and underflow. Of course, buffer regulation is also considered by the buffer control or rate control mechanism in the encoder. The video, audio, and data information are multiplexed according to the system syntax by inserting time stamps for decoding, presenting, and delivering the coded audio, video, and other data. It should be noted that both the program stream and the transport stream are packet-oriented multiplexing. Before we explain these streams, we first give a set of parameter definitions used in the system documents. Then, we describe the overall picture regarding the basic multiplexing approach for single video and audio elementary streams.

### 20.2.1   Major Technical Definitions in the MPEG-2 System Document

In this section, the technical definitions that are often used in the system document are provided. First, the major packet- and stream-related definitions are given.

> **Access unit:**  A coded representation of a presentation unit. In the case of audio, an access unit is the coded representation of audio frame. In the case of video, an access unit indicates all the coded data for a picture, and any stuffing that follows it, up to but not including the start of the next access unit. In other words, the access unit begins with the first byte of the first start code. Except for the end of sequence, all bytes between the last byte of the coded picture and the sequence end code belong to the access unit.
>
> **DSM-CC:**  Digital storage media command and control.
>
> **Elementary stream (ES):**  A generic term for one of the coded video, coded audio, or other coded bitstreams in PES packets. One elementary stream is carried in a sequence of PES

packets with one and only one stream identification. This implies that one elementary stream can only carry the same type of data, such as audio or video.

**Packet:** A packet consists of a header followed by a number of contiguous bytes from an elementary data stream.

**Packet identification (PID):** A unique integer value used to associate elementary streams of a program in a single- or multiprogram transport stream. It is a 13-bit field, which indicates the type of data stored in the packet payload.

**PES packet:** The data structure used to carry elementary stream data. It contains a PES packet header followed by PES packet payload.

**PES stream:** A PES stream consists of PES packets, all of whose payloads consist of data from a single elementary steam, and all of which have the same stream identification. Specific semantic constraints apply.

**PES packet header:** The leading fields in the PES packet up to and not including the PES packet data byte fields. Its function will be explained in the section on syntax description.

**System target decoder (STD):** A hypothetical reference model of a decoding process used to describe the semantics of the MPEG-2 system-multiplexed bitstream.

**Program-specific information (PSI):** PSI includes normal data that will be used for demultiplexing of programs in the transport stream by decoders. One case of PSI, the nonmandatory network information table, is privately defined.

**System header:** The leading fields of program stream packets.

**Transport stream packet header:** The leading fields of program stream packets.

The following definitions are related to timing information:

**Time stamp:** A term that indicates the time of a specific action such as the arrival of a byte or the presentation of a presentation unit.

**System clock reference (SCR):** A Time stamp in the program stream from which decoder timing is derived.

**Elementary stream clock reference (ESCR):** A time stamp in the PES stream from which decoders of the PES stream may derive timing information.

**Decoding time stamp (DTS):** A time stamp that may be presented in a PES packet header used to indicate the time when an access unit is decoded in the system target decoder.

**Program clock reference (PCR):** A time stamp in the transport stream from which decoder timing is derived.

**Presentation time stamp (PTS):** A time stamp that may be presented in the PES packet header used to indicate the time that a presentation unit is presented in the system target decoder.

## 20.2.2 TRANSPORT STREAMS

The transport stream is a stream definition that is designed for communicating or storing one or more programs of coded video, audio, and other kinds of data in lossy or noisy environments where significant errors may occur. A transport stream combines one or more programs with one or more time bases into a single stream. However, there are some difficulties with constructing and delivering a transport stream containing multiple programs with independent time bases such that the overall bit rate is variable. As in other standards, the transport stream may be constructed by any method that results in a valid stream. In other words, the standards just specify the system coding syntax. In this way, all compliant decoders can decode bitstreams generated according to the standard syntax. However, the standard does not specify how the encoder generates the bitstreams. It is possible to generate transport streams containing one or more programs from elementary coded data streams, from program streams, or from other transport streams, which may themselves contain

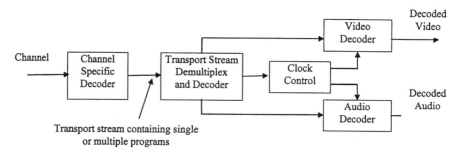

**FIGURE 20.2**  Example of transport demultiplexing and decoding. (From ISO/IEC 13818-1, 1996. With permission.)

one or more programs. An important feature of a transport stream is that the transport stream is designed in such a way that makes the following operations possible with minimum effort. These operations include several transcoding requirements, including the following:

- Retrieve the coded data from one program within the transport stream, decode it, and present the decoded results. In this operation, the transport stream is directly demultiplexed and decoded. The data in the transport stream are constructed in two layers: a system layer and a compression layer. The system decoder decodes the transport streams and demultiplexes them to the compressed video and audio streams that are further decoded to the video and audio data by the video decoder and the audio decoder, respectively. It should be noted that nonaudio/video data is also allowed. The function of the transport decoder includes demultiplexing, depacketization, and other functions such as error detection, which will be explained in detail later. This procedure is shown in Figure 20.2.
- Extract the transport stream packets from one program within the transport stream and produce as the output a new transport stream that contains only that one program. This operation can be seen as system-layer transcoding that converts a transport stream containing multiple programs to a transport stream containing only a single program. In this case, the remultiplexing operation may need the correction of PCR values to account for changes in the PCR locations in the bitstream.
- Extract the transport stream packets of one or more programs from one or more transport streams and produce as output of a new transport stream. This is another kind of transcoding that converts selected programs of one transport stream to a different one.
- Extract the contents of one program from the transport stream and produce as output another program stream. This is a transcoding that converts the transport program to a program stream for certain applications.
- Convert a program stream to a transport stream that can be used in a lossy communication environment.

To answer the question of how to define the transport stream and then make the above transcoding simpler and more efficient, we will begin by describing the technical detail of the systems specification in the following section.

### 20.2.2.1  Structure of Transport Streams

As described earlier, the task of the transport stream coding layer is to allow one or more programs to be combined into a single stream. Data from each elementary stream are multiplexed together with timing information, which is used for synchronization and presentation of the elementary

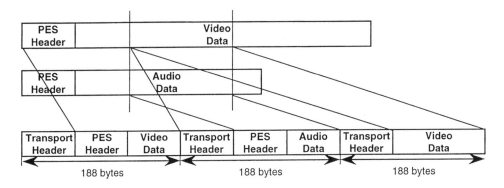

**FIGURE 20.3** Structure of transport stream containing only PES packets. (From ISO/IEC 13818-1, 1996. With permission.)

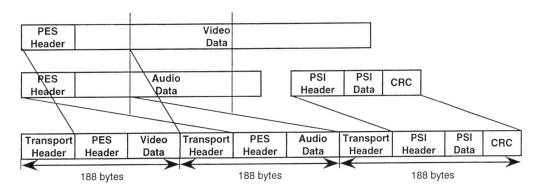

**FIGURE 20.4** Structure of transport stream containing both PES packets and PSI packets.

stream during decoding. Therefore, the transport stream consists of one or more programs such as audio, video, and data elementary stream access units. The transport stream structure is a layered structure. All the bits in the transport stream are packetized to the transport packets. The size of transport packet is chosen to be 188 bytes, among which 4 bytes are used as the transport stream packet header. In the first layer, the header of the transport packets indicates whether or not the transport packet has an adaptation field. If there is no adaptation field, the transport payload may consist of only PES packets or consist of both PES packets and PSI packets. Figure 20.3 illustrates the case of containing PES packets only. If the transport stream carries both PES and PSI packets, then the structure of transport stream is as shown in Figure 20.4 would result. If the transport stream packet header indicates that the transport stream packet includes the adaptation field, then the construct is as shown in Figure 20.5.

In Figure 20.5, the appearance of the optional field depends on the flag settings. The function of adaptation field will be explained in the syntax section. Before we go ahead, however, we should give a little explanation regarding the size of the transport stream packet. More specifically, why is a packet size of 188 bytes chosen? Actually, there are several reasons. First, the transport packet size needs to be large enough so that the overhead due to the transport headers is not too significant. Second, the size should not be so large that the packet-based error correction code becomes inefficient. Finally, the size 188 bytes is also compatible with ATM packet size which is 47 bytes; one transport stream packet is equal to four ATM packets. So the size of 188 bytes is not a theoretical solution but a practical and compromised solution.

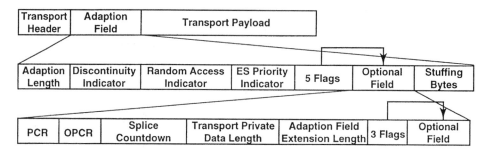

**FIGURE 20.5**    Structure of transport stream whose header contains an adaptation field.

### 20.2.2.2    Transport Stream Syntax

As we indicated, the transport stream is a layered structure. To explain the transport stream syntax we start from the transport stream packet header. Since the header part is very important, it is the highest layer of the stream. We describe it in more detail. For the rest, we do not repeat the standard document and just indicate the important parts that we think may cause some confusion for readers. The detail of other parts that are not covered here can be found from the MPEG standard document (ISO/IEC, 1996).

**Transport stream packet header** — This header contains four bytes that are assigned as eight parts:

| Syntax | No. of bits | Mnemonic |
|---|---|---|
| sync_byte | 8 | bslbf |
| transport_error_indicator | 1 | bslbf |
| payload_unit_start_indicator | 1 | bslbf |
| transport_priority | 1 | bslbf |
| PID | 13 | uimsbf |
| transport_scrambling_control | 2 | bslbf |
| adaptation_field_control | 2 | bslbf |
| continuity_counter | 4 | uimsbf |

The mnemonic in the above table means:

| | |
|---|---|
| bslbf | Bitstream left bit first |
| unimsbf | Unsigned integer, most significant bit first |

- The sync_byte is a fixed 8-bit field whose value is 0100 0111 (hexadecimal 47 = 71).
- The transport_error_indicator is a 1-bit flag, when it is set to 1, it indicates that at least 1 uncorrectable bit error exists in the associated transport stream packet. It will not be reset to 0 unless the bit values in error have been corrected. This flag is useful for error concealment purpose, since it indicates the error location. When an error exists, either resynchronization or another concealment method can be used.
- The payload_unit_start_indicator is a 1-bit flag that is used to indicate whether the transport stream packets carry PES packets or PSI data. If it carries PES packets, then the PES header starts in this transport packet. If it contains PSI data, then a PSI table starts in this transport packet.
- The transport_priority is a 1-bit flag which is used to indicate that the associated packet is of greater priority than other packets having the same PID which do not have the flag

bit set to 1. The original idea of adding a flag to indicate the priority of packets comes from video coding. The video elementary bitstream contains mostly bits that are converted from DCT coefficients. The priority indicator can set a partitioning point that can divide the data into a more important part and a less important part. The important part includes the header information and low-frequency coefficients, and the less important part includes only the high-frequency coefficients that have less effect on the decoding and quality of reconstructed pictures.

- PID is a 13-bit field that provides information for multiplexing and demultiplexing by uniquely identifying which packet belongs to a particular bitstream.
- The transport_scrambling_control is a 2-bit flag. 00 indicates that the packet is not scrambled, the other three (01, 10, and 11) indicate that the packet is scrambled by a user-defined scrambling method. It should be noted that the transport packet header and adaptation field (when it is present) should not be scrambled. In other words, only the payload of transport packets can be scrambled.
- The adaptation_field_control is a 2-bit indicator that is used to inform whether or not there is an adaptation field present in the transport packet. 00 is reserved for future use: 01 indicates no adaptation field; 10 indicates that there is only an adaptation field and no payload. Finally, 11 indicates that there is an adaptation field followed by a payload in the transport stream packet.
- The continuity_counter is a 4-bit counter which increases with each transport stream packet having the same PID.

From the header of the transport stream packet we can obtain information about future bits. There are two possibilities; if the adaptation field control value is 10 or 11, then the bits following the header are adaptation field; otherwise, the bits are payload. The information contained in the adaptation field is described as follows.

**Adaptation field** — The structure of the adaptation field data is shown in Figure 20.5. The functionality of these headers is basically related to the timing and decoding of the elementary bit steam. Some important fields are explained below:

- Adaptation field length is an 8-bit field specifying the number of bytes immediately following it in the adaptation field including stuffing bytes.
- Discontinuity indicator is 1-bit flag which when it is set to 1 indicates that the discontinuity state is true for the current transport packet. When this flag is set to 0, the discontinuity is false. This discontinuity indicator is used to indicate two types of discontinuities, system time-base discontinuities and continuity-counter discontinuities. In the first type, this transport stream packet is the packet of a PID designed as a PCR-PID. The next PCR represents a sample of a new system time clock for the associated program. In the second type, the transport stream packet could be any PID type. If the transport stream packet is not designated as a PCR-PID, the continuity counter may be discontinuous with respect to the previous packet with the same PID or when a system time-base discontinuity occurs. For those PIDs that are not designated as PCR-PIDs, the discontinuity indicator may be set to 1 in the next transport stream packet with the same PID, but will not be set to 1 in three consecutive transport stream packet with the same PID.
- Random access indicator is a 1-bit flag that indicates the current and subsequent transport stream packets with the same PID, containing some information to aid random access at this point. Specifically, when this flag is set to 1, the next PES packet in the payload of the transport stream packet with the current PID will contain the first byte of a video sequence header or the first byte of an audio frame.

- Elementary stream priority indicator is used for data-partitioning application in the elementary stream. If this flag is set to 1, the payload contains high-priority data, such as the header information, or low-order DCT coefficients of the video data. This packet will be highly protected.
- PCR flag and OPCR flag: If these flags are set to 1, it means that the adaptation field contains the PCR data and original PCR data. These data are coded in two parts.
- Splicing point flag: When this flag is set to 1, it indicates that a splice-countdown field will be present to specify the occurrence of a splicing point. The splice point is used to splice two bitstreams smoothly into one stream. The Society of Motion Picture and Television Engineers (SMPTE) has developed a standard for seamless splicing of two streams (SMPTE, 1997). We will describe the function of splicing later.
- Transport private flag: This flag is used to indicate whether the adaptation field contains private data.
- Adaptation filed extension-flag: This flag is used to indicate whether the adaptation field contains the extension field that gives more-detailed splicing information.

**Packetized elementary stream** — It is noted that the elementary stream data is carried in PES packets. A PES packet consists of a PES packet header followed by packet data, or payload. The PES packet header begins with a 32-bit start-code that also identifies the stream or stream type to which the packet data belong. The first byte of each PES packet header is located at the first available payload location of a transport stream packet. The PES packet header may also contain decoding time stamps (DTS), presentation time stamps (PTS), elementary stream clock reference (ESCR), and other optional fields such as DSM trick-mode information. The PES packet data field contains a variable number of contiguous bytes from one elementary stream. Readers can learn this part of the syntax in the same way as described for the transport packet header and adaptation field.

**Program-specific information** — PSI includes both MPEG-2 system-compliant data and private data. In the transport streams, the program-specific information is classified into four table structures: program association table, program map table, conditional access table, and network information table. The network information table is private data and the other three are MPEG-2 system compliant data. The program associate table provides the information of program number and the PID value of the transport stream packets. The program map table specifies PID values for components of one or more programs. The conditional access (CA) table provides the association between one or more CA systems, their entitlement management messages (EMM), and any special parameters associated with them. The EMM are private conditional-access information that specify the authorization levels or the services of specific decoders. They may be addressed to a single decoder or groups of decoders. The network information table is optional and its contents are private. Its contents provide physical network parameters such as FDM frequencies, transponder numbers, etc.

## 20.2.3 Transport Stream Splicing

The operation of bitstream splicing is switching from one source to another according to the requirements of the applications. Splicing is the most common operation performed in TV stations today (Hurst, 1997). The examples include inserting commercials into programming, editing, inserting, or replacing a segment into an existing stream, and inserting local commercials or news into a network feed. The most important problem for bitstream splicing is managing the buffer fullness at the decoder. Usually, the encoded bitstream satisfies the buffer regulation with a buffer control algorithm at the encoder. During decoding, this bitstream will not cause the decoder buffer to suffer from buffer overflow and underflow. A typical example of buffer fullness trajectory at the decoder is shown in Figure 20.6. However, after bitstream splicing, the buffer regulation is not

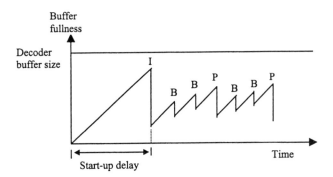

**FIGURE 20.6**   Typical buffer fullness trajectory at the decoder.

guaranteed depending on the selection of splicing point and the bit rate of the new bitstream. It is necessary to have a rule for selecting the splicing point.

The committee on packetized television technology, PT20 of SMPTE, has proposed a standard that deals with the splice point for MPEG-2 transport streams (SMPTE, 1997). In this standard, two techniques have been proposed for selecting splicing points. One is the seamless splicing and the other is nonseamless splicing. The seamless splicing approach can provide clean and instant switching of bitstreams, but it requires careful selection of splicing points on video bitstreams. The nonseamless splicing approach inserts a "drain time" that is a period of time between the end of an old stream and the start of a new stream to avoid overflow in the decoder buffer. The "drain time" assures that the new stream begins with an empty buffer. However, the decoder has to freeze the final presented picture of the old stream and wait for a period of start-up delay while the new stream is initially filling the buffer. The difference between seamless splicing and nonseamless splicing is shown in Figure 20.7.

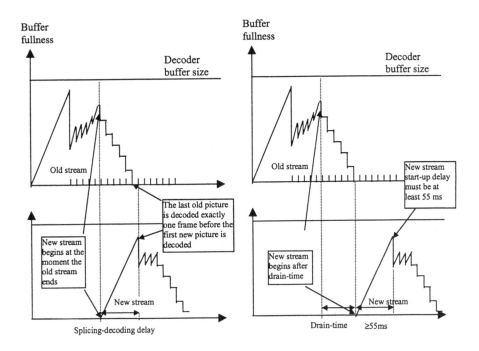

**FIGURE 20.7**   Difference between seamless splicing and nonseamless splicing: (a) the VBV buffer behavior of seamless splicing, (b) the VBV buffer behavior of nonseamless buffer behavior.

In the SMPTE proposed standard (SMPTE, 1997), the optional indicator data in the PID streams (all the packets with the same PID within a transport stream) are used to provide important information about the splice for the applications such as inserting commercial programs. The proposed standard defines a syntax that may be carried in the adaptation field in the packets of the transport stream. The syntax provides a way to convey two kinds of information. One type of information is splice-point information that consists of four splicing parameters: drain time, in-point flag, ground id and picture-param-type. The other types of information are splice point indicators that provide a method to indicate application-specific information. One such application example is the insertion indicator for commercial advertisements. This indicator includes flags to indicate that the original stream is obtained from the network and that the splice point is the time point where the network is going out or going in. Other fields give information about whether it is scheduled, how long it is expected to last, as well as an ID code. The detail about splicing can be found in the proposed standard (SMPTE, 1997).

Although the standard provides a tool for bitstream splicing, there are still some difficulties for performing bitstream splicing in practice. One problem is that the selection of a splicing point has to consider that the bitstream contains video that has been encoded by a predictive coding scheme. Therefore, the new stream should begin from the anchor picture. Other problems include uneven timing frames and splicing of bitstreams with different bit rates. In such cases, one needs to be aware of any consequences related to buffer overflow and underflow.

## 20.2.4  PROGRAM STREAMS

The program stream is defined for the multiplexing of audio, video, and other data into a single stream for communication or storage application. The essential difference between the program stream and transport stream is that the transport stream is designed for applications with noisy media, such as in terrestrial broadcasting. Since the program stream is designed for applications in the relatively error-free environment, such as in the digital video disk (DVD) and digital storage applications, the overhead in the program stream is less than in the transport stream.

A program stream contains one or more elementary streams. The data from elementary streams are organized in the form of PES packets. The PES packets from different elementary streams are multiplexed together. The structure of a program stream is shown in Figure 20.8.

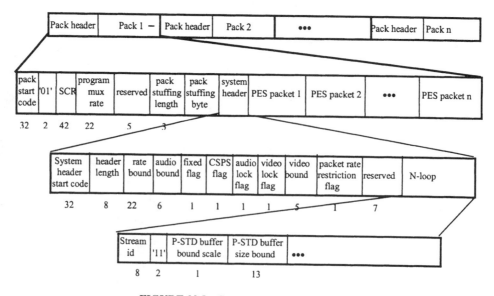

**FIGURE 20.8**  Structure of program stream.

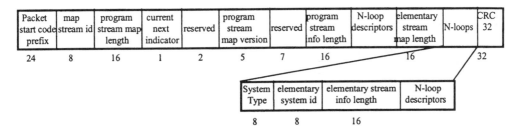

| Packet start code prefix | map stream id | program stream map length | current next indicator | reserved | program stream map version | reserved | program stream info length | N-loop descriptors | elementary stream map length | N-loops | CRC 32 |
|---|---|---|---|---|---|---|---|---|---|---|---|
| 24 | 8 | 16 | 1 | 2 | 5 | 7 | 16 | | 16 | | 32 |

| System Type | elementary system id | elementary stream info length | N-loop descriptors |
|---|---|---|---|
| 8 | 8 | 16 | |

**FIGURE 20.9** Data structure of program stream map.

A program stream consists of packs. A pack begins from a pack header followed by PES packets. The pack header is used to carry timing and bit-rate information. It begins with a 32-bit start-code followed by system clock reference (SCR) information, program muxing rate, and stuffing bits. The SCR indicates the intended arrival time of the byte that contains the last bit of SCR base at the input of the decoder. The program muxing rate is a 22-bit integer that specifies the rate at the decoder. The value of this rate may vary from pack to pack. The stuffing bits are inserted by the encoder to meet channel requirements. The pack header may contain a system header that may be repeated optionally. The system header contains the summary of the system parameters, such as header length, rate bound, audio bound, video bound, stream id, and other system parameters. The rate bound is used to indicate the maximum rate in any pack of the program stream, and it may be used to assess whether the decoder is capable of decoding the entire stream. The audio bound and video bound are used to indicate the maximum values of audio and video in the program stream. There are some other flags that are used to give some system information. A PES packet consists of a PES packet header followed by packet data. The PES packets have the same structure as in the transport stream.

A special type of PES packet is the program stream map; it is present when the stream id value is $0 \times BC$. The program stream map provides a description of the elementary streams in the program stream and their relationship to one another. The data structure of program stream map is shown in Figure 20.9.

Other special types of PES packets include program stream directory and program element descriptors. The major information contained in the program stream directory includes the number of access units, packet stream id, and presentation time stamp (PTS). The program and program descriptors provide the coding information about the elementary streams. There are a total of 17 descriptors including video descriptor, audio descriptor, and hierarchy descriptor. For the detail on these descriptors, the reader is referred to the standard document (ISO/IEC, 1996).

### 20.2.5 TIMING MODEL AND SYNCHRONIZATION

The principal function of the MPEG system is to define the syntax and semantics of the bitstreams that allow the system decoder to perform two operations among multiple elementary streams: demultiplexing and resynchronization. Therefore, the system encoder has to add the timing information to the program streams or transport streams during the process of multiplexing the coded video, audio, and data elementary streams to a single stream or multiple streams. System, video, and audio all have a timing model in which the end-to-end delay from the signal input to an encoder to the signal output from a decoder is a constant. The delay is the sum of encoding, encoder buffering, multiplexing, transmission or storage, demultiplexing, decoding buffering, decoding, and presentation delays. The buffering delays could be variable, while the sum of total delays should be constant.

In the program stream, the timing information for a decoding system is the SCR; in the transport stream, the timing information is given by the PCR. The SCR and PCR are time stamps that are used to encode the timing information of the bitstream itself. The 27 MHz SCR is the kernel time base for the entire system. The PCR is 90 kHz, which is $\frac{1}{300}$ of the SCR. In the transport stream, the PCR is encoded with 33 bits and is contained in the adaptation field of the transport stream. The PCR can be extended to the SCR with an additional 9 bits in the adaptation field. For the program stream, the SCR is directly encoded with 42 bits and it is located in the pack header of the program stream. The synchronization among multiple elementary streams is accomplished with a PTS in the program and transport streams. The PTS is 90 kHz and represented with a 33-bit number coded in three separate parts contained in the PES packet header. In the case of audio, if a PTS is present, it will refer to the first access unit commencing in the PES packet. An audio access unit starts in a PES packet if the first byte of the audio access unit is present in the PES packet. In the case of video, if a PTS occurs in the PES packet header, it refers to the access-unit containing the first picture start-code that commences in this PES packet. A picture start-code commences in the PES packet if the first byte of the picture start-code is present in the PES packet. In an MPEG-2 system, the system clock reference is specified to satisfy the following conditions:

27 MHz to 810 Hz $\leq$ SCR $\leq$ 27 MHz + 810 Hz
Change rate of SCR $\leq 75 \times 10^{-3}$ Hz/second

In the encoder, the SCR or PCR are encoded in the bitstream at intervals up to 100 ms in the transport stream and up to 700 ms in the program stream. As such, they can be used to reconstruct the system time clock in the decoder with sufficient accuracy for all identified applications. The decoder has its own system time clock (STC) with the same frequency, 90 kHz for the transport stream and 27 MHz for the program stream. In a correctly constructed MPEG-2 system bitstream, each SCR arrives at the decoder at precisely the time indicated by the value of that SCR. If the decoder's clock frequency matches the one in the encoder, the decoding and presentation of video and audio will automatically have the same rate as those in the encoder; then the end-to-end delay will be constant. However, the STC in the decoder may not exactly match the one in the encoder due to the independent oscillators. Therefore, a decoder's system clock frequency may not match the encoder's system clock frequency that is sampled and indicated in the SCR. One method is to use a free run 27 MHz in the decoder. The mismatch between the encoder's system time clock and the decoder's system time clock is handled by skipping or repeating frames. Another method to handle the mismatch is to use the received SCRs (which occur at least once in the intervals of 100 ms for the transport stream and 700 ms for the program stream). In this way, the decoder's system time clock is a slave to the encoder's system time clock. This can be implemented with a phase-locked loop (PLL) as shown in Figure 20.10.

The synchronization among multiple elementary streams can be achieved by adjusting the decoding of streams to a common master time base rather than by adjusting the decoding of one

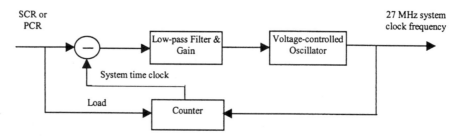

**FIGURE 20.10**   System time clock recovery using PLL.

stream to match that of another. The master time base may be one of the many decoder clocks, the clock of the data source, or some external clock. Each program in a transport stream, which may contain multiple programs, may have its own time base. The time bases of different programs within a transport stream may be different.

In the digital video systems, the 13.5 MHz sampling rate of the luminance signal and 6.25 MHz chrominance signals of the CCIR601 digital video are all synchronized to 27 MHz system time clock. The NTSC or PAL TV signals are also phase-locked to the same 27 MHz clock such that the horizontal and vertical synchronous signals and the color burst clock are all locked to the 27-MHz system time clock.

In the TV studio applications, the entire TV studio equipment is synchronized to the same time base as a composite horizontal and vertical synchronization signal in order to perform the seamless video source switching and editing. It should be noted that this time base is definitely not synchronized to the PCRs from various remote encoder sites. The 27-MHz local decoder's system time clock is locked to the same studio composite horizontal and vertical synchronization signal. The 33 bits of video STC counter is initialized by the latest video PTS, then calibrated using the 90-kHz clock derived from the 27-MHz system clock in the decoder. If the 27-MHz system clock in the decoder is synchronized with the system clock on the transmitting end, the STC counter will be always the same as the incoming PTS numbers. However, there may be some mismatch between the system clocks. As each new PTS arrives, the PTS will be compared with the STC counter. If the PTS is larger than the STC plus half of the duration of the PTS, it means that the 27-MHz decoder clock is too slow and the bit buffer may overflow. In this case, the decoder should skip some of the current data to search for the next anchor frame so that decoding can be continued. If the PTS is less than the STC minus half of the duration of the PTS, the bit buffer may underflow. The decoding will halt and repeatedly display the current frame. The audio decoder will also be locked on the same 27-MHz system clock, where similar skipping and repeating of audio data are used to handle the mismatch.

In the low-cost consumer "set-top box" (STB) applications, a simple free-run 27-MHz decoder system clock with the skipping and repeating frame scheme can provide pretty good results. In fact, the skipping or repeating frame may happen once in a 2- or 4-hour period with a free-run 27-MHz crystal clock. The STC counter will be set by the latest PTS, then count on the 90-kHz STC clock derived from the free-run 27-MHz crystal clock. The same skipping or repeating display control as the TV studio will be used.

For a complex STC solution, a phase-locked loop with VCXO (voltage controlled crystal oscillator) in the decoder is used to synchronize the incoming PCR data. The 33-bit decoder's PCR counter is initialized by the latest PCR data, then the 27-MHz system clock is calibrated. If the decoder's system clock is synchronized with the encoder's remote 27-MHz system clock, every incoming PCR data will be the same as the decoder's PCR counter or have small errors from PCR jitter. The difference between the decoder's PCR counter and the incoming PCR data indicates this frequency jitter or drift. As long as the decoder's 27-MHz system clock is locked on the PCR data, the STC counter will be initialized by the latest PTS, then calibrated using the 90-kHz clock. The similar skipping and repeating frame scheme will be used again, but the 27-MHz system clock in the decoder is synchronized with the incoming PCR. As long as the decoder's 27 MHz is locked on the encoder's 27 MHz, there will be no skipping or repeating of frames. However, if the PCR PLL is not working properly, the skipping or repeating of frames will occur more often than when the free-run 27-MHz system clock is used.

Finally, it should be noted that the PTS-DTS flag is used to indicate whether or not the PTS and DTS or both of them (decoding time stamp) will be present in the PES packet header. The DTS is a 33-bit number coded in three separate fields in the PES packet header. It is used to indicate the decoding time.

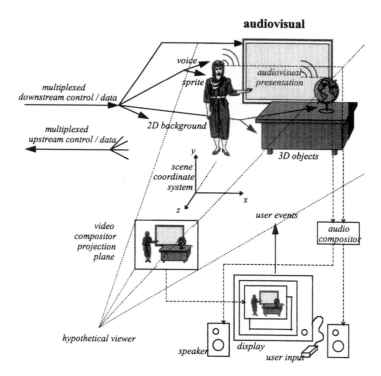

**FIGURE 20.11**   An example of MPEG-4 audiovisual scene. (From ISO/IEC 14496-1, 1998. With permission.)

## 20.3   MPEG-4 SYSTEM

This section describes the specification of the MPEG-4 system or ISO/IEC 14496-1.

### 20.3.1   OVERVIEW AND ARCHITECTURE

The specification of the MPEG-4 system (ISO/IEC, 1998) is used to define the requirements for the communication of interactive audiovisual scenes. An example of such a scene is shown in Figure 20.11 (ISO/IEC, 1998). The overall operation of this system can be summarized as follows. At the encoder, the audio, visual, and other data information is first compressed and supplemented with synchronization timing information. The compressed data with timing information are then passed to a delivery layer that multiplexes these data into one or more coded binary streams for storing or transmission. At the decoder, these streams are first demultiplexed and decompressed. The reconstructed audio and visual objects are then composed according to the scene description and synchronization information. The composed audiovisual objects are then presented to the end user. The important feature of the MPEG-4 standard is that the end user may have the option to interact with this presentation since the compression is performed on the object or content basis. The interaction information can be processed locally or transmitted back to the encoder. The scene information is contained in the bitstreams and used in the decoding processes.

The system part of the MPEG-4 standard specifies the overall architecture of a general receiving terminal. Figure 20.12 shows the basic architecture of the receiving terminal. The major elements of this architecture are delivery layer, sync layer (SL), and compression layer. The delivery layer consists of the FlexMux and TransMux. At the encoder, the coded elementary streams, which include the coded video, audio, and other data with the synchronization and scene description information, are multiplexed to the FlexMux streams. The FlexMux streams are transmitted to the

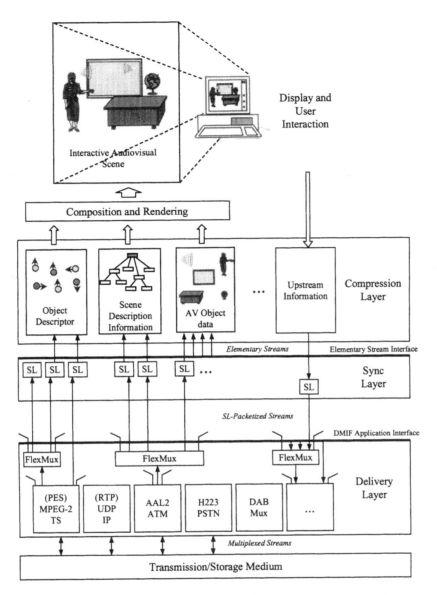

**FIGURE 20.12**  The MPEG-4 system terminal architecture. (From ISO/IEC 14496-1, 1998. With permission.)

TransMux of the delivery layer from the network. The function of TransMux is not within the scope of the system standard and it can be any of the existing transport protocols such as MPEG-2 transport stream, RTP/UDP/IP, AAL5/ATM, and H223/Mux.

Only the interface to the TransMux layer is part of the standard. Usually, the interface is the DMIF application interface (DAI), which is not specified in the system part, but in Part 6 of the MPEG-4 standard. The DAI specifies the data that need to be exchanged between the SL and the delivery layer. The DAI also defines the interface for signaling information required for session and channel setup as well as teardown. For some simple applications, it does not require full functionality of the system specification. A simple multiplexing tool, FlexMux, with low delay and low overhead, is defined in the system part of MPEG-4. The FlexMux tool is a flexible multiplexer that accommodates the interleaving of SL-packetized streams with varying instantaneous bit rates. The FlexMux packet has a variable length which may contain one or more SL packets. Also, the

FlexMux tool provides the identification for the SL packets to indicate which elementary stream they come from. FlexMux packets with data from different SL-packetized streams, therefore, can be arbitrarily interleaved.

The SL specifies the syntax for packetizing the elementary streams to the SL packets. The SL packets contain an SL packet header and an SL packet payload. The SL packet header provides the information for continuity checking in case of data loss and also carries the timing and synchronization information as well as fragmentation and random access information. The SL packet does not contain its length information. Therefore, SL packets must be framed by the FlexMux tool. At the decoder, the SL packets are demultiplexed to the elementary streams in the SL. At the same time, the timing and the synchronization information as well as fragmentation and random access information are also extracted for synchronizing the decoding process and subsequently for composition of the elementary streams.

At the compression layer, the encoded elementary streams are decoded. The decoded information is then used for the reconstruction of audiovisual information. The operation of the reconstruction includes composition, rendering, and presentation with the timing synchronization information.

### 20.3.2 Systems Decoder Model

The systems decoder model (SDM) is a conceptual model that is used to describe the behavior of decoders complying with MPEG-4 systems. It may be used for the encoder to predict how the decoder or receiving terminal will behave in terms of buffer management and synchronization during the process of decoding, reconstructing, and composing of audiovisual objects. The systems decoder model includes a system timing model and a system buffer model. These models specify the interfaces for accessing demultiplexed data streams, decoding buffers for each elementary stream, the behavior of elementary stream decoders, composition memory for decoded data from each decoder, and the output behavior of composition memory toward the compositor. The systems decoder model is shown in Figure 20.13.

The timing model defines the mechanisms that allow a decoder or receiving terminal to process time-dependent objects. This model also allows the decoder or receiving terminal to establish mechanisms to maintain synchronization both across and within particular media types as well as with user interaction events. In order to facilitate these functions at the decoder or receiving terminal, the timing model requires that the transmitted data streams contain implicit or explicit timing information. There are two sets of timing information that are defined in the MPEG-4 system. One indicates the periodic values of the encoder clock that is used to convey the encoder's time base to the decoder or the receiving terminal, while the other is the desired presentation timing for each audiovisual object. For real-time applications, the end-to-end delay from the encoder input to the decoder output is constant. The delay is equal to the sum of the delay due to the encoding process, buffering, multiplexing at the encoder, the delay due to the delivery layer and demultiplxing, decoder buffering, and decoding process at the decoder.

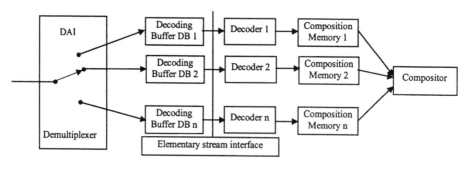

**FIGURE 20.13** Block diagram of systems decoder model. (From ISO/IEC 14496-1, 1998. With permission.)

The buffer model is used for the encoder to monitor and control the buffer resources that are needed for decoding each elementary stream at the decoder. The information of the buffer requirements is transmitted to the decoder by descriptors at the beginning of the decoding process. The decoder can then decide whether or not it is capable of handling this particular bitstream. In summary, the buffer model allows the encoder to schedule data transmission and to specify when the bits may be removed from these buffers. Then the decoder can choose proper buffers so that the buffers will not overflow or underflow during the decoding process.

### 20.3.3  SCENE DESCRIPTION

In multimedia applications a scene may consist of audiovisual objects that include the objects of natural video, audio, texture, 2-D or 3-D graphics, and synthetic video. Since MPEG-4 is the first object-based coding standard, reconstructing or composing a multiple audiovisual scene is quite new. The decoder not only needs the elementary streams for the individual audiovisual objects, but also synchronization timing information and the scene structure. This information is called the scene description and it specifies the temporal and spatial relationships between the objects or scene structures. The information of the scene description can be defined in the encoder or interactively determined by the end user and transmitted with the coded objects to the decoder. The scene description only describes the scene structure. The action of assembling these audiovisual objects to a scene is called composition. The action of transmitting these objects from a common representation space to a specific presentation device is called rendering.

The MPEG-4 system defines the syntax and semantics of a bitstream that can be used to describe the relationships of the objects in space and time. However, for visual data, the system standard does not specify the composition algorithms. Only for audio data the composition process is specified in a normative manner. In order to allow the operations of authoring, editing, and interaction of visual objects at the decoder the scene descriptions are coded independently from the audiovisual media. This allows the decoder to modify the scene according to the requirements of the end user. Two kinds of user interaction are provided in the system specification. One is client-side interaction that involves object manipulations requested in the end user's terminal. The manipulation includes the modification of attributes of scene objects according to specified user actions. The other type of manipulation is the server-side interaction that the standard does not deal with.

The scene description is a hierarchical structure that can be represented as a graph. The example of the audiovisual scene in Figure 20.11 can be represented as in Figure 20.14. The scene description

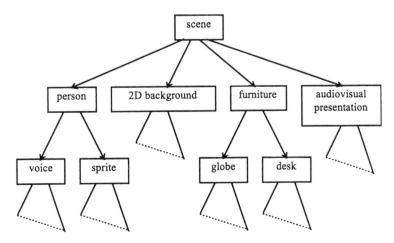

**FIGURE 20.14**  Hierarchical graph representation of an audiovisual scene. (From ISO/IEC 14496-1, 1998. With permission.)

is represented by a parametric approach, the binary format for scenes (BIFS). The description consists of an encoded hierarchical tree of nodes with attributes and other information. In this tree, the leaf nodes correspond to the elementary audiovisual objects and information for grouping, transformation, and other operation.

### 20.3.4 OBJECT DESCRIPTION FRAMEWORK

The elementary streams carry data for audio or visual objects as well as for the scene description itself. The purpose of the object description framework is to provide the link between the elementary streams and the audiovisual scene description. The object description framework consists of a set of descriptors that allow identifying, describing, and appropriately associating elementary streams to each other and to audiovisual objects used in the scene description. Each object descriptor is a collection of one or more elementary stream descriptors that are associated to a single audiovisual object or a scene description. Object descriptors are themselves conveyed in elementary streams. Each object descriptor is assigned an identifier (object descriptor ID), which is unique within a defined name scope. This identifier is used to associate audiovisual objects in the scene description with a particular object descriptor, and thus the elementary streams related to that particular object. Elementary stream descriptors include information about the source of the stream data, in the form of a unique numeric identifier (the elementary stream ID) or a URL pointing to a remote source for the stream. Elementary stream descriptors also include information about the encoding format, configuration information for the decoding process and the SL packetization, as well as quality-of-service requirements for the transmission of the stream and intellectual property identification. Dependencies between streams can also be signaled within the elementary stream descriptors. This functionality may be used, for example, in scalable audio or visual object representations to indicate the logical dependency of an enhancement layer stream to a base-layer stream. It can also be used to describe alternative representations for the same content (e.g., the same speech content in various languages).

The object description framework provides the hooks to implement intellectual property management and protection (IPMP) systems. IPMP descriptors are carried as part of an object descriptor stream. IPMP elementary streams carry time-variant IPMP information that can be associated to multiple object descriptors. The IPMP system itself is a nonnormative component that provides intellectual property management and protection functions for the terminal. The IPMP system uses the information carried by the IPMP elementary streams and descriptors to make protected IS 14496 content available to the terminal. An application may choose not to use an IPMP system, thereby offering no management and protection features.

## 20.4 SUMMARY

In this chapter, the MPEG system issues are discussed. Two typical systems, MPEG-2 and MPEG-4, are introduced. The major task of the system layer is to multiplex and demultiplex video, audio, and other data to a single bitstream with synchronization timing information. For MPEG-4 systems, there are additional issues addressed, such as interface with network applications.

## 20.5 EXERCISES

20-1. What are two major system streams provided by MPEG-2 system? Describe some application examples for these two streams and explain the reasons.

20-2. The MPEG-2 system bitstream is a self-contained bitstream to facilitate synchronous playback of video, audio, and related data. Describe what kind of timing signals are contained in the bitstream to achieve the goal of synchronization.

**20-3.** How does the MPEG-2 system deal with different system clocks between the encoder and decoder? Describe what a system may do when the decoder clock is running too slow or too fast?

**20-4.** Why is the 27-MHz system clock in MPEG-2 represented in two parts: 33 + 9 bit extension?

**20-5.** What is bitstream splicing of a transport stream? Give several application examples of bitstream splicing and indicate the problems that may arise.

**20-6.** Describe the difference between the MPEG-2 system and the MPEG-4 system.

## REFERENCES

Hurst, N. Splicing — high definition broadcasting technology, year 1 demonstration, Meeting talk, 1997.

ISO/IEC 13818-1: 1996 Information Technology — Generic Coding of Moving Pictures and Associated Audio Information, 1996.

ISO/IEC 14496-1: 1998 Information Technology — Coding of Audio-Visual Objects, 1998.

SMPTE, 1997, Proposed SMPTE Standard for Television — Splice Points for MPEG-2 Streams, PT20.02, April 4, 1997.

# Index

## A

ACATS (Advisory Committee on Advanced Television
    Service), 367-368
Access unit, 450
Accuracy
    half-pixel, 434-435
    matching, 234, 246
    one-pixel, 234
AC prediction, 409-410, 426
Action units, 308
Activity masking, 12-13
Adaptation fields, 453-456
Adaptive error concealment, 396-397
Adaptive prediction, 64-65, 68, 72
Adaptive quantization, 45-49, 51-52
Additive white Gaussian noise (AWGN), 11-12, 20
Advanced Television Standards Committee (ATSC),
    368-371. See also ATSC DTV standards
Advanced Television Test Center (ATTC), 368
Advisory Committee on Advanced Television Service
    (ACATS), 367-368
A-law characteristic, 45
Alpha planes, 413
Alternative inter-VLC (AIV) mode, 446
Amplitude distribution, 98
Anandan method, 277-278
Aperture problem, 266-268, 305
AR (autoregressive) model, 133-134, 301
Arbitrarily slice ordering (ASO), 443
Arithmetic coding, 107, 119-127, 163, 438
ARQ (automatic repeat request) system, 138
ASCII code, 108-109
ASO (arbitrarily slice ordering), 443
ATM networks, 360, 362, 372, 453
ATSC (Advanced Television Standards Committee),
    368-371. See also ATSC DTV standards
ATSC DTV standards, 367-371
    compression layer, 370
    digital television and, 328
    history of, 367-368
    picture layer, 369-370
    technical overview, 368-371
    transcoding with bitstream scaling, 371-379
    transmission layer, 371
    transport layer, 370-371
ATTC (Advanced Television Test Center), 368
Audio visual objects (AVOs), 403
Audiovisual scene, 465
Autocorrelation, 5, 7, 62
Automatic repeat request (ARQ) system, 138

Autoregressive (AR) model, 133-134, 301
AVOs (audio visual objects), 403
AWGN (additive white Gaussian noise), 11-12, 20

## B

BAB (binary alpha block), 413, 426
Backward temporal masking, 14
Basis images, 83-84, 86-87, 103
Basis vectors, 79-80, 83, 103
Bergmann algorithms, 260, 262
Binary alpha block (BAB), 413, 426
Binary alpha planes, 413
Binary format for scenes (BIFS), 466
Binary shape coding, 414-415
Binary splitting, 188
Binary tree search quantization, 188
Bit, origin of term, 25
Bit allocation, 95-102, 354-356, 378
Bit error probability, 65
Bit-plane coding, 181
Bit rate
    constant, 350, 372
    conversion, 372
    definition, 3
    example requirement, 4
    MPEG, 363
    optical flow, 265, 301
    reducing, 374-375
    variable, 350, 372
Bitstreams. See also Bitstream scaling
    compressed, 371-372
    editing, 404
    error concealment, 392
    layered, 439
    splicing, 456-458
    syntax and semantics, 420-421
Bitstream scaling, 371-379
    analysis, 378-379
    architectures of, 374-378
    background, 371-373
    principles of, 373-374
Block artifacts, 235-236
Block layer, 432
Block matching, 221-247
    block size, 222, 229
    coarse-fine three-step search, 226
    conjugate direction search, 226-227
    conservation information, 306
    generalized, 350

Block matching *(cont.)*
    hierarchical, 236-239, 246, 349
    limitations of, 235-236, 246
    2-D logarithmic search, 224-225, 348
    matching accuracy, 234, 246
    matching criteria, 222-224, 246
    MPEG, 335, 347-348
    multigrid, 238-242, 246
    multiresolution, 227, 246
    overlapped, 244-245, 247
    overview, 221-222
    performance comparison, 313-315
    predictive motion field segmentation, 242-244, 247, 313
    as region-based approach, 311
    searching procedures, 224-234, 246
    subsampling, 227
    thresholding multigrid, 239-241
    thresholding multiresolution, 229-234
Block overlapping, 102, 105
Block quantization. *See* Transform coding
Block truncation coding (BTC), 191
Brightness invariance equations, 270-271, 278, 312
Buffer fullness trajectory, 456-457
Buffer model, 465

## C

CAE (content-based arithmetic encoding), 413-415
Cafforio and Rocca algorithm, 261
Carry-over, 126
CBR (constant bit rate) encoders, 360-362
CCIR, 19, 327, 334
CCITT, 139-140, 330
CDF (cumulative distribution function), 121
Channel capacity, 26-27, 325
Channel encoding, 65
Chebyshev polynomials, 89
Chrominance, 15-17, 100, 396, 434, 436-437, 446
CIE (Commission Internationale de l'Eclairage), 15
CIF (common intermediate format), 327-328, 429-430
CMV (concealment motion vectors), 344
Code alphabets, 107-108
Codebooks, 117-119, 185, 187-191
Codes
    binary, 108
    block, 108-110, 120
    compact, 111
    definition, 108
    discrete memoryless, 111-114
    instantaneous, 111, 114-115
    stream-based, 120, 128
    uniquely decodable, 108-111
Codewords. *See also* Coding algorithms
    basics of, 107-108
    coding theorem, 25-26
    domain error concealment, 392-393
    length, 119-120
    transform coding, 102

Coding algorithms
    adaptive arithmetic, 418
    area, 139
    arithmetic, 107, 119-127
    dictionary, 140-149, 151-152
    efficiency, 404
    fractal image, 185, 193-197
    frame-difference, 70, 209
    hierarchical mode, 157-158, 166-167, 349
    Huffman, 101, 107, 114-120, 125, 127
    information sources, 107-108
    lossless bilevel still image compression, 150
    lossless JPEG, 157-158, 164-166
    lossless multilevel still image compression, 150-151
    model-based, 185, 197-198
    motion-compensated, 209-211, 245
    nonstandard, 185-198
    progressive DCT-based mode, 157-158, 163-165
    raster, 139
    run-length, 32, 134-139, 151
    sequential DCT-based mode, 157-163
    vector quantization, 185-193
Coding control, 431
Coding rate. *See* Bit rate
Coding redundancy, 4, 8-9, 26
Coding theorems, 25-28
Color masking, 14-17
Color representation, 327
Commission Internationale de l'Eclairage (CIE), 15
Common intermediate format (CIF), 327-328, 429-430
Companding quantization, 44-46, 51
Compression characteristic, 45
Compression layer, 464
Concealment motion vectors (CMV), 344
Conditional replenishment, 68-71, 208
Conservation information, 279-280, 293-300, 306
Constant bit rate (CBR) encoders, 360-362
Content-based arithmetic encoding (CAE), 413-415
Content-based coding, 403. *See also* MPEG-4
Content-based efficient compression, 404-405
Content-based interactivity, 404
Contour curves, 254
Contractive affine transformations, 194
Contrast, optical flow, 296
Contrast masking, 10-12, 18
Contrast sensitivity function, 10-11, 18
Convergence speed, 255-256, 258, 261-262
Correlation-feedback technique, 281-293
Correlation measure, 223
Costs, 19-20
Covariance matrix, 76-79
Cumulative distribution function (CDF), 121
Cumulative probability (CP), 121, 123, 126-127

## D

DAI (DMIF application interface), 463
DC prediction, 409-410, 426
DCT. *See* Discrete cosine transform

DCT-to-spatial transformation, 388-389
Deblocking filters (DF), 441-443
Decision levels, 34, 45-46
Decoding
    arithmetic coding, 123-125
    block diagram, 32-33, 48
    dictionary coding, 142-147, 149
    down-conversions, 217, 379-391
    full-memory, 385-387
    full-resolution, 380
    H.261, 431
    half-memory, 385-387
    independent segmentation, 444
    low-resolution, 380
    MPEG-2, 452
    MPEG-4, 426-427
    quarter-memory, 385-387
    reduced-resolution update, 444-445
    scalable, 385-388
    systems decoder model, 464-465
    system target, 451
    variable-length, 340
Decoding time stamps (DTS), 451, 456, 461
Delivery layer, 462
Delta modulation (DM), 65-68, 72
Demultiplexing, 452
Descent methods, 251-258. *See also* Pel recursive techniques
    conditions for, 252-253
    convergence speed, 255-256
    Newton-Raphson's method, 257-258, 260
    steepest, 256-257, 261, 306
    strategy, 253-255
Detail dependence, 12-13
Detelecine processing, 347
Deterministic *vs.* stochastic methods, 308
DFD (displaced frame difference), 251, 259, 262, 282
DF (deblocking filters), 441-443
DFT (discrete Fourier transform), 82, 84-86, 90-92
DHT (discrete Hadamard transform), 87-88
Differential coding, 55-72. *See also* Quantization
    adaptive prediction, 64-65, 72
    conditional replenishment, 68-71
    1-D, 2-D, and 3-D, 63-64, 69-70, 72
    delta modulation, 65-68, 72
    DPCM coding of wavelet coefficients, 418
    general DPCM systems, 58-60, 72
    history of, 60
    implementation, 62-65
    information-preserving, 71-72
    interframe, 68-69
    motion-compensated predictive, 64-65
    optimum DPCM system, 62-63
    optimum linear prediction, 59-62, 64, 69-70
    order of predictors, 64, 72
    pixel-to-pixel DPCM, 55-59, 63
    step size, 67
    transform coding and, 103
    transmission errors, 65, 72
Differential pulse code modulation (DPCM), 55-60, 69-70, 72, 103, 206. *See also* Differential coding

Digital halftones, 150-151
Digital television (DTV), 328, 367-370. *See also* ATSC DTV standards
Digital Video Disk (DVD), 331, 458
Digram coding, 141
Dirty window effect, 209
Discontinuity indicators, 455
Discrete cosine transform (DCT)
    DCT-to-spatial, 388-389
    differential coding, 60
    field-frame syntax, 342-343
    frequency masking, 14
    motion estimation, 315-318
    MPEG and, 334, 359
    progressive DCT-based mode, 157-158, 163-165
    pseudophases, 315-317
    run-length coding and, 134
    sequential DCT-based mode, 157-163
    shape-adaptive, 412-413
    thresholding, 97-98
    transform coding, 88-92, 95, 104
    video coding, 206
    zigzag scans, 100
    zonal coding, 95
Discrete Fourier transform (DFT), 82, 84-86, 90-92
Discrete Hadamard transform (DHT), 87-88
Discrete memoryless codes, 111-114
Discreteness, 25-26
Discrete sine transform (DST), 315-317
Discrete Walsh-Hadamard transform (DWHT), 87
Discrete Walsh transform (DWT), 86-88
Discrimination threshold, 12
Disocclusion, 306-307
Displaced frame difference (DFD), 251, 259, 262, 282
Displacement vectors, 251. *See also* Pel recursive techniques
Dissimilarity metric, 224
Distortion
    H.263, 354, 356-357
    MPEG, 354, 356-357
    quantization, 35-37
    rate distortion function, 27-28, 324-326
Dithering, 50
DM (delta modulation), 65-68, 72
DMIF application interface (DAI), 463
Down-conversion decoders, 379-391
    background, 379-381
    DCT-to-spatial transformation, 388-389
    frequency synthesis, 381-383
    motion-compensated, 217, 383-385
    structures, 380
    summary, 388
    three-layer scalable, 385-388
DPCM (differential pulse code modulation), 55-60, 69-70, 72, 103, 206. *See also* Differential coding
DSM-CC, 450
DST (discrete sine transform), 315-317
DTS (decoding time stamps), 451, 456, 461
DTV (digital television), 328, 367-370. *See also* ATSC DTV standards

Dual prime prediction, 342
DVD (Digital Video Disk), 331, 458
DWHT (discrete Walsh-Hadamard transform), 87
DWT (discrete Walsh transform), 86-88

## E

Early vision, 269
EBCOT (embedded block-based coding with optimized
    truncation), 180-182
EDTV (extended definition television), 367-368
Efficiency, coding schemes, 26, 96
Eigenvector transform. *See* Hotelling transform
Elementary stream clock reference (ESCR), 451, 456
Elementary streams (ES), 450-451, 466
Element difference, 69-70
Embedded block-based coding with optimized truncation
    (EBCOT), 180-182
Embedded zerotree technique (EZW), 176-177
EMM (entitlement management messages), 456
Encoder structure
    block diagrams, 32-33, 48, 75
    H.261, 430-431
    MPEG, 335-339
    MPEG-4, 422-425
    quantization, 31-33
Encoding delays, 25-26
End-of-band (EOB), 164, 177
End-of-line (EOL), 134, 138
End-to-end delay, 430
Enhancement, 215-217, 446
Entitlement management messages (EMM), 456
Entropy
    average information per symbol, 25
    block matching, 233-234, 241
    coding efficiency, 26
    discrete memoryless source, 112-113
    DPCM, 70
    Huffman coding, 117, 120, 127
    information measure, 24-25
    Markov source, 132-133
    optical flow, 297
    rate distortion function, 326
    run-length coding, 135
EOB (end-of-band), 164, 177
EOL (end-of-line), 134, 138
Error concealment, 391-400
    adaptive, 396-397
    algorithms, 392-397
    directional interpolation, 397-398
    domain, 392-393
    I-picture motion vectors, 398
    MPEG-4, 420
    spatial scalable, 399-400
    summary, 400
Errors. *See also* Error concealment
    angular, 287-288
    coding, 59-60
    displacement vectors, 210
    mean square error, 20

mean square prediction, 61
mean square quantization, 36, 39-40, 51
mean square reconstruction, 78, 93-95, 102
minimization, 272
prediction, 55, 59, 62, 241
quantization, 35, 38-39, 43, 57-60, 62, 379
resilience, 419-420
transmission, 65, 72, 95, 102, 138-139
ESCR (elementary stream clock reference), 451, 456
ES (elementary streams), 450-451, 466
Extended definition television (EDTV), 367-368
Extrapolation, low-pass, 412
EZW (embedded zerotree technique), 176-177

## F

Facilitation, 10. *See also* Masking
Facsimile coding, 134, 139-140, 149. *See also* Run-length
    coding
False contouring, 13, 50
FCC, 368
Feature correspondence, 212
Field difference, 70
Field-frame prediction mode, 341-342
Filters
    deblocking, 441-443
    down-conversion, 382-385
    Gabor energy, 308-311, 315
    Gaussian, 274
    Kalman, 306-307
    loop, 431
    low-pass, 228, 236, 274
    motion compensation, 390
    noise, 216
    sine-phase, 310
    temporal, 216
Finite precision, 126-127
FlexMux, 462-464
FMD (full-memory decoder), 385-387
Fourier transforms, 82, 84-86, 90-92, 169-174
Fractal image coding, 185, 193-197
Frame-difference coding, 70, 209
Frames
    prediction, 431
    replenishment, 208-209, 211, 218
    temporal sampling rate, 323
FRD (full-resolution decoder), 380
Freeze requests, 446
Frequency masking, 13-14
Frequency synthesis, 381-383
Full-memory decoder (FMD), 385-387
Full-resolution decoder (FRD), 380

## G

Gabor energy filters, 308-311, 315
Gabor transform, 169-174
Gamma correction, 15
Gamma input distribution, 40-42

Gaussian filters, 274
Gaussian input distribution, 40-43, 46, 84
Gaussian pyramids, 228-230, 277-278
Gaussian Seidel iterative method, 273
General switched telephone network (GSTN), 134
GOB (group of block) structure, 420, 432
GOV (group of video object plane), 420-421
Gradient method, 256-257. *See also* Descent methods
Grand Alliance, 368
Granular noise, 36, 38, 67
Gray level discrimination, 10-11
Gray-level quantization, 323
Gray scale shape coding, 415-416
Greedy algorithm, 358-359
Greedy parsing, 141
Group of block (GOB) structure, 420, 432
Group of video object plane (GOV), 420-421
GSTN (general switched telephone network), 134

# H

H.26L standard, 330, 446-447
H.261 standard, 329-330, 429-432
H.263 standard
    completion time, 330
    overview, 433-434
    scalability, 439-440
    status of, 329-330
    technical features, 434-438
    version 2, 439-446
H.263++ standard, 446-447
Haar wavelet, 170
Half-memory decoder (HMD), 385-387
Hardware complexity, 377
HDTV. *See* High definition television
Hessian matrix, 253
High definition television (HDTV). *See also* ATSC DTV
    standards
    block matching, 245
    digital formats, 328
    down-conversion, 217, 380
    formats, 369
    history of, 367-369
High-frequency coefficients, 374-375, 378-379
Histogram equalization, 15
HMD (half-memory decoder), 385-387
Horizontal coding mode, 138
Horizontal edgeness, 296
Horn and Schunck method, 269-275, 282, 284-285, 293, 300
Hotelling transform
    basis vector interpretation, 79-80
    covariance matrix, 75-77
    energy compaction, 94
    geometrical interpretation, 78-79
    inverse, 76-77, 83
    statistical interpretation, 77-78
HSI model, 15-16
Hue, 15-16
Huffman coding
    algorithm, 115-117, 127

arithmetic coding and, 125
in JPEG algorithms, 160-163
limitations, 120
modified, 117-119
in MPEG standards, 338
optimum instantaneous codes, 114-115
run-length coding and, 135, 137
transform coding, 101
Human visual system (HVS)
    color masking, 14-17
    differential sensitivity, 17-18, 23-24
    frequency masking, 13-14
    luminance masking, 10-12, 17
    motion perception, 212
    normalization factors, 99
    objective quality assessment and, 21-24
    perception, 9-10
    psychovisual redundancy, 9
    rigid and nonrigid motion, 307-308
    temporal masking, 14
    texture masking, 12-13
Hybrid coding, 103

# I

IFS (iterated function systems), 185, 194-197
IGS (improved gray-scale) quantization, 13-14, 21
Image and video data compression
    content-based, 404-405
    criteria for, 18-19
    definition and importance of, 3
    feasibility of, 4-18
    practical needs for, 4
    steps in, 4
    visual storage systems, 31-32
Image compression, 205
Image frames, 205
Image intensity, 5-6, 15, 55, 296-297
Image sampling, 323
Image sequences, 203-205, 211-217, 245
Image storage, 3
Image transformation, 80-83
Imaging space, 204, 206
Improved gray-scale (IGS) quantization, 13-14, 21
Independent segmentation decoding (ISD), 444
Information. *See also* Information theory
    coding algorithms, 107-108
    conservation, 279-280, 293-300, 306
    definition, 3
    discrete memoryless, 25-27
    measure, 24-25
    neighborhood, 280, 306, 311
    preserving, 71-72
    program-specific, 451, 456
    side, 235-236
    spatial, 23
    temporal, 23
Information theory. *See also* Variable-length coding
    dictionary coding, 140-149, 151-152
    entropy, 24-26

Information theory *(cont.)*
    Markov source model, 131-134
    rate distortion function, 27-28, 324-326
    run-length coding, 32, 134-139, 151
    Shannon's noiseless source coding theorem, 25-26
    Shannon's noisy channel coding theorem, 26-27
Information transmission theorem, 27
Input-output characteristics, 34-35, 37-38, 46, 66, 99
Intellectual property management and protection (IPMP), 466
Intensity, 5-6, 15, 55, 296-297
Interframe coding, 7-8, 103, 105
Interframe correlation, 205-208
Interframe predictive coding, 7-8
Interframe redundancy, 5-8, 206, 208
Interlaced video coding, 417
Internet videophony, 447
Interpixel correlation. *See* Differential coding; Transform coding
Interpixel redundancy, 4, 8
Interpolation
    block matching, 237-238
    directional, 397-398
    linear, 214
    motion-compensated, 214-215
    Pel recursive algorithm, 259
Intracoding, 441
Intraframe redundancy, 5-6, 55
Inverse Hotelling transform, 76-77
I-picture motion vectors, 398
IPMP (intellectual property management and protection), 466
ISD (independent segmentation decoding), 444
Iterated function systems (IFS), 185, 194-197
ITU-T, 429

**J**

JBIG, 128, 150-151
Joint probability density functions, 84
JPEG, 157-167
    arithmetic coding, 128
    color masking, 17
    completion time, 330
    differential coding, 60
    hierarchical mode, 157-158, 166-167
    lossless algorithms, 149-151, 157-158, 164-166
    motion, 206
    normalization factors, 99
    progressive DCT-based mode, 157-158, 163-165
    sequential DCT-based mode, 157-163
    status of, 329
    threshold coding, 96
JPEG-2000, 179-182, 330

**K**

Kalman filtering, 306-307
Karhunen-Loeve transform (KLT), 77, 91, 94-95, 104

**L**

Laplacian input distribution, 40-42
Laplacian pyramids, 277-278
LBG algorithm, 187-188
Level curves, 254
Linear block code, 108
Linear prediction, 59-62, 64, 69-70
Linear transforms, 80-83
Line difference, 70
Lloyd clustering (LBG) algorithm, 187-188
Lloyd-Max quantizers, 42
Logarithmic quantization, 44-46, 51
Logarithmic search procedures, 224-225, 348
Lossless still image compression standards, 149-151
Lossy compression, 3, 32, 324
Lossy shape coding, 414-415
Low-pass extrapolation, 412
Low-resolution decoders (LRDs), 380
Lucas and Kanade method, 275
Luminance, 15-17, 100, 396, 434, 436-437, 446
Luminance masking, 10-12, 18
LZ77 algorithms, 142-145, 151-152
LZ78 algorithms, 145-147, 151-152
LZW algorithm, 147-149, 152

**M**

Macroblock layer, 432-433, 438
MAD (mean absolute difference), 223-224, 231-232, 245
Mallat wavelet, 180
Mapper, 31-32, 55
Markov source model, 131-134, 152
Masking, 10-12, 21
Matrices
    basis, 83-84, 86-87, 103
    covariance, 76-79
    Hessian, 253
    quantization, 343
    transform coding, 82
MC coding. *See* Motion-compensated (MC) coding
Mean absolute difference (MAD), 223-224, 231-232, 245
Mean square error (MSE), 20, 187, 223-224
Mean square prediction error, 61-62, 64
Mean square quantization error, 36, 39-40, 51
Mean square reconstruction error, 78, 93-95, 102
Message codes, 107
Message ensemble, 108
Midrise quantizers, 34
Midtread quantizers, 33-34, 36, 38
Minimization, 272, 280-281
Minimum redundancy code, 111
Mismatches, 46
Model-based coding, 185, 197-198
Monochrome image encoding, 55
Moore-Penrose inverse, 384
Morlet wavelet, 170
Motion, 265-266

Motion analysis
  computer vision, 212-213
  definition, 213
  frame replenishment, 208-209, 211
  human vision, 212
  interframe correlation, 205-208
  motion-compensating coding, 209-211, 245
  region-based *vs.* gradient-based, 311-312
  signal processing, 213-214
Motion-compensated (MC) coding. *See also* Motion
  compensation
  block matching, 245
  efficiency, 218
  MPEG, 336
  predictive, 71, 103
  video coding, 60
Motion compensation. *See also* Motion-compensated (MC)
  coding
  enhancement, 215-217
  error concealment, 393-395
  full-resolution, 389-391
  H.261, 430-431
  half-pixel accuracy, 434-435
  HRTV, 380
  interpolation, 214-215
  low-resolution, 383-385
  MPEG-2, 347-350
  MPEG-4, 407-411
  overlapped, 408-409, 436
  scalable decoders and, 385-388
Motion estimation
  DCT-based, 315-318
  deterministic *vs.* stochastic, 308
  forward *vs.* backward, 312-313
  frequency domain, 349-350
  hierarchical, 349
  integer pixel, 435
  MPEG-2, 347-350
  MPEG-4, 407-411
  occlusion and disocclusion, 306-307
  overlapped, 349
  reduced data set, 349
  rigid and nonrigid, 307-308
  spatial *vs.* frequency domain, 308-311
Motion vectors
  advanced prediction mode, 435-437
  I-picture, 398
  motion estimation, 407
  substitution, 399
  unrestricted, 434-435
MPEG. *See also* ATSC DTV standards; MPEG-1;
  MPEG-2; MPEG-4; Multiplexing
  bitstream scaling, 371-379
  color masking, 17
  down-conversion decoders, 217, 379-391
  error concealment, 391-400
  H.261 and, 429-430
  MPEG-7, 330
  MPEG96/M960, 423
  MPEG97/M3147, 423

  optimum mode decision, 354-360
  rate control, 350-355, 363
  statistical multiplexing operations, 360-365
  wavelet transform, 179
MPEG-1
  bitstream structure, 339-340
  completion time, 330
  decoding process, 340
  encoder structure, 335-339
  features, 333-340
  H.261 and, 429-430
  layered structure, 334-335
  status of, 329
MPEG-2
  adaptation fields, 453-456
  completion time, 330
  concealment motion vectors, 344
  features, 340-346
  field-frame prediction mode, 341-342
  field-frame syntax, 342-343
  I-picture motion vectors, 398
  matching criteria, 347-348
  motion estimation, 347-350
  multiblock coding, 354-356
  pan and scan, 343-344
  preprocessing, 346-347
  program streams, 457-459
  quantization matrix, 343
  rate control, 346, 350-354, 363
  scalability, 344-346, 372, 399-400
  searching algorithm, 348-349
  status of, 329, 331
  technical definitions, 450-451
  test model, 421
  timing model and synchronization, 459-461
  transport streams, 451-456
  transport stream splicing, 456-458
  video encoding, 346-350
MPEG-4, 403-427
  completion time, 330
  decoder structure, 426-427
  encoder structure, 422-425
  error resilience, 419-420
  generalized scalability, 418-419
  goals of, 403
  half-pixel accuracy, 434
  interlaced video coding, 417
  motion estimation and compensation, 407-409
  multiple VO rate control, 425
  object description framework, 466
  overview and architecture, 462-464
  requirements and functionalities, 404-406
  scene description, 465-466
  shape coding, 413-416
  sprite coding, 416-417, 424
  status of, 329
  systems decoder model, 464-465
  texture coding, 409-413
  texture decoding, 426
  verification model, 421-427

MPEG-4 *(cont.)*
    video overview, 406
    visual bitstream syntax and semantics, 420-421,
      464-465
    wavelet-based texture coding, 417-418
MPEG-7, 330
MPEG96/M960, 423
MPEG97/M3147, 423
MRA (multiresolution analysis), 173
MSE (mean square error), 20, 187, 223-224
μ-law compression characteristic, 45
Multimedia applications. *See* MPEG-4
Multiple VO rate control, 425
Multiplexing
    MPEG-2, 450-461
    MPEG-4, 462-466
    standards, 449
Multiplication, 126
Multiresolution analysis (MRA), 173
Multiresolution block matching, 227
Muxing rates, 458

**N**

Nagel method, 276
National Television Systems Committee (NTSC) systems,
    16, 368-369, 429-430, 461
Natural binary code (NBC), 9, 32, 39
Neighborhood information, 280, 306, 311
Netravali and Robbins algorithm, 251, 256, 258-260, 262,
    306. *See also* Descent methods
Newton-Raphson's method, 257-258, 260, 262
Noise filtering, 216
Noiseless source coding theorem, 3, 25-26, 112-114
Noisy channel coding theorem, 26-27
Nonsingular code, 108-110
Normalization, 98-100, 104
NTSC (National Television Systems Committee) systems,
    16, 368-369, 429-430, 461

**O**

Object description framework, 466
Occlusion, 306-307
Occurrence probability, 25
Optical flow, 265-304
    Anandan method, 277-278
    aperture problem, 266-268, 305
    classification, 269
    computer vision and, 212-213
    correlation-based approach, 276-293, 301
    correlation-feedback technique, 281-293, 295
    definition, 266
    experiments, 284-291, 299-300
    fundamentals, 265-269
    Gabor energy filters, 308-311
    gradient-based, 269-279, 286, 288, 300, 312
    Horn and Schunck method, 269-275, 282, 284-285,
      293, 300
    ill-posed problems, 267-269, 300, 305
    Lucas and Kanade method, 275
    multiple image attributes, 293-300
    Nagel method, 276
    normal, 267
    Pan, Shi, and Shu method, 281-293, 295, 301
    performance comparison, 313-315
    Singh method, 278-281, 284-285, 292-293, 301
    Uras, Girosi, Verri, and Torre method, 276
    Weng, Ahuja, and Huang method, 294-295
    Xia and Shi method, 296-300
Optimization, 252-255, 354-360
Optimum code, 111
Orthogonality, 61, 83, 317
Overload noise, 37, 48
Oversampling technique, 67

**P**

Px64 (H.261) standard, 329-330, 429-432
Packet identification (PID), 451, 456-457
Packetization, 370-371
Packetized elementary stream (PES) packet header, 451,
    456
Packets, 451
Packs, 458
Pairwise nearest neighbor (PNN) algorithm, 188
PAL system, 16, 429-430, 461
PAM (pulse amplitude modulation), 49
Pan, Shi, and Shu method, 281-293, 295, 301
Pan-scan, 343-344
Parsing strategies, 141
Partitioned IFS (PIFS), 196
Pass coding mode, 136
PB-frames, 438, 440-441
PCM (pulse code modulation), 49-50, 52, 55
PCR (program clock reference), 451, 455, 460-461
Peak signal-to-noise ratio (PSNR), 21, 231, 242
Pel recursive techniques, 251-262
    Bergmann algorithms, 260, 262
    Cafforio and Rocca algorithm, 261
    conservation information, 306
    descent methods, 251-258
    as gradient-based approach, 311-312
    Netravali and Robbins algorithm, 251, 256, 258-260,
      262, 306
    performance comparison, 261-262, 313-315
    problem formulation, 251-252
    segmentation, 243
    Walker and Rao algorithm, 261
PES (packetized elementary stream) packet header, 451,
    456
PES packets, 451, 453, 458-461
PES stream, 451
Phase-locked loops (PLL), 460-461
Picture aspect ratio, 430
Picture autocorrelators, 5
Picture layer, 369-370, 432, 446
PID (packet identification), 451, 456-457
PIFS (partitioned IFS), 196
Pixels
    binary shape coding, 414-415

changing, 208-209, 218
correlation, 135-136
differencing, 23, 31
intensity, 5, 55
redundancy, 4, 8
PLL (phase-locked loops), 460-461
PNN (pairwise nearest neighbor) algorithm, 188
Postfiltering, 102, 105
Power spectrum, 5, 8
PPM (pulse position modulation), 49
Prediction
    AC, 409-410, 426
    adaptive, 64-65, 68, 72
    advanced, 435-438
    DC, 409-410, 426
    drift, 381
    dual prime, 342
    errors, 55, 59, 62, 241
    field-frame, 341-342
    frame, 431
    linear, 59-62, 64, 69-70
    optimum linear, 59-62, 64, 69-70
Prediction drift, 381
Predictive coding. *See* Differential coding
Predictive motion field segmentation, 242-244
Prefix condition, 111
Presentation time stamps (PTS), 451, 456, 459, 461
Present switch telephone network (PSTN), 4
Probabiloscopes, 5
Program clock reference (PCR), 451, 455, 460-461
Program-specific information (PSI), 451, 456
Program streams, 457-459
Pseudophases, 315-317
PSNR (peak signal-to-noise ratio), 21, 231, 242
PSTN (present switch telephone network), 4
Psychovisual redundancy, 8, 9-19, 27
PTS (presentation time stamps), 451, 456, 459, 461
Pulse amplitude modulation (PAM), 49
Pulse code modulation (PCM), 49-50, 52, 55
Pulse position modulation (PPM), 49
Pulse width modulation (PWM), 49

**Q**

QCIF (quarter-CIF), 429-430
QDC (quantized DC), 409-410
QM-coder, 127
QMD (quarter-memory decoder), 385-387
Quad-tree embedded algorithms, 181, 228
Quad-tree segmentation, 241
Quality of reconstructed image
    bitstream scaling, 378
    objective based on subjective, 21-24, 28
    objective evaluation, 20-24, 36
    requirements, 3
    subjective evaluation, 19-20, 35-36, 400
Quantization, 31-52. *See also* Uniform quantization; Vector
        quantization
    adaptive, 45-49, 51-52, 353-354
    binary tree search, 188

bitstream scaling, 376
block. *See* Transform coding
companding, 43-46, 51
definition, 31-32, 50
distortion, 35-37
error, 35, 38-39, 43, 57-60, 62, 379
gray-level, 13-14, 21, 323
H.263, 442, 446
JPEG algorithms, 160
logarithmic, 44-46, 51
luminance masking and, 11, 18
matrices, 343
MPEG, 337, 343, 354, 361
noise, 36, 38
nonuniform, 40-46
optimum, 39-40, 42-43, 51
pulse code modulation, 49-50, 52
source encoders and, 31-33
step size, 39-40, 48, 50, 67
switched, 48-49, 52
texture masking, 12-13
uniform, 33-41, 50
wavelet coefficients, 418
Quantized DC (QDC), 409-410
Quantizers
    design of, 37
    midrise, 33-34
    midtread, 33-34, 36, 38, 99
    optimum, 37-40
Quantizing (reconstruction) levels, 36-37, 50, 66
Quarter-CIF (QCIF), 429-430
Quarter-memory decoder (QMD), 385-387

**R**

Random access, 404, 455
Rate bounds, 458
Rate buffer feedback, 102, 104
Rate control
    bitstream scaling, 376
    MPEG, 350-355, 363
    multiple VO, 425
Rate distortion function, 27-28, 35, 324-326, 354, 364-365,
        424
Reaction parameter, 352
READ code, 135, 150
Reconstructed data, 394-395, 420
Reconstruction levels, 36-37, 50, 66
Rectangular slice (RS) submode, 443
Recursions, 251-252
Reduced-resolution update (RRU), 444-445
Reencoding, 376-377
Reference picture resampling (RPR), 444
Reference picture selection (RPS), 443-444
Restoration, 217, 398
Resynchronization, 138, 419-420
Reversible variable-length codes (RVLC), 420
RGB model, 15
RLC (run-length coding), 32, 134-139, 151
Root mean square error (RMSE), 20, 39, 196

Roundoff, 98-100, 104
RPR (reference picture resampling), 444
RPS (reference picture selection), 443-444
RRU (reduced-resolution update), 444-445
RS (rectangular slice) submode, 443
Run-length coding (RLC), 32, 134-139, 151
RVLC (reversible variable-length codes), 420


**S**

SA-DCT (shape-adaptive discrete cosine transform), 412-413
SAD (sum of absolute difference), 407-408
Sampling
    block matching, 227
    image, 323
    oversampling technique, 67
    reference picture resampling, 444
    threshold, 96-102, 104
    zonal, 95-96, 104
Sampling rate, 323, 461
Satellite broadcasting, 360
Saturation, 15-16
Scalability
    H.263 version 2, 439-440
    MPEG-2, 344-346
    MPEG-4, 418-419
    multilayer, 439-441
    spatial, 344-345, 418-419, 440
    temporal, 345, 418-419, 439
Scalar quantization, 32
Scene description, 465-466
SCR (system clock reference), 451, 458, 460
SDM (systems decoder model), 464-465
SDTV (Standard Definition Television), 370-371
Segmentation, 147, 241-244, 313, 423-424, 444
Separability, 81
Set partitioning in hierarchical trees ( SPIHT), 176-178
Set-top box (STB), 461
Shannon-Fano coding, 120
Shannon rate distortion theory, 324, 326
Shannon's noiseless source coding theorem, 25-26, 112-114
Shannon's noisy channel coding theorem, 26-27
Shannon's source coding theorem, 27
Shape-adaptive discrete cosine transform (SA-DCT), 412-413
Shape coding, 413-416
Shifting, 97-98
Short-Time Fourier Transform (STFT), 169-174
Side information, 235-236
SIF (source input format), 327, 334
Signal processing, 31-32, 55
Signal-to-noise ratio (SNR), 20-21, 39, 345, 439-440
Significance maps, 177-178
Similarity measure, 223
Singh method, 278-281, 284, 292, 294, 301
Sinusoidal orthogonal principle, 317
SI (spatial information), 23
Skipped frames, 430
Slice structured mode, 443

Sliding window algorithms, 142-145, 151-152
Slope overload error, 67
Smoothness constraint, 271-272
SNHC (synthetic and natural hybrid coding), 404
SNR (signal-to-noise ratio), 20-21, 39, 345, 439-440
Source alphabets, 107-108, 112-113, 116-117, 121-122, 127
Source coding theorem, 27
Source decoder. *See* Decoding
Source encoder. *See* Encoder structure
Source input format (SIF), 327, 334
Source symbols, 120-121
Spatial frequency, 18
Spatial image sequencing, 204, 206
Spatial information (SI), 23
Spatial masking, 12-13
Spatial orientation trees, 178
Spatial redundancy, 5-6, 55
Spatial scalability, 344-345, 418-419, 440
Spatial segmented wavelet transform (SSWT), 180
Spatial *vs.* frequency domain methods, 308-311
Speech signals, 44-45
SPIHT (set partitioning in hierarchical trees), 176-178
Splicing, 456-458
Sprite coding, 416-417, 424
SSD (sum of the squared difference), 223, 277, 282-283
SSE (sum of the squared error), 223, 239
SSWT (spatial segmented wavelet transform), 180
Stability range, 261-262
Standard Definition Television (SDTV), 370-371
Standards. *See also* ATSC DTV standards; H.263 standard; JPEG; MPEG
    digital coding, 328-331
    facsimile coding, 139-140, 149
    H.26L standard, 330
    H.261 standard, 329-330, 429-432
    H.263++ standard, 446-447
    JPEG-2000, 179-180
    lossless still image compression, 149-151
    multiplexing, 449
    wavelet transform, 330
State diagrams, 132
Statistical multiplexing, 360-365
Statistical redundancy, 4-9
StatMux, 360-365
STB (set-top box), 461
STC (system time clock), 460-461
STD (system target decoder), 451
STFT (Short-Time Fourier Transform), 169-174
Still image coding, 157-167. *See also* JPEG
Stochastic methods, 308
Strings, arithmetic coding, 121, 124-125
Subband coding, 84
Sub-block partitioning, 181
Suboptimum transforms, 94
Sum of absolute difference (SAD), 407-408
Sum of the squared difference (SSD), 223, 277, 282-283
Sum of the squared error (SSE), 223, 239
Symmetry, 82
Synchronization, 459-461
Sync layer, 462-464

Synthetic and natural hybrid coding (SNHC), 404
Synthetic video, 403
System clock reference (SCR), 451, 458, 460
System header, 451
Systems decoder model (SDM), 464-465
System target decoder (STD), 451
System time clock (STC), 460-461

**T**

Target bit allocation, 350-351
TC. *See* Transform coding
Television. *See also* ATSC DTV standards; High definition
    television
  cable, 360
  digital, 328, 367-370
  extended definition, 367-368
  interframe correlation, 207-208
  interframe differential coding, 68-69
  motion-compensated interpolation, 214-215
  MPEG-2, 347
  National Television Systems Committee systems, 16,
      368-369, 429-430, 461
  PAL system, 16, 429-430, 461
  power spectrum, 5, 8
  progressive and interlaced, 327
  SECAM system, 16
  spatial redundancy, 5
  Standard Definition, 370-371
  studios, 461
  synchronization, 461
Temporal filtering, 216
Temporal image sequencing, 203-204, 206
Temporal information (TI), 23
Temporal interpolation, 215
Temporal masking, 14
Temporal redundancy, 5-8, 206, 208
Temporal scalability, 345, 418-419, 439
Texture coding, 409-413, 417-418
Texture decoding, 426
Texture masking, 12-13
Threshold coding, 96-102, 104
Thresholding multigrid block matching, 239-241
Thresholding multiresolution block matching,
    229-234
Time stamp, 451
Timing models, 459-461, 464
TI (temporal information), 23
Transcoding, 371-379
Transformations
  contractive affine, 194
  DCT-to-spatial, 388-389
  image, 80-83
  kernels, 82, 85-90
  overview, 31-32, 55
Transform coding (TC), 75-105. *See also* Transform
    coefficients
  basis vector interpretation, 79-80
  bit allocation, 95-102
  block diagram, 80

computational complexity, 95
discrete cosine transform, 14, 60, 88-92
discrete Fourier transform, 82, 84-86, 90-92
discrete Hadamard transform, 87-88
discrete Walsh transform, 86-87
DPCM and, 103
energy compaction, 92-93
frequency masking and, 14
geometrical interpretation, 78-79
Hotelling transform, 75-79
hybrid coding, 103
inverse, 76-77, 83
linear transforms, 80-84
matrix form, 82
mean square reconstruction error, 78, 93-95, 102
orthogonality, 83
performance comparison, 92-95
procedures, 80
separability, 81
statistical interpretation, 77-78
subimage size, 84, 105
symmetry, 82
truncation, 95-102, 104
zonal coding, 95-96
Transform coding gain, 93
Transform coefficients
  basis images, 83, 103
  energy compaction, 93
  mean square reconstruction error, 93-94
  zonal coding, 95-96
Transmission errors, 65, 95, 102
TransMux, 462-463
Transport stream packet header, 451, 454-455
Transport streams, 451-456
Tree search quantization, 188
Trellis coders, 190
Trick-play tracks, 373
Truncation, 95-102, 104

**U**

Uniform quantization
    definitions, 33-35
    different input distributions, 40-41
    distortion, 35-37
    overview, 50
    uniformly distributed input, 37-39
Unitary transform, 79, 82, 104
Uras, Girosi, Verri, and Torre method, 276

**V**

Variable bit rate (VBR) coding, 361-363, 372
Variable-length coding (VLC), 107-128
    alternative inter-VLC, 446
    arithmetic codes, 107, 119-127, 438
    coding redundancy, 9
    H.261, 431
    Huffman codes, 101, 107, 114-120, 125, 127

Variable-length coding (VLC) *(cont.)*
   noiseless source coding theorem and, 113-114
   reversible, 420
VBR (variable bit rate) coding, 361-363, 372
VCXO (voltage controlled crystal oscillators), 461
Vector quantization (VQ), 185-193
   block truncation coding, 191
   classified, 189-190
   codebook generation, 185, 187-191
   finite state, 190
   lattice, 191-193
   overview, 32
   predictive, 190
   quantization, 188
   residual, 189
   training set generalization, 187
   transform domain, 190
   vector formation, 186-187
Vertical coding mode, 136-138
Vertical edgeness, 296
Very large scale integrated (VLSI) circuits, 4
Vestigial sideband (VSB) modulation, 371
Video coding, 323-331
   block matching, 240-241
   digital coding standards, 328-331
   digital formats, 327-328
   digital representation, 323
   motion-compensated, 265-266
   optical flow, 301
   rate distortion function, 27-28, 324-326
Video compression
   color masking, 17
   definition, 205
   error concealment, 392
   interframe correlation, 205-206
   motion-compensated, 209-211, 214-215, 218
Videoconferencing. *See* H.261 standard; H.263 standard
Video delivery systems, 391-392
Video objects (VO), 405-406, 420-421, 425
Video object layers (VOL), 420-421
Video object planes (VOP)
   binary shape coding, 414-415
   content-based coding, 405-406
   encoding and decoding process, 422-427
   generalized scalability, 418-419
   motion estimation/compensation, 410-413
   sprite coding, 416-417, 424
   in syntax hierarchy, 420-421
   texture coding, 409, 411-413
Video-on-demand (VOD), 372-373
Videophony, 3, 308. *See also* H.261 standard; H.263 standard
Video segmentation, 423-424
Video sequences, 205-207
Video session (VS), 421
Video tape recorders (VTRs), 373
Video verification model, 421-427
Virtual buffers, 352-353
Visual bitstream syntax and semantics, 420-421

Visual quality measurement
   objective, 20-24, 36
   objective based on subjective, 21-24, 28
   subjective, 19-20, 35-36
VLC. *See* Variable-length coding
VLSI (very large scale integrated) circuits, 4
VOD (video-on-demand), 372-373
Voltage controlled crystal oscillators (VCXO), 461
VOL (video object layer), 420-421
Von Koch curves, 193-194
VOP. *See* Video object planes
VO (video objects), 405-406, 420-421, 425
VQ. *See* Vector quantization
VSB (vestigial sideband) modulation, 371
VS (video session), 421
VTRs (video tape recorders), 373

# W

Walker and Rao algorithm, 261
Wavelet-based texture coding, 417-418
Wavelet transform, 169-182
   definition, 169-170
   digital, 174-179
   discrete, 172-174
   embedded image coding, 176-179
   short-time Fourier transform and, 169-172
   standards, 330
   theory, 174-176
Weber's law, 10-11
Weighted least-square method, 279-280, 283, 297-298
Weng, Ahuja, and Huang method, 294-295
Window functions, 245, 247
Wrapping, 444

# X

Xia and Shi method, 296-300

# Y

YCbCr model, 17
YDbDr model, 16-17
YIQ model, 16
Yule-Walker equations, 62
YUV model, 16

# Z

Zero run length, 101
Zerotree coding, 176-177
Zigzag scans, 100-101, 161-162, 338, 343
Zonal coding, 95-96, 104